Key Concepts of Physical Chemistry

Key Concepts of Physical Chemistry

Edited by
Jamie Langdon

Willford Press
www.willfordpress.com

Published by Willford Press,
118-35 Queens Blvd., Suite 400,
Forest Hills, NY 11375, USA

ISBN: 978-1-68285-435-8

Cataloging-in-Publication Data

Key concepts of physical chemistry / edited by Jamie Langdon.
p. cm.
Includes bibliographical references and index.
ISBN 978-1-68285-435-8
1. Chemistry, Physical and theoretical. 2. Chemistry. I. Langdon, Jamie.
QD453.3 .K49 2018
541--dc23

For information on all Willford Press publications
visit our website at www.willfordpress.com

Willford Press

Printed in China.

Contents

Preface

Physical chemistry blends the principles and techniques of physics to the study of chemical phenomena. It has a number of allied fields like electrochemistry, photochemistry, thermochemistry, spectrochemistry, chemical kinetics, surface chemistry, etc. This book unravels the recent studies in the field of physical chemistry. It is a vital tool for all researching or studying physical chemistry, as it gives incredible insights into emerging trends and concepts. A number of latest researches have been included to keep the readers up-to-date with the global concepts in this area of study.

The information contained in this book is the result of intensive hard work done by researchers in this field. All due efforts have been made to make this book serve as a complete guiding source for students and researchers. The topics in this book have been comprehensively explained to help readers understand the growing trends in the field.

I would like to thank the entire group of writers who made sincere efforts in this book and my family who supported me in my efforts of working on this book. I take this opportunity to thank all those who have been a guiding force throughout my life.

Editor

1

Chirality-selected phase behaviour in ionic polypeptide complexes

Sarah L. Perry[1,*], Lorraine Leon[1,2,*], Kyle Q. Hoffmann[1,3], Matthew J. Kade[1,2], Dimitrios Priftis[1], Katie A. Black[4], Derek Wong[1], Ryan A. Klein[1], Charles F. Piercelll[1], Khatcher O. Margossian[1], Jonathan K. Whitmer[1,2], Jian Qin[1], Juan J. de Pablo[1,2] & Matthew Tirrell[1,2]

Polyelectrolyte complexes present new opportunities for self-assembled soft matter. Factors determining whether the phase of the complex is solid or liquid remain unclear. Ionic polypeptides enable examination of the effects of stereochemistry on complex formation. Here we demonstrate that chirality determines the state of polyelectrolyte complexes, formed from mixing dilute solutions of oppositely charged polypeptides, via a combination of electrostatic and hydrogen-bonding interactions. Fluid complexes occur when at least one of the polypeptides in the mixture is racemic, which disrupts backbone hydrogen-bonding networks. Pairs of purely chiral polypeptides, of any sense, form compact, fibrillar solids with a β-sheet structure. Analogous behaviour occurs in micelles formed from polypeptide block copolymers with polyethylene oxide, where assembly into aggregates with either solid or fluid cores, and eventually into ordered phases at high concentrations, is possible. Chirality is an exploitable tool for manipulating material properties in polyelectrolyte complexation.

[1] Institute for Molecular Engineering, University of Chicago, Chicago, Illinois 60637, USA. [2] Argonne National Laboratory, Argonne, Illinois 60439, USA. [3] Department of Chemical Engineering, University of Wisconsin, Madison, Wisconsin 53706, USA. [4] Department of Bioengineering, University of California at Berkeley, Berkeley, California 94720, USA. * These authors contributed equally to this work. Correspondence and requests for materials should be addressed to M.T. (email: mtirrell@uchicago.edu).

Polyelectrolyte complexation has a long history of utility in encapsulation applications[1,2], produces compartmentalization that is implicated in development of certain biological assemblies[3–12] and in origin of life scenarios[13,14], and is accelerating new synthetic materials applications from surface coatings to self-assembled structures[3–12,15]. It is an entropically driven process, where the initial electrostatic attraction between oppositely charged polyelectrolytes is followed by the release of small, bound counter-ions and the restructuring of water molecules[4,16,17]. Studies on understanding and controlling polyelectrolyte complexation and self-assembly have focused almost exclusively on charge-driven phenomena, such as charge density, pH and ionic strength[4,5,18–23]. The resulting complexes can form either solid precipitates or liquid complex coacervates[4]. Wang and Schlenhoff[24] give an excellent account of the progression from precipitate to coacervate to solution in a particular system, but the underlying causes and general understanding of phase selection between solid and fluid state complexes are obscure.

While precipitates result in dense, solid, hydrated materials, fluid coacervate phases retain significant amounts of water (up to 90% in some cases)[18,19], display viscoelastic properties[16,22,25,26] and have very low surface tension with water[20,23] Complexation can be coupled with molecular design, linking polyelectrolyte domains to hydrophilic, neutral polymer blocks to stabilize microphase separation and drive the assembly of ordered phases associated with traditional block copolymers[3–6,8,10,12,27,28].

A delicate balance of forces determines whether complexation yields a liquid or a solid. This process occurs cooperatively, with the initial electrostatic interaction between the oppositely charged segments of two polymer chains nucleating phase formation[17]. The strength of these interactions can be mediated by charge screening and by parameters such as the acidity or basicity of the charged groups and the density of charges along the polymer[4,23]. Higher polyelectrolyte charge density provides an increased stability against salt-induced dissolution of the complexes and enables phase separation with shorter polyelectrolyte chains[18,22,23]. Higher charge density and strongly charged polyelectrolytes tend to favour solid precipitates over liquid coacervates[4,16]. The strength of electrostatic interaction appears to be the one recognized determinant of phase selection.

Various polyelectrolytes have been used in studies of complexation, ranging from charged particles and micelles[29,30] to proteins[4–6], biologically derived polymers such as gelatin, chitosan and heparin[5,31,32], synthetic polymers such as poly(acrylic acid) and poly(allylamine)[3,5,6,21,22] and biomimetic polymers such as polypeptides[5,6,8,12,14,17,19,20]. Ionic polypeptides are distinctive owing to the chirality of the monomers and the resulting potential to form secondary structures based on hydrogen bonding. While repulsion among like-charged groups favours the random coil structure of individual polypeptide chains, both α-helical and β-sheet structures can form on charge neutralization or screening[33,34]. However, the preponderance of studies on the complexation of polypeptides have used at least one polyelectrolyte lacking the capacity to form secondary structure, such as achiral polymers, polysaccharides or α/β-random co-polypeptides[5,6,8,12,16,17,19,20,32,35]. The effect of the charged side chains in complexation has been well studied in a variety of systems[4–6,8,12,17,19,20]; however, the effect of backbone chirality on phase selection and materials design in polyelectrolyte complexes has not been tested.

In a broader sense, the stereochemistry of the polymer backbone can have a direct influence on the formation of semicrystalline or amorphous domains. However, these interactions are driven largely by stereoselective van der Waals interactions[36]. Here we examine the chirality effects during polyelectrolyte complexation, taking advantage of not only electrostatic and stereospecific van der Waals forces, but also hydrogen-bonding interactions between polypeptide chains. This diverse set of orthogonal interactions enables the independent tuning of material properties without altering the chemical identity and demonstrates that chirality presents unique opportunities for the manipulation of physical properties in material systems built on polyelectrolyte complexation.

Results

Characterization of chirality on bulk polyelectrolyte complexes. We present data here on the complexation of poly(lysine) (pK, using the single-letter abbreviation for amino acids,) with poly(-glutamic acid) (pE) while systematically varying the polymer chirality (homochiral; p*L*K, p*D*K, p*L*E and p*D*E and racemic; p(*D*,*L*)K and p(*D*,*L*)E, see Supplementary Table 1 and Supplementary Figs 1 and 2 for details on characterization). Throughout this article, pK stands for polylysine and pE stands for polyglutamic acid; *D* and *L* indicate chirality, with *D*,*L* used to designate racemic polymers. Complexes were formed by mixing stoichiometric amounts of polyanion and polycation (that is, charge-matched conditions), in dilute solution in the presence of varying amounts of NaCl. Visual identification of liquid coacervates or solid precipitates was made using optical microscopy (Leica DMI6000 B). Liquid coacervates appear as small, spherical fluid droplets, while precipitates form amorphous solid clusters. Liquid coacervates were observed exclusively when at least one of the polypeptides present was racemic (Fig. 1). Although the majority of the work done here utilized a 50/50 random copolymer of *D* and *L* monomers, we have also observed the formation of liquid coacervates when a sequence-controlled polymer of alternating *D* and *L* monomers was used (Supplementary Fig. 3).

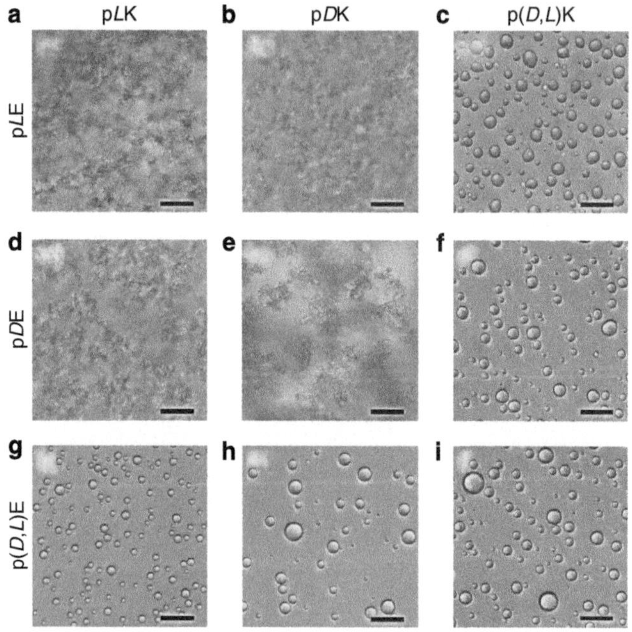

Figure 1 | Optical micrographs of polyelectrolyte complexes. Bright-field optical micrographs showing the liquid coacervates or solid precipitates resulting from the stoichiometric electrostatic complexation of *L*, *D*, or racemic (*D*,*L*) poly(lysine) with *L*, *D* or racemic (*D*,*L*) poly(glutamic acid) at a total residue concentration of 6 and 100 mM NaCl. Complexes are formed from (**a**) p*L*K + p*L*E, (**b**) p*D*K + p*L*E, (**c**) p(*D*,*L*)K + p*L*E, (**d**) p*L*K + p*D*E, (**e**) p*D*K + p*D*E, (**f**) p(*D*,*L*)K + p*D*E, (**g**) p*L*K + p(*D*,*L*)E, (**h**) p*D*K + p(*D*,*L*)E, (**i**) p(*D*,*L*)K + p(*D*,*L*)E. Liquid coacervate droplets are only observed during complexation involving a racemic polymer. Scale bars, 25 μm.

In contrast, complexes composed only of homochiral polymers (that is, pLK + pLE, pLK + pDE, pDK + pLE, pDK + pDE) formed solid precipitates, regardless of salt concentration.

To understand the secondary structure of the polypeptides within the resultant solid or fluid complexes, we employed transmission Fourier transform infrared spectroscopy (FTIR, Bruker Vertex 70 spectrometer). The location of the amide I carbonyl stretching vibration provides a characterization of polypeptide secondary structure[33,34,37]. As can be seen in Fig. 2a, all samples (that is, individual polypeptides, liquid coacervates and solid precipitates) display a peak at 1,644 cm^{-1}, characteristic of a random coil chain configuration (complete FTIR data and discussion are available in Supplementary Fig. 4 and Supplementary Note 1, respectively)[33,34,37]. This random coil structure was expected for the individual, charged polypeptides, and is consistent with previous characterization of the polymer structure in liquid coacervates closely approximating that of an ideal Gaussian chain[26] For the solid precipitates, we observe additional peaks that are characteristic of β-strands and amyloids[33,34,37]. This signal was present at 1,611 cm^{-1} for solid precipitates formed from polypeptides with matching chirality (pLK + pLE, pDK + pDE) and at 1,613 cm^{-1} for complexes formed from polypeptides with opposite chirality (pLK + pDE, pDK + pLE). An additional, low-intensity peak near 1,680 cm^{-1} is also attributable to the presence of β-sheet structure. The blue-shift observed in the main amide I band has been attributed to a decrease in the number of peptide-water contacts, implying that complexes with opposite chirality exclude more water[34,37,38].

These results demonstrate the correlation between polypeptide chirality, polymer conformation and the fluid or solid nature of the resulting complex. While the random coil structure of an individual polypeptide homopolymer is a consequence of electrostatic repulsion from the charged side chains, neutralization of these charges allows attractive hydrogen-bonding interactions to dominate, resulting in the formation of close-packed polypeptide secondary structure (that is, α-helix or β-sheet). For the case of homochiral polypeptides, the electrostatic interaction of oppositely charged side chains facilitates alignment of the peptide backbone and the formation of hydrogen-bonded β-strand structures where backbone hydrogen bonds replace surface-bound water that is released during the aggregation of the polymer[33]. However, the presence of a racemic polypeptide prevents the formation of compact protein-like secondary structures and appears to limit both the extent of complexation and the expulsion of water from the polymer rich phase, while maintaining a Gaussian coil conformation[26].

Stability of polyelectrolyte complexes against salt and urea. We also investigated the stability of the various solid and fluid polyelectrolyte complexes with respect to salt. Turbidimetric measurements for liquid coacervates showed a characteristic increase in the amount of complex formation at low salt concentrations, followed by a decrease in complex formation up to a critical salt concentration, above which no phase separation is observed (Fig. 3a, Supplementary Fig. 5a and Supplementary Note 2). Taking into account the dependence of polymer chain length on the critical salt concentration (see Supplementary Table 2 and Supplementary Note 2)[18,19], we observed a significant decrease in stability against salt (lower stability means that complexes dissociate and dissolve at lower salt concentrations) when fully racemic complexes were formed, as opposed to complexes with only a single racemic polymer (for example, p(D,L)K + p(D,L)E versus p(D,L)K + p(D)E). However, coacervates were significantly less stable in salt as compared with

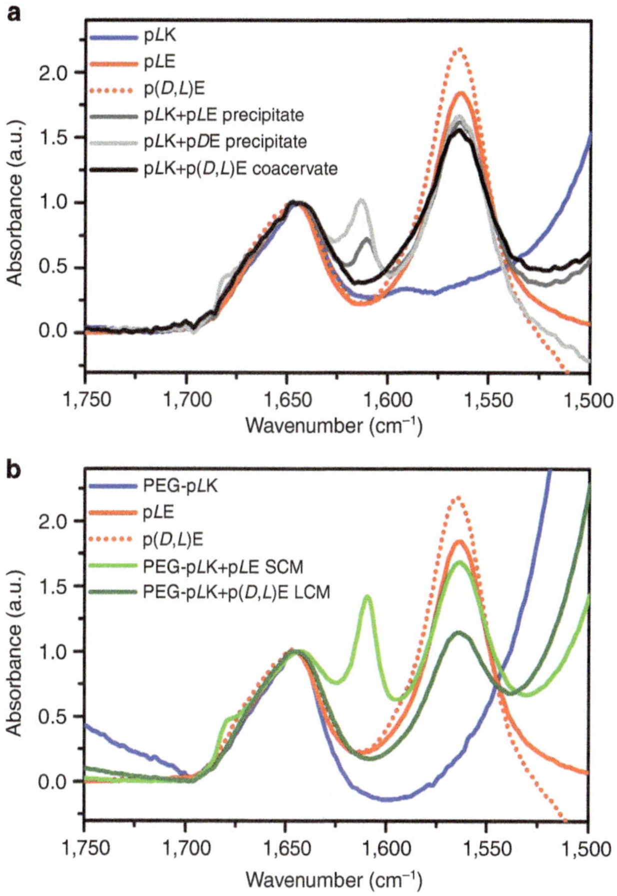

Figure 2 | FTIR of polyelectrolyte complexes. FTIR spectra showing the amide I region for (**a**) individual polypeptides and the resulting liquid coacervates and solid precipitates, as well as (**b**) the polypeptides and block copolymers involved in the formation of related LCM and SCM. All samples were prepared in D_2O. Polypeptides were analysed at a concentration of 10 mM with respect to monomer, liquid coacervates and solid precipitates at a concentration of 6 mM with respect to monomer and 100 mM NaCl, while micellar complexes were prepared at a concentration of 0.186 mM polymer, with no salt. All materials show a peak at 1,644 cm^{-1}, characteristic of random coil polypeptide structure. However, the additional peaks associated with aggregated β-strands are present for the solid precipitates and SCMs. For solid precipitates formed from polypeptides with matching chirality (pLK + pLE, pDK + pDE), the main peak is located at 1,611 cm^{-1} and is shifted to 1,613 cm^{-1} for opposite chirality (pLK + pDE, pDK + pLE). For SCMs, this peak is located at 1,610 cm^{-1}. An additional low-intensity peak is also present near 1,680 cm^{-1}. The signal for the carbonyl stretching of the glutamic acid can also be observed at 1,564 cm^{-1}. Micelles were prepared using a polyethylene glycol-pLK block copolymer with an average $N = 50$ and either pLE with $N = 50$ for SCMs or $N = 100$ for LCMs.

solid precipitates, and solid complexes formed from polypeptides with opposite chirality showed even higher salt stability than those of matched chirality (for example, pLK + pLE, pLK + pDE) (Fig. 3a, Supplementary Fig. 5b and Supplementary Note 2). These variations as a function of polypeptide chirality suggest that electrostatic interactions are not solely responsible for the stability of the complex, but also van der Waals stereoregular interactions and hydrogen bonding.

In polypeptides, short-range hydrogen bonding plays a critical role in defining stable structural motifs. Therefore, we also investigated the stability of our complexes in urea, a denaturant that interacts preferentially with the peptide backbone and disrupts hydrogen bonds[39]. While urea did not affect the formation or stability of liquid coacervates, even at very high concentrations, a much more dramatic effect was observed for solid precipitates. The addition of sufficient quantities of urea resulted in the melting of solid precipitates into a coacervate-like liquid (Fig. 3b, Supplementary Fig. 6 and Supplementary Note 2), a transition that had only been previously observed in relation to charge-driven phenomena in polyelectrolyte complexes[24]. Furthermore, the trends in stability observed as a function of urea matched those observed for salt (Supplementary Table 2), implying that the variations in salt stability are attributable to the strength of hydrogen-bonding effects. This conclusion is supported by the observed differences in polypeptide secondary structure, where hydrogen-bonded β-sheets provide increased stability against salt dissolution for solid complexes as compared with liquid coacervates. Furthermore, the differences in stability observed for solid complexes of differing chirality are supported by the FTIR data (Fig. 2a and Supplementary Fig. 4d), suggesting that complexes with opposite chirality incorporate less water[34,37,38], and thus denaturant, from their surfaces, enhancing the stability of hydrogen-bonding interactions.

Chirality effects in micellar polyelectrolyte complexes. On the basis of experience with bulk complexes and previous reports on polyelectrolyte complex micelles formed using racemic polypeptides[12], we extended our investigation to include micellar complexes formed from the complexation of a homopolymer polypeptide with a diblock copolymer with one peptide block (Supplementary Fig. 2). In agreement with our results from bulk experiments, micelles formed from electrostatic complexation involving at least one racemic polypeptide (for example, PEG-p*L*K with p(*D*,*L*)E) resulted in the formation of coacervate- or liquid-core micelles (LCMs), as described in the literature[6,8,12]. These LCMs are highly monodisperse ($R_h = 27.1$ nm, polydispersity = 0.074), as characterized by dynamic light scattering (DLS, Brookhaven Instruments BI-200SM, see Supplementary Table 3) and also similar in size to polypeptide-based polyelectrolyte complex micelles of comparable polymer chain lengths observed previously[40]. Secondary structure analysis via both FTIR (Fig. 2b) and circular dichroism spectroscopy (CD, Fig. 4a) shows a random coil conformation, in agreement with the results obtained for bulk complexes. However, the structures formed from the complexation of homochiral polypeptides (for example, PEG-p*L*K with p*L*E) resulted in micelles that were larger, more polydisperse ($R_h = 32.8$ nm, polydispersity = 0.117), and displayed β-sheet character (Figs 2b and 4a). We designate these complexes as solid-core micelles (SCMs) based on the behaviour of similar complexes in bulk. Interestingly, while the FTIR spectra for both LCMs and SCMs display the same features as the analogous bulk materials, the location of the β-strand peak for SCMs is slightly red-shifted compared with the bulk, from 1,611 to 1,610 cm^{-1}. This shift suggests that the confinement imposed by the micellar structure could enhance packing of the β-sheet network, compared with the bulk, because of forced chain alignment at the polyelectrolyte core-PEG corona interface[37].

CD enabled characterization of the urea-induced polypeptide structural changes in the micellar systems (Fig. 4b). While our

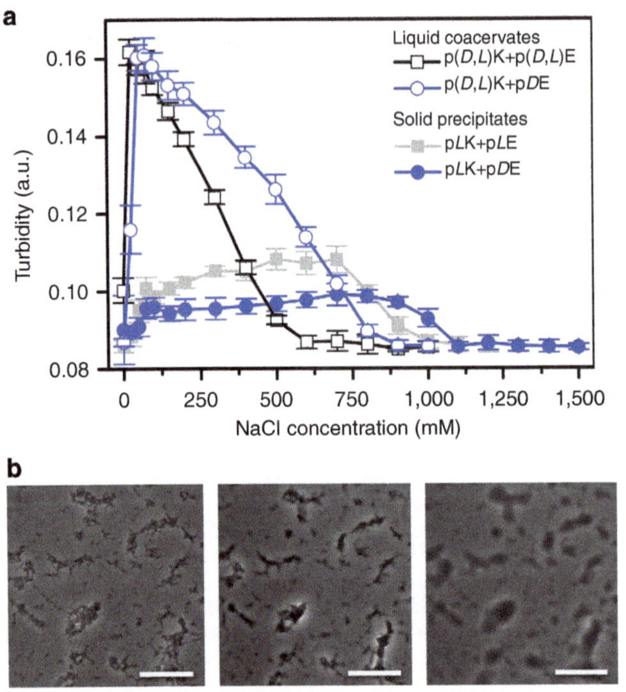

Figure 3 | Salt and urea stability of polyelectrolyte complexes. (**a**) Turbidity as a function of NaCl concentration for various liquid coacervates (open symbols) and solid precipitates (solid symbols) prepared at 1 mM total residue concentration and pH = 7.0. Error bars are the s.d. from triplicate measurements. (**b**) Optical micrographs showing the transition from solid precipitate to liquid coacervate for p*L*K + p*L*E complexes with increasing urea concentration. Scale bars, 25 μm.

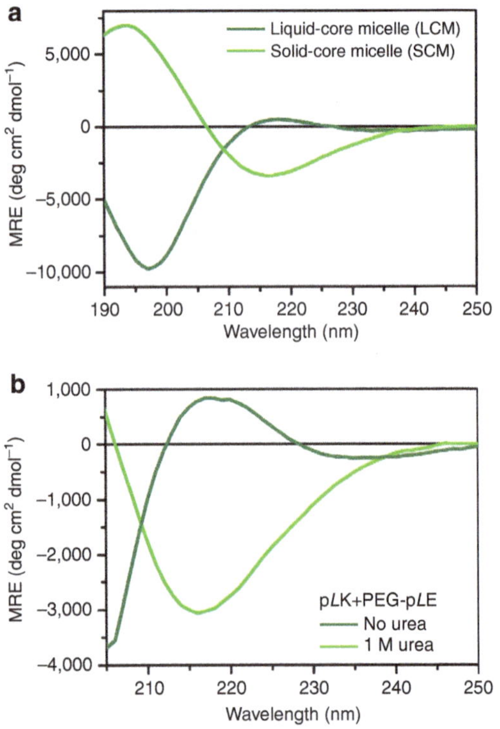

Figure 4 | Secondary structure of micellar complexes. CD spectra for (**a**) LCM (dark green) and SCM (light green) micelles formed from PEG-p*L*K + p(*D*,*L*)E and PEG-p*L*K + p*L*E, respectively, and (**b**) SCMs in the absence (light green) and presence of 1 M urea (dark green), respectively. LCMs show a random coil structure, while SCMs display β-sheet character in the absence of urea, but convert to a random coil structure in the presence of 1 M urea, suggestive of a LCM. LCMs were prepared at a polymer concentration of 0.01 mM total polymer concentration, while SCMs were prepared at 0.0125 mM with an average $N = 100$ for the micelles in **a**, and both LCMs and SCMs were prepared at a total polymer concentration of 0.04 mM with an average $N = 50$ for **b**.

initial SCM sample displayed 100% β-sheet character, as calculated using CD basis spectra and evidenced by the characteristic minimum at 215 nm (Supplementary Fig. 7a), the addition of 1 M urea caused an unfolding transition, suggested by the appearance of a maximum near 220 nm and a minimum below 205 nm. Spectral fitting suggested a micelle with 99.8% random coil structure (Supplementary Fig. 7c,d, Supplementary Note 3). As in the case of our bulk complexes, the observed transition to a random coil structure is reminiscent of liquid complexes or LCMs (Fig. 4a and Supplementary Fig. 7b), suggesting that hydrogen bonding is a key determinant of the solid or liquid character of polypeptide-based polyelectrolyte complexes. Furthermore, this provides a way of independently manipulating the solid or liquid character of complexes in bulk or in domains. The moduli and self-healing character of ordered phases based on polyelectrolyte complexation will depend on the phase selection of the domains[10].

Commensurate with phase selection and materials properties, solid and liquid complexes also display differences in the kinetics of assembly. Light scattering measurements were utilized to track the formation and equilibration of both LCMs and SCMs as a function of time (Fig. 5). The dynamic nature of the liquid coacervate core enabled the fast equilibration of LCMs. After 1 h, LCMs had reached 72% of their equilibrium size. This is in contrast to SCMs, where the hydrogen-bonded, β-sheet solid core is more resistant to chain rearrangement and was only able to attain 25% of its equilibrium size after 1 h. We confirmed the dominant effect of hydrogen bonding in preventing chain rearrangement by further examining a presumed SCM sample that was formed in the presence of 4 M urea. Here disruption of the hydrogen bonds by urea enabled recovery of the liquid-like timescales for chain rearrangement, attaining 74% of the equilibrium value after 1 h. These data provide further support for the hypothesis that the physical state of the system (that is, SCM or LCMs) is determined by the presence of hydrogen bonds, which favour a more solid-like structure by creating a higher barrier for chain rearrangement and thus slower equilibration kinetics.

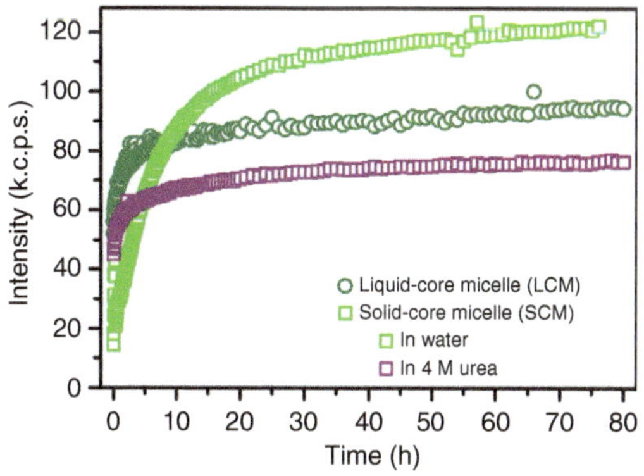

Figure 5 | Kinetics of micelle formation. Total scattered intensity as a function of time for LCMs (circles) and SCMs (squares) depicting the timescale for equilibration. LCMs in water equilibrate relatively quickly, reaching 72% of their equilibrium value within 1 h, while SCMs in water show much slower kinetics, on the order of days; only reaching 25% of their equilibrium value in 1 h. However, the addition of 4 M urea to a SCM sample to disrupt hydrogen bonding between the polypeptides enables fast equilibration, similar to that of LCMs (74% of equilibrium value in 1 h). Micelles were prepared at a polymer concentration of 0.01 mM total polymer concentration using a polyethylene glycol-p*L*E block copolymer with an average $N = 50$ and either p*L*K with $N = 100$ for SCMs or p(*D,L*)K with $N = 100$ for LCMs.

Molecular dynamics investigation of chirality effects. To further investigate the different intermolecular interactions that give rise to the various solid and liquid complexes, we performed molecular dynamics (MD) calculations. Multiple independent simulations were run for a 10-residue p*L*K interacting with a 10-residue p(*D,L*)E, while one simulation was run for a 10-residue p*L*K interacting with a 10-residue p*L*E. All simulations were performed in 173 mM NaCl over the course of 1,000 ns. In all systems, the peptides formed an electrostatic complex in less than 10 ns, adopting a disordered structure. Around 150 ns, the p*L*K + p*L*E system began to form a parallel β-sheet in the centre of the peptides, which was fully formed by 200 ns (Fig. 6). This structure remained stable throughout the duration of the simulation, with the exception of small fluctuations at the termini. The formation of a stable β-sheet structure allowed the peptides to bind more tightly, reducing the centre of mass distance to about 0.2 nm. However, the p*L*K + p(*D,L*)E system shown in Fig. 6 formed a less compact structure with a centre of mass distance varying between 0.3 and 0.9 nm. These peptides remained in mostly coil, bend and turn conformations, with only rapidly transient β-sheet formation observed for durations of less than 100 ns, and mostly in a region of the p(*D,L*)E peptide where a sequence of three consecutive *L* amino acids was present (Fig. 6). Furthermore, the analysis of backbone hydrogen-bonding interactions indicated that the unstructured p*L*K + p(*D,L*)E system, corresponding to the liquid coacervate complexes, showed a preferential interaction with water and formed very few inter-peptide hydrogen bonds (Fig. 7a). In contrast, the dense β-sheet p*L*K + p*L*E complex satisfied nearly all of its backbone hydrogen bonds through peptide-peptide interaction (Fig. 7b). Averaging multiple independent enactments of these 1,000-ns simulations with some variability of initial conditions, the p*L*K + p*L*E system had an average β-sheet content over all residues of 61.3%, compared with 11.4% for the average of the p*L*K + p(*D,L*)E simulations. These results provide direct support of our experimental observations, confirming the structural modes of interaction between the various oppositely charged polypeptides and demonstrating the key role that chirality and hydrogen bonding play in determining the structural and physical state of polypeptide-based polyelectrolyte complexes.

Discussion

In summary, the polypeptide chirality not only determines the physical state of the resulting polyelectrolyte complexes (that is, liquid or solid), but also defines the strength of intermolecular interactions, and thus the material properties. While electrostatic interactions act over long distances, the shorter-range nature of polar hydrogen-bonding forces, combined with steric packing and hydration, provide additional methods for controlling self-assembly. The effects we have presented here raise obvious follow-on questions about the effects of sequence distribution within globally achiral polymers or designed effects that can be created by tailoring sequences of chiral peptides, which we are now pursuing. While sequence specificity in biology controls the three-dimensional assembly of proteins, we propose that patterns of chirality could have significant implications for tailoring of material properties without otherwise altering the chemical composition of polypeptide-based materials. For instance, this type of control could be utilized to tailor the rheological response of bulk materials and formulate delivery systems with controlled water content. Furthermore, coupling this type of polar and electrostatically guided self-assembly with more complex molecular architectures, as in the block copolymer systems described here, enables the creation of interesting classes of new materials with novel self-assembling structures, functionality and

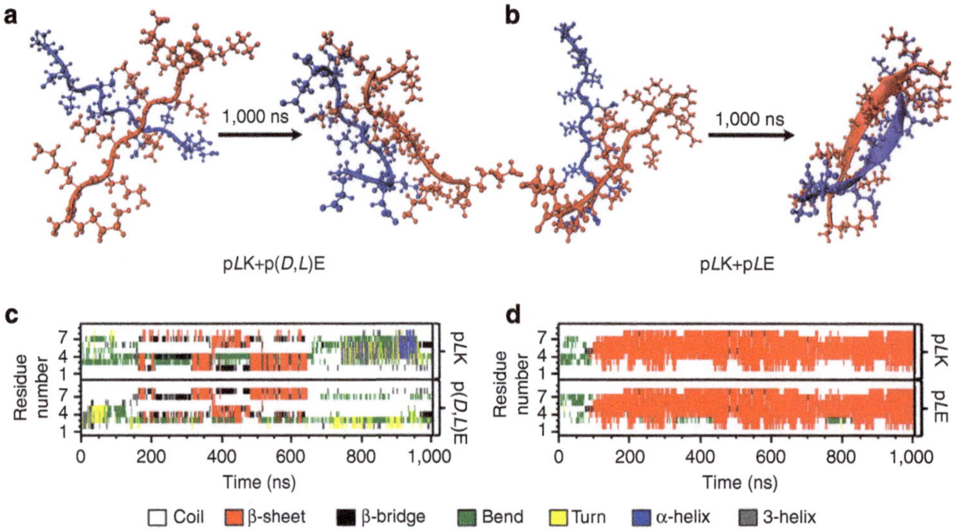

Figure 6 | MD simulations of polyelectrolyte complexes. Visualization and residue maps indicating polypeptide secondary structure from representative MD simulations for two pairs of poly(lysine) and poly(glutamic acid) peptides, $N = 10$. Polypeptides are initially equilibrated in a random coil conformation and then allowed to complex for 1,000 ns. (**a**) A representative simulation of homochiral p*L*K complexing with racemic p(*D*,*L*)E indicates preservation of a mostly random coil structure, as would be expected for liquid coacervates, while (**b**) homochiral polypeptides p*L*K with p*L*E shows the evolution of β-strand structure expected for a solid precipitate. Map of secondary structure as a function of time for (**c**) p*L*K + p(*D*,*L*)E and (**d**) p*L*K + p*L*E.

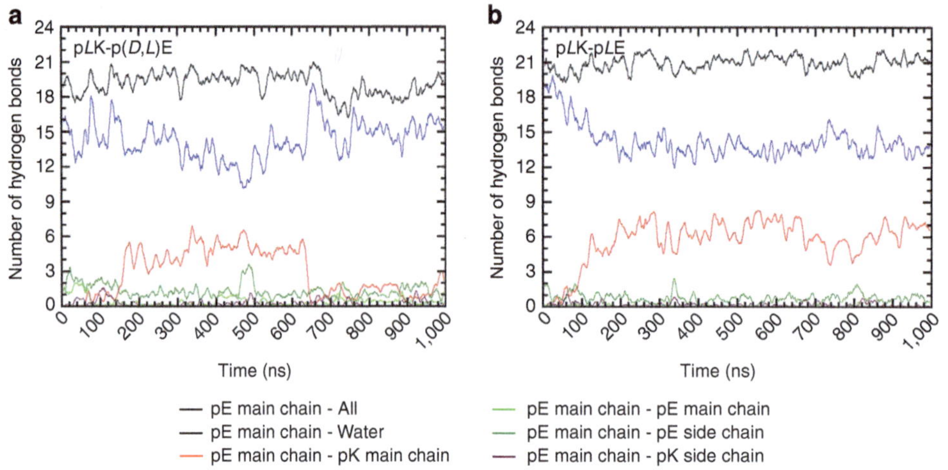

Figure 7 | Hydrogen bonding in polyelectrolyte complexes. Quantification of the number of hydrogen bonds formed as a function of time during a 1,000 ns MD simulation of complex formation between (**a**) p*L*K + p(*D*,*L*)E and (**b**) p*L*K + p*L*E.

responsiveness[41]. We note, too, that while this work clarifies the role of chirality on phase selection in complexes of charged polypeptides, other polyelectrolyte complex systems also form both solid and fluid phases[24]. We suggest that the causes of solid phase formation in polyelectrolyte complexes are to be found in short-range forces, which may be influenced by tacticity, hydration packing and other factors, acting in concert with longer range electrostatic forces.

Methods

Materials. Polypeptides were obtained either from Alamanda Polymers Inc. and used directly, without further purification, or were synthesized in house using *N*-carboxyanhydride polymerization[42]. In a previously published article[19], we reported results on a polypeptide that we said was p*L*E; after communication with the supplier (Alamanda Polymers Inc.) and performing further characterization, it came to our attention that the polymers in this previous work were not optically pure *L* but contained a number of *D* repeating units and should therefore have been referred to as p(*D*,*L*)E. A correction of this error has now been published[43] Identification of this issue led to the current work in which we have retested a wide range of conditions and have been unable to formulate liquid coacervates using homochiral polypeptides in any instance.

The degree of polymerization for the prepared polymers was obtained via ^{1}H NMR. Gel permeation chromatography (Waters) coupled with refractive index (Optilab UT-rex, Wyatt Technologies) and light scattering detectors (miniDAWN Treos, Wyatt Technologies) was used to characterize the polydispersity of the samples (Supplementary Table 1). CD (Jasco J-815 CD Spectrometer) was used to confirm both the random coil structure and the homochiral or racemic composition of the individual polypeptides (Supplementary Figs 1 and 2).

Here we abbreviate poly(glutamic acid) as pE in general, or the different chiral polymers as p*L*E, p*D*E, p(*D*,*L*)E in specific, taking advantage of the single-letter abbreviation strategy for amino acids. Similarly, we refer to poly(lysine) as pK in general, or p*L*K, p*D*K, p(*D*,*L*)K in specific. This naming convention also allows referring to *D*-amino acids using a lower-case letter (that is, pk rather than p*D*K). This strategy will be useful for the investigation of sequence-controlled polymers of *D* and *L* amino acids (for example, $(Kk)_{14}W$ as in Supplementary Fig. 3), but for the current discussion of homochiral or random copolymers we will explicitly specify the *D* or *L* chirality of the polymer for clarity.

pE is a weak acid with a pKa around pH 4.3, while pK is a weak base with a pKa around pH 10.0 (refs 44,45). We make the assumption that, at solution conditions at least two pH units away from the pKa, the polypeptides are fully charged. Stock solutions were prepared gravimetrically using MilliQ water (resistivity of 18.2 MΩ-cm, Millipore) at a concentration of 10 mM with respect to the number of monomers (that is, the number of acid or base groups) present in solution and then adjusted to pH 7.0 using concentrated solutions of HCl and NaOH, as needed. Monomer concentration was chosen as the experimental basis to easily enable direct stoichiometric comparison of the number of positively and negatively charged units present in solution. Stock solutions of 2 M sodium chloride (NaCl, ACS reagent, ≥99%, Acros Organics) and 8 M urea (Bioreagent, Sigma) were prepared gravimetrically and adjusted to pH 7.0, as above.

Preparation of bulk polyelectrolyte complexes. Complexation was performed using stoichiometric quantities of positive and negatively charged polypeptides at a total residue concentration of 1 mM at pH 7.0, in the presence of varying concentrations of NaCl and urea. Under these conditions, it is a reasonable approximation to describe all of the residues on both polypeptides as charged. All polymers used for the preparation of bulk complexes have an approximate degree of polymerization $N = 100$. Samples were prepared by first mixing a concentrated solution of NaCl with MilliQ water in a microcentrifuge tube (1.5 ml, Eppendorf). Other additives such as concentrated urea were also added at this stage, unless otherwise specified. The polyanion (pE) was then added to this mixture, followed by the polycation (pK) to a final volume of 500 µl. The mixture was vortexed for 5 s immediately after the addition of each component to ensure fast mixing. For all experiments, samples were prepared to a final concentration of 1 mM monomers (total cation and anion). Unless otherwise indicated, all bulk complexes (that is, liquid coacervates and solid precipitates) were prepared in 100 mM NaCl. The effect of salt was examined over the range of 0 to 1.5 M NaCl. The effect of urea was examined from 0 to 6.8 M. All samples were prepared immediately before analysis and studied at room temperature (25 °C).

Preparation of micellar polyelectrolyte complexes. Micelles were formed by mixing either PEG-p*LK* with pE or PEG-p*LE* with pK at an equal charge molar ratio, in water, using the order of operations described above for the homopolymers. Micellar solutions for the DLS experiments were made using PEG-p*LK* and pE, in which the average charged polypeptide segments of each molecule was $N = 100$. The micelle formed using PEG-p*LK* and p(*D,L*)E was made at a total polymer concentration of 0.07 mM, which was subsequently diluted to 0.01 mM when measured using CD. The micelle formed using PEG-p*LK* and p*LE* was made at a total polymer concentration of 0.05 mM, which was diluted to 0.0125 mM when measured using CD. The micellar urea experiments were performed by complexing PEG-p*LE* with pK at a total polymer concentration of 0.04 mM, with all the charged segments containing an average $N = 50$. Micellar solutions for the FTIR experiments were made at a total polymer concentration of 0.19 mM in D_2O using a PEG-p*LK* block copolymer with an average $N = 50$ and either p*LE* with $N = 50$ for SCMs or $N = 100$ for LCMs.

Visual characterization of complexes. The resulting complexes formed from homopolymers were imaged within 1 h of preparation. Bright-field and phase contrast optical microscopy (Leica DMI 6000B using Leica AFC image acquisition software) were used to identify the formation of liquid coacervates or solid precipitates, to identify the critical salt concentration, above which no phase separation occurs, and the minimum urea concentration necessary to trigger the transformation between solid precipitates and liquid coacervates. Liquid coacervates appear as small spherical droplets in solution, while precipitates form amorphous solid clusters (Figs 1 and 3 and Supplementary Figs 3 and 6). Imaging was performed using both ultra-low attachment 96-well plates (Costar, Corning Inc.) and glass slides (Fisherbrand).

Turbidimetry. Turbidity was used to qualitatively measure the extent of complex formation as a function of charge stoichiometry and salt concentration. Turbidity measurements were made using a plate reader equipped with a ultraviolet spectrophotometer (Tecan, Infinite M200 P80). Turbidity was measured at a wavelength of 550 nm and a temperature of 25 °C. None of the polymers absorb light at this wavelength. Turbidity is defined as $-\ln(I/I_o)$, with I_o = incident light intensity and I = intensity of light passed through the sample volume, and is measured in absorption units (a.u.). After preparation, 100 µl of sample was pipetted in triplicate into a 96-well plate (black/clear with lid, BD Falcon) for analysis. Triplicate measurements were made for each well, and all experiments were repeated three times. Error bars on turbidity plots represent the calculated s.d. of the data.

Characterization of polypeptide secondary structure. Analysis of the secondary structure of the various individual polypeptides, liquid coacervates, solid precipitates and micellar complexes was performed using transmission FTIR spectroscopy. The instrument used was a Bruker Vertex 70 FTIR spectrometer with a DTGS detector. The sample was prepared in D_2O and held between two 1-mm thick CaF_2 windows with a 50-µm PTFE spacer. The sample chamber was purged with N_2 for 10 min before data collection. One hundred and twenty-eight scans were collected from 4,000 to 800 cm^{-1} at a spectral resolution of 2 cm^{-1}. Spectra were obtained using a D_2O sample as a background and were dynamically corrected for atmospheric water by subtracting the signal obtained for the open chamber (exposed to air), taken with a N_2 background and scaled by a factor of ~ -0.018. Data were then normalized such that the minimum baseline signal above the amide I signal (in the range of 1,755–1,700 cm^{-1}) was set to zero, and the height of the random coil peak near 1,644 cm^{-1} was set to 1. The location of individual peaks was determined using non-linear least-squares fitting of Gaussian peaks to the features of the spectrum, with baseline correction where necessary.

CD (Jasco J-815 CD Spectrometer) was used to confirm the random coil structure, the chiral nature of the individual polypeptide solutions (Supplementary Figs 1 and 2) and the secondary structure of micellar solutions (Supplementary Fig. 7). All the CD data presented are an average of five scans collected between 250 and 190 nm at room temperature, in a 0.1-cm cuvette. In experiments involving urea, the baseline measurements of the solvent were subtracted from the micellar spectra. The fitting of the CD data was done using a linear combination of poly(lysine) basis spectra[46] resulting in specific percentages of α-helix, β-sheet and random coil secondary structure.

Dynamic light scattering. DLS was measured at 90° using a BI-200SM goniometer containing a red laser diode with a wavelength of 637 nm and a TurboCorr digital correlator (Brookhaven Instruments, Holtsville, NY). Brookhaven Instruments DLS software was used to analyse the intensity autocorrelation function using the cumulant method[47] to obtain characteristic decay rates from which apparent diffusion coefficients were calculated. The Stokes–Einstein equation was used to convert diffusion coefficients to hydrodynamic radii. Polydispersity was obtained from the second order cumulant term.

MD simulations. MD simulations were performed for 10-residue peptides of poly(glutamic acid) (pE) and poly(*L*-lysine) (p*LK*). Two versions of pE were used: a homochiral *L* version (p*LE*) and a racemic version (p(*D,L*)E) with the sequence LLDDDLLLDD, which was chosen by shuffling a sequence with five *L* and five *D* chiral centres. The side chains were fully charged with the N terminus and C terminus consisting of a charged NH_3^+ and COO^- group, respectively. These were initialized into a β-sheet conformation. One pE chain and one p*LK* chain were placed perpendicular to each other ~1 nm apart in a 6-nm dodecahedral box. The system was then solvated with 4889 TIP3P waters and 16 Na^+ and Cl^- ions (that is, 104 mM NaCl), modelled with the ion parameters from Joung *et al.*[48] Molecular simulations were run using the GROMACS 4.6.3 simulation package[49]. The energy of the system was minimized for 500 steps without constraints and then for an additional 10,000 steps with all bonds constrained using the steepest descent algorithm. A MD simulation was then run using a step size of 2 fs and the CHARMM22* force field[50,51] for 100 ps at constant volume and a temperature of 298 K. This was continued for 1,000 ns at a constant pressure of 1 bar using the Nose–Hoover[52] and Parrinello–Rahman[53] coupling schemes to maintain temperature and pressure, respectively. Electrostatic interactions were performed using a particle mesh Ewald method[54] with a cutoff of 0.9 nm and a Fourier spacing of 0.33 nm. Hydrogen bonds were constrained using the LINCS[55] algorithm. The secondary structure was determined using the DSSP (Define Secondary Structure of Proteins) criteria[56] over an average of four replicate simulations for p(D,L)E with pLK and one simulation for pLE with pLK.

References

1. Madene, A., Jacquot, M., Scher, J. & Desobry, S. Flavour encapsulation and controlled release—a review. *Int. J. Food Sci. Technol.* **41,** 1–21 (2006).
2. Augustin, M. A. & Hemar, Y. Nano- and micro-structured assemblies for encapsulation of food ingredients. *Chem. Soc. Rev.* **38,** 902–912 (2009).
3. Hunt, J. N. *et al.* Tunable, high modulus hydrogels driven by ionic coacervation. *Adv. Mater.* **23,** 2327–2331 (2011).
4. van der Gucht, J., Spruijt, E., Lemmers, M. & Cohen Stuart, M. A. Polyelectrolyte complexes: bulk phases and colloidal systems. *J. Colloid Interface Sci.* **361,** 407–422 (2011).
5. Pergushov, D. V., Müller, A. H. E. & Schacher, F. H. Micellar interpolyelectrolyte complexes. *Chem. Soc. Rev.* **41,** 6888–6901 (2012).
6. Voets, I. K., de Keizer, A. & Stuart, M. A. C. Complex coacervate core micelles. *Adv. Colloid Interface Sci.* **147-148,** 300–318 (2009).
7. Eulalio, A., Behm-Ansmant, I. & Izaurralde, E. P bodies: at the crossroads of post-transcriptional pathways. *Nat. Rev. Mol. Cell Biol.* **8,** 9–22 (2007).
8. Kataoka, K. *et al.* Spontaneous formation of polyion complex micelles with narrow distribution from antisense oligonucleotide and cationic block copolymer in physiological saline. *Macromolecules* **29,** 8556–8557 (1996).
9. Hyman, A. A. & Simons, K. Beyond oil and water—phase transitions in cells. *Science* **337,** 1047–1049 (2012).
10. Krogstad, D. V. *et al.* Effects of polymer and salt concentration on the structure and properties of triblock copolymer coacervate hydrogels. *Macromolecules* **46,** 1512–1518 (2013).
11. Hwang, D. S. *et al.* Viscosity and interfacial properties in a mussel-inspired adhesive coacervate. *Soft Matter* **6,** 3232–3236 (2010).
12. Harada, A. & Kataoka, K. Chain length recognition: core-shell supramolecular assembly from oppositely charged block copolymers. *Science* **283,** 65–67 (1999).
13. Keating, C. D. Aqueous phase separation as a possible route to compartmentalization of biological molecules. *Acc. Chem. Res.* **45,** 2114–2124 (2012).
14. Koga, S., Williams, D. S., Perriman, A. W. & Mann, S. Peptide–nucleotide microdroplets as a step towards a membrane-free protocell model. *Nat. Chem.* **3,** 720–724 (2011).

15. Hammond, P. T. Engineering materials layer-by-layer: challenges and opportunities in multilayer assembly. *AIChE J.* **57**, 2928–2940 (2011).
16. de Kruif, C. G., Weinbreck, F. & de Vries, R. Complex coacervation of proteins and anionic polysaccharides. *Curr. Opin. Colloid Interface Sci.* **9**, 340–349 (2004).
17. Priftis, D., Laugel, N. & Tirrell, M. Thermodynamic characterization of polypeptide complex coacervation. *Langmuir* **28**, 15947–15957 (2012).
18. Spruijt, E., Westphal, A. H., Borst, J. W., Cohen Stuart, M. A. & van der Gucht, J. Binodal compositions of polyelectrolyte complexes. *Macromolecules* **43**, 6476–6484 (2010).
19. Priftis, D. & Tirrell, M. Phase behaviour and complex coacervation of aqueous polypeptide solutions. *Soft Matter* **8**, 9396–9405 (2012).
20. Priftis, D., Farina, R. & Tirrell, M. Interfacial energy of polypeptide complex coacervates measured via capillary adhesion. *Langmuir* **28**, 8721–8729 (2012).
21. Perry, S. L., Li, Y., Priftis, D., Leon, L. & Tirrell, M. The effect of salt on the complex coacervation of vinyl polyelectrolytes. *Polymers* **6**, 1756–1772 (2014).
22. Priftis, D. *et al.* Ternary, tunable polyelectrolyte complex fluids driven by complex coacervation. *Macromolecules* **47**, 3076–3085 (2014).
23. Qin, J. *et al.* Interfacial tension of polyelectrolyte complex coacervate phases. *ACS Macro Lett.* **3**, 565–568 (2014).
24. Wang, Q. & Schlenoff, J. B. The polyelectrolyte complex/coacervate continuum. *Macromolecules* **47**, 3108–3116 (2014).
25. Spruijt, E., Sprakel, J., Lemmers, M., Stuart, M. A. C. & van der Gucht, J. Relaxation dynamics at different time scales in electrostatic complexes: time-salt superposition. *Phys. Rev. Lett.* **105**, 208301 (2010).
26. Spruijt, E. *et al.* Structure and dynamics of polyelectrolyte complex coacervates studied by scattering of neutrons, X-rays, and light. *Macromolecules* **46**, 4596–4605 (2013).
27. Koide, A. *et al.* Semipermeable polymer vesicle (PICsome) self-assembled in aqueous medium from a pair of oppositely charged block copolymers: physiologically stable micro-/nanocontainers of water-soluble macromolecules. *J. Am. Chem. Soc.* **128**, 5988–5989 (2006).
28. van der Kooij, H. M. *et al.* On the stability and morphology of complex coacervate core micelles: From spherical to wormlike micelles. *Langmuir* **28**, 14180–14191 (2012).
29. Kizilay, E., Kayitmazer, A. B. & Dubin, P. L. Complexation and coacervation of polyelectrolytes with oppositely charged colloids. *Adv. Colloid Interface Sci.* **167**, 24–37 (2011).
30. Pawar, N. & Bohidar, H. B. Anisotropic domain growth and complex coacervation in nanoclay-polyelectrolyte solutions. *Adv. Colloid Interface Sci.* **167**, 12–23 (2011).
31. Bungenberg de Jong, H. G. in *Colloid Science* (ed. Kruyt, H.) 2, 335–432 (Elsevier Publishing Company, 1949).
32. Chu, H., Gao, J., Chen, C.-W., Huard, J. & Wang, Y. Injectable fibroblast growth factor-2 coacervate for persistent angiogenesis. *Proc. Natl Acad. Sci. USA* **108**, 13444–13449 (2011).
33. Fandrich, M. & Dobson, C. M. The behaviour of polyamino acids reveals an inverse side chain effect in amyloid structure formation. *EMBO J.* **21**, 5682–5690 (2002).
34. Dzwolak, W., Ravindra, R., Nicolini, C., Jansen, R. & Winter, R. The diastereomeric assembly of polylysine is the low-volume pathway for preferential formation of β-sheet aggregates. *J. Am. Chem. Soc.* **126**, 3762–3768 (2004).
35. Conrad, U., Fahr, A. & Scriba, G. K. E. Kinetics of aspartic acid isomerization and enantiomerization in model aspartyl tripeptides under forced conditions. *J. Pharm. Sci.* **99**, 4162–4173 (2010).
36. Slager, J. & Domb, A. J. Biopolymer stereocomplexes. *Adv. Drug Deliv. Rev.* **55**, 549–583 (2003).
37. Baiz, C. R., Reppert, M. & Tokmakoff, A. in *Ultrafast Infrared Vibrational Spectroscopy* (ed. Fayer, M. D.) 361–404 (CRC Press, 2012).
38. Fuhrhop, J.-H., Krull, M. & Büldt, G. Precipitates with β-pleated sheet structure by mixing aqueous solutions of helical poly(D-lysine) and poly(L-lysine). *Angew. Chem. Int. Ed.* **26**, 699–700 (1987).
39. Auton, M., Holthauzen, L. M. F. & Bolen, D. W. Anatomy of energetic changes accompanying urea-induced protein denaturation. *Proc. Natl Acad. Sci. USA* **104**, 15317–15322 (2007).
40. Harada, A. & Kataoka, K. Effect of charged segment length on physicochemical properties of core – shell type polyion complex micelles from block ionomers. *Macromolecules* **36**, 4995–5001 (2003).
41. Hamley, I. *Block Copolymers in Solution: Fundamentals and Applications* (John Wiley & Sons, Ltd., 2005).
42. Kramer, J. R. & Deming, T. J. General method for purification of α-amino acid- N-carboxyanhydrides using flash chromatography. *Biomacromolecules* **11**, 3668–3672 (2010).
43. Priftis, D. & Tirrell, M. Correction: phase behaviour and complex coacervation of aqueous polypeptide solutions. *Soft Matter* **11**, 422 (2014).
44. Harada, A. & Kataoka, K. Formation of polyion complex micelles in an aqueous milieu from a pair of oppositely-charged block copolymers with poly (ethylene glycol) segments. *Macromolecules* **28**, 5294–5299 (1995).
45. Liu, Y., Li, C., Wang, H.-Y., Zhang, X. Z. & Zhuo, R.-X. Synthesis of thermo- and pH-sensitive polyion complex micelles for fluorescent imaging. *Chem.-Eur. J.* **18**, 2297–2304 (2012).
46. Greenfield, N. J. & Fasman, G. D. Computed circular dichroism spectra for the evaluation of protein conformation. *Biochemistry* **8**, 4108–4116 (1969).
47. Koppel, D. E. Analysis of macromolecular polydispersity in intensity correlation spectroscopy: The method of cumulants. *J. Chem. Phys.* **57**, 4814 (1972).
48. Joung, I. S. & Cheatham, T. E. Determination of alkali and halide monovalent ion parameters for use in explicitly solvated biomolecular simulations. *J. Phys. Chem. B* **112**, 9020–9041 (2008).
49. Hess, B., Kutzner, C., van der Spoel, D. & Lindahl, E. GROMACS 4: algorithms for highly efficient, load-balanced, and scalable molecular simulation. *J. Chem. Theory Comput.* **4**, 435–447 (2008).
50. Bjelkmar, P., Larsson, P., Cuendet, M. A., Hess, B. & Lindahl, E. Implementation of the CHARMM force field in GROMACS: Analysis of protein stability effects from correction maps, virtual interaction sites, and water models. *J. Chem. Theory Comput.* **6**, 459–466 (2010).
51. Piana, S., Lindorff-Larsen, K. & Shaw, D. E. How robust are protein folding simulations with respect to force field parameterization? *Biophys. J.* **100**, L47–L49 (2011).
52. Hoover, W. G. Canonical dynamics: equilibrium phase-space distributions. *Phys. Rev. A* **31**, 1695–1697 (1985).
53. Parrinello, M. & Rahman, A. Polymorphic transitions in single crystals: a new molecular dynamics method. *J. Appl. Phys.* **52**, 7182–7190 (1981).
54. Darden, T., York, D. & Pedersen, L. Particle mesh Ewald: an N log(N) method for Ewald sums in large systems. *J. Chem. Phys.* **98**, 10089–10092 (1993).
55. Hess, B., Bekker, H., Berendsen, H. J. C. & Fraaije, J. G. E. M. LINCS: a linear constraint solver for molecular simulations. *J. Comput. Chem.* **18**, 1463–1472 (1997).
56. Kabsch, W. & Sander, C. Dictionary of protein secondary structure: pattern recognition of hydrogen-bonded and geometrical features. *Biopolymers* **22**, 2577–2637 (1983).

Acknowledgements

We thank Dr Carlos Baiz and Dr Denise Schach from the Tokmakoff lab for their assistance and guidance in collecting and analysing the FTIR data. This work also benefited from extensive discussions with Professor Tobin Sosnick. This work was supported by the U.S. Department of Energy Office of Science, Program in Basic Energy Sciences, Materials Sciences and Engineering Division.

Author contributions

S.L.P. and L.L. contributed equally to this work. S.L.P. designed and performed the experiments involving bulk polyelectrolyte complexes. L.L. designed and performed the experiments involving micellar polyelectrolyte complexes. S.L.P., L.L., K.Q.H. and M.T. wrote the article. All authors discussed and contributed to the interpretation of the data. M.J.K., C.F.P., L.L. and R.A.K. synthesized the racemic polycations. D.P. and K.A.B. performed proof of concept work. R.A.K. and K.O.M. performed characterization experiments on bulk polyelectrolyte complexes using optical microscopy and turbidity measurements. D.W. performed characterization experiments using CD and DLS on the micellar polyelectrolyte complexes. K.Q.H. designed and performed MD simulations. S.L.P., L.L, D.P., K.Q.H., J.Q., J.K.W., J.J.d.P. and M.T. edited the manuscript. J.J.d.P. supervised the computational work. M.T. supervised the experimental work.

Additional information

Competing financial interests: The authors declare no competing financial interests.

2

Atomic-scale disproportionation in amorphous silicon monoxide

Akihiko Hirata[1], Shinji Kohara[2,3,4,5,6], Toshihiro Asada[7], Masazumi Arao[7], Chihiro Yogi[7], Hideto Imai[7], Yongwen Tan[1,8], Takeshi Fujita[1] & Mingwei Chen[1,8,9]

Solid silicon monoxide is an amorphous material which has been commercialized for many functional applications. However, the amorphous structure of silicon monoxide is a long-standing question because of the uncommon valence state of silicon in the oxide. It has been deduced that amorphous silicon monoxide undergoes an unusual disproportionation by forming silicon- and silicon-dioxide-like regions. Nevertheless, the direct experimental observation is still missing. Here we report the amorphous structure characterized by angstrom-beam electron diffraction, supplemented by synchrotron X-ray scattering and computer simulations. In addition to the theoretically predicted amorphous silicon and silicon-dioxide clusters, suboxide-type tetrahedral coordinates are detected by angstrom-beam electron diffraction at silicon/silicon-dioxide interfaces, which provides compelling experimental evidence on the atomic-scale disproportionation of amorphous silicon monoxide. Eventually we develop a heterostructure model of the disproportionated silicon monoxide which well explains the distinctive structure and properties of the amorphous material.

[1] WPI Advanced Institute for Materials Research, Tohoku University, Sendai 980-8577, Japan. [2] Quantum Beam Unit, National Institute for Materials Science (NIMS), 1-1-1 Kouto, Sayo, Hyogo 679-5148, Japan. [3] Information Integrated Materials Research Unit, Research Center for Information Integrated Materials, NIMS, 1-2-1 Sengen, Tsukuba, Ibaraki 305-0047, Japan. [4] Division of Research & Utilization , Japan Synchrotron Radiation Research Institute, Hyogo 679-5198, Japan. [5] Schools of Materials Science, Japan Advanced Institute of Science and Technology, Nomi, Ishikawa 923-1291, Japan. [6] JST, PRESTO, 4-1-8 Honcho, Kawaguchi, Saitama 332-0012, Japan. [7] Device-functional Analysis Department, NISSAN ARC Ltd., 1 Natsushima, Yokosuka 237-0061, Japan. [8] State Key Laboratory of Metal Matrix Composites and School of Materials Science and Engineering, Shanghai Jiao Tong University, Shanghai 200030, China. [9] JST, CREST, 4-1-8 Honcho, Kawaguchi, Saitama 332-0012, Japan. Correspondence and requests for materials should be addressed to A.H. (email: hirata@wpi-aimr.tohoku.ac.jp) or to M.C. (email: mwchen@wpi-aimr.tohoku.ac.jp).

Silicon monoxide (SiO), the most common oxide of Si (ref. 1), was first reported by Charles Mabery[2] and has been widely used as surface coatings, insulating layers in integrated circuits, dielectric material in capacitors and anode materials for Li-ion batteries[3–7]. However, since it was discovered, amorphous SiO has been the subject of much investigation and controversy[1,2]. In particular, the atomic structure of SiO has been debated for nearly a century despite numerous experimental and theoretical efforts devoted to this problem[8–12]. The uncommon +2 valence state of Si in the amorphous material cannot be described by either Si–4Si or Si–4O tetrahedra that have been successfully employed in the continuous random network models of amorphous Si and SiO_2 (refs 13–19). Although a random bonding Si-($Si_{4-x}O_x$) tetrahedral configuration can satisfy the requirements in Si valence state and stoichiometry of SiO (ref. 9), experiments and theoretical calculations have suggested that it is inherently unstable and undergoes an unusual disproportionation by forming amorphous Si- and SiO_2-like clusters[8,12]. Local chemical analyses by transmission electron microscopy (TEM)[11] and atom probe tomography[7] have supported the existence of Si clusters in the Si–O matrix. However, it has been long known that the heat of combustion of amorphous SiO is significantly higher than that of an equilibrium mixture of amorphous Si and SiO_2 (ref. 8) and, importantly, the X-ray diffraction (XRD) patterns of amorphous SiO cannot be interpreted by the summation of amorphous Si and SiO_2 spectra[10]. Apparently, amorphous SiO is not a simple composite of amorphous Si and SiO_2 clusters but may have a unique atomic structure, possibly, in the interfacial regions between Si and SiO_2 domains as suggested by Hohl and co-authors[12]. Nevertheless, in spite of extensive investigations by XRD, X-ray photoelectron spectroscopy, X-ray Raman scattering, small-angle X-ray scattering and so on[12,20–22], these techniques only provide average or spectroscopic information on the structure of the amorphous SiO. The unique and well-defined local atomic configurations of SiO have not been directly realized by experiments mainly because of the limitation in spatial resolution of conventional diffraction methods.

In this study we employ our recently developed angstrom-beam electron diffraction (ABED) method[23,24] to investigate the local structure of amorphous SiO, which is supplemented by synchrotron high-energy XRD (HEXRD) and computational simulations based on molecular dynamics (MD) and reverse Monte Carlo (RMC) calculations. The ABED patterns from Si and SiO_2-like nanoscale regions in amorphous SiO are obtained, as well as those from the Si/SiO_2 interfaces. The diffraction intensity profile from the interfaces shows a unique feature different from either Si or SiO_2. On the basis of ABED and HEXRD experiments, we develop a heterostructure model of the disproportionated amorphous SiO which well explains the structure and properties of the amorphous material.

Results

High-energy X-ray diffraction. Commercially available amorphous SiO, fabricated by a vacuum sublimation, was used in this study (Osaka Titanium Technologies Co., Ltd.). Figure 1a shows the X-ray structure factor S(Q) of amorphous SiO (red curve) obtained by HEXRD, together with those of amorphous Si (black curve) and SiO_2 (orange curve) reported in the literature[15,19]. The S(Q) of amorphous SiO appears to be intermediate between those of amorphous Si and SiO_2. However, the summation curve (blue) of amorphous Si and SiO_2 is not fully consistent with that of SiO, agreeing with previous observations[10]. In particular, the first sharp diffraction peak of the summation data splits into two sub-peaks, which cannot reproduce the first peak at $Q \sim 1.8$ Å^{-1} of amorphous SiO. The good agreement between the summation curve and SiO at the high Q portion indicates that the short-range structure of amorphous SiO could be similar to those of Si–4Si and Si–4O tetrahedra in amorphous Si and SiO_2.

Transmission electron microscopy. The amorphous nature of SiO was verified by aberration-corrected TEM and selected area electron diffraction (Fig. 2a and b). The high-resolution TEM image shows typical maze-like contrast of amorphous materials. Neither a nanocrystalline phase nor inhomogeneity can be seen from the phase-contrast micrograph. On the other hand, the high-angle annular dark field scanning TEM (HAADF-STEM) image in Fig. 2c exhibits a faint inhomogeneous contrast with dark and bright nano-sized domains. Since HAADF-STEM is sensitive to the local density and chemistry, the contrast variation may result from the disproportionation of SiO. Separate STEM

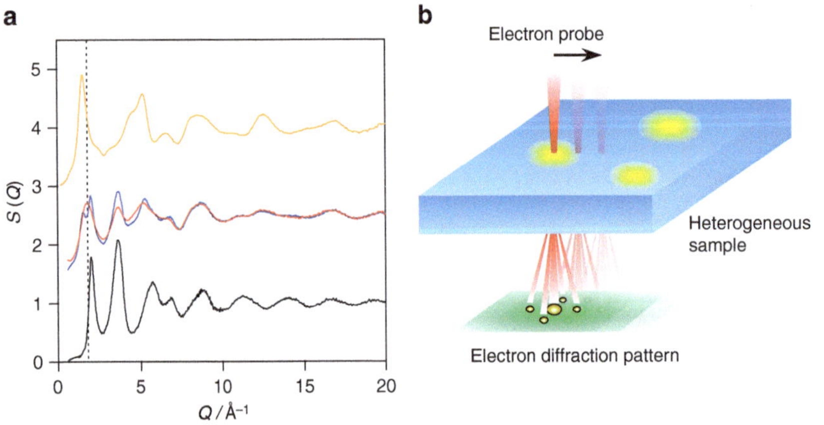

Figure 1 | HEXRD and ABED experiments of amorphous SiO. (**a**) X-ray total structure factors *S(Q)* of amorphous SiO, Si and SiO_2. The X-ray *S(Q)* of amorphous SiO obtained from HEXRD (red) is shown together with the *S(Q)* data of amorphous Si (black) and SiO_2 (orange) reported in the refs 15 and 19, respectively. The *S(Q)* profile for the summation of amorphous Si and SiO_2 is shown as a blue curve. The dotted line is a guide to the eyes. The weighting factors of Si-Si, Si-O and O-O pairs for X-rays used in the RMC modelling are plotted in Supplementary Fig. 10, together with those for electrons. (**b**) The schematic diagram of the ABED measurements of amorphous SiO with nanoscale structural heterogeneity. Nano-regions indicated by yellow colour are structurally different from those indicated by blue colour. Diffraction patterns from yellow and blue nano-regions, and also their interface can be acquired by using ABED technique.

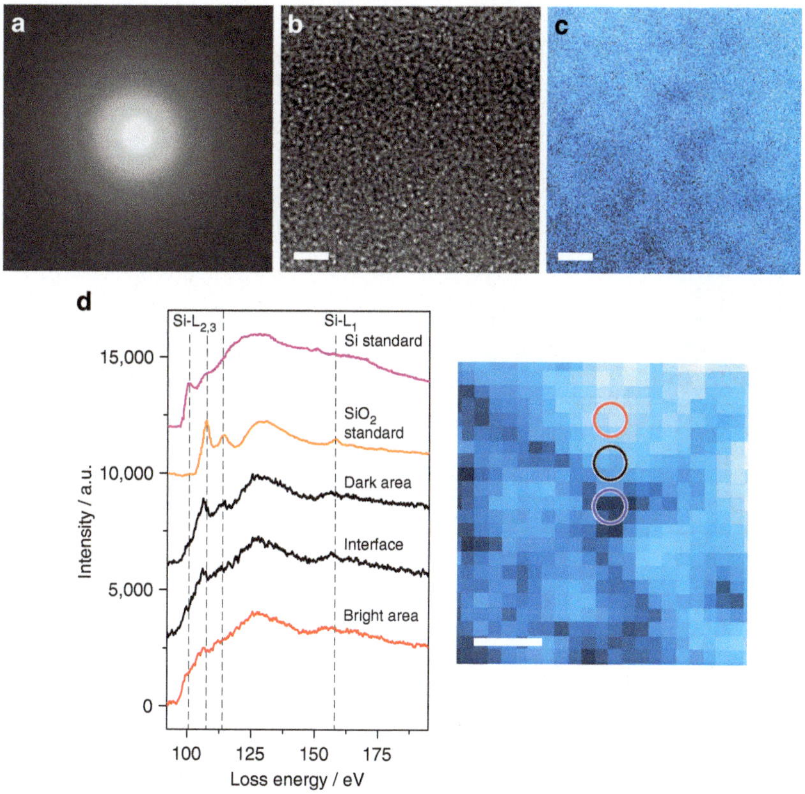

Figure 2 | TEM/STEM micrographs and EELS spectra of amorphous SiO. (**a**) Selected area electron diffraction pattern; (**b**) High-resolution TEM image; and (**c**) HAADF-STEM image of the amorphous SiO. Scale bar, 2 nm. (**d**) EELS profiles (Si-*K* edge) taken from the dark (blue circle), bright (red circle) and interfacial (black circle) regions in the HAADF-STEM image. Standard EELS profiles of amorphous Si and SiO_2 were extracted from the literature[11]. Scale bar, 2 nm.

electron energy loss spectroscopy (EELS) spectra of Si-*L* edges taken from the dark, bright and interfacial regions in the HAADF-STEM image show that the Si bonding states in the dark and bright regions are different (Fig. 2d). The local EELS spectra for the dark regions are similar to those of amorphous SiO_2 while the ones from the bright regions are analogous to amorphous Si. The EELS spectra of amorphous Si and SiO_2 were taken from the standard samples for comparison[11].

Angstrom-beam electron diffraction. The local atomic structure of amorphous SiO was investigated by ABED as illustrated in Fig. 1b. The full width at half maximum electron probe was set as ~0.8 nm to match the size of short-range order in the amorphous material. Different from conventional nano-beam electron diffraction, the convergence angle of the electron beam for ABED is as small as 1.0 mrad to form a nearly parallel electron probe. Figure 3a–c shows three typical ABED patterns obtained from dark, bright and interface regions. The dash lines marked in the ABED patterns represent Debye-Scherrer rings, corresponding to the major peaks found in $S(Q)$ of amorphous SiO, measured by HEXRD and most ABED spots locate at the well-defined rings. Thus the diffraction data from global HEXRD and local ABED are qualitatively consistent with each other. On the basis of the Q values of each diffraction spot and the angles between the diffraction vectors, the corresponding local structures of those ABED patterns can be determined. It is found that the diffraction patterns in Fig. 3a and c are akin to those of amorphous Si and SiO_2 with the smallest Q values close to the first sharp diffraction peaks of Si ($Q \sim 2.0\,\text{Å}^{-1}$) and SiO_2 ($Q \sim 1.5\,\text{Å}^{-1}$) in HEXRD. For amorphous Si-like pattern, even the feature Q values close to the second peak of Si ($Q \sim 3.6\,\text{Å}^{-1}$) can be seen. In contrast, the ABED patterns taken from the interface region (Fig. 3b) cannot be interpreted by either amorphous Si or SiO_2. Interestingly, the smallest Q values ($\sim 1.8\,\text{Å}^{-1}$) in the ABED patterns are very close to the first sharp diffraction peak of SiO ($Q \sim 1.8\,\text{Å}^{-1}$). It is worth noting that the first sharp diffraction peak of SiO is in the middle between the Si ($Q \sim 2.0\,\text{Å}^{-1}$) and SiO_2 ($Q \sim 1.5\,\text{Å}^{-1}$). This indicates that the interface regions may be suboxide SiO_x with a variety of atomic coordinates between Si–4Si and Si–4O. Indeed, the simulated ABED patterns based on suboxide-type tetrahedra match well with the experimental one. The simulated ABED patterns of amorphous Si, SiO_x and SiO_2 clusters are shown in Fig. 3d–f, together with the corresponding atomic models (Fig. 3g–i), which are the typical configurations found in the large atomic models (MD model of pure amorphous Si (Supplementary Fig. 1a), MD–RMC model of amorphous SiO (Fig. 4a) and MD model of pure amorphous SiO_2 (Supplementary Fig. 1b)). Although the atomic arrangements of amorphous SiO are highly disordered, symmetric patterns can be frequently obtained by ABED experiments and by rotating the structural models of the amorphous clusters to match the experimental results.

While it is a powerful approach to determine the local structure of amorphous SiO by carefully analysing individual ABED patterns as shown in Fig. 3, this method may only offer qualitative structure information for highly disordered and heterogeneous materials, such as amorphous SiO because of a large number of possible local atomic configurations. To overcome this shortage of ABED, we selectively integrated ABED patterns taken from different regions (dark, bright and interface), respectively, and transferred the two-dimensional diffraction pattern into intensity profiles to expose the structure of these heterogeneous domains in a statistic manner (Supplementary

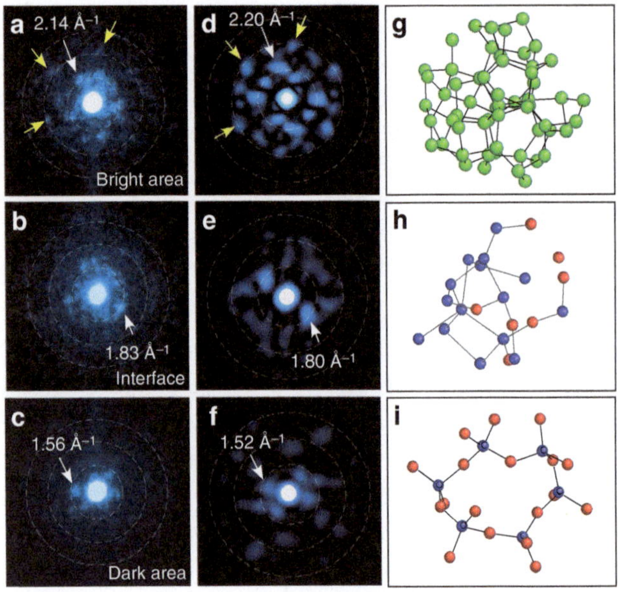

Figure 3 | Experimental/simulated ABED from SiO local structures. (a–c) Typical ABED patterns taken from bright, interface and dark regions in the HAADF-STEM image shown in Fig. 2c. **(d–f)** The simulated ABED patterns based on the atomic models of amorphous Si, interfacial suboxide-type tetrahedra and amorphous SiO_2 in **g–i**. The consistence between the experimental and simulated ABED patterns was achieved by systematically rotating the atomic models to proper orientations that agree with the experimental ABED patterns. The *Q* values as three dashed white rings shown in **a–c** are ~1.8, 3.6 and 5.2 $Å^{-1}$, which correspond to those of the first three peaks in X-ray total structure factor *S*(*Q*) of amorphous SiO shown in Fig. 1a. The atomic models shown in **g–i** were typical configurations found in the large models (MD model of pure amorphous Si (Supplementary Fig. 1a), MD-RMC model of amorphous SiO (Fig. 4a), MD model of pure amorphous SiO_2 (Supplementary Fig. 1b) for **g–i**, respectively).

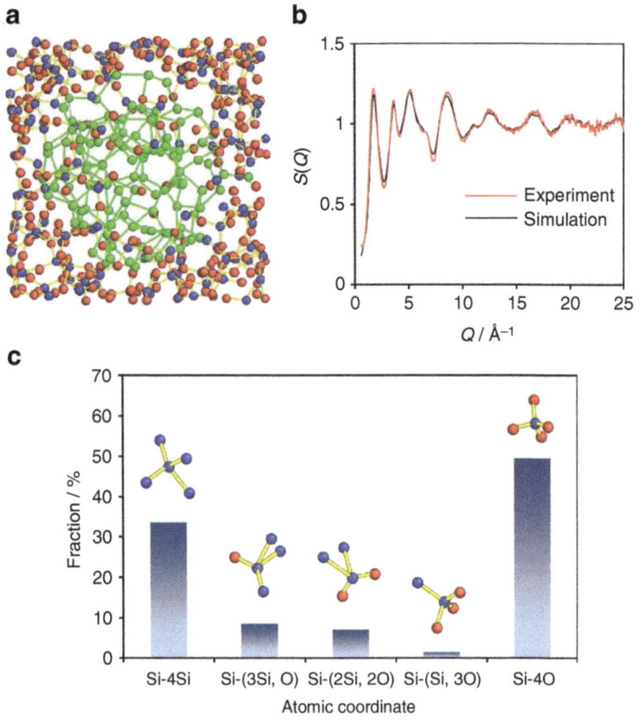

Figure 4 | Atomic model of disproportionated amorphous SiO. **(a)** Reconstructed heterostructure model of amorphous SiO. The inner part corresponds to an amorphous Si cluster and the outer part is amorphous SiO_2 matrix. The blue, red and green circles denote Si and O in amorphous SiO_2 and Si in the Si cluster, respectively. **(b)** Experimental and simulated X-ray total structure factor *S*(*Q*) curves. The red curve denotes the experimental *S*(*Q*) data obtained from the HEXRD measurements. The black curve shows the *S*(*Q*) data obtained from heterostructure model after RMC refinement. **(c)** Fractions of the five atomic coordinates found in amorphous SiO. Si-4Si and Si-4O are from the Si cluster and SiO_2 matrix while Si-(3Si, O) Si-(2Si, 2O) and Si-(Si, 3O) appear at the interfacial regions between the Si cluster and amorphous SiO_2 matrix.

Fig. 2). Apparently, the normalized intensity profiles of ABED from the dark, bright and interface regions are observably different (Fig. 5). In fact, the profiles from dark and bright regions are similar to the normalized intensity profiles of amorphous SiO_2 and Si obtained by selected area electron diffraction while the one from the interface region has unique diffraction features which the amorphous Si and SiO_2 do not have. The statistical analysis further demonstrates the nanoscale disproportionation of amorphous SiO by forming amorphous Si and SiO_2-like clusters as well as unique interface domains.

Structural modelling. On the basis of the ABED and HEXRD results, we constructed an atomic model of amorphous SiO by the combination of MD simulations and RMC modelling. In this approach, the structure of homogeneous amorphous Si and amorphous SiO_2 were first generated by MD simulations and verified by ABED (Fig. 3) and HEXRD. The heterogeneous structure revealed by ABED and HAADF-STEM was constructed by embedding an amorphous Si cluster into the amorphous SiO_2 surrounding (Fig. 4a). The heterostructure was relaxed to reduce the total energy by MD simulations using variable-charge atomic potentials[25,26]. The relaxed model was further fitted to the experimental structure factor *S*(*Q*) by RMC modelling (Fig. 4b and Supplementary Fig. 3)[27]. The final structure model is shown in Fig. 4a, in which the coordinates of Si–4Si and Si–4O tetrahedral from the amorphous Si and SiO_2 regions are still visible. Additionally, the suboxide-type tetrahedral coordinates (Si-(3Si, O), Si-(2Si, 2O) and Si-(Si, 3O)) revealed by ABED are formed at the Si/SiO_2 interface regions during the structure relaxation (Supplementary Fig. 4). We also calculated the effective charges using Voronoi prescription in which atomic cells and volumes are based on geometry and the corresponding charges are calculated from the electron density enclosed inside the cell[28,29]. The electron density of the model was calculated using density functional theory (DFT). The electron charge of Si should be 0 and $+4$ in pure Si and SiO_2 regions, respectively. In the interface regions between Si and SiO_2, on the other hand, the charge of Si varies in a range from 0 to $+4$, depending on the surrounding atoms. The calculations confirm the presence of the suboxide-type tetrahedra in the interface regions (Supplementary Fig. 5). These suboxide-type tetrahedra keep the random networks of amorphous Si and SiO_2 continuously across the interface regions. The number fractions of the five typical atomic coordinates are shown in Fig. 4c. There are considerable amounts of suboxide-type tetrahedra of Si-(3Si, O), Si-(2Si, 2O) and Si-(Si, 3O) from interface regions. These suboxide-type tetrahedra as the transition layers bridge amorphous Si and amorphous SiO_2 by preserving the continuity of the random networks of Si and SiO_2 (Supplementary Fig. 4). Therefore, the formation of the conspicuous interface regions with distinctive suboxide Si–O coordinates appear to be the structural origins of the amorphous

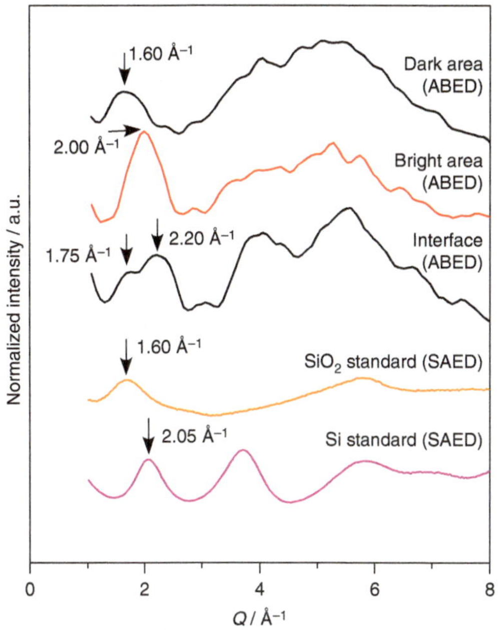

Figure 5 | Integrated intensity profiles of ABED from different domains. The normalized intensity profiles were constructed by integrating 26, 35 and 14 ABED patterns obtained from dark, bright, and interface areas in HAADF-STEM images, respectively. The detailed procedure is described in Supplementary Fig. 2. For comparison, the selected area electron diffraction (SAED) profiles of amorphous SiO_2 and Si are also shown here.

SiO different from the simple mixture of amorphous Si and SiO_2 in the thermal properties and X-ray diffraction.

Raman spectroscopy and XANES. We further conducted Raman spectroscopy and X-ray absorption near-edge structure (XANES) analyses to obtain vibrational and electronic features of the amorphous SiO. In the Raman spectrum of amorphous SiO (Supplementary Fig. 6), besides the vibration bands similar to those of amorphous Si and SiO_2, there are two characteristic peaks at ca. 510 and 645 cm^{-1}. Although further theoretical analysis is required to explain the origins of the bands, based on the local diffraction data achieved in this study, it is most likely that the characteristic Raman bands are the suboxide-type vibration modes from the interfacial regions. Moreover, we measured XANES spectra of as-prepared and annealed SiO specimens (Supplementary Fig. 7). Annealing especially at 1050 °C completely led to a formation of ($Si + SiO_2$) two phases, because the XANES spectrum can be understood as a simple summation of those of Si and SiO_2. On the other hand, the spectrum of as-formed amorphous SiO is quite different from the summation. The unique peak ~1843.5 eV for amorphous SiO found in between the peaks from Si and SiO_2 is presumably due to the suboxide-type tetrahedra at the interface. This result is consistent well with the local EELS experiment shown in Fig. 2.

Discussion

The fundamental challenge in structural characterization of disordered materials is that structural models with different heterogeneities can yield almost identical structure factors because X-ray and neutron diffractions only provide information about averaged structure. Indeed, one can construct a homogeneous model of amorphous SiO with basic units of fourfold silicon (total coordination number of silicon and/or oxygen atoms around a silicon atom is four) by RMC modelling (Supplementary Fig. 8), which is similar to a theoretical homogeneous model reported recently[30]. In principle, RMC modelling can fit any given diffraction data by introducing highly disordered uniform structures although these uniform configurations may not be thermodynamically realistic[31]. In contrast, it is difficult to construct an inhomogeneous amorphous structure only using conventional RMC modelling or MD simulations. In this study, we developed a new approach to construct an amorphous heterostructure model with the appropriate degree of inhomogeneity (~1–2 nm) based on very local ABED and the combination of variable-charge MD simulation and RMC modelling. The experiment-based atomic structure can fully reproduce the experimental $S(Q)$ obtained from the HEXRD measurements in all Q space. Moreover, DFT calculations demonstrate that the structural model of the amorphous SiO is energetically more stable in comparison with the homogeneous atomic model predicted by RMC and the composite model of amorphous Si and SiO_2 constructed only by MD (Supplementary Fig. 9). Although the RMC approach, which is known to fail in determining definite local atomic structure, is also used to generate the initial structure of amorphous SiO, the uncertainness of the RMC can be overcome by the successful utilization of ABED which provides well-defined local atomic configurations in the heterogeneous SiO. Therefore, the experiment-based MD–RMC approach developed in this study could be a powerful and generic method to model disordered materials with structural heterogeneity based on both local and global diffraction data.

In summary, we systematically investigated the atomic structure of amorphous SiO by utilizing state-of-the-art ABED, complemented with synchrotron HEXRD and MD-RMC simulations. The ABED experiment provides direct evidence on the atomic-scale disproportionation in amorphous SiO, predicted by theoretical calculations. The distinctive interfacial structure between amorphous Si and SiO_2 clusters, revealed by the sub-nanoscale electron diffraction, uncovers the structural origins of amorphous SiO, different from the simple mixture of Si and SiO_2. The heterostructure model of disproportionated amorphous SiO resolves the long-standing question on the structure of amorphous SiO. Moreover, the MD–RMC procedure based on both local and global diffraction could be a generic approach to model atomic structure of inhomogeneous amorphous materials.

Methods

Angstrom-beam electron diffraction. We employed a JEOL JEM-2100F TEM/STEM system with double Cs-correctors (operated at 200 kV) for the measurements. All the ABED patterns were recorded with a charge-coupled device camera (Gatan, US1000). A coherent electron beam was produced using a specially designed small-condenser aperture with a diameter of 3.5 μm (Daiwa Techno Systems Co., Ltd.). The convergence angle was estimated to be 1.0 mrad. The instrumental parameters such as the spherical aberration coefficient, defocus and astigmatism were precisely measured using the Ronchigram method[32]. By using a scanning function of the STEM system, we were able to obtain a large number of ABED patterns (more than 10,000 frames) from a thin area near the specimen edge. The specimens with a thickness <5 nm were prepared by the conventional crushing method for the TEM/STEM observations. The ABED patterns were calculated using the multislice simulation software, which has been detailed in ref. 33.

High-energy X-ray diffraction. The HEXRD measurements were performed on powder samples at room temperature using the BL04B2 beamline of SPring-8 (ref. 34). A two-axis diffractometer dedicated to the study of glass, liquids and amorphous materials was employed to measure the diffraction spectra.
The energy of the incident X-rays was 61.4 keV and the diffraction patterns were measured in the transmission geometry. The intensity of the incident X-rays was monitored in an ionization chamber filled with Ar gas and the scattered X-rays were detected using three CdTe detectors. A vacuum chamber was used to suppress air scattering around the sample. The collected data were corrected using a standard programme[34].

Computational simulations. The models of amorphous Si and SiO_2 were constructed by using MD methods (Supplementary Fig. 1). Three-body Tersoff-type[16,35] and two-body Born–Mayer–Huggins-type[36] atomic potentials were employed for amorphous Si and SiO_2, respectively. The MD simulations were performed under constant-NVT conditions. First, 1,000 Si atoms for amorphous Si and 216 Si, and 432 O atoms for SiO_2 were placed randomly in cubic MD cells with periodic boundary conditions. The densities used for amorphous Si and SiO_2 were 2.33 and 2.30 g cm^{-3}, respectively. To construct the model of SiO, we embedded a spherical amorphous Si cluster into the SiO_2 matrix with a density of 2.15 g cm^{-3} and then relaxed at 1,000 K for short periods (5–100 fs) using variable-charge potentials where the electron charges are variable based on the atomic coordination environments[25,26]. The fractions of amorphous Si and SiO_2 in the initial configuration, containing 128 Si atoms of amorphous Si as well as 128 Si and 256 O atoms for amorphous SiO_2, were determined by the stoichiometry of SiO.

The RMC refinement was performed using the RMC+ + code[37]. The atomic configuration produced by the MD simulation was employed as the starting structure. During the simulation based on the X-ray total structure factor, $S(Q)$, the coordination number distribution of oxygen around silicon and that of silicon around silicon, as well as the bond angle distribution of O–Si–O, were constrained to the values obtained from the MD simulation to avoid producing unreasonable disordered local structure.

A simulation box containing 552 atoms obtained by MD and MD-RMC simulations on heterostructure models, and by RMC homogeneous modelling, were used for DFT calculations. The CP2K programme was used in the DFT mode. CP2K employs two representations of the electron density: localized Gaussian and plane wave basis sets. For the Gaussian-based (localized) expansion of the Kohn–Sham orbitals, we used a library of contracted molecularly optimized valence double-zeta plus polarization basis sets[38], and the complementary plane wave basis set had a cutoff of 400 Rydberg for the electron density. The valence electron–ion interaction was based on the norm-conserving and separable pseudopotentials of the analytical form derived by Goedecker, Teter and Hutter[39], and the generalized gradient corrected approximation of Perdew, Burke and Ernzerhof was adopted for the exchange-correlation energy function[40].

References

1. Jutzi, P. & Schubert, U. *Silicon Chemistry: From The Atom To Extended Systems* (Wiley-VCH, 2003).
2. Mabery, C. F. On the composition of certain products from the Cowles electrical furnace. *Am. Chem. J.* **9,** 11–15 (1887).
3. Yang, J. *et al.* SiO_x-based anodes for secondary lithium batteries. *Solid State Ion.* **152–153,** 125–129 (2002).
4. Nagao, Y., Sakaguchi, H., Honda, H., Fukunaga, T. & Esaka, T. Structural analysis of pure and electrochemically lithiated SiO using neutron elastic scattering. *J. Electrochem. Soc.* **151,** A1572–A1575 (2004).
5. Park, C. M. Characterizations and electrochemical behaviors of disproportionated SiO and its composite for rechargeable Li-ion batteries. *J. Mater. Chem.* **20,** 4854–4860 (2010).
6. Yamada, M. *et al.* Reaction mechanism of "SiO"-carbon composite-negative electrode for high-capacity lithium-ion batteries. *J. Electrochem. Soc.* **159,** A1630–A1635 (2012).
7. Sepehri-Amin, H. *et al.* Evidence for nano-Si clusters in amorphous SiO anode materials for rechargeable Li-ion batteries. *Scr. Mater.* **69,** 92–95 (2013).
8. Mellor, J. W. *A Comprehensive Treatise on Inorganic and Theoretical Chemistry* (John Wiley & Sons, 1972).
9. Greaves, G. N. EXAFS and the structure of glass. *J. Non Cryst. Solids* **71,** 203–217 (1985).
10. Yasaitis, J. A. & Kaplow, R. Structure of amorphous silicon monoxide. *J. Appl. Phys.* **43,** 995–1000 (1972).
11. Schulmeister, K. & Mader, W. TEM investigation on the structure of amorphous silicon monoxide. *J. Non Cryst. Solids* **320,** 143–150 (2003).
12. Hohl, A. *et al.* An interface clusters mixture model for the structure of amorphous silicon monoxide (SiO). *J. Non Cryst. Solids* **320,** 255–280 (2003).
13. Zachariasen, W. H. The atomic arrangement in glass. *J. Am. Chem. Soc.* **54,** 3841–3851 (1932).
14. Car, R. & Parrinello, M. Structural, dynamical, and electronic properties of amorphous silicon: An ab initio molecular-dynamics study. *Phys. Rev. Lett.* **60,** 204–207 (1988).
15. Laaziri, K. *et al.* High resolution radial distribution function of pure amorphous silicon. *Phys. Rev. Lett.* **82,** 3460–3463 (1999).
16. Ishimaru, M. Molecular-dynamics study on atomistic structures of amorphous silicon. *J. Phys. Condens. Matter* **13,** 4181–4189 (2001).
17. Mazzi, R. L. & Warren, B. E. The structure of vitreous silica. *J. Appl. Crystallogr.* **2,** 164–172 (1969).
18. Vashishta, P., Kalia, R. K. & Rino, J. P. Interaction potential for SiO_2: a molecular-dynamics study of structural correlations. *Phys. Rev. B* **41,** 12197–12209 (1990).
19. Kohara, S. & Suzuya, K. Intermediate-range order in vitreous SiO_2 and GeO_2. *J. Phys. Condens. Matter* **17,** S77–S86 (2005).
20. Sternemann, C. *et al.* X-ray Raman scattering at the Si LII, III-edge of bulk amorphous SiO. *J. Phys. Chem. Solids* **66,** 2277–2280 (2005).
21. Sakko, A. *et al.* Suboxide interface in disproportionating a-SiO studied by x-ray Raman scattering. *Phys. Rev. B* **81,** 205317 (2010).
22. Feroughi, O. M. *et al.* Phase separation and Si nanocrystal formation in bulk SiO studied by x-ray scattering. *Appl. Phys. Lett.* **96,** 081912 (2010).
23. Hirata, A. *et al.* Direct observation of local atomic order in a metallic glass. *Nat. Mater.* **10,** 28–33 (2011).
24. Hirata, A. *et al.* Geometric frustration of icosahedron in metallic glasses. *Science* **341,** 376–379 (2013).
25. Rappé, A. K. & Goddard, III W. A. Charge equilibration for molecular dynamics simulations. *J. Phys. Chem.* **95,** 3358–3363 (1991).
26. Streitz, F. H. & Mintmire, J. W. Electrostatic potentials for metal-oxide surfaces and interfaces. *Phys. Rev. B* **50,** 11996–12003 (1994).
27. Mcgreevy, R. L. & Pusztai, L. Reverse Monte Carlo simulation: a new technique for the determination of disordered structures. *Mol. Simul.* **1,** 359–367 (1988).
28. Bickelhaupt, F. M., van Eikema Hommes, N. J. R., Guerra, C. F. & Baerends, E. J. The carbon-lithium electron pair bond in $(CH_3Li)_n$ ($n = 1, 2, 4$). *Organometallics* **15,** 2923–2931 (1996).
29. Guerra, C. F., Handgraaf, J. W., Baerends, E. J. & Bickelhaupt, F. M. Voronoi deformation density (VDD) charges: assessment of the mulliken, bader, hirshfeld, weinhold, and VDD methods for charge analysis. *J Comput. Chem.* **25,** 189–210 (2004).
30. Alkaabi, K., Prasad, D. L. V. K., Kroll, P., Ashcroft, N. W. & Hoffmann, R. Silicon monoxide at 1 atm and elevated pressures: crystalline or amorphous? *J. Am. Chem. Soc.* **136,** 3410–3423 (2014).
31. McGreevy, R. L. Reverse Monte Carlo modelling. *J. Phys. Condens. Matter* **13,** R877–R913 (2001).
32. Lin, J. A. & Cowley, J. M. Calibration of the operating parameters for an HB5 STEM instrument. *Ultramicroscopy* **19,** 31–42 (1986).
33. Kirkland, E. J. *Advanced Computing in Electron Microscopy* (Plenum, 1998).
34. Kohara, S. *et al.* Structural studies of disordered materials using high-energy x-ray diffraction from ambient to extreme conditions. *J. Phys. Condens. Matter* **19,** 506101–506115 (2007).
35. Tersoff, J. Empirical interatomic potential for silicon with improved elastic properties. *Phys. Rev. B* **38,** 9902–9905 (1988).
36. Fumi, F. G. & Tosi, M. P. Ionic sizes and Born repulsive parameters in the NaCl-type alkali halides I. *J. Phys. Chem. Solids* **25,** 31–43 (1964).
37. Gereben, O., Jóvári, P., Temleitner, L. & Pusztai, L. A new version of the RMC+ + Reverse Monte Carlo programme, aimed at investigating the structure of covalent glasses. *J. Optoelectron. Adv. Mater.* **9,** 3021–3027 (2007).
38. VandeVondele, J. & Hutter, J. Gaussian basis sets for accurate calculations on molecular systems in gas and condensed phases. *J. Chem. Phys.* **127,** 114105–114109 (2007).
39. Goedecker, S., Teter, M. & Hutter, J. Separable dual-space Gaussian pseudopotentials. *Phys. Rev. B* **54,** 1703–1710 (1996).
40. Perdew, J. P., Burke, K. & Ernzerhof, M. Generalized gradient approximation made simple. *Phys. Rev. Lett.* **77,** 3865–3868 (1996).

Acknowledgements

This work was sponsored by JST-CREST 'Phase Interface Science for Highly Efficient Energy Utilization', JST (Japan), and the Fusion Research Funds from 'World Premier International (WPI) Research Centre Initiative for Atoms, Molecules and Materials' programme of the MEXT of Japan, the Grant-in-Aid for Exploratory Research (No. 15K14116) from Japan Society for the Promotion of Science (JSPS) and Cross-ministerial Strategic Innovation Promotion Programme (SIP, Structural Materials for Innovation D72) of the Ministry of Agriculture, Forestry and Fisheries of Japan, JST-PRESTO "Materials research by Information Integration" Initiative (MI2I) project of the Support Program for Starting Up Innovation Hub from JST and the MEXT Project of the SR Centre in Ritsumeikan University for the XAFS measurement (Proposal No. R1405). This work was also sponsored by MOST 973 of China (Grant No. 2015CB856800) and National Natural Science Foundation of China (Grant No. 11327902, 51271113). The synchrotron radiation experiments were performed at the beamline BL04B2 of SPring-8 with the approval of the Japan Synchrotron Radiation Research Institute (JASRI; Proposal No's. 2013B1541 and 2014A1545). We thank Dr Naoto Kitamura at Tokyo University of Science and Mr. Ichiro Ishikawa at NISSAN ARC Ltd. for their technical assistance. The authors gratefully acknowledge Prof. Sjoerd Roorda at Université de Montréal and Prof. Laszlo Pusztai at Wigner Research Centre for Physics for providing us with diffraction data of amorphous silicon.

Author contributions

A.H. and M.C. conceived this study. A.H. performed ABED, STEM and EELS experiments and analyses. M.C. contributed to the data analysis and the technical development of ABED. S.K., T.A. and H.I. conducted to HEXRD experiments and analyses. C.Y. and H.I. implemented XANES experiments. Y.T. and T.F. performed Raman scattering experiments. A.H. and S.K. contributed to the structural modelling using computational methods. S.K. performed DFT calculations. T.A., M.A. and H.I. prepared the samples. A.H., S.K. and M.C. wrote the manuscript and all the authors discussed the results and commented on the manuscript.

Additional information

Competing financial interests: The authors declare no competing financial interests.

3

Structural control of mixed ionic and electronic transport in conducting polymers

Jonathan Rivnay[1,†], Sahika Inal[1], Brian A. Collins[2,3], Michele Sessolo[4], Eleni Stavrinidou[1], Xenofon Strakosas[1], Christopher Tassone[5], Dean M. Delongchamp[2] & George G. Malliaras[1]

Poly(3,4-ethylenedioxythiophene) doped with poly(styrenesulfonate), PEDOT:PSS, has been utilized for over two decades as a stable, solution-processable hole conductor. While its hole transport properties have been the subject of intense investigation, recent work has turned to PEDOT:PSS as a mixed ionic/electronic conductor in applications including bioelectronics, energy storage and management, and soft robotics. Conducting polymers can efficiently transport both holes and ions when sufficiently hydrated, however, little is known about the role of morphology on mixed conduction. Here, we show that bulk ionic and electronic mobilities are simultaneously affected by processing-induced changes in nano- and meso-scale structure in PEDOT:PSS films. We quantify domain composition, and find that domain purification on addition of dispersion co-solvents limits ion mobility, even while electronic conductivity improves. We show that an optimal morphology allows for the balanced ionic and electronic transport that is critical for prototypical mixed conductor devices. These findings may pave the way for the rational design of polymeric materials and processing routes to enhance devices reliant on mixed conduction.

[1] Department of Bioelectronics, Ecole Nationale Supérieure des Mines, CMP-EMSE, MOC, Gardanne 13541, France. [2] Material Science and Engineering Division, National Institute of Standards and Technology (NIST), Gaithersburg, Maryland 20899, USA. [3] Department of Physics and Astronomy, Washington State University, Pullman, Washington 99164, USA. [4] Instituto de Ciencia Molecular, Universidad de Valencia, Paterna 46980, Spain. [5] Stanford Synchrotron Radiation Lightsource (SSRL), SLAC National Accelerator Laboratory, Menlo Park, California 94025, USA. † Present address: PARC, a Xerox Company, 3333 Coyote Hill Road, Palo Alto, California 94304, USA. Correspondence and requests for materials should be addressed to J.R. (email: rivnay@gmail.com).

Mixed electronic/ionic transport in conducting polymers (CPs) is attracting a great deal of attention due to a host of new devices that leverage the coupling of these carriers to enable new modes of operation. One example is the organic electrochemical transistor (OECT), which consists of a CP layer in which ions injected from an electrolyte modify hole conductivity. Owing to their high transconductance[1], these devices have been shown to be efficient transducers of biological activity, with applications ranging from neural interfacing[2,3] to biosensing[4]. The inverse mode of operation, where a CP film injects ions into an electrolyte, has been used for drug delivery with exceptional spatio-temporal control[5,6]. In addition to applications in bioelectronics, OECTs are being developed for printed electronics[7,8], and more recently, neuroinspired electronics[9–11]. Mixed transport in CPs is also leveraged in energy applications including batteries/supercapacitors[12], electrochromic windows[13] and in electro-mechanical actuators for soft robotics[14,15]. Despite the many opportunities enabled by mixed conduction in CPs, the lack of fundamental understanding of combined ionic/electronic transport and its connection to morphology hinders rational materials and device design.

A prototypical CP used in many mixed conductor devices is poly(3,4-ethylenedioxythiophene) doped with poly(styrene sulfonate); (PEDOT:PSS). This material has attracted interest for its high hole conductivity (> 1,000 S cm^{-1}), high stability and for its commercial availability as a dispersion for solution processing[16]. PEDOT:PSS dispersions are typically described as being composed of gel-like particles comprising a polyanion (PSS^{-})-rich shell which helps to stabilize the PEDOT-enriched particles in aqueous solvents[17,18]. PEDOT oligomers are believed to polymerize onto the PSS template (Fig. 1a), meaning that PEDOT itself is inherently ill-characterized as far as molecular weight and polydispersity. The gel particles (Fig. 1b) deposit in what has been described as a pancake-like morphology[19–21], and show signs of π-stacking[22–25] (Fig. 1c,d). PEDOT:PSS structure has been intensely studied by a variety of techniques including electron microscopy, X-ray scattering, elemental and surface-sensitive scanning probes[17,22–26], largely with the goal of understanding and enhancing hole transport to make better electrodes for optoelectronic devices. Changes in formulation content, processing or post-processing have been shown to lead to enhancements in hole conductivity by inducing morphological re-arrangements, minimizing excess dopant phase (that is, insulating PSS) or both. A thorough review of the methods and suggested mechanisms to enhance conductivity can be found in ref. 27.

Ion transport in PEDOT:PSS (and in CPs, in general) has been explored only on a limited basis. The presence of electronic carriers in the film renders most techniques used to study ion conduction inutile. Nevertheless, the electrochromic nature of PEDOT:PSS, which gives rise to changes in film absorption on electrochemical (de)doping, allows us to follow ion motion in the film[28–30], and to extract ion-drift mobility[31]. Similar to ion-exchange membranes such as Nafion[32], hydration of PEDOT:PSS was found to establish ion-transport pathways, leading to a high ion mobility[31]. The nature of these ion pathways and their connection to film structure and morphology is not well-understood. Moreover, unlike the case of hole transport, attempts to modify the structure of PEDOT:PSS in a systematic manner and to monitor the effect on ion transport are lacking. It is important, for example, to deduce if hole and ion transport can be improved simultaneously, and, if not, whether an optimal morphology exists wherein device performance is maximized. This calls for a comprehensive study where both ion and hole transport are measured in films processed to have different structure and morphology. Thorough quantification of the latter is particularly important, as there have been differing reports on the influence of the same additive on film morphology[24,25]. This is a challenging proposition, as PEDOT and PSS have similar densities and bonding environments (limiting X-ray scattering contrast), and are disordered on multiple length scales.

Here, we investigate the microstructure and morphology of PEDOT:PSS films modified with varying amounts of co-solvent dispersion additive, ethylene glycol (EG), and relate these findings to measurements of electronic and ionic transport. To overcome the low contrast and a relatively disordered film microstructure, we use Synchrotron radiation and resonant soft X-ray scattering (rSoXS) and investigate the multi-scale microstructure of PEDOT:PSS films. The combined use of carbon-edge X-ray scattering and spectroscopy allows the deduction of the absolute nano-scale domain composition of the films. We find that the PEDOT:PSS-rich cores and PSS-rich matrix are surprisingly impure even with a large dispersion content of EG, but that the co-solvent enhances film heterogeneity. We then investigate both ionic and electronic (hole) transport in the same formulations. Ionic transport is probed using a spectroscopic analysis of the electrochromic moving front, revealing an ion motion heavily affected by the meso-scale domain size and composition. We find that ion and hole transport are oppositely affected by the addition of EG—a finding that is corroborated by the structural work herein. We show that this suggests an optimal morphology exists, enabling enhanced ionic mobility and hole conductivity, and leading to improved transconductance in OECTs. Therefore, this

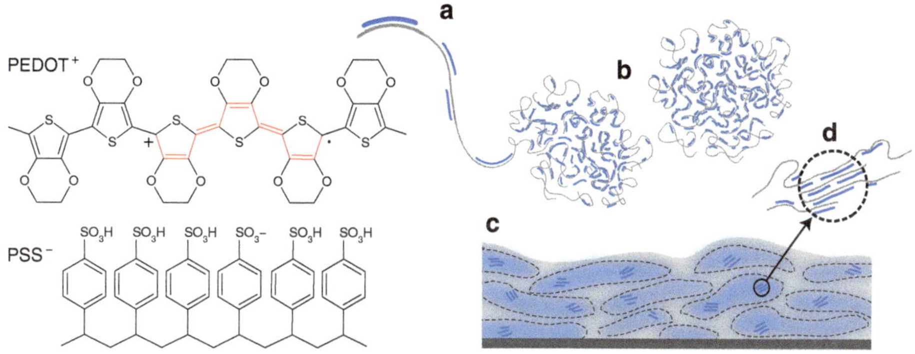

Figure 1 | PEDOT:PSS structure and morphology. The chemical structure of PEDOT:PSS and commonly described microstructure of the CP system (**a**) synthesis onto PSS template, (**b**) formation of colloidal gel particles in dispersion and (**c**) resulting film with PEDOT:PSS-rich (blue) and PSS-rich (grey) phases. (**d**) Aggregates/crystallites support enhanced electronic transport.

work presents a study of the bulk composition and microstructure of an organic electronic system, and its effect on mixed ionic and electronic transport.

Results

Structure in PEDOT:PSS films. The low-density contrast in PEDOT:PSS largely limits the utility of traditional electron or X-ray methods, however, the use of synchrotron radiation, as well as rSoXS allows for multi-scale structure determination. Wide-angle grazing incidence scattering (GIXS) of PEDOT:PSS films allows one to focus on the relative molecular packing arising from aggregation/crystallization[25,33]. Figure 2a and Supplementary Fig. 1 show the changes in wide-angle scattering on addition of EG as co-solvent to the PEDOT:PSS formulation. Dispersion co-solvents should not be present within the film (after film formation, annealing and vacuum treatment)[34,35]. π-stacking scattered intensity at $Q = 1.83\,Å^{-1}$ is enhanced with higher EG formulation content. This enhancement is accompanied by a narrower peak width, indicating an increase in coherence length from $\cong 1$ to 3 nm (due to larger crystallite/aggregate size, or a higher degree of aggregate ordering)[36]. A tendency for the π-stacked aggregates to slightly favour a face-on texture is observed, similar to previous reports[25,33].

Meso-scale structure of PEDOT:PSS films is investigated using C-edge rSoXS. The optical constants are determined from films of Na:PSS and PEDOT:Cl, which are dominated by the bonding of the polymeric/oligomeric species in the soft X-ray regime. The Na^+ and Cl^- ions contribute a linear background absorbance but do not participate in the energetic transitions at the C-edge, enabling facile removal of their signals. The calculated materials and vacuum contrast functions can be computed[37], and are shown in Supplementary Fig. 2. Two energies are selected for investigation, the pre-edge energy 270 eV, and the 1 s to π^* C=C resonance at 285.1 eV. Figure 2b shows scattering profiles of PEDOT:PSS samples in the transmission geometry. The scattered intensity at $Q < 0.01\,nm^{-1}$ ($2\pi/Q > 600$ nm) matches well between the two incident energies tested for all samples measured, indicating a predominant contribution from vacuum contrast (that is, film roughness). The scattering feature in the $Q \cong 0.1\,nm^{-1}$ range is enhanced at resonance owing to materials (PEDOT to PSS) contrast (see Supplementary Fig. 2) and is thus a measure of the bulk meso-scale domain morphology. The upturn at the highest Q-values is primarily attributed to X-ray fluorescence (scaled by Q^2), since it follows the energy dependence of absorbance and approaches a constant $I(Q \to \infty)$. Some of the scattering at the highest Q-values may also originate from nanocrystals within PEDOT:PSS-rich domains. Previous works suggest the overall film morphology consists of closely packed PEDOT-rich domains in a PSS-rich matrix (Fig. 1c)[17,20,21,26] thus, to quantify how the average meso-scale domain size and purity varies with EG, the features on a Lorentz plot $IQ^2(Q)$ are fit to a log-normal peak with a cubic background (Supplementary Fig. 3). The peak intensity (area) is found to scale linearly by one order of magnitude when EG content ranges from 0 to 50 vol%. This suggests a substantial increase in either domain purity or volume fraction, in concordance with the Porod Invariant formalism (see Supplementary Note 1 for specifics of the calculation)[38]. The feature position (Fig. 2c, top) is indicative of domain spacing (gel-like particle spacing in plane), and indicates the size of particles increases from 16 to 42 nm over the same range of EG concentrations, a size scale range in accordance with others[23,26]. Taken together, these data strongly suggest an increase of film heterogeneity with the addition of EG rather than having a homogenizing/inversion effect.

Compositional analysis using transmission NEXAFS of the same samples (same X-ray beam on the same spot as scattering) reveals only minimal variation in average film composition[39], and corroborates the expectation that EG is effectively absent in the films[34,35]. The linear combination of the background-subtracted PEDOT:Cl and Na:PSS NEXAFS spectra is used to determine film thickness and composition. The PEDOT to PSS content in the film varies around 1:1.5 to 1:1.8 by weight, without a consistent trend with EG concentration, and is slightly lower than the as-received content of 1:2.5. Since there is not a large difference in the scattering population, the increased scattering intensity must arise from increased contrast between the scatterers. These findings confirm the domain purification on addition of the co-solvent, rather than loss of overall PSS content during the casting process.

Combining the NEXAFS and rSoXS analyses, we calculate an estimate of the absolute PEDOT concentration of the PEDOT:PSS-rich phase and surrounding PSS-rich matrix (Fig. 2c, bottom). The composition analysis (detailed in the

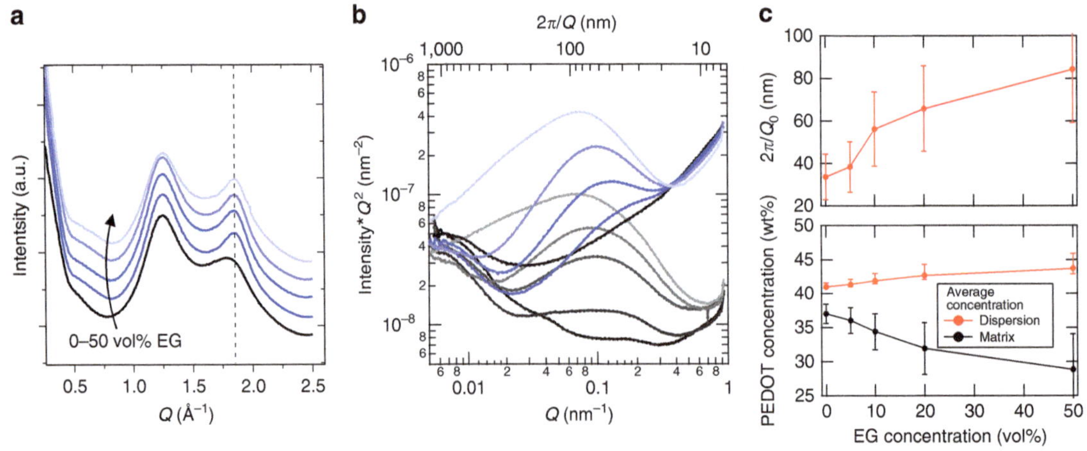

Figure 2 | Microstructure and morphology of PEDOT:PSS from X-ray scattering. (**a**) Linearly offset wide-angle GIXS of PEDOT:PSS films cast from dispersions with EG concentration of 0, 5, 10, 20 and 50 vol% (varying from dark to light colours). The dotted vertical line is the position of the PEDOT:PSS π-stacking scattered intensity (020). (**b**) C-edge rSoXS of PEDOT:PSS films at pre-edge energy (270 eV, black) and at the C 1s to π^* resonance (285.1 eV, blue). (**c**) Domain spacing from peak position ($2\pi/Q_0$) of the features observed at resonance in **b** and quantification of the absolute domain purity (as PEDOT concentration) from peak area. Uncertainties for purity are derived in Supplementary Fig. 3 (mainly from uncertainty in domain volume fraction), while uncertainties for spacing are calculated using the peak width to demonstrate the distribution of domain spacings.

Supplementary Figs 3–5) reveals that neither phase in the film is pure. As EG content in the dispersion is increased, the PEDOT concentration in the film's collapsed gel particles increases mildly from 41 to 44 wt%. Considering our X-ray diffraction evidence of PEDOT chain packing along the π-stacking direction, such a low PEDOT concentration suggests the aggregates that lie within these domains consist of alternating PEDOT and PSS chains along the non-π-stacking direction (similar to that of PEDOT: tosylate)[40], rather than a crystal of pure PEDOT as suggested by others[24]. The PEDOT concentration varies more significantly in the matrix, where it decreases from 37 to 29 wt% with increasing EG co-solvent. These calculations assume a constant 0.75 volume fraction of the PEDOT enriched gel particles—an assumption based on cross-sectional AFMs reported by Nardes *et al.*[23]—which show a skin layer around the gel particles in this approximate volume ratio. The possibility of a changing volume ratio is included in the uncertainty calculations (along with other possible sources—see Supplementary Fig. 3). It is unlikely that the marked increase in scattering intensity, ΔTSI, originates from a change in domain volume fraction since our calculations show that such a ΔTSI would require an unrealistic volume fraction of >0.9. Indeed, with rSoXS results indicating a purification and rarefication of PEDOT: PSS-rich domains with the EG co-solvent, a pure skin layer covering these domains would be expected to pose a significant barrier to hole conductivity. Although our measurements cannot address connectivity directly, the calculated PEDOT concentration in this skin layer above 20 wt% indicates a possible means for percolation of charges between the PEDOT:PSS-rich particles.

To the best of our knowledge, this is the first time an analysis of absolute nano-scale domain composition in an all-polymer blend has been achieved using a combination of NEXAFS and X-ray scattering at a resonant photon energy of a molecule. It should be noted that the resonant scattering experiments described above are performed in transmission, with the incident X-ray beam normal to the plane of the film. Therefore, only the lateral meso-scale structure is quantified. Previous works show the collapsed gel particles (the pancake-like morphology) possess an approximate aspect ratio of 0.1–0.5 (ref. 41), with the out-of-plane direction possessing a Q-value out of the range of the current experimental setup. Despite this known anisotropic morphology, the trends in domain composition should not be significantly affected, and are discussed in detail in Supplementary Fig. 4.

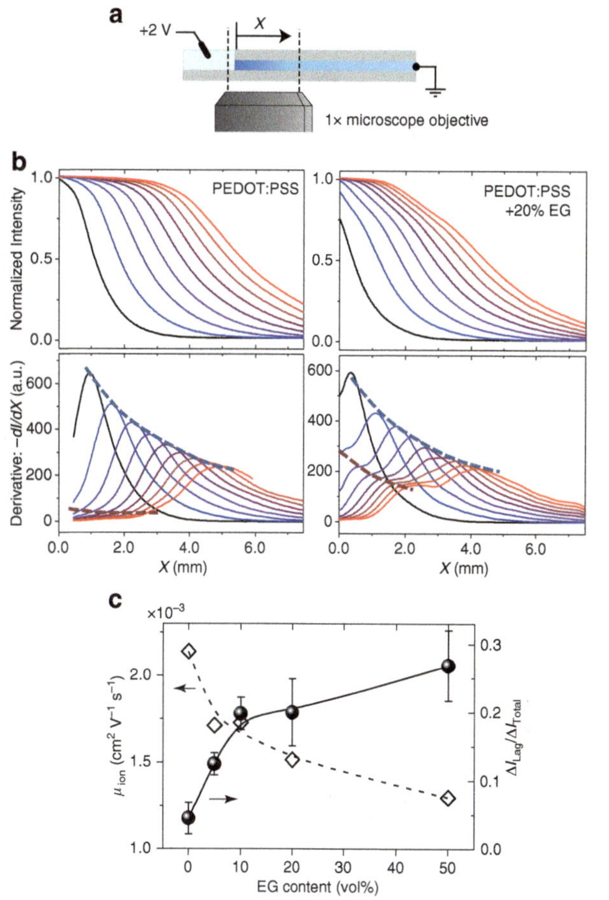

Figure 3 | Co-solvent influences ion transport. (**a**) Temporal characteristics of spatial moving front profiles due to ion transport. (**b**) 1D normalized-intensity profiles as a function of distance, X, from the electrolyte-PEDOT:PSS interface (from $t=5$ s, blue, to $t=45$ s, red; $\Delta t=5$ s) are obtained from time-lapse optical micrographs. Use of EG as a co-solvent affects the shape of the moving front profiles, which is highlighted by investigating the spatial derivative of the intensity profiles, revealing a 'leading' and 'lagging' front (blue and red dotted lines, respectively). (**c**) The effective ion mobility as derived from the moving front data (open symbols/dotted line), and the fractional contribution of the lagging front ($\Delta I_{Lag}/\Delta I_{Total}$) (solid symbols/line), as a function of the EG content in the dispersion. Error bars were determined from the standard deviation of $\Delta I_{Lag}/\Delta I_{Total}$ from 5-10 consecutive spatial derivative profiles in the $t=5$-50 s range (the inclusion of a specific profile in quantification depended on the ability to fit both leading and lagging contributions within the experimentally restricted spatial range, X).

Ionic transport in PEDOT:PSS. We probed ion transport in PEDOT:PSS films with a one-dimensional version of the 'moving front' experiment[31]. The method involves monitoring visible light transmission through an electrochromic film as it is (de)doped due to lateral injection of ions from a planar junction with an electrolyte (Fig. 3a). The drift mobility of ions is then determined from the time-dependent position of the moving front, by assuming a simple equivalent circuit model[42]. Applying this methodology to our EG-containing PEDOT:PSS formulations allows us to relate the meso-scale structure of films, discussed above, with ion motion therein.

The electrochromic moving front is represented as a 1D normalized-intensity profile at each time point (Fig. 3b, top and Supplementary Fig. 6). The effective mobility of K^+ ions, μ_{ion}, drops consistently and in a reproducible manner from $2.2\times10^{-3}\,cm^2\,V^{-1}\,s^{-1}$ to $1.3\times10^{-3}\,cm^2\,V^{-1}\,s^{-1}$ for films prepared with 0 and 50 vol% EG, respectively (Fig. 3c). The EG-induced reduction in effective ionic mobility is most likely due to the changes in meso-scale arrangements.

The subtle change in the shape of the moving front profiles on addition of EG reveals that the ion motion can be described by two moving fronts. This behaviour is more clearly observed in the derivative of the front profiles (Fig. 3b, bottom and Supplementary Fig. 6, bottom). The derivative profiles can be used to deconvolute the moving front into a 'leading' front and a 'lagging' front (Supplementary Fig. 7), whereby the fraction of PEDOT:PSS de-doping attributed to each wave of effective ion motion is deduced. With 50 vol% EG, up to 30% of the complete film de-doping is due to the lagging front alone—a contribution which is negligible when no co-solvent is used (Fig. 3c). The evolution of the lagging front is correlated with the position and breadth of the leading front, ruling out the possibility of different (impurity) cations.

To further deduce the origin of different moving fronts observed in EG-modified films, we add an element of absorption spectroscopy to the lateral moving front experiment and monitor

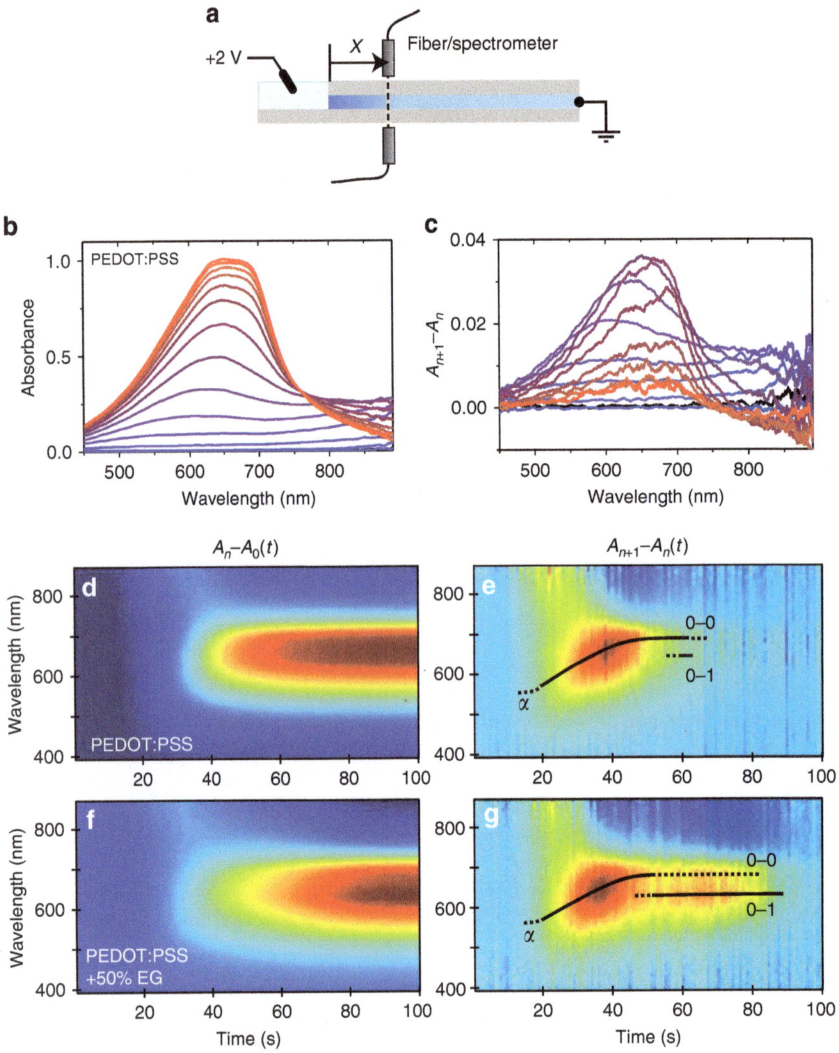

Figure 4 | Spectro-temporal characteristics of ion transport and the effect of EG content. (**a**) Spectro-temporal data are collected with a white-light fibre spectrometer at a fixed position ($X \cong 4$ mm) from the edge of the well. (**b**) Absorbance data ($A_n - A_0$, where the subscripts represent time $t = n$ and $t = 0$ s) of a hydrated PEDOT:PSS film (0 vol% EG) as the de-doping front moves past the position of the optical fibre ($t = 0$ s, blue to $t = 60$ s, red). (**c**) Temporal derivative spectra ($A_{n+1} - A_n$) of the same film revealing the variable electrochromic signatures associated with K^+ transport as the moving front passes. (**d**,**e**) Colour plot representations of the data in **b**,**c**. (**f**,**g**) Colour plot spectra of PEDOT:PSS with 50 vol% EG in the formulation. Note that the time to reach saturation increases with EG formulation content; the added time is due to a lengthy tail of aggregate absorption, as noted by the 0-0 and 0-1 vibronic features (**e**,**g**).

the time-resolved ion motion through the local microstructure. When a cation effectively de-dopes a PEDOT chromophore, the associated change in the visible absorption spectrum of the film can reveal not only the presence of ions, but also the local molecular order of the chromophore in question. The higher energy feature of a polythiophene absorption spectrum is typically associated with amorphous molecular arrangements, as in solution phase, while the low-energy vibronic progression is attributed to weakly interacting H- and J-aggregates[43]. The intensity and ratios of the vibronic peak progression are correlated with the degree of local order of the conjugated backbone[44].

A fibre light source/spectrometer is mounted at a distance of $X \cong 4$ mm from the edge of the CP film/electrolyte interface, in the same configuration as that of the microscopy measurements (Fig. 4a). The ultraviolet–visible spectra are recorded in 1 s intervals as a +2 V bias is applied across the film (Fig. 4b,d,f). To deduce the time course of the instantaneous optical signature and to better understand the local ionic environment, we plot the absorption of individual 1 s intervals (Fig. 4c,e,g). For all samples, we observe first the arrival of the front at $t \cong 10$ s, after which the spectra evolve from a broad peak absorption at ~550–600 nm, to A_{0-1} and A_{0-0} vibronic features at 625 and 685 nm, respectively, over a 40–50 s time span. The sustained addition of vibronic absorption persists in EG-rich samples (Fig. 4g) for an additional 30–40 s (for 50 vol% EG), and is responsible for the delay in reaching saturation in Fig. 4f. The spectroscopy data thus suggest that the rapidly moving leading front is due to ions passing through regions of disordered PEDOT chromophores (most likely those in the PSS-rich matrix), while the lagging front is dictated by ion penetration into aggregate-rich domains such as those in the PEDOT:PSS-densified cores. The persistence of the aggregate absorption recovery within films cast from a higher EG content directly relates to the existence of a lagging front, and thus suggests that aggregates and domain content dictate effective ion mobility.

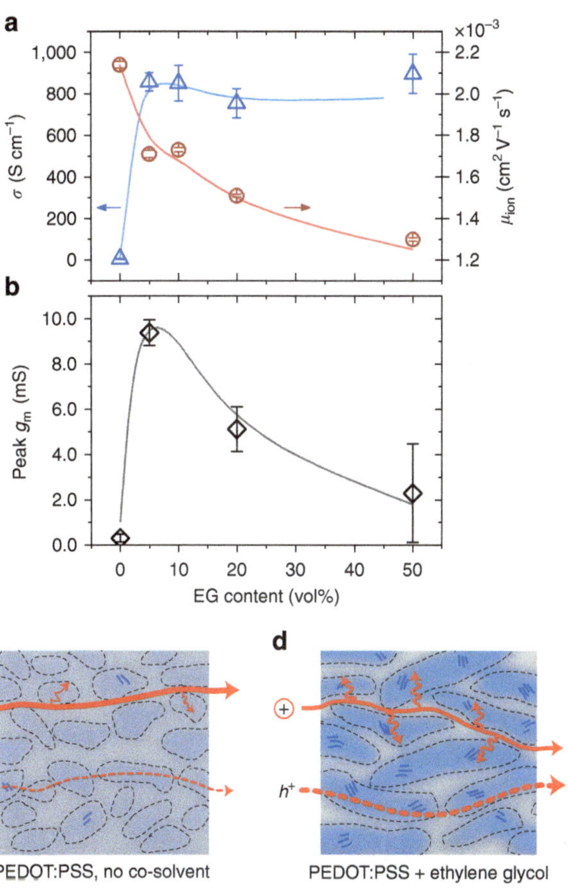

Figure 5 | Effect of mixed conduction in PEDOT:PSS on OECT performance. (**a**) Electronic conductivity (blue), K^+ ion mobility (red) and (**b**) OECT peak transconductance (g_m) at $V_D = -0.6$ V as a function of EG formulation content. Devices have dimensions $W/L = 50\,\mu m/50\,\mu m$ and an equivalent thickness of $d = 200$ nm. Solid lines are guides to the eye. Note that neither efficient electronic nor ionic transport alone is sufficient for achieving the highest g_m. A schematic summary of the morphological changes and associated transport of ions and holes viewed in the cross section of bulk film of PEDOT:PSS (**c**), and PEDOT:PSS with EG (**d**). Addition of EG varies the PEDOT content (shade of blue), size/purity of PEDOT:PSS-rich phases, and amount of PEDOT:PSS aggregates/crystallites. Error bars in **a** represent standard deviations of conductivity values from $n = 8$ measurements, while those in **b** are from OECT measurements of four transistors for each EG condition.

Implications for device performance. Figure 5a shows that the conductivity improves from 6 to 800 S cm^{-1} upon addition of just 5 vol% EG, in agreement with previous reports[45,46]. Increasing the EG content >10 vol% does not lead to a notable change in conductivity of the films. The trends in effective ion mobility in films of the same formulation are shown overlayed in Fig. 5a. The film cast from PEDOT:PSS dispersion without EG displays maximum ionic mobility but poor electrical conductivity, while on addition of 50 vol% EG to the dispersion, the electrical conductivity is maximized at the expense of ionic mobility. Figure 5b shows that neither of these cases is optimal, but it is the 5 vol% EG formulation which yields OECTs with the highest transconductance (g_m) (Supplementary Fig. 8). The formulation with 0 vol% EG leads to an $\cong 30\times$ lower g_m despite showing the best ionic mobility, and the one with 50 vol% EG shows $\cong 4\times$ lower g_m despite displaying the highest electrical conductivity (Fig. 5b).

The difference in effective ionic mobility between samples with low (5 vol%) and high (50 vol%) EG content may seem small in comparison with the observed variation in OECT transconductance. The ion mobilities determined from the moving front displacement are, in fact, dominated by lateral drift of ions, whereas operation of the OECT implies initial through-plane penetration of ions, such that a direct correlation of ionic mobility to device performance is not straightforward. Especially for the films prepared from the dispersions of high EG content, the fractional contribution of the lagging front on the total moving front increases substantially (see Fig. 3c). Furthermore, due to the larger lateral extent of domains, these films may show a higher degree of anisotropy, with more tortuous pathways in the out-of-plane direction through the PSS-rich regions. Taken together, ion mobilities determined from lateral moving front experiments, in particular for the films cast from high-EG-fraction dispersions, are likely overestimations of the through-plane ion transport.

Discussion

Correlating device performance with transport and ultimately film microstructure/morphology is the primary route by which to design new formulations and materials. Addition of the higher boiling point polar solvent, EG, in the casting dispersion leads to changes in the film structure at both the molecular and meso-scale (see Fig. 5c,d). EG leads to an increase in PEDOT aggregation, and slightly tighter π-stacking, in agreement with the work of Palumbiny *et al.*[25,33] At the same time, the domain size of the gel-like particles is known to grow[26,47], increasing by a factor of 3 in the in-plane direction in this work. The growth of domain size with higher EG content is accompanied by a purification of the PEDOT:PSS-rich cores (towards the ultimate limit of 1:1 PEDOT:PSS), and of the PSS matrix ($\sim$30% reduction in PEDOT; a lower purity than that observed in drop casting studies)[48].

The coarsened, heterogeneous film morphology of films cast with EG is responsible for the observed dual moving front profiles in Fig. 3b, and leads to lower effective ion mobility. The spectroscopic investigations show that films cast from formulations with higher EG content exhibit a longer, lagging tail of visible absorption recovery (de-doping) attributed to the interaction of ions with aggregated PEDOT chromophores. This aggregate absorption is responsible for the enhanced contribution of the lagging front in the 50 vol% EG film—up to 27% of the total extent of de-doping. We propose, therefore, that the transport of ions takes place through the PSS-rich regions (via the initial leading front), but that the effective ionic mobility is limited by the subsequent diffusion of ions into the densified PEDOT:PSS-rich domains (which slows the total de-doping front).

The observed improvement of hole transport and deterioration of ion transport on the addition of EG thus agree with changes in the multi-scale morphology. At the inter-molecular level, the quality of hopping between molecules is aided by tighter π-stacking and enhanced aggregation[33], largely improving electronic conductivity via an enhancement in hole mobility. At the meso-scale, the growth and purification of PEDOT:PSS-rich domains should aide in electronic transport, however, the percolative nature of the network—the connectivity of the collapsed gel-like particles—cannot be probed with the scattering techniques used. The increase in PEDOT content in the cores, due to potential recruitment of previously inactive PEDOT into the electrically percolating network, may serve to further enhance the film conductivity by increasing hole concentration. Ultimately, the same π–π aggregation and densification of the PEDOT:PSS cores that aides in improving electronic transport, slows down the effective ion motion, leading to opposite trends in conductivity and ion mobility.

The trend in OECT performance, and its similarity to the convoluted trends in electronic and ionic transport give credence to the importance of mixed conduction for device operation. OECT operation relies on efficient lateral bulk electronic transport, and good vertical ion penetration into the film. As can be seen here, these requirements are often contradictory with electron transport being supported by rigid backbones with π-stacked crystallites, while ion transport would benefit from an open, hydrogel-like morphology. Balancing these needs will lead to efficient mixed conduction-based devices. It has been shown that the product of electronic charge carrier mobility (μ), and the volumetric capacitance (C^*) is the material's figure-of-merit for high transconductance devices[3]. Indeed, a high C^* requires that hydrated ions readily access the largest possible percolating conjugated polymer (PEDOT) content, placing greater importance on the lagging front (ions arriving at PEDOT aggregates).

OECT device operation, especially for applications in bioelectronics, necessitates an aqueous environment. Inherently, some of the techniques used here are limited in their ability to probe the films in their fully hydrated state. X-ray scattering and conductivity measurements were performed in the dry state, while the films were in contact with aqueous media during the moving front experiments, the latter most resembling device operation. While film hydration may affect the meso-scale spacings/relative domain sizes, the transport trends observed with EG formulation content are preserved. This verification was performed by extracting the hole mobility of PEDOT:PSS films from OECT operation (see the 'Methods' section) for different EG casting conditions, and comparing the trend with the conductivity/mobility obtained for dry films (Supplementary Fig. 9). Hydrated films were found to have a slightly higher mobility, with larger discrepancy noted for the pristine PEDOT:PSS films. Estimates of hole mobility from Hall effect measurements in this work, for example, $\mu = 1.3\,\text{cm}^2\,\text{V}^{-1}\,\text{s}^{-1}$ for the 20 vol% EG-cast film, are in agreement with previous reports[49], and are on par with OECT-determined hole mobility for the same formulation ($\mu = 2.7\,\text{cm}^2\,\text{V}^{-1}\,\text{s}^{-1}$). This result is in agreement with Wang *et al.*[50] who reported that relative humidity affects the electronic conductivity of PEDOT-based films only when the ionic transport governs the electrical properties. Accordingly, environmental humidity is crucial for poorly electrically conducting films, such as melanin, where conductivity is dominated by proton hopping[51]. Taken together, the agreement of mobility between dry and hydrated PEDOT:PSS films, and the conservation of the general trends observed, confirm that structure–electronic property correlations can be made for the present system independently of the polymer hydration.

By revealing the multi-scale morphological changes triggered by the addition of a commonly used co-solvent, we are able to establish structure-property relations describing mixed conduction in a prototypical CP. We extend the state of understanding of PEDOT:PSS film structure by quantifying domain composition using NEXAFS and resonant scattering techniques. Even with a large volume fraction of EG in the casting dispersion, the phases are seen to trend towards purification, with the PSS phase being far from pure. The observed phase purification enhances electrical conductivity while adversely affecting effective ion mobility as observed in electrochromic moving front studies. Here, the co-solvent EG is used as a means to tune both electronic and ionic transport, and to maximize the performance of OECTs. Finding routes to simultaneously improve ionic and electronic conductivity of CPs and their devices is relevant for many rising materials for applications in bioelectronics, electronics, energy storage and management, and robotics. Enhancing ion transport by approaching hydrogel-like structures without limiting long-range electronic transport remains a materials design hurdle in CP systems. Nevertheless, the structure-mixed conduction property work presented here lays the foundation for future materials design of organic mixed conductors.

Methods

Materials and film formation. Films for structural, electronic and planar junction ion-transport measurements were spun from aqueous dispersion formulation onto glass, Si or Si/SiN window substrates (cleaned by ultra-sonication in dilute soap and acetone/IPA, and/or rinsed in DI water before drying). Substrates are cleaned for 2 min in oxygen plasma before spin casting the dispersions: PEDOT:PSS (Clevios PH-1000 from Heraeus Holding, GmbH) with dodecyl benzene sulfonic acid (DBSA; 0.002 vol%) is mixed with EG (between 0–50% by volume of dispersion as noted in the text/figures) before ultra-sonication (10–15 min) and filtration (through 1.2 μm cellulose acetate filters). Unless otherwise noted, all chemicals are purchased from Sigma Aldrich. Formulations are spun at 1,000–3,000 r.p.m., 1–3 times, soft baked after each spin (90 °C, 60 s), followed by a hard bake (140 °C, 40 min).

X-ray scattering. Grazing incidence wide-angle X-ray scattering was carried out on films spun on native oxide silicon and was performed at the Stanford Synchrotron Radiation Lightsource (SSRL) on beamline 11-3 (2D scattering with an area detector, MAR345 image plate, at grazing incidence), with incident energy of 12.7 keV. The incidence angle was slightly larger than the critical angle, ensuring that we sampled the full film depth, and measurements were performed in a helium-filled chamber to minimize beam damage and reduce air scattering. Data are expressed as a function of the scattering vector $Q = 4\pi\sin(\theta)/\lambda$, where θ is half the scattering angle and λ is the wavelength of the incident radiation. Wide-angle GIXS data were distortion corrected and integrated from $\chi = 0$ to 90°, where χ is defined as the angle between the orientation of a scatterer and the surface normal. Films for rSoXS were solution cast onto 1.5 mm × 1.5 mm, 75 nm thick Si_3N_4 membranes (Norcada, Inc.). Scattering measurements were acquired in transmission geometry onto a CCD at beamline 11.0.1.2 of the Advanced Light Source (ALS, Lawrence Berkeley National Lab) using previously published methods[52].

Near-edge absorption fine structure. NEXAFS spectra for optical constants were measured in transmission mode via a phosphor-coated PMT in a scanning transmission X-ray microscope at beamline 5.3.2.2 of the ALS. The PEDOT:Cl (as in ref. 53) and Na:PSS (from Polymer Souce, spin cast from $1\,\text{mg}\,\text{ml}^{-1}$ solution) films were cast onto Si_3N_4 membranes. The measurement and data processing followed previously published methods[37]. NEXAFS spectra for average film composition were measured at beamline 11.0.1.2 of the ALS in transmission mode using a photodiode and illuminating the same film spot under investigation for scattering. NEXAFS measurements before and after the scattering experiment were used to confirm no sample damage occurred during the measurement.

Moving front characterization. Planar junctions for electrochromic moving front experiments were prepared as described previously[31]. Briefly, The PEDOT:PSS formulations described above are spun cast on 2-μm thick parylene-C (SCS Labcoater 2) on a 25 × 75 mm glass substrate, before vapour depositing a gold electrode on one side of the film. A ≅ 40-μm-thick SU-8 film was spin-coated, patterned and baked on top of the PEDOT:PSS film to serve as an ion barrier. A polydimethylsiloxane rim was placed on top of the SU-8 well to confine 1.5–2 ml of electrolyte. The length of PEDOT:PSS film between the electrolyte and the Au contact was 3.2 cm. A Ag/AgCl electrode (Warner Instruments) was immersed in the 100 mM KCl in DI water electrolyte. De-doping was achieved by applying a + 2 V bias to the Ag/AgCl electrode; the device was then short-circuited to return to its doped state (this defined one cycle). Measurements were reproducible after 2–3 cycles. Spatio-temporal measurements were conducted on an inverted Carl Zeiss Axio-Observer Z1 in the bright-field mode with a 1 × objective. Voltage was applied, and current measured using a Keithley 2612A source-measure unit connected to a computer running custom LabView software. Grey-level profiles, front shape normalization and deconvolution were calculated using a custom Matlab program. UV-visible spectro-temporal measurements were carried out by mounting the planar junction devices in a 3D-printed holder with fibres for the light source (Hamamatsu) and detector (Ocean Optics USB4000) aligned across from each other at a fixed X distance from the edge of the well. Time-lapse spectra were taken every second using Ocean Optics OceanView software. $A_n - A_0$, $A_{n+1} - A_n$ spectra, normalization and spectral-time colour plots were obtained through custom Matlab software.

Electrical characterization. Conductivity of the polymer thin films were measured with a Van der Pauw Ecopia HMS-3000 Hall Measurement System. PEDOT:PSS formulations with increasing EG concentrations were coated on $10 \times 10\,\text{mm}^2$ glass slides with patterned ITO contacts on the corners. IV scans confirmed the ohmic behaviour of the sample contacts. Measurements were performed at room temperature using a current source of 1 mA and a magnetic field strength of 0.56 T.

OECTs were fabricated in a cleanroom and patterned photolithographically, as previously described[1]. They consisted of a PEDOT:PSS channel (formulation described above, with 0.1 wt% of (3-Glycidyloxypropyl)trimethoxysilane to stabilize the film in aqueous) with Au source and drain electrodes and interconnect lines. Parylene-C was used to insulate the Au interconnects from the electrolyte. A Ag/AgCl pellet (Warner Instruments) was used as the gate electrode. Electrical measurements followed previously defined protocols[1]. Device characterization/analysis were carried out with custom LabView and MATLAB software. Hole mobilities from working OECTs were determined using drain current transients at different constant gate currents, which allows for extraction of hole transit times. The approximate electric field at a low drain bias along with the hole velocity allows for mobility estimation, as shown in refs 3 and 54.

References

1. Khodagholy, D. *et al.* High transconductance organic electrochemical transistors. *Nat. Commun.* **4,** 2133 (2013).
2. Khodagholy, D. *et al. In vivo* recordings of brain activity using organic transistors. *Nat. Commun.* **4,** 1575 (2013).
3. Rivnay, J. *et al.* High-performance transistors for bioelectronics through tuning of channel thickness. *Sci. Adv.* **1,** e1400251 (2015).
4. Lin, P., Yan, F., Yu, J., Chan, H. L. W. & Yang, M. The application of organic electrochemical transistors in cell-based biosensors. *Adv. Mater.* **22,** 3655–3660 (2010).
5. Jonsson, A. *et al.* Therapy using implanted organic bioelectronics. *Sci. Adv.* **1,** e1500039 (2015).
6. Williamson, A. *et al.* Controlling epileptiform activity with organic electronic ion pumps. *Adv. Mater.* **27,** 3138–3144 (2015).
7. Kim, S. H. *et al.* Electrolyte-gated transistors for organic and printed electronics. *Adv. Mater.* **25,** 1822–1846 (2013).
8. Kaihovirta, N. *et al.* Printed all-polymer electrochemical transistors on patterned ion conducting membranes. *Org. Electron.* **11,** 1207–1211 (2010).
9. Gkoupidenis, P., Schaefer, N., Garlan, B. & Malliaras, G. G. Neuromorphic functions in PEDOT:PSS organic electrochemical transistors. *Adv. Mater.* **27,** 7176–7180 (2015).
10. Tarabella, G. *et al.* A hybrid living/organic electrochemical transistor based on the *Physarum polycephalum* cell endowed with both sensing and memristive properties. *Chem. Sci.* **6,** 2859–2868 (2015).
11. Josberger, E. E., Deng, Y., Sun, W., Kautz, R. & Rolandi, M. Two-terminal protonic devices with synaptic-like short-term depression and device memory. *Adv. Mater.* **26,** 4986–4990 (2014).
12. Malti, A. *et al.* An organic mixed ion–electron conductor for power electronics. *Adv. Sci.* **3,** 1500305 (2016).
13. Mortimer, R. J., Dyer, A. L. & Reynolds, J. R. Electrochromic organic and polymeric materials for display applications. *Displays* **27,** 2–18 (2006).
14. Haines, C. S. *et al.* Artificial muscles from fishing line and sewing thread. *Science* **343,** 868–872 (2014).
15. Smela, E. Conjugated polymer actuators for biomedical applications. *Adv. Mater.* **15,** 481–494 (2003).
16. Elschner, A., Kirchmeyer, S., Lövenich, W., Merker, U. & Reuter, K. *PEDOT: Pinciples and Applications of an Intrinsically Conductive Polymer* (CRC Press, 2010).
17. Crispin, X. *et al.* The origin of the high conductivity of Poly(3,4-ethylenedioxythiophene) – Poly(styrenesulfonate) (PEDOT – PSS) plastic electrodes. *Chem. Mater.* **18,** 4354–4360 (2006).
18. Greczynski, G., Kugler, T. & Salaneck, W. R. Characterization of the PEDOT-PSS system by means of X-ray and ultraviolet photoelectron spectroscopy. *Thin Solid Films* **354,** 129–135 (1999).
19. Lang, U., Müller, E., Naujoks, N. & Dual, J. Microscopical investigations of PEDOT:PSS thin films. *Adv. Funct. Mater.* **19,** 1215–1220 (2009).
20. Nardes, A. M. *et al.* Microscopic understanding of the anisotropic conductivity of PEDOT:PSS thin films. *Adv. Mater.* **19,** 1196–1200 (2007).
21. Kemerink, M., Timpanaro, S., de Kok, M. M., Meulenkamp, E. A. & Touwslager, F. J. Three-dimensional inhomogeneities in PEDOT:PSS films. *J. Phys. Chem. B* **108,** 18820–18825 (2004).
22. Alemu Mengistie, D., Wang, P.-C. & Chu, C.-W. Effect of molecular weight of additives on the conductivity of PEDOT:PSS and efficiency for ITO-free organic solar cells. *J. Mater. Chem. A* **1,** 9907–9915 (2013).
23. Nardes, A. M., Janssen, R. A. J. & Kemerink, M. A morphological model for the solvent-enhanced conductivity of PEDOT:PSS thin films. *Adv. Funct. Mater.* **18,** 865–871 (2008).
24. Takano, T., Masunaga, H., Fujiwara, A., Okuzaki, H. & Sasaki, T. PEDOT nanocrystal in highly conductive PEDOT:PSS polymer films. *Macromolecules* **45,** 3859–3865 (2012).
25. Palumbiny, C. M. *et al.* Molecular reorientation and structural changes in cosolvent-treated highly conductive PEDOT:PSS electrodes for flexible indium tin oxide-free organic electronics. *J. Phys. Chem. C* **118,** 13598–13606 (2014).
26. Timpanaro, S., Kemerink, M., Touwslager, F. J., De Kok, M. M. & Schrader, S. Morphology and conductivity of PEDOT/PSS films studied by scanning-tunneling microscopy. *Chem. Phys. Lett.* **394,** 339–343 (2004).
27. Shi, H., Liu, C., Jiang, Q. & Xu, J. Effective approaches to improve the electrical conductivity of PEDOT:PSS: a review. *Adv. Electron. Mater.* **1,** 1500017 (2015).
28. Aoki, K., Aramoto, T. & Hoshino, Y. Photographic measurements of propagation speeds of the conducting zone in polyaniline films during electrochemical switching. *J. Electroanal. Chem.* **340,** 127–135 (1992).
29. Johansson, T., Persson, N.-K. & Inganäs, O. Moving redox fronts in conjugated polymers studies from lateral electrochemistry in polythiophenes. *J. Electrochem. Soc.* **151,** E119–E124 (2004).
30. Wang, X., Shapiro, B. & Smela, E. Visualizing ion currents in conjugated polymers. *Adv. Mater.* **16,** 1605–1609 (2004).
31. Stavrinidou, E. *et al.* Direct measurement of ion mobility in a conducting polymer. *Adv. Mater.* **25,** 4488–4493 (2013).
32. Mauritz, K. A. & Moore, R. B. State of understanding of nafion. *Chem. Rev.* **104,** 4535–4585 (2004).
33. Palumbiny, C. M. *et al.* The crystallization of PEDOT:PSS polymeric electrodes probed *in situ* during printing. *Adv. Mater.* **27,** 3391–3397 (2015).
34. Jönsson, S. K. M. *et al.* The effects of solvents on the morphology and sheet resistance in poly(3,4-ethylenedioxythiophene)–polystyrenesulfonic acid (PEDOT–PSS) films. *Synt. Met.* **139,** 1–10 (2003).
35. Huang, J. *et al.* Investigation of the Effects of doping and post-deposition treatments on the conductivity, morphology, and work function of Poly(3,4-ethylenedioxythiophene)/Poly(styrene sulfonate) films. *Adv. Funct. Mater.* **15,** 290–296 (2005).
36. Rivnay, J., Mannsfeld, S. C. B., Miller, C. E., Salleo, A. & Toney, M. F. Quantitative determination of organic semiconductor microstructure from the molecular to device scale. *Chem. Rev.* **112,** 5488–5519 (2012).
37. Yan, H., Wang, C., McCarn, A. R. & Ade, H. Accurate and facile determination of the index of refraction of organic thin films near the carbon 1s absorption edge. *Phys. Rev. Lett.* **110,** 177401 (2013).
38. Roe, R.-J. *Methods of X-Ray and Neutron Scattering in Polymer Science* (Oxford University Press, 2000).
39. Collins, B. A. & Ade, H. Quantitative compositional analysis of organic thin films using transmission NEXAFS spectroscopy in an X-ray microscope. *J. Electron. Spectrosc. Relat. Phenom.* **185,** 119–128 (2012).
40. Aasmundtveit, K. E. *et al.* Structural aspects of electrochemical doping and dedoping of poly(3,4-ethylenedioxythiophene). *Synt. Met.* **113,** 93–97 (2000).
41. Nardes, A. M., Kemerink, M. & Janssen, R. A. J. Anisotropic hopping conduction in spin-coated PEDOT:PSS thin films. *Phys. Rev. B* **76,** 085208 (2007).
42. Stavrinidou, E. *et al.* A simple model for ion injection and transport in conducting polymers. *J. Appl. Phys.* **113,** 244501 (2013).
43. Clark, J., Silva, C., Friend, R. & Spano, F. Role of intermolecular coupling in the photophysics of disordered organic semiconductors: aggregate emission in regioregular polythiophene. *Phys. Rev. Lett.* **98,** 206406 (2007).
44. Spano, F. C. Modeling disorder in polymer aggregates: the optical spectroscopy of regioregular poly(3-hexylthiophene) thin films. *J. Chem. Phys.* **122,** 234701 (2005).
45. Kim, Y. H. *et al.* Highly conductive PEDOT:PSS electrode with optimized solvent and thermal post-treatment for ITO-free organic solar cells. *Adv. Funct. Mater.* **21,** 1076–1081 (2011).
46. Wei, Q., Mukaida, M., Naitoh, Y. & Ishida, T. Morphological change and mobility enhancement in PEDOT:PSS by adding co-solvents. *Adv. Mater.* **25,** 2831–2836 (2013).
47. Thomas, J. P., Zhao, L., McGillivray, D. & Leung, K. T. High-efficiency hybrid solar cells by nanostructural modification in PEDOT:PSS with co-solvent addition. *J. Mater. Chem. A* **2,** 2383–2389 (2014).
48. Ouyang, L., Musumeci, C., Jafari, M. J., Ederth, T. & Inganäs, O. Imaging the phase separation between PEDOT and polyelectrolytes during processing of highly conductive PEDOT:PSS films. *ACS Appl. Mater. Interfaces* **7,** 19764–19773 (2015).
49. Tsai, T.-C., Chang, H.-C., Chen, C.-H. & Whang, W.-T. Widely variable Seebeck coefficient and enhanced thermoelectric power of PEDOT:PSS films by blending thermal decomposable ammonium formate. *Org. Electron.* **12,** 2159–2164 (2011).
50. Wang, H., Ail, U., Gabrielsson, R., Berggren, M. & Crispin, X. Ionic Seebeck effect in conducting polymers. *Adv. Energy Mater.* **5,** 1500044 (2015).
51. Mostert, A. B. *et al.* Role of semiconductivity and ion transport in the electrical conduction of melanin. *Proc. Natl Acad. Sci. USA* **109,** 8943–8947 (2012).
52. Gann, E. *et al.* Soft x-ray scattering facility at the advanced light source with real-time data processing and analysis. *Rev. Sci. Instrum.* **83,** 045110 (2012).
53. Winther-Jensen, B. *et al.* Order-disorder transitions in poly(3,4-ethylenedioxythiophene). *Polymer* **49,** 481–487 (2008).
54. Bernards, D. A. & Malliaras, G. G. Steady-state and transient behavior of organic electrochemical transistors. *Adv. Func. Mater.* **17,** 3538–3544 (2007).

Acknowledgements

J.R. acknowledges support from a Marie Curie post-doctoral fellowship (FP7). M.S. acknowledges the support by the Fundación BBVA, Spain. We acknowledge Prof. B. Winther-Jensen (Monash) for providing vapour-phase polymerized PEDOT:Cl films, Dr E. Gann (Monash) for NEXAFS measurements, Dr M. Ramuz (EMSE) for 3D printing efforts and I. Uguz (EMSE) for film thickness measurements. We additionally thank Prof. C. Silva (Montreal) for fruitful discussions. Portions of this research were carried out at the Stanford Synchrotron Radiation Laboratory (SSRL), a national user facility operated by the Stanford University on behalf of the US Department of Energy, Office of Basic Energy Sciences.

Author contributions

J.R. conceived the research. B.A.C. and J.R. prepared samples and performed resonance scattering experiments. C.T. performed GIXD experiments. B.A.C. and J.R. analysed and interpreted all X-ray scattering data with input from D.M.D. J.R., S.I., and E.S. prepared samples, performed and analysed all electrochromic moving front experiments. M.S., S.I., and X.S. characterized electrical properties including 4-point probe and Hall measurements. J.R., G.G.M., oversaw the research. J.R., S.I. and G.G.M. wrote the manuscript, and all authors provided input in manuscript preparation. Certain commercial equipment or materials are identified in this paper to specify the experimental procedure adequately. Such identification is not intended to imply recommendation or endorsement by the National Institute of Standards and Technology, nor is it intended to imply that the materials or equipment identified are necessarily the best available for the purpose.

Additional information

Competing financial interests: The authors declare no competing financial interests.

4

Atomistic understanding of cation exchange in PbS nanocrystals using simulations with pseudoligands

Zhaochuan Fan[1], Li-Chiang Lin[1], Wim Buijs[1], Thijs J.H. Vlugt[1] & Marijn A. van Huis[2]

Cation exchange is a powerful tool for the synthesis of nanostructures such as core-shell nanocrystals, however, the underlying mechanism is poorly understood. Interactions of cations with ligands and solvent molecules are systematically ignored in simulations. Here, we introduce the concept of pseudoligands to incorporate cation-ligand-solvent interactions in molecular dynamics. This leads to excellent agreement with experimental data on cation exchange of PbS nanocrystals, whereby Pb ions are partially replaced by Cd ions from solution. The temperature and the ligand-type control the exchange rate and equilibrium composition of cations in the nanocrystal. Our simulations reveal that Pb ions are kicked out by exchanged Cd interstitials and migrate through interstitial sites, aided by local relaxations at core-shell interfaces and point defects. We also predict that high-pressure conditions facilitate strongly enhanced cation exchange reactions at elevated temperatures. Our approach is easily extendable to other semiconductor compounds and to other families of nanocrystals.

[1] Process and Energy Department, Delft University of Technology, Leeghwaterstraat 39, 2628CB Delft, The Netherlands. [2] Soft Condensed Matter, Debye Institute for Nanomaterials Science, Utrecht University, Princetonplein 5, 3584CC Utrecht, The Netherlands. Correspondence and requests for materials should be addressed to M.A.v.H. (email: M.A.vanHuis@uu.nl).

To synthesize novel nanocrystals (NCs) and heteronanocrystals (HNCs) with designed shape and structure, cation exchange (CE) has recently been extensively explored and applied[1–11]. This process involves colloidal ionic NCs whereby the cations in the NCs are (partially) replaced by the cations from the surrounding solution. The CE technique has greatly boosted the syntheses of nanostructures with various morphologies. Starting from elementary NCs that are synthesized using well-known synthesis routes such as hot injection[12], the CE technique can be used to modify nanostructures into other novel nanostructures or heterostructures such as binary rods[13] and core/shell structures[10,14]. The CE technique can be extremely powerful when combined with other advanced (post)syntheses techniques such as oriented attachment and seeded growth, which leads to many novel nanostructures including nanooctapods[8], nanorod couples[15] and 2D honeycomb superlattices[16].

Pb→Cd exchange (that is, replacing Pb^{2+} by Cd^{2+}) in lead chalcogenide NCs is a typical CE process that produces PbE/CdE (E = S, Se, Te) core/shell HNCs[14,17,18]. This CE process for colloidal PbS NCs can be expressed as follows:

$$Cd^{2+}_{liquid} + PbS_{solid} \rightarrow Pb^{2+}_{liquid} + CdS_{solid} \quad (1)$$

A typical experimental route of this CE consists of two steps[14,19]: (1) synthesis of the PbS parent NCs, and (2) immersing the parent NCs in Cd-oleate solutions and applying heat for 0.5–10 h at 373–473 K. PbS has a rocksalt (RS) crystal structure where the atoms have a sixfold coordination, whereas CdS has a zinc blende (ZB) crystal structure where the atoms have a fourfold coordination. The lattice mismatch between these two structures is <2% (ref. 20). Pb→Cd CE in PbE NCs has often been described as a self-limiting process given that a core/shell structure is generally found in the product. Complete conversion is usually difficult to achieve[18,19,21]. Compared to other types of CE[4,22] (for example, Cd→Ag) that take place spontaneously at ambient temperature and at a shorter timescale, Pb→Cd CE requires a relatively high temperature and a long time to overcome a large energy barrier[1].

Although the Pb→Cd CE process has been studied by modern experimental techniques such as energy dispersive X-ray spectrometry, high-resolution transmission electron microscopy (HRTEM) and 3D electron tomography[17,19,21,23,24], a detailed understanding of the atomistic mechanism is still missing. Theoretical investigation of this CE process using molecular simulations also remains a great challenge. Understanding the thermodynamics and kinetics of CE at the atomic scale is crucial for the future development of the precise control of this technique. In this study, classical molecular dynamics (MD) simulations are used to study Pb→Cd CE in PbS NCs within a solution. Our results indicate that the rate of CE can be controlled by changing the ligand type, or by adjusting the temperature and/or pressure. By changing the ligand type, the direction of CE can also be controlled, which enables the reverse CE process to take place. Furthermore, we reveal the atomistic mechanism of ion transportation in the NCs. During CE, Pb ions are kicked out by exchanged Cd interstitials and migrate through interstitial sites. Local structural distortions and point defects present at the PbS/CdS interface and in the CdS shell region assist the Pb jumps by locally expanding the voids in Cd sublattice and by reducing the Coulombic repulsion between Pb and Cd ions.

Results

Thermodynamic driving force. As schematically shown in Fig. 1a, CE not only involves parent NCs, but also the surrounding solution, consisting of exchanging cations, ligands and solvent molecules. CE is initiated at the interface between the NCs and the solution and then proceeds to occur within the NCs. Modelling the whole system atomistically requires large system sizes and long simulation times, which is not computationally feasible. To overcome this, we developed a coarse-grained model to mimic the cation–ligand solution in which the effects of ligands and solvent are incorporated into negatively charged large spherical particles (Fig. 1b, the orange circles). These particles are denoted by pseudoligands (PSLs). PbS–CdS systems are modelled using our newly developed all-atom force field[20], which accurately describes the crystal structures, elastic properties, polymorphic stabilities, and surface energies of PbS and CdS solids and $Pb_xCd_{1-x}S$ mixed phases. Details about the coarse-grained PSL model and the all-atom force field are provided in Methods section, Supplementary Methods, Supplementary Discussion, Supplementary Figs 1,2 and Supplementary Table 1.

The thermodynamic driving force for Pb→Cd CE is controlled by two main factors: the free energy difference between the two materials forming the NCs and the solubility of both cations in the ligand solution[1,2]. The former depends on several factors, such as the lattice energy, surface energy, size and shape of the NCs, the interfacial energy when forming HNCs and the entropy of mixing. The dominating factor is the lattice energy difference between ZB-CdS and RS-PbS, as the difference in surface energies and the lattice mismatch between ZB-CdS and RS-PbS are relatively small[20]. During CE, the change in $T\Delta S$ is typically one or two orders of magnitude smaller than that from the enthalpy change (ΔH) at ~500 K (ref. 24), and thus the entropy of mixing is also not a dominating factor[21]. The lattice energy of

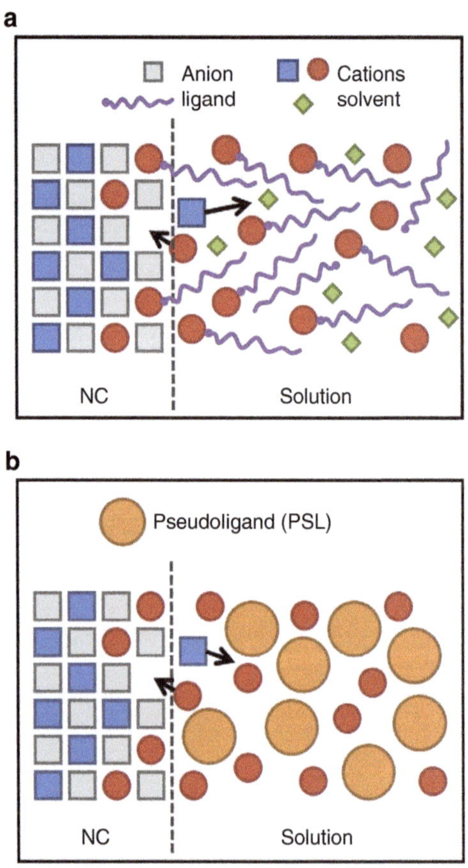

Figure 1 | Schematic representations of a CE process. (a) CE in experiments. **(b)** CE in MD simulations with a coarse-grained PSL model. The dashed lines indicate the NC/solution boundaries. The black arrows indicate the motion of the cations during CE.

ZB–CdS computed using our force field is about 0.3 eV per formula unit (eV/f.u.) less than that of RS–PbS[20]. Therefore, if the PSL–Cd and PSL–Pb interatomic interactions are equivalent (that is, using non-preferential PSLs, see Methods section) and when the temperature is sufficiently high, the relatively low-lattice energy of ZB–CdS will promote the Pb→Cd CE process in PbS NCs. MD simulations of a 4.7-nm-sized PbS nanocrystal immersed in a non-preferential PSL solution were carried out at 500 K. Pb→Cd CE was indeed observed in our simulations. Figure 2a shows that, as a function of simulation time, the number of Pb cations in the parent NC decreases, whereas the number of Cd cations increases, clearly demonstrating the exchange of cations. We also observed that the number of S anions remained nearly constant, which indicates that, as the anionic sublattice is more stable, the anionic sublattice is preserved during the CE process. The CE process found in the MD simulations is in excellent agreement with experimental observations[21]: the exchanging Cd ions in the solutions initially cover the surface of the parent NC (that is, corresponding to the sharp increase of the number of in-going Cd ions during the first few nanoseconds as shown in Fig. 2a), while the surface Pb ions in the parent NC are dissolved into the solution (Supplementary Movies 1 and 2). Subsequently, the Cd ions diffuse inwards into the NCs. When the Cd ions occupy the sites of the first few outermost monolayers (MLs), the exchange slows down, forming a PbS/CdS core/shell HNC (see Fig. 2e; Supplementary Movies 1 and 2). The formation of the CdS shell impedes the outward diffusion of the Pb ions from the PbS core. As the exchange slows down, the number of in-going Cd (N_{Cd}^{NC}), the number of remaining Pb in the NC (N_{Pb}^{NC}) and the exchange ratio (ρ_{Cd}), defined as the number of in-going Cd ions divided by the total number of cations in the NC ($\rho_{Cd}=N_{Cd}^{NC}/(N_{Cd}^{NC}+N_{Pb}^{NC})$, grey circles in Fig. 2b,c), become nearly constant after a few tens of nanoseconds. The PbS/CdS core/shell HNCs with higher exchange ratios (for example, the final configuration of an exchanged HNC at 600 K in a MD simulation of 100 ns in Fig. 2i) have a structure and morphology, which is very similar to those of experimentally observed core/shell HNCs, as shown in HRTEM images of PbS/CdS (Fig. 2j), PbSe/CdSe (Fig. 2k) and PbTe/CdTe (Fig. 2l) configurations[17,25,26].

The preferential dissolution of the cations in solutions with different ligands has a crucial role in determining the direction and exchange ratio of CE. Using the same system and temperature (that is, the 4.7-nm-sized PbS NC immersed in a Cd–PSL solution at 500 K), we investigated the influence of different ligand solutions on the CE process by considering different PSL–cation interactions. For Pb-preferred, non-preferential and Cd-preferred PSL solutions (see Methods section), the exchange ratio ρ_{Cd} as a function of simulation time is shown in Fig. 2b. As shown in Fig. 2b, Pb→Cd CE can be significantly accelerated while a Pb-preferred solution is used, and it can also be completely prohibited using a Cd-preferred solution. Note that a small amount of Cd ($\rho_{Cd}\sim 20\%$) was found in the NC with the Cd-preferred PSLs. These Cd ions did not penetrate into the NC but were adsorbed at the NC surface (Fig. 2f). We also found that, by immersing a ZB-CdS NC in a solution containing Pb ions and Cd-preferred PSLs, the system underwent a reverse Cd→Pb CE

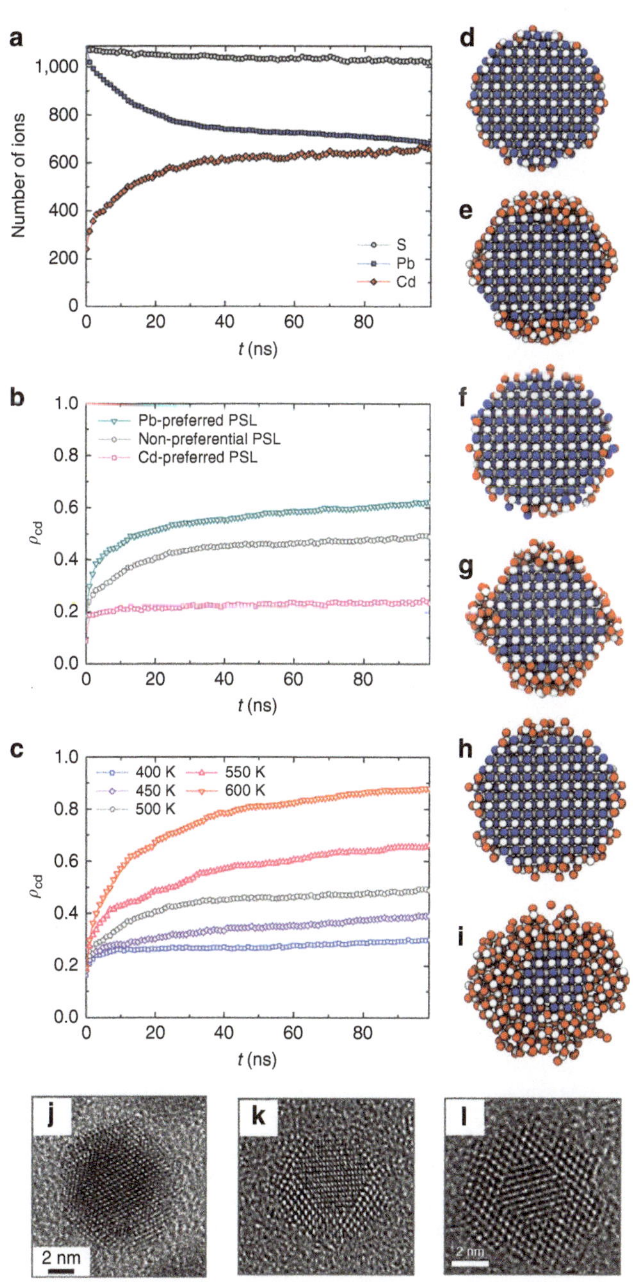

Figure 2 | The Pb→Cd CE process in a 4.7-nm-sized PbS NC using different PSLs at different temperatures. (**a**) Number of in-going Cd ions (red diamonds), remaining Pb ions (blue squares) and S ions (grey circles) in the parent PbS NCs as a function of time. MD simulations were performed at 500 K with non-preferential PSLs. (**b**,**c**) Time evolution of exchange ratio ρ_{Cd} (the ratio of in-going Cd ions over the total number of cations in the NC). (**b**) Simulations performed at 500 K with different PSLs. A constant value of ρ_{Cd} (~0.2) for simulation with Cd-preferred PSLs indicates that CE did not take place. The low fraction of Cd ions corresponds to those Cd ions adsorbed at the NC surface. (**c**) Simulation performed using non-preferential PSLs at different temperatures. (**d**-**i**) Snapshots of the RS-(110) sections of the NCs in 100-ns MD simulations: (**d**) initial configuration; (**e**-**i**) final configurations of 100-ns simulations at 500 K with non-preferential PSLs (**e**), Cd-preferred PSLs (**f**), and Pb-preferred PSLs (**g**) and those with non-preferential PSLs at 400 K (**h**) and 600 K (**i**). The white, blue and red spheres represent S, Pb and Cd, respectively. (**j**-**l**) HRTEM images of PbE/CdE core/shell HNCs achieved by experimental CE, in a [110] projection. (**j**) A 9-nm-sized PbS/CdS core/shell HNC, image adapted with permission from (ref. 25) (copyright 2014 American Chemical Society). (**k**) A 5.6-nm-sized PbSe/CdSe core/shell HNC, image adapted with permission from (ref. 26) (copyright The Royal Society of Chemistry 2011). (**l**) A 6.6-nm-sized PbTe/CdTe core/shell HNC, image adapted with permission from (ref. 17) (copyright 2009 American Chemical Society).

(see Supplementary Methods and Supplementary Fig. 3). The reverse CE did not take place in a non-preferential or a Pb-preferred PSL solution. As shown experimentally, a way to achieve both the forward and backward Pb⇔Cd CE is to change the type of ligand: Cd-oleates are often chosen as cation–ligands solutions to achieve Pb→Cd CE[14,19,21], whereas aqueous Pb nitrate solutions, $PbCl_2$–oleylamine and Pb–oleates–oleylamine have been used to achieve the reverse CE process[27–30].

The temperature is another key factor for determining the equilibrium and the exchange rate of CE. MD simulations for PbS nanocrystals immersed in non-preferential PSL solutions were performed at temperatures ranging from 400 to 600 K. The CE process was found to be faster at higher temperatures with a more complete Pb→Cd replacement (Fig. 2c). At 600 K, nearly 90% of the cations in the NC was replaced by Cd within 100 ns, and a CdS shell with a thickness of about four MLs was formed (Fig. 2c,i). In experiments, high temperatures often lead to severe damage and distortion of the NCs (among others, Ostwald ripening can take place)[19,21]. In the simulations, we also observed that at 700 K the whole NC dissolved in the solution. Note that temperatures used in the MD simulations should not be directly compared with those in experiments. The timescale in our MD simulations is generally of the order of hundreds of nanoseconds,

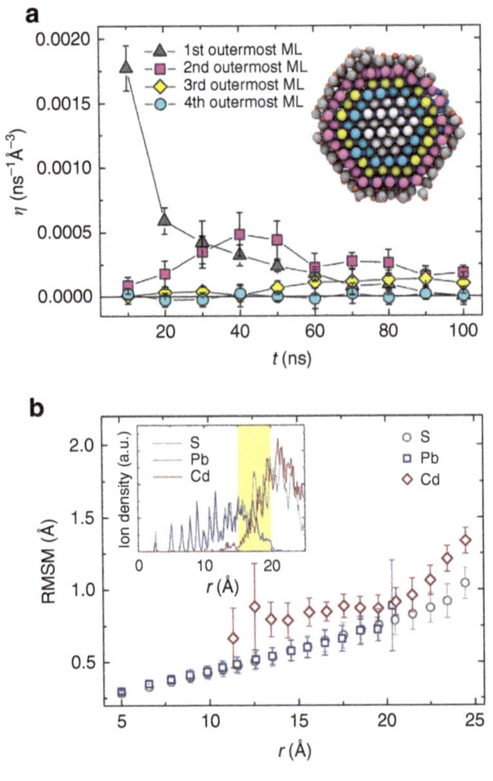

Figure 3 | Exchange rate and atomic mobility analysis of Pb→Cd CE. (**a**) Time evolution of the averaged volume-scaled exchange rate η (see Methods section) for the four outermost atomic MLs. Data were collected from 10 independent 100-ns MD simulations of a PbS NC at 550 K. The inset shows the final configuration of the RS-(110) section from one of the MD simulations. The S ions are coloured according to the colour code of the different outermost MLs. The small red and blue spheres are Cd and Pb, respectively. (**b**) Averaged RMSM (see Methods section) of S (grey circles), Pb (blue squares) and Cd (red diamonds) as a function of r (r is the distance between the ion and the centre of the NC). The inset shows the ion density as a function of r. The yellow region labels the $Cd_xPb_{1-x}S$ mixed phase domain between the PbS core and the CdS shell. Data were sampled during the last 100 ps of the ten independent 100-ns MD simulations.

which is much shorter than a typical timescale of the Pb→Cd CE experiments (mins–hr)[14,17–19,21]. Accordingly, a higher temperature is required in our simulations in order to accelerate the CE process and so that it falls within the timescale accessible by MD.

Self-limiting exchange. One of the appealing features of CE at the nanoscale is the fast exchange rate compared to the process at the macroscopic scale. Nonetheless, Pb→Cd CE is often reported to be self-limiting. Although complete Pb→Cd exchange has been experimentally reported[16,18,21], accurate controls, high temperatures and long times are required. Figure 3a shows the time evolution of the averaged volume-scaled exchange rate η (see Methods section) at the first four outermost MLs of the PbS NC from the MD simulations at 550 K with non-preferential PSLs. The CE process takes place in a layer-by-layer fashion. The exchange front moves from the first outermost ML to the second ML in 20∼40 ns and further proceeds to the third outermost ML after ca. 50 ns onward (Fig. 3a; Supplementary Movies 1 and 2; Supplementary Fig. 4). The exchange rate of the fourth layer is always zero, suggesting that the exchange front has not yet reached the fourth ML in 100 ns. The overall exchange rate η significantly decreases as the exchange front moves inward. This is consistent with the average root mean square motion (RMSM, see Methods) calculated for different ionic species as a function of the distance from the initial position of an ion to the centre of the NC (denoted by r) as shown in Fig. 3b. To clarify the ionic positions in the PbS/CdS core/shell structure, the radial densities of each ionic species are plotted in the inset of Fig. 3b. The RMSM of all types of ions increases as the ions are closer to the NC surface, indicating a faster CE process at the surface than inside the NC. Obviously, the large surface-to-volume ratio of nanostructures leads to a swift CE process at the nanoscale, but the spherical morphology of the PbS parent NCs and the layer-by-layer mode of exchange in the Pb→Cd CE process determine its self-limiting nature. This has also been reported experimentally for an analogous Cd→Zn CE process in CdSe NCs[10].

It is instructive to point out that the radial density of each atom type (the inset of Fig. 3b) indicates a pure PbS domain ($r \leq 15$ Å), a $Cd_xPb_{1-x}S$ mixed phase domain (15 Å $< r <$ 20 Å) and a pure CdS domain (20 Å $\leq r <$ 24 Å). Considering the inhomogeneity in PbS/CdS HNCs, the mixed phase layer with a thickness of 5 Å is relatively thin. It has been also shown experimentally that PbS/CdS HNCs obtained by CE have a sharp PbS/CdS boundary[17,19]. Unlike systems containing materials with the same crystal structure, which easily form mixed phases or solid solutions (for example, the CdSe/ZnSe[10] and CdSe/CdS/ZnS[31] systems), PbS and CdS in HNCs undergo a phase separation even at relatively high temperatures[24,26].

Kinetic mechanism. Based on a variety of experimental data, two mechanisms have been proposed for describing the kinetics of CE: a 'vacancy-mediated' mechanism[18,21,24] and a 'kick-out' mechanism[32]. The 'vacancy-mediated' mechanism suggests that Pb ions from the PbS core migrate through Cd vacancies[18,21,24] (Fig. 4a). These Cd vacancies form at the surface of NCs and migrate further towards the PbE/CdE interface, transporting the Pb ions through the CdS shell into the solution. In a recent paper, Bothe *et al.*[32] challenged the 'vacancy-mediated' mechanism by pointing out that the formation energy of Cd vacancies is rather high, and that therefore, cations migration through vacancies is unlikely. These authors proposed that the Pb ions are kicked out of the PbS core by exchanged Cd ions (the 'kick-out' mechanism) and migrate trough interstitial sites[32] (Fig. 4b). From an energetic point of view, the 'vacancy-mediated' mechanism suggests that the Pb jumps in the CE process are assisted by 'energy traps' (vacancies) and thermal fluctuations (Fig. 4a), whereas the 'kick-out' mechanism suggests that the Pb jumps are stimulated by Coulombic repulsion between the Pb ions and exchanged Cd interstitials (Fig. 4b). Due to the limitations in resolution of state-of-the-art experimental techniques, it is difficult to determine which mechanism is correct for the Pb→Cd CE process. By analysing the atomic trajectories in our simulations, we are able to unveil the mechanism for CE. We divide the CE process into two steps: (1) CE at the PbS/CdS interface and (2) the migration in the CdS shell region. We frequently observed the 'kick-out' mechanism[32] at the PbS/CdS interface in our simulations. Figure 4c shows a typical 'kick-out' motion: two MD snapshots showing configurations before and after the jump of a (green coloured) Pb ion out of the PbS core. An exchanged Cd ion (the pink sphere) located at an interstitial site in the PbS core region is present in the pre-jump configuration. The exchanged Cd interstitial kicks the Pb ion out of the PbS core by Coulombic repulsion and then occupies its site. We identified many more Pb jumps at the PbS/CdS interfaces from five independent 200 ns MD simulations on which we performed statistical analysis (see Supplementary Methods and Supplementary Table 2). Exchanged Cd interstitials near the jumping Pb ions can be found in about 47% of the identified Pb jumps, strongly supporting the 'kick-out' mechanism.

Another difference between the 'vacancy-mediated' and 'kick-out' mechanisms is whether the Pb ions migrate through vacancies or interstitial sites during CE. We examined the pre-jump atomistic environment (that is, the distribution of anions and cations) around the post-jump locations of the jumping Pb ions (see Supplementary Methods). If a Pb ion jumps into a ZB-CdS domain, it is mostly located at one of the two different cation sites: the tetrahedral cation site (Fig. 5a,b) or the octahedral cation site (Fig. 5c). The former is a tetrahedrally coordinated site with four nearest neighbour anions whereby the Pb ion can be either located at an interstitial (Fig. 5a) or at a substitutional site (Fig. 5b). The octahedrally coordinated site has six nearest neighbour anions so that the Pb ion can only be an interstitial (Fig. 5c). By counting the coordination numbers of the nearest neighbour anions around the positions in which the Pb ions jumped, the tetrahedral and octahedral cation sites can be distinguished (see Supplementary Methods). Figure 5d shows the coordination numbers for the identified Pb jumps in the interfacial and shell regions. The results indicate that at the PbS/CdS interface, the Pb ions show a strong preference to jump to octahedral cation sites (coordination number 6) rather than to tetrahedral cation sites (coordination number 4). A slight

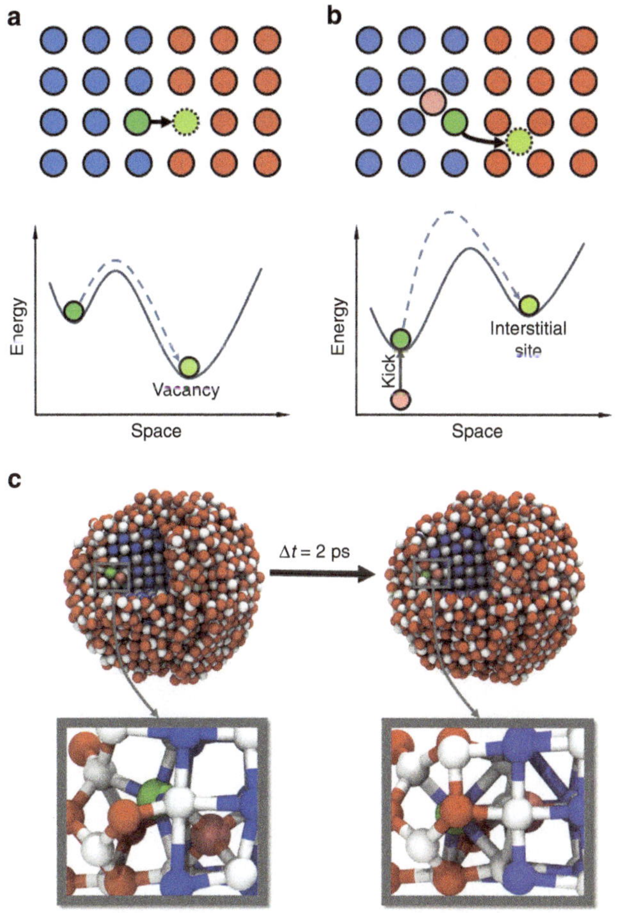

Figure 4 | 'Vacancy-mediated' mechanism versus 'kick-out' mechanism. (**a**,**b**) Schematic representations of the 'vacancy-mediated' mechanism (**a**) and the 'kick-out' mechanism (**b**). The blue and red circles represent Pb and Cd, respectively. The green circles denote the Pb ions migrating at the PbS/CdS interface. The pink circle is a Cd interstitial kicking the Pb ion out of the PbS core region. The lower panels of (**a**,**b**) schematically show the energy landscapes of these two different mechanisms. (**c**) Typical snapshots from MD simulations showing the 'kick-out' mechanism. Left and right parts of (**c**) are the configurations of the NC before and after the jump of a Pb ion (the green sphere) out of the PbS core, respectively. The Pb jump is initiated by a Cd interstitial (the pink sphere) in the PbS region. The blue, red and white spheres are Pb, Cd and S, respectively. The lower panels of (**c**) are magnifications of the region where the 'kick-out' motion takes place.

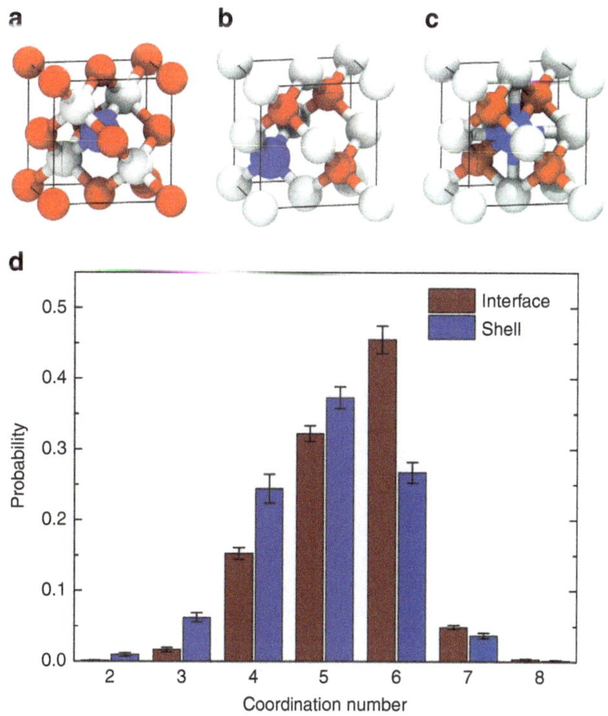

Figure 5 | Statistical analysis of different Pb jumps in the Pb→Cd CE. (**a-c**) Three different sites for a Pb ion in ZB-CdS: (**a**) a tetrahedrally coordinated interstitial site; (**b**) a tetrahedrally coordinated substitutional site; (**c**) an octahedrally coordinated interstitial site. The blue, red and white spheres are Pb, Cd and S, respectively. (**d**) Distribution of the coordination numbers of the jumping Pb ions after the jumps at the PbS/CdS interface and in the CdS shell region, based on an analysis of about 10,000 Pb jumps identified from five independent 200-ns MD simulations.

preference for the octahedral cation sites can be found when the Pb ions are migrating in the CdS shell region. We calculated the averaged distances between the four nearest anions and the jumping Pb ions ($\bar{r}_a$), and those between the four nearest cations and the jumping Pb ions ($\bar{r}_c$) in the post-jump configurations (see Supplementary Table 3). The distance analysis included the Pb jumps at the PbS/CdS interface and in the CdS shell region with anionic coordination numbers 4 and 6. Note that coordination numbers 3, 5 and 7 correspond to local structural distortions or off-site Pb ions, which were excluded from the distance calculation. The results show that the values of $\bar{r}_a$ are in a narrow range of 3.0–3.1 Å, regardless of the tetrahedral/octahedral site type and the region (interface or shell) where the Pb jumps took place. The values of $\bar{r}_a$ are consistent with the Pb–S bond length in the bulk crystal PbS (3.0 Å), indicating that the locations of the Pb ions in the CdS domains are mainly determined by Pb–S interactions. For the Pb ions located at the tetrahedral cation sites, the ratio of $\bar{r}_c/\bar{r}_a$ is about 1.13 (see Supplementary Table 3), which suggests that these Pb ions are indeed interstitials (the value of $\bar{r}_c/\bar{r}_a$ for a tetrahedrally coordinated Pb interstitial is 1.15) rather than substitutional ions (the value of $\bar{r}_c/\bar{r}_a$ for a substitutional Pb is 1.63). For the Pb ions located at the octahedral cations sites, $\bar{r}_c/\bar{r}_a$ is about 1.08 (see Supplementary Table 3). This value is significantly larger than the value of ideal Pb interstitials at octahedral cation sites (0.87). The mismatch can only be explained by local structural distortions and the formation of Cd vacancies. The statistical analysis of our simulations supports the 'kick-out' mechanism[32] whereby Pb ions migrate through interstitial sites and not through vacancies during CE. However, the migration of the Pb ions through octahedral sites needs to be assisted by Cd vacancies.

Discussion

The exchange rate in our simulations is faster than that in experiments, and this Pb→Cd CE reaction rate may be attributed to the higher reaction temperature and the higher concentrations of Cd in the solution compared to those in experiments. In Pb→Cd CE experiments of PbS NCs, the reaction temperature is typically in the range of 100–200 °C (373–473 K). In our simulations, the highest temperature is 600 K in order to accelerate the CE process, considering that the timescale of our simulations is several orders of magnitude shorter than of experiments. Such a high temperature can increase the probability of cation jumps by two to six orders of magnitude. Raising the temperature in the simulations bears the risk that the kinetic mechanism found in our simulations may differ from that in experiments at a lower temperature. However, as the atomistic configurations of the exchanged nanocrystals in our simulations are very similar to those observed in experiments (Fig. 2e–l), it is anticipated that the mechanisms are the same. The coarse-grained PSL model was used to mimic the ligand and solvent molecules. The size of a PSL is smaller than that of actual ligands used in CE experiments, and the number of the guest cations (Cd) is set equal to the number of PSLs. Therefore, the concentration of guest cations in the solution is higher in our simulations in comparison to experiments (Supplementary Discussion), which may also result in a faster exchange in the simulations. It should be noted that our simulations are only able to show the initial stages (the first few hundreds of nanoseconds) of the CE process. Compared to the exchanged NCs in experiments that has formed CdS shells with a thickness of 1–2 nm (Fig. 2j–l), the exchanged NC at 400 K (within the temperature range used in CE experiments) only has formed an incomplete and thin layer of CdS (thickness of the shell <0.3 nm) after 100 ns (Fig. 2h), thus the CE took place to a lesser extent. To show a CE process for this NC in a simulation at 400 K with a similar degree of exchange as those in experiments, the required simulation time may be extremely long (>1 s), which is not computationally feasible.

Increasing the temperature is an effective way to accelerate the CE process, however, it has been reported that Ostwald ripening takes place by heating colloidal PbS NCs at 473 K, thus destroying the monodispersity of the NCs[19]. A multiple step heating scheme[18,19] has been therefore used for CE experiments at high temperatures in which the NCs are protected by CdS shells that form at a lower temperature. We did find significant dissolution of S ions in our simulations at high temperatures (Supplementary Movies 1 and 2; Supplementary Fig. 5), however, Ostwald ripening could not be directly observed in our simulations since a single NC was considered. The dissolved S anions may also recrystallize on other parts of the same NC with the cations in the solution, thus enabling massive migration of anions and reconstruction of PbE/CdE HNCs. This phenomenon becomes more important at high temperatures and/or with active ligands, in some extreme cases PbE NCs or PbE/CdE core/shell HNCs may transform to a Janus-like heterostructure[26,29,33].

Besides the multiple step heating scheme, we found that increasing the pressure could be an alternative approach to preserve NCs at elevated temperatures in order to accelerate the CE processes and to achieve complete compositional conversions. The Pb→Cd CE process cannot be accelerated by merely applying a hydrostatic pressure to the solution (Supplementary Methods and Supplementary Fig. 6), however, an increased hydrostatic pressure is able to increase the melting point of ionic semiconductor solids[34] and their stability at high temperatures. For example, only five Pb ions remained in the exchanged NC ($\rho_{Cd}=0.997$) in a MD simulation of 100 ns at 750 K and a hydrostatic pressure of 1.00 GPa, and a PbS NC completely transformed into a CdS NC in 70 ns at 800 K and 1.0 GPa (Supplementary Fig. 6). At a standard pressure the NC would have been severely damaged upon reaching a temperature of 700 K, whereas the exchanged NC remains solid phase and retains a quasi-spherical morphology at high-temperature–high-pressure conditions ($p=1.00$ GPa, $T\leq 800$ K, Supplementary Fig. 6). Our simulations therefore suggest that a high-pressure–high-temperature approach is a promising method to control the rate and degree of the Pb→Cd CE process.

The exchange process is initiated by Cd interstitials in the PbS domain, whereby these Cd interstitials kick Pb ions out of the PbS core. A high temperature is needed to promote the penetration of these guest cations into the NC, and this explains why some types of CE processes (for example, Pb→Cd and Zn→Cd) only take place upon annealing. Although it has been shown that Pb ions in Pb→Cd CE migrate through interstitial sites, it is important to note that Cd vacancies, or most likely, Cd Frenkel pairs (a Cd ion moves to a neighbouring interstitial site, thereby forming a vacancy and an interstitial) play a role during the exchange process, especially at the PbS/CdS interface. Since PbS has a sixfold RS structure, there should be a strong preference for Pb ions to occupy octahedral cation sites. However, in CdS domains, Pb ions are prevented from jumping into octahedral interstitial sites due to strong Coulombic repulsion between the Pb ions and the densely packed Cd ions around the octahedral interstitial sites. Only when assisted by Cd vacancies, which are present due to the formation of Cd Frenkel pairs, Pb ions can jump into octahedral cation sites. It is well known that structural instability will be introduced by dipole moments at the polar/non-polar interface or the polar/polar interface with different crystal structures[35]. Therefore, formation of Cd Frenkel pairs and local structural distortion occurs more easily and more frequently at the PbS/CdS interface than in the CdS domain at high temperatures. This conclusion agrees with the results from our

MD simulations showing that the Pb ions show a stronger preference for octahedral cation sites while jumping at the PbS/CdS interface. In general, the CE processes take place more easily and faster in the systems forming interfaces of two different structures (for example, PbE/CdE[14,17,18], Ag_2E/CdE[22] and Cu_2E/CdE[36]) than those having the same structure (for example, CdE/ZnE[10,31]). In addition, we point out here that the 'vacancy-mediated' mechanism wrongly guessed the detailed kinetics of this particular CE process. Cd vacancies do not migrate back and forth between the NC surface and the PbS/CdS interface, and Pb ions do not migrate through Cd vacancies. The formation of Cd vacancies (without Cd Frenkel pairs) is indeed difficult[32], especially at high temperatures whereby vacancies tends to be annealed out[36]. Pb substitutions in the CdS shell region have been rarely observed in our simulations.

As CdS and PbS have different crystal structures, a structural transition must take place during a CE process. Two most stable phases of CdS, wurtzite (WZ) and ZB, have almost the same cohesive energy (the difference between the cohesive energies of these two CdS polymorphs is about 1 meV/f.u. according to density functional theory (DFT) calculations[20]). During the CE process of Pb and Cd chalcogenides nanostructures, however, the RS-to-ZB transition has been more frequently observed experimentally than the RS-to-WZ transition[16–18,23,24] (see Fig. 2j–l). The latter requires an extra step of anionic sublattice slips[37] as the former only requires slips of the cationic sublattice[20]. Therefore, the RS-to-ZB transition has a lower energy barrier than the RS-to-WZ transition. In our simulations, almost equal amounts of the WZ and ZB domains were found in the CdS shell of an exchanged HNC (Supplementary Fig. 7). This discrepancy between experiments and simulations may be attributed to the adopted force field for CdS as the stability of ZB phase is slightly underestimated[20]. However, we note that the competition between the RS-to-ZB and RS-to-WZ transitions in Pb→Cd CE also depends on many other factors, including the morphology of initial parent NCs, types of ligand and solvent, temperatures and nucleation energies. For instance, the WZ stacking faults and even direct RS-to-WZ transition have been observed in CE experiments[16–18,23,24]. In our simulations, most of the Cd ions were located at the tetrahedral cation sites (ZB or WZ), and only those at or near the PbS/CdS interfaces may jump between tetrahedral and octahedral sites. It has been reported that CdS shells may remain in a metastable RS phase during CE processes[25], which has not been observed in our simulations.

In summary, we used classical MD simulations to investigate Pb→Cd CE in colloidal PbS NCs. The cation-ligands solutions were simulated using a coarse-grained PSL model, allowing us to study this large system at a long-time scale. The main driving force of the Pb→Cd CE process is the competition between the ionic crystal lattice energies and the cations solvent solubility. By adjusting the PSL–cation interatomic interactions (analogous to changing the cation–ligand solutions in experiments) and the temperature, the direction and rate of the CE process can be controlled. Our simulation results reveal that the CE process is initiated by 'kick-out' motions and that Pb ions migrate through interstitial sites. Highly mobile Cd ions at the PbS/CdS interface facilitate the formation of Cd Frenkel pairs, which enables the jumps of cations and accelerates the CE process. Our study provides a profound insight into the physics underlying the Pb→Cd CE at the atomic level. The hybrid all-atom/coarse-graining approach is able to effectively reveal both the thermodynamics and kinetics of the Pb→Cd CE, and, more importantly, this model may also be adapted to investigate other types of CE (for example, Cd→Cu and Zn→Ag) and other synthesis techniques (for example, hot injection, seeded growth and vapour–liquid–solid growth). Finally, our simulations also predict that high-temperature–high-pressure conditions can greatly enhance the CE rate. To date, several key questions remain unclear, such as: Why does CE behave very differently in different materials? What is the optimal ligand solution, temperature and pressure to yield complete Pb→Cd CE? How can the thickness and uniformity of the CdE shell be completely controlled? Our approach provides opportunities to further address these questions, and we anticipate that the coarse-grained PSL model will be extended and further improved.

Methods

Colloidal PbS NC systems and MD simulations. A spherical PbS NC with a diameter of 4.7 nm containing 1,088 {Pb–S} pairs was cut from a RS–PbS matrix with a lattice parameter of 6.0 Å. The PbS NC was placed in the centre of a cubic simulation box with a size of 10 nm. The simulation box was then randomly filled with 4,352 {Cd-PSL} pairs (to avoid overlaps, we ensured that the distance between any two ions is larger than 3.5 Å). Periodic boundary conditions were applied in all directions.

In MD simulations, the PSL–PSL or PSL–cation interatomic interactions consist of two parts: long-ranged Coulombic interactions and short-ranged interactions described by a Buckingham potential. We assume that electron transfer does not play an important role during Pb→Cd CE, that is, the charges of the cations, anions and anionic ligands in the solid and liquid phases, and at the solid/liquid interface, do not change. The main function of our PSL model is to create a solution in which the guest cations are dissociated and CE takes place at the NC/solution interface (see Supplementary Discussion). The parameters in the Buckingham potentials can be altered to create Pb-preferred, Cd-preferred or non-preferential PSLs (Pb/Cd-preferred means that the Pb/Cd–PSL interactions are stronger than the Cd/Pb–PSL interactions; non-preferential means that the Cd–PSL and Pb–PSL interactions are equivalent, see Supplementary Discussion, Supplementary Table 1 and Supplementary Fig. 1), so that the influence of the cation-ligand interactions on CE can be quantitatively investigated.

All MD simulations were carried out using the LAMMPS code[38] (http://lammps.sandia.gov). The particle-particle-particle-mesh method[39] was used to calculate the Coulombic interactions and a cut-off radius of 10 Å was set for all short-ranged interactions. The equations of motion were integrated using the velocity–Verlet algorithm with a time step of 1 fs. Four S ions in the centre of the NC were constrained at fixed positions to prevent net motion and rotation of the NCs. Simulations of a few hundreds ns were carried out in the NPT ensemble and the first 100 ps was used for preliminary equilibration at the desired temperatures and zero pressure. The longest simulation time was 300 ns.

Volume-scaled exchange rate. Several MLs were defined in a PbS NC (one ML has a thickness of 3.5 Å, see the inset of Fig. 3a). The exchange rate in the jth ML and at time t, $\eta_j(t)$, was defined as:

$$\eta_j(t) = \frac{(N_{Cd}^{o} - N_{Cd}^{i}) - (N_{Pb}^{o} - N_{Pb}^{i})}{\Delta t \Delta V_j} \tag{2}$$

where N is the number of Cd or Pb ions passing through the surface o or i during a time Δt. The superscripts o and i represent the outer and the inner surfaces of the jth ML, respectively. A positive value of N indicates the cations are moving inward to a given ML, while a negative value indicates that the cations are moving outward. Δt is a small time interval ($\Delta t = 1$ ps) and ΔV_j is the volume of the jth ML. Using this definition, a positive value of η indicates a Pb→Cd CE process and a negative value indicates the reverse Cd→Pb CE. We sampled the η_j every 100 ps for the first four outermost MLs from simulations for a PbS NC with non-preferential PSLs at 550 K. Each data point in Fig. 3a corresponds to an averaged η_j over a time interval of 10 ns from 10 independent MD simulations.

Root mean square motion. The RMSM was used to describe the vibration and mobility of ions. The RMSM was defined as:

$$\mathrm{RMSM} = \sqrt{\frac{1}{t}\sum_{t_j=1}^{t}\left(\mathbf{x}_i(t_j) - \bar{\mathbf{x}}_i\right)^2} \tag{3}$$

where t is the time to sample the data, which is the last 100 ps of 10 independent 100- ns simulations. $\mathbf{x}_i(t_j)$ is the position of ion i at time t_j, $\bar{\mathbf{x}}_i$ is the time-averaged position of the ion i. The time interval between two samplings is 1 ps. The RMSM of each ion and its initial distance r to the centre of the NC were calculated. The RMSM was further averaged over the same types of ions with similar values of r (± 0.5 Å).

References

1. Beberwyck, B. J., Surendranath, Y. & Alivisatos, A. P. Cation exchange: a versatile tool for nanomaterials synthesis. *J. Phys. Chem. C* **117,** 19759–19770 (2013).

2. Rivest, J. B. & Jain, P. K. Cation exchange on the nanoscale: an emerging technique for new material synthesis, device fabrication, and chemical sensing. *Chem. Soc. Rev.* **42,** 89–96 (2013).
3. Gupta, S., Kershaw, S. V. & Rogach, A. L. 25th anniversary article: ion exchange in colloidal nanocrystals. *Adv. Mater.* **25,** 6923–6944 (2013).
4. Son, D. H., Hughes, S. M., Yin, Y. & Alivisatos, A. P. Cation exchange reactions in ionic nanocrystals. *Science* **306,** 1009–1012 (2004).
5. Robinson, R. D. *et al.* Spontaneous superlattice formation in nanorods through partial cation exchange. *Science* **317,** 355–358 (2007).
6. Li, H. *et al.* Sequential cation exchange in nanocrystals: preservation of crystal phase and formation of metastable phases. *Nano Lett.* **11,** 4964–4970 (2011).
7. Zhang, B., Jung, Y., Chung, H.-S., Vugt, L. V. & Agarwal, R. Nanowire transformation by size-dependent cation exchange reactions. *Nano Lett.* **10,** 149–155 (2009).
8. Miszta, K., Dorfs, D., Genovese, A., Kim, M. R. & Manna, L. Cation exchange reactions in colloidal branched nanocrystals. *ACS Nano* **5,** 7176–7183 (2011).
9. Jain, P. K., Amirav, L., Aloni, S. & Alivisatos, A. P. Nanoheterostructure cation exchange: anionic framework conservation. *J. Am. Chem. Soc.* **132,** 9997–9999 (2010).
10. Groeneveld, E. *et al.* Tailoring ZnSe—CdSe colloidal quantum dots via cation exchange: from core/shell to alloy nanocrystals. *ACS Nano* **7,** 7913–7930 (2013).
11. Wark, S. E., Hsia, C.-H. & Son, D. H. Effects of ion solvation and volume change of reaction on the equilibrium and morphology in cation-exchange reaction of nanocrystals. *J. Am. Chem. Soc.* **130,** 9550–9555 (2008).
12. Murray, C. B., Norris, D. J. & Bawendi, M. G. Synthesis and characterization of nearly monodisperse CdE (E = sulfur, selenium, tellurium) semiconductor nanocrystallites. *J. Am. Chem. Soc.* **115,** 8706–8715 (1993).
13. Sadtler, B. *et al.* Selective facet reactivity during cation exchange in cadmium sulfide nanorods. *J. Am. Chem. Soc.* **131,** 5285–5293 (2009).
14. Pietryga, J. M. *et al.* Utilizing the lability of lead selenide to produce heterostructured nanocrystals with bright, stable infrared emission. *J. Am. Chem. Soc.* **130,** 4879–4885 (2008).
15. Jia, G. *et al.* Couples of colloidal semiconductor nanorods formed by self-limited assembly. *Nat. Mater.* **13,** 301–307 (2014).
16. Boneschanscher, M. P. *et al.* Long-range orientation and atomic attachment of nanocrystals in 2D honeycomb superlattices. *Science* **344,** 1377–1380 (2014).
17. Lambert, K., Geyter, B. D., Moreels, I. & Hens, Z. PbTe|CdTe core|shell particles by cation exchange, a HR-TEM study. *Chem. Mater.* **21,** 778–780 (2009).
18. Casavola, M. *et al.* Anisotropic cation exchange in PbSe/CdSe core/shell nanocrystals of different geometry. *Chem. Mater.* **24,** 294–302 (2011).
19. Zhao, H., Chaker, M., Wu, N. & Ma, D. Towards controlled synthesis and better understanding of highly luminescent PbS/CdS core/shell quantum dots. *J. Mater. Chem.* **21,** 8898–8904 (2011).
20. Fan, Z. *et al.* A transferable force field for CdS-CdSe-PbS-PbSe solid systems. *J. Chem. Phys.* **141,** 244503 (2014).
21. Justo, Y. *et al.* More cation exchange and the chemistry of the nanocrystal surface. *ACS Nano* **8,** 7948–7957 (2014).
22. Chan, E. M. *et al.* Millisecond kinetics of nanocrystal cation exchange using microfluidic X-ray absorption spectroscopy. *J. Phys. Chem. A* **111,** 12210–12215 (2007).
23. Bals, S. *et al.* Three-dimensional atomic imaging of colloidal core-shell nanocrystals. *Nano Lett.* **11,** 3420–3424 (2011).
24. Yalcin, A. O. *et al.* Atomic resolution monitoring of cation exchange in CdSe-PbSe heteronanocrystals during epitaxial solid-solid-vapor growth. *Nano Lett.* **14,** 3661–3667 (2014).
25. Lechner, R. T. *et al.* Crystal phase transitions in the shell of PbS/CdS core/shell nanocrystals influences photoluminescence intensity. *Chem. Mater.* **26,** 5914–5922 (2014).
26. Grodzińska, D., Pietra, F., van Huis, M. A., Vanmaekelbergh, D. & de Mello Donegá, C. Thermally induced atomic reconstruction of PbSe/CdSe core/shell quantum dots into PbSe/CdSe Bi-hemisphere hetero-nanocrystals. *J. Mater. Chem.* **21,** 11556–11565 (2011).
27. Zhou, H. S., Honma, I., Komiyama, H. & Haus, J. W. Coated semiconductor nanoparticles; the CdS/PbS system's synthesis and properties. *J. Phys. Chem.* **97,** 895–901 (1993).
28. Zhang, J. *et al.* PbSe quantum dot solar cells with more than 6% efficiency fabricated in ambient atmosphere. *Nano Lett.* **14,** 6010–6015 (2014).
29. Zhang, J. *et al.* Preparation of Cd/Pb chalcogenide heterostructured janus particles via controllable cation exchange. *ACS Nano* **9,** 7151–7163 (2015).
30. Lee, D. *et al.* Direct Cd-to-Pb exchange of CdSe nanorods into PbSe/CdSe axial heterojunction nanorods. *Chem. Mater.* **27,** 5295–5304 (2015).
31. Yalcin, A. O. *et al.* Heat-induced transformation of CdSe—CdS—ZnS Core—multishell quantum dots by Zn diffusion into inner layers. *Chem. Commun.* **51,** 3320–3323 (2015).
32. Bothe, C. *et al.* Solid-state chemistry on the nanoscale: ion transport through interstitial sites or vacancies? *Angew. Chem. Int. Edit* **54,** 14183–14186 (2015).
33. Habinshuti, J. *et al.* Phase transformation of PbSe/CdSe nanocrystals from core–shell to janus structure studied by photoemission spectroscopy. *Phys. Rev. B* **87,** 184102 (2013).
34. Li, D., Zhang, P. & Yan, J. *Ab initio* molecular dynamics study of high-pressure melting of beryllium oxide. *Sci. Rep.* **4,** 4707 (2014).
35. Harrison, W. A., Kraut, E. A., Waldrop, J. R. & Grant, R. W. Polar heterojunction interfaces. *Phys. Rev. B* **18,** 4402 (1978).
36. Lesnyak, V., Brescia, R., Messina, G. C. & Manna, L. Cu vacancies boost cation exchange reactions in copper selenide nanocrystals. *J. Am. Chem. Soc.* **137,** 9315–9323 (2015).
37. Fan, Z. *et al.* From sphere to multipod: thermally induced transitions of CdSe nanocrystals studied by molecular dynamics simulations. *J. Am. Chem. Soc.* **135,** 5869–5876 (2013).
38. Plimpton, S. Fast parallel algorithms for short-range molecular dynamics. *J. Comput. Phys.* **117,** 1–19 (1995).
39. Darden, T., York, D. & Pedersen, L. Particle mesh Ewald: an Nlog(N) method for Ewald sums in large systems. *J. Chem. Phys.* **98,** 10089–10092 (1993).

Acknowledgements

Z.F. acknowledges Dengpan Dong from University of Utah for useful discussions of the kinetic mechanism. This work is part of the research program of the Foundation for Fundamental Research on Matter (FOM), which is part of the Netherlands Organisation for Scientific Research (NWO). This work was sponsored by NWO Exacte Wetenschappen (Physical Sciences) for the use of supercomputer facilities, with financial support from the Nederlandse Organisatie voor Wetenschappelijk Onderzoek (Netherlands Organisation for Scientific Research, NWO). M.A.v.H. acknowledges NWO for a VIDI grant. The data that support the findings of this study will be stored in a repository, and are available from the authors upon request.

Author contributions

Z.F., T.J.H.V. and M.A.v.H. conceived the study. Z.F. developed the simulation model and performed the simulations. Z.F., L.-C.L. and W.B. performed the analyses. All authors discussed the results and participated in writing the manuscript.

Additional information

Competing financial interests: The authors declare no competing financial interests.

5

An unforeseen polymorph of coronene by the application of magnetic fields during crystal growth

Jason Potticary[1], Lui R. Terry[1], Christopher Bell[2], Alexandros N. Papanikolopoulos[3], Peter C.M. Christianen[3], Hans Engelkamp[3], Andrew M. Collins[4], Claudio Fontanesi[5,6], Gabriele Kociok-Köhn[7], Simon Crampin[5], Enrico Da Como[5] & Simon R. Hall[1]

The continued development of novel drugs, proteins, and advanced materials strongly rely on our ability to self-assemble molecules in solids with the most suitable structure (polymorph) in order to exhibit desired functionalities. The search for new polymorphs remains a scientific challenge, that is at the core of crystal engineering and there has been a lack of effective solutions to this problem. Here we show that by crystallizing the polyaromatic hydrocarbon coronene in the presence of a magnetic field, a polymorph is formed in a β-herringbone structure instead of the ubiquitous γ-herringbone structure, with a decrease of 35° in the herringbone nearest neighbour angle. The β-herringbone polymorph is stable, preserves its structure under ambient conditions and as a result of the altered molecular packing of the crystals, exhibits significant changes to the optical and mechanical properties of the crystal.

[1] Complex Functional Materials Group, School of Chemistry, University of Bristol, Bristol BS8 1TS, UK. [2] School of Physics, HH Wills Physics Laboratory, Tyndall Avenue, Bristol BS8 1TL, UK. [3] High Field Magnet Laboratory (HFML-EMFL), Radboud University, Toernooiveld 7, 6525 ED Nijmegen, The Netherlands. [4] Bristol Centre for Functional Nanomaterials, HH Wills Physics Laboratory, Tyndall Avenue, Bristol BS8 1TL, UK. [5] Department of Physics, University of Bath, Claverton Down, Bath BA2 7AY, UK. [6] Dipartimento di Ingegneria Enzo Ferrari, Universita' di Modena e Reggio Emilia, Via Vivarelli 10, 41125 Modena, Italy. [7] Department of Chemistry, University of Bath, Claverton Down, Bath BA2 7AY, UK. Correspondence and requests for materials should be addressed to S.R.H. (email: simon.hall@bristol.ac.uk).

The ability to discover new phenomena and properties in materials depends on our ability to synthesize new structures. The discovery of new structures, however, should rely on variables that transcend the typical thermodynamic parameters of temperature and pressure. In the field of functional molecular materials, polymorphism, the presence of different crystal structures of the same molecular system, can be an opportunity to discover novel phenomena[1] and tune properties[2]. For example, charge carrier mobility in organic semiconductors can be increased by crystallizing molecules under pressure, resulting in shorter intermolecular distances favourable for transport[3]. One successful tool at the scientist's disposal to control crystal structure has been to perform crystallization in the presence of electric fields. This is a well-known process and is used extensively to prepare nonlinear optical materials through electric poling. Crystal growth in the presence of an external magnetic field is much less explored, although it has been proven to be efficacious in the melt texturing of alloys[4] or in cases where controlled convection of the crystallizing solution is required, for example, in the growth of high-quality protein crystals[5], chiral aggregates[6] or in the alignment of liquid crystal and block copolymer arrays[7,8]. A magnetic field has even been demonstrated to be able to separate polymorphs of crystals post-synthetically[9] and in one case to preferentially nucleate a monoclinic form of terpyridine[10]. Although it is known that magnetic forces can have an effect on solidification and subsequent physical behaviour in crystals, their use to create previously unknown polymorphs in single crystals has never been reported.

Polyaromatic hydrocarbons (PAHs) are commonly researched molecules due to their rigid planar structure, high stabilities and characteristic optical and electronic behaviour[10,11]. As molecular solids, PAHs crystallize in four basic structure types according to well-defined geometric and energetic considerations[12]. These are the herringbone structure, the gamma-herringbone (γ-) structure, the sandwich-herringbone (SHB) structure and the beta-herringbone (β-) structure (Supplementary Fig. 1). Comprehensive studies of PAHs have shown that the adoption of one of the four structure types depends ultimately on the relative strength of nonbonded C···C and C···H interactions[12-14]. Polymorphism in PAHs has been demonstrated previously by growing thin films on substrates, where variations in processing conditions direct the formation of different polymorphs[15,16], although in single crystals at ambient pressure, polymorphism in PAHs is rare with perylene and pyrene being two notable cases exhibiting both herringbone and SHB polymorphs[17]. In terms of applications, single crystals of PAHs are to be desired, however, as they have distinct physical advantages over thin-film versions of the same material. Most importantly, for conductive applications such as transistors, or optical applications in solar cells, single crystals typically have higher carrier mobilities than their thin-film analogues[18,19], which have to be defect free to achieve mobilities of the same order of magnitude.

Here we show that by crystallizing the PAH coronene in the presence of a magnetic field, it can be made to form as a new β-herringbone polymorph instead of the ubiquitous γ-herringbone form, with a change as large as 35° to the herringbone nearest neighbour angle. The β polymorph is stable and can preserve its structure in ambient conditions and zero magnetic field. Dispersion corrected density functional theory (DFT-D) calculations indicate that the new form is energetically favoured at low temperature. Furthermore, we demonstrate how the new supramolecular structure generates remarkable changes of the electronic, optical and mechanical properties in the crystal.

Results

Crystal growth of a new polymorph. Coronene is a PAH composed of six aromatic rings arranged in a planar discoidal geometry (Supplementary Fig. 2). The high molecular symmetry (D_{6h}) and 24 electron π-system has made coronene an ideal model system for the study of graphene, due to it being large enough to display exotic electronic behaviour, but not so large that contortion becomes a complicating factor[20]. Centimetre-long crystals of coronene (typically ~0.75 cm) were grown from a supersaturated solution of the molecules in toluene cooled slowly ($0.04\,K\,min^{-1}$) from 328 K to 298 K over a period of 12 h (Fig. 1a, yellow crystal labelled 'γ'). The very slow cooling rate was chosen to eliminate changes in solution shearing which have been shown to influence nucleation and growth with concomitant changes to the resultant crystal structure[21]. Single crystal X-ray diffraction (XRD) of these crystals indicated that the structure is the conventional γ-polymorph; $a = 10.02$ Å, $b = 4.67$ Å, $c = 15.60$ Å, $\beta = 106.7°$, $Z = 2$, space group $P2_1/n$, with an inter-planar distance (d_π) of 3.43 Å, which is consistent with parallel π-stacking[12] (Fig. 2a,b,e,g).

In the presence of an external magnetic field of 1 Tesla, however, significantly longer (typically ~2.5 cm) coronene crystals were grown from the same supersaturated solution and exhibit a different colour to normal γ-coronene crystals (Fig. 1a orange crystal labelled 'β'). Single crystal XRD indicates that the unit cell of these crystals are consistent with a new β-coronene structure; $a = 10.39$ Å, $b = 3.84$ Å, $c = 17.23$ Å, $\beta = 96.24°$, $Z = 2$, space group $P2_1/n$ and an inter-planar distance (d_π) of 3.48 Å (Fig. 2c,d,f,h). As a previously unreported polymorph of coronene, the crystallographic data has been lodged with the Cambridge Crystallographic Data Centre (deposition number CCDC 1409823).

By comparing the two crystal forms in Fig. 2c, it can be seen that the short axis, b, is substantially decreased in length for the β-coronene crystallized in the magnetic field when compared with the γ-coronene. This new polymorph also has a significantly smaller nearest neighbour herringbone angle of 49.71° compared with 95.86° in γ-coronene. The identification of the new crystal polymorph as a β-structure can be made by reference to a plot of the inter-planar angle of nearest neighbour molecules versus the unit cell short axis[12] (Fig. 3). From this, it can be seen that the new polymorph sits squarely with other PAHs identified as

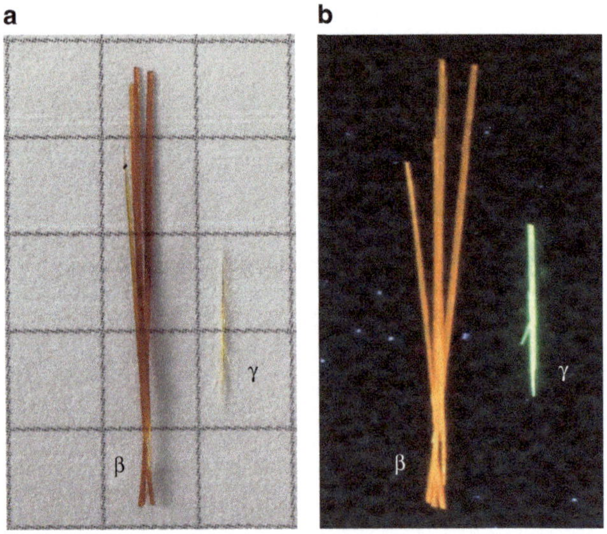

Figure 1 | Optical images of the β- and γ- polymorphs of coronene. (**a**) in daylight and (**b**) under ultraviolet ($\lambda = 365$ nm) illumination to show fluorescence. The squares on the grid in **a** are $0.5 \times 0.5\,cm^2$.

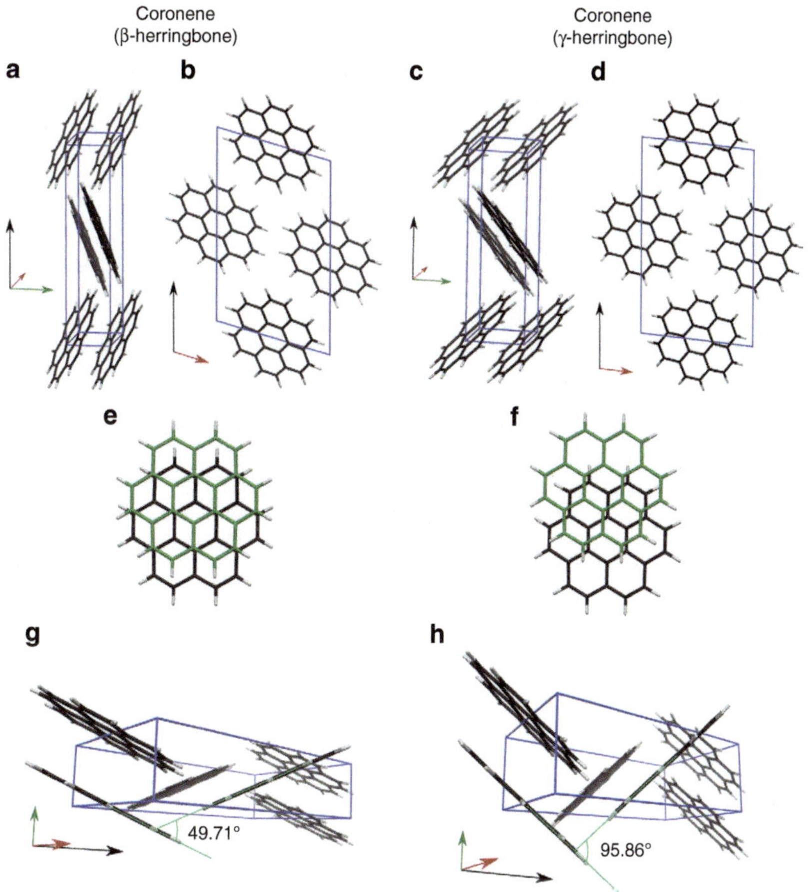

Figure 2 | Representation of the β- and γ- polymorphs of coronene. Differing perspectives of both unit cells (blue boxes) viewed slightly offset from along the *a*-axis (**a** and **c**) and along the *b*-axis (**b** and **d**). The relative shift of the molecules along the stacks are shown for β- (**e**) and γ- (**f**). (**g**) and (**h**) show an orientation of the unit cell clearly demonstrating the difference in nearest neighbour angle between the two polymorphs. Red green and black arrows indicate the direction of the *a*-, *b*- and *c*-axis respectively.

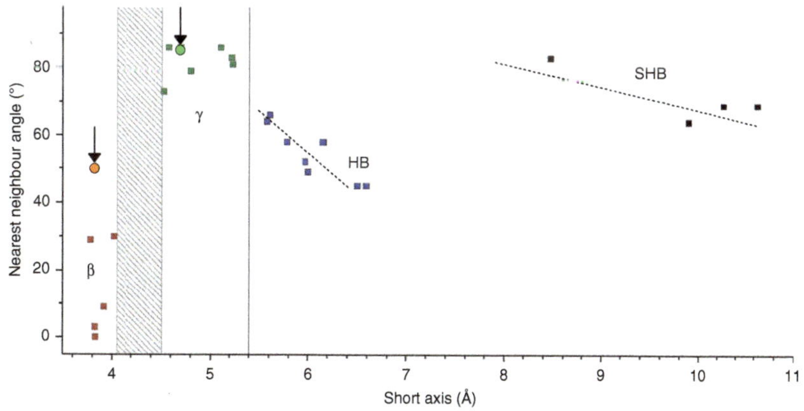

Figure 3 | Grouping of PAHs into structure types. Plotted on the graph are PAHs that adopt the herringbone structure (blue squares), the gamma-herringbone (γ-) structure (green squares), the SHB structure (black squares) and the beta-herringbone (β-) structure (red squares), according to the crystallographic short axis and nearest neighbour herringbone angle. The positions of both γ- and β-coronene polymorphs are indicated by circles and marked with arrows. Adapted from Desiraju and Gavezzotti[12], in which the names of the crystals corresponding to all of the marked squares can be found. Reproduced with permission of the International Union of Crystallography, http://journals.iucr.org/.

having the β-herringbone structure. Under an applied magnetic field of 1 T, the new β-coronene polymorph grows exclusively, is reproducible (>10 times to-date), stable under ambient conditions and also forms in the non-aromatic solvent hexane.

To determine the mechanism of crystal growth, we followed the inception of crystallization of coronene at 1 T and 0 T via *in situ* UV–vis spectrometry. At 358 nm, there are no peaks due to molecules of coronene in solution, nor in the solid state (as crystals). However, a peak will be observed due to the formation of nucleation clusters, which we determined via a single-point energy calculation using time-dependent self-consistent field density functional theory, using basis set

6-31G(D). Supplementary Figure 3 shows that there is a stark difference between the growth of coronene crystals under 1 T of applied field and at 0 T. At 1 T, there is suppression of nucleation which results in the crystallization of the least stable polymorph, that is, beta-coronene. This is entirely consistent with previously published work on the formation of meta-stable polymorphs and is epitomized in Ostwald's rules of stages[22]. Our experimental data therefore demonstrates that the magnetic field is achieving the effect of polymorphic control through suppression of nucleation.

As to the question of why an applied magnetic field of 1 T is having this effect in coronene and not in other molecules at this field strength, it is likely to be intimately linked to the magnetic susceptibility of the molecule. Based on inductively coupled plasma atomic absorption spectroscopy, we note that the level of magnetically active impurities such as cobalt, iron and nickel are at the parts per billion level ($Co = 0.138 \pm 0.98$ ppb; $Fe = 33.81 \pm 0.56$ ppb; and $Ni = 13.46 \pm 0.65$ ppb), and so are unlikely to play any role in polymorph selection. Instead we note that coronene has a very high diamagnetic susceptibility (χ) of -243×10^{-6} emu mol^{-1} (refs 23,24). Crystal growth performed at weaker field strengths of 0.2, 0.5 and 0.8 T resulted in crystals of the γ – polymorph (Supplementary Fig. 4), suggesting that 1 T is close to a threshold for energetic selection between the two forms. In the case of other PAHs such as pyrene, which do not have such a strong diamagnetic susceptibility, our attempts to control polymorphism at 1 T were unsuccessful. It would appear that for molecules with lower diamagnetic susceptibilities, higher fields would be required. Indeed, work reported in a thesis in 2013 has shown that a simple organic dye molecule with a lower diamagnetic susceptibility than coronene, namely an isoxazolone derivative, can be crystallized in a different known polymorph at fields >2.5 T (ref. 25).

Optical behaviour. Figure 1b shows optical images of the two different polymorphs taken under UV irradiation ($\lambda = 365$ nm). The remarkably different colours of the fluorescence of γ-coronene and β-coronene crystals is further confirmation that the molecular packing in the crystal has been transformed, leading to altered electronic behaviour. To quantify this, we measured the absorption spectrum of two single crystals of the two polymorphs by shining unpolarized light perpendicular to the *a*–*b* plane of the crystals, that is, light propagation parallel to [001]. As shown in Fig. 4a (green plot), the γ-coronene single crystal is characterized by a first absorption resonance at 468 nm assigned to the free exciton in coronene, in agreement with previous studies[26]. The β-coronene spectrum is by stark contrast almost featureless, with an absorption onset at 780 nm and a maximum at ~500 nm (Fig. 4a, orange plot). This is a remarkable change in the optical properties of this material. These data can be plotted as the extinction molar coefficient

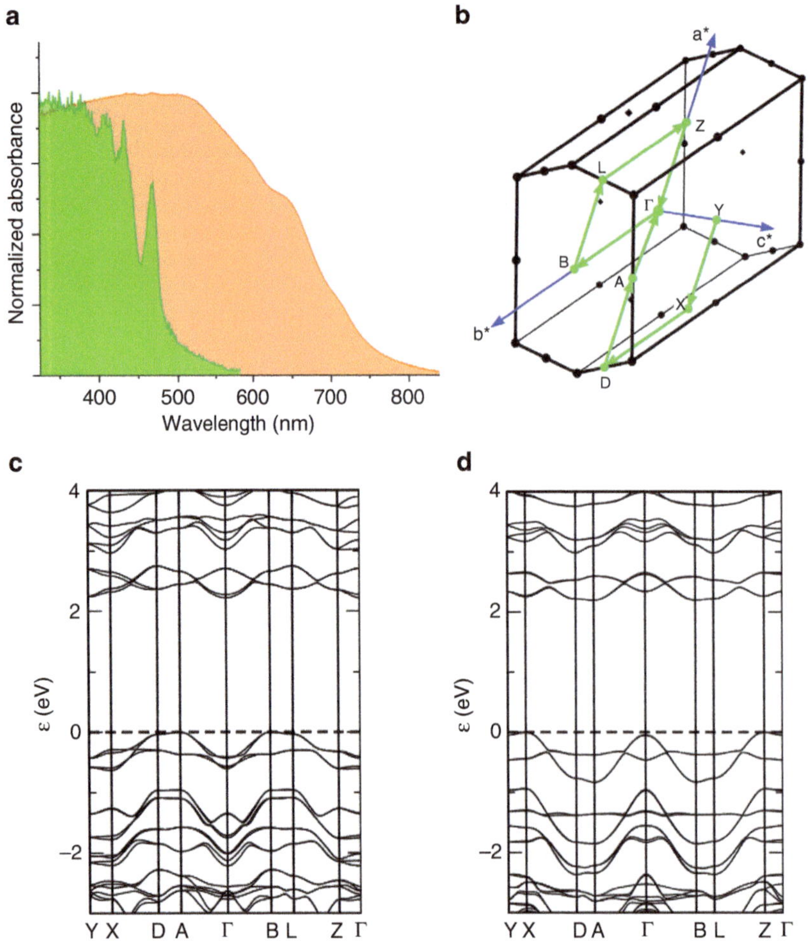

Figure 4 | Electronic structure of γ- and β-coronene. (**a**) Absorption spectra of γ-coronene (green) and β-coronene (orange) single crystals. Unpolarized light was irradiated perpendicular to the *a*-*b* plane at room temperature; (**b**) Brillouin zone with reciprocal lattice vectors and high symmetry points; (**c**) and (**d**) band dispersion along high symmetry points in γ- and β-coronene, respectively.

versus wavelength, to decouple any effects due to differences in the thickness of the crystals of the two polymorphs (50 microns and 76 microns for γ-coronene and β-coronene, respectively; Supplementary Fig. 5). From these spectra, it can be seen clearly that β-coronene is a more strongly absorbing polymorph than γ-coronene. Organic materials tend to be sensitive to a particular range of wavelengths, which can be seen in the features of the γ-coronene absorption spectrum and the lack of absorption of wavelengths >583 nm. In β-coronene, however, the crystal strongly absorbs over a wide band of radiation, from the UV into the near-IR (320–847 nm).

DFT-D calculations correctly describe the existence of the new stable polymorph and indicate that no appreciable difference in the indirect bandgap (Fig. 4b–d) exists between the two polymorphs, suggesting that the large shift in the light absorption onset is instead related to a change in the fundamental photoexcitations in the two structures[27,28]. In β-coronene it is difficult to identify sharp resonances from Frenkel excitons as observed for the γ-polymorph. Instead, the structureless absorption band of the β − polymorph originates from charge transfer excitons that are favoured by the larger overlap between the molecular π orbitals. Moreover, the transition dipole moment of a charge transfer exciton is likely to be oriented parallel to the *a*–*b* plane, which would ensure optimal coupling with the electromagnetic radiation in our experiment.

Structural stability. The new β-polymorph structure is notable for the large change in the herringbone angle. We can gain insights into the relative stability of the β-coronene structure when compared with the γ-polymorph, by consideration of these changes in the CH···π hydrogen bonding motif. It is known that the stronger the hydrogen bond, the stronger the trend for linearity[29,30] depending on the strength of the proton donor. From XRD, we can see that the CH···π angle in γ-coronene of 95.86° suggests a strong, almost linear CH···π hydrogen bond, whereas in β-coronene, this angle becomes 49.71° (viz. Fig. 2). This change in angle would suggest a weakening of the hydrogen bonding in β-coronene, which is concomitant with the change in estimated CH···π hydrogen bond length[31] that increases from 2.5 Å in the γ-polymorph to 3.0 Å in the β-polymorph. This weakening of the CH···π hydrogen bonding should therefore manifest itself as a diminution of the physicomechanical properties of β-coronene. To confirm this, we have determined the melting point (T_M) and elastic modulus (*E*—the modal value measured on the ($\bar{1}0\underline{1}\underline{1}$ and $10\bar{1}\bar{1}$ crystal faces) of both polymorphs and find that in γ-coronene, $T_M = 436.14 \pm 0.01$ °C and $E = 227$ GPa, whereas for β-coronene $T_M = 435.48 \pm 0.01$ °C and $E = 92$ GPa (Supplementary Figs 6 and 7). These data suggest weaker intermolecular forces in β-coronene at room temperature.

DFT-D calculations show that β-coronene is actually the more stable of the two polymorphs at 0 K, by 3.8 kJ mol^{-1} based on lattice energy differences (Supplementary Table 1). Confirmation that this is the case comes from powder XRD of polycrystalline γ-coronene recorded through a temperature cycle from room temperature to 12 K. On cooling through a temperature of 150 K, three new peaks emerge which cannot be indexed to the γ − polymorph (JCPDS card number 12-1611; Supplementary Fig. 8). These emergent peaks correspond to the (002), (101) and (112; $2\theta = 10.55°$, 10.67° and 27.72° respectively) reflections of the new polymorph, β-coronene. Thus β- and γ-coronene are enantiotropic polymorphs, with a critical temperature between 100 K and 150 K (Fig. 5). In general, in an enantiotropic pair of polymorphs, the more intrinsically stable member will have the greater specific enthalpy of fusion. As the differential scanning calorimetry (DSC) experiment to determine melting point (Supplementary Fig. 6) is conducted at far higher temperatures than 150 K, it is the γ-coronene polymorph which will be the more stable at these temperatures and therefore be expected to have the greater specific enthalpy of fusion. From Supplementary Fig. 5, we can calculate the values for the heat of fusion for each of the polymorphs as 20.78 kJ mol^{-1} and 25.26 kJ mol^{-1} for β- and γ-coronene, respectively, confirming that these are enantiotropic polymorphs.

In summary, through judicious application of a magnetic field, we have demonstrated that a well-studied PAH can be grown as a new polymorph and confirmed that a range of physical properties have been significantly altered. The optical properties of the β-coronene crystal have been transformed to such an extent that it absorbs panchromatically from the UV into the near-IR. This new single crystal may therefore be of great interest in organic optoelectronics and photovoltaics.

Methods

Coronene. Coronene crystals (Sigma Aldrich, purity 97%) were recrystallized from toluene and purity assayed by nuclear magnetic resonance (C^{13} and H^+). Trace metals were tested for using inductively coupled plasma—atomic emission spectroscopy (Agilent 710 ICP-AES) after digestion in 3% nitric acid (Sigma Aldrich, ≥99.999%, trace metal basis). Structural analysis was conducted by optical microscopy and single crystal XRD. Analysis of the single crystal data of non-magnetic field grown crystals suggests favourable growth along the *b*-axis, which form the characteristic needle shaped crystals.

Crystal growth under 1 T of magnetic field. A supersaturated solution of coronene (2.5 mg ml^{-1}) in toluene was prepared and stored in an oven at 93 °C. The solution was then passed through a 0.22-μm PTFE filter directly into a 5-mm quartz cuvette with a stopper. Once sealed, the cuvette was placed in the magnetic cavity (Supplementary Fig. 9) and the whole system maintained at 93 °C for 4 h post cuvette insertion. The oven was then programmed to cool to 83, 73 and 63 °C then finally 50 °C, with a 24-h dwell at each temperature, before cooling to room temperature. Crystals of β-coronene are the sole polymorph to grow under the magnetic field and to-date the experiment has been repeated >10 times.

Single crystal X-ray crystallography. Intensity data for all coronene structures were collected at different temperatures on an Agilent SuperNova-E Dual diffractometer equipped with an Oxford Cryosystem, using CuKα radiation ($\lambda = 1.5418$ Å). Data were processed using the CrysAlisPro software (CrysAlisPro, Agilent Technologies, Version 1.171.37.35 (release 13-08-2014 CrysAlis171 .NET; compiled on 13 Aug 2014). For all structures a symmetry-related (multi-scan) absorption correction was applied. Crystal parameters are provided in Supplementary Table 2. See also Supplementary Data 1 and 2. Structure solution, followed by full-matrix least squares refinement was performed using the WINGX-v2014.1 suite of programs throughout.

Physical characterization. Optical absorption spectra were recorded with a ultraviolet–visible–near infrared spectrometer Agilent Cary 5000 measuring in transmission configuration. The spectrometer can measure absorbance up an optical density of 10. Single crystals of β- or γ-coronene were suspended in the sample beam path. Unpolarized light was shone perpendicular to the *a*–*b* plane of the unit cells. High-quality crystals with flat surfaces were carefully selected to avoid light scattering effects. The beam size was narrowed to half the lateral size of crystals. All absorption spectra were recorded at room temperature in air. Elastic moduli were determined using an Asylum Research MFP-3D Infinity AFM operating in AMFM viscoelastic imaging mode, using an AC160TS-R3 silicon tip (9 ± 2 nm radius). Freshly cleaved mica was used as a calibration standard (measured at 178 GPa). Melting point determination was done using a TA-Instruments Q100 DSC with temperature ramp of 1 °C min^{-1} between 35 and 460 °C with 2.2 mg of coronene hermetically sealed under a nitrogen atmosphere in an aluminium pan.

Computational calculations. Computational calculations have been performed with density functional codes CASTEP (ref. 32) and VASP (ref. 33), using the Perdew–Burke–Enzerhof exchange correlation functional with semi-empirical dispersion corrections (DFT-D) to account for van der Waals interactions. CASTEP calculations (version 7.03) used in-built ultra-soft pseudopotentials for C and H atoms, a plane wave cutoff of 600 eV, and Monkhorst-Pack[34] k-point samplings of 3 × 4 × 2. Cell parameters and atomic coordinates were fully relaxed using the Broyden–Fletcher–Goldfarb–Shanno (BFGS) method, halting when residual forces fell below 1 meV Å^{-1}. Changing the cutoff energy from 450 to 600 eV caused structural parameters to change by <0.001 nm or <0.01°. Increasing k-point sampling to 6 × 8 × 4 changed energies by <1 meV. Robustness

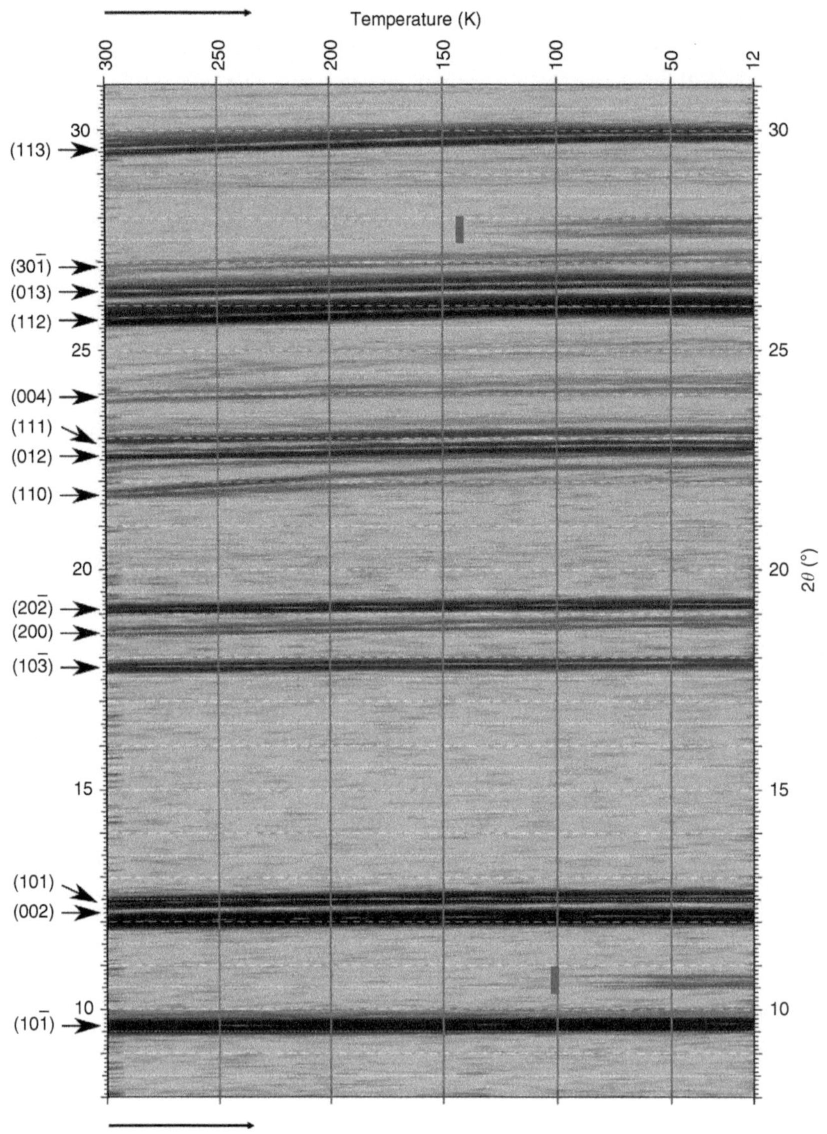

Figure 5 | 2D map of XRD peak shifts as a function of temperature in γ-coronene. Temperature is decreased from 300 to 12 K (left to right). Blue markers indicate the emergence of the new peaks due to β-coronene formation. The blue line near 10° 2θ encompasses two emergent peaks.

of results to choice of semi-empirical dispersion correction was assessed through the use[35] of using the Grimme scheme[36] with both default vdW radii ($R_H = 1.001$, $R_C = 1.452$) and experimental ($R_H = 1.090$, $R_C = 1.750$), as previously employed by Fedorov *et al.*[37] for γ-coronene. VASP calculations (version 5.3.3) use PAW potentials[38] with 500 eV energy cutoff, $3 \times 4 \times 2$ k-point sampling and the Tkatchenko-Scheffler[39,40] dispersion correction. Geometry optimization was performed starting from experimental cell parameters and atom coordinates, aboth with and without symmetry constraint. Stability of optimized geometries was verified by re-optimizing after randomly displacing atoms by 0.05 Å in *x*, *y* and *z*.

References

1. Ganin, A. Y. *et al.* Polymorphism control of superconductivity and magnetism in Cs_3C_{60} close to the Mott transition. *Nature* **466,** 221–225 (2010).
2. Bernstein, J. *Polymorphism in Molecular Crystals* (Oxford University Press, 2002).
3. Giri, G. *et al.* Tuning charge transport in solution-sheared organic semiconductors using lattice strain. *Nature* **480,** 504–508 (2011).
4. Beaugnon, E. *et al.* Material processing in high static magnetic field. A review of an experimental study on levitation, phase separation, convection and texturation. *J. Phys. I France* **3,** 399–421 (1993).
5. Nakamura, A. *et al.* Improvement in quality of protein crystals grown in a high magnetic field gradient. *Cryst. Growth Des.* **12,** 1141–1150 (2012).
6. Micali, N. *et al.* Selection of supramolecular chirality by application of rotational and magnetic forces. *Nat. Chem.* **4,** 201–207 (2012).
7. Boamfa, M. I., Lazarenko, S. V., Vermolen, E. C. M., Kirilyuk, A. & Rasing, T. Magnetic field alignment of liquid crystals for fast display applications. *Adv. Mater.* **17,** 610–614 (2005).
8. Gopinadhan, M. *et al.* Thermally switchable aligned nanopores by magnetic-field directed self-assembly of block copolymers. *Adv. Mater.* **26,** 5148–5154 (2014).
9. Atkinson, M. B. J. *et al.* Using magnetic levitation to separate mixtures of crystal polymorphs. *Angew. Chem. Int. Ed.* **52,** 10208–10211 (2013).
10. Honjo, S., Yokota, M., Doki, N. & Shimizu, K. Magnetic field influence on the crystal structure of 2,2':6',2''-terpyridine. *Kagaku Kogaku Ronbun.* **34,** 383–387 (2008).
11. Rieger, R. & Müllen, K. Forever young: polycyclic aromatic hydrocarbons as model cases for structural and optical studies. *J. Phys. Org. Chem.* **23,** 315–325 (2010).
12. Desiraju, G. R. & Gavezzotti, A. Crystal structures of polynuclear aromatic hydrocarbons. Classification, rationalization and prediction from molecular structure. *Acta Crystallogr. Sect. B Struct. Sci.* **45,** 473–482 (1989).
13. Loots, L. & Barbour, L. J. A simple and robust method for the identification of π–π packing motifs of aromatic compounds. *CrystEngComm* **14,** 300–304 (2014).

14. Schatschneider, B., Phelps, J. & Jezowski, S. A new parameter for classification of polycyclic aromatic hydrocarbon crystalline motifs: a Hirshfeld surface investigation. *CrystEngComm* **11,** 7216–7223 (2011).
15. Mattheus, C. C. *et al.* Identification of polymorphs of pentacene. *Synth. Metals* **138,** 475–481 (2003).
16. Hiszpanski, A. M. *et al.* Tuning polymorphism and orientation in organic semiconductor thin films via post-deposition processing. *J. Am. Chem. Soc.* **136,** 15749–15756 (2014).
17. Botoshansky, M., Herbstein, R. H. & Kapon, M. Towards a complete description of a polymorphic crystal: the example of perylene redetermination of the structures of the (Z = 2 and 4) polymorphs. *Helv. Chim. Acta* **86,** 1113–1128 (2003).
18. Hu, W.-S., Weng, S.-Z., Tao, Y.-T., Liu, H.-J. & Lee, H.-Y. Oriented growth of rubrene thin films on aligned pentacene buffer layer and its anisotropic thin-film transistor characteristics. *Org. Electron.* **9,** 385–395 (2008).
19. Hamaguchi, A. *et al.* Single-crystal-like organic thin-film transistors fabricated from dinaphtho[2,3- b :2′,3′- f]thieno[3,2- b]thiophene (DNTT) precursor–polystyrene blends. *Adv. Mater.* **27,** 6606–6611 (2015).
20. Zhao, Y. & Truhlar, D. A prototype for graphene material simulation: structures and interaction potentials of coronene dimers. *J. Phys. Chem. C* **112,** 4061–4067 (2008).
21. Giri, G. *et al.* Effect of solution shearing method on packing and disorder of organic semiconductor polymers. *Chem. Mater.* **27,** 2350–2359 (2015).
22. Ostwald, W. Studien über die bildung und umwandlung fester körper. *Z. Phys. Chem.* **22,** 289–330 (1897).
23. Steiner, E., Fowler, P. W. & Jenneskens, L. W. Counter-rotating ring currents in coronene and corannulene. *Angew. Chem. Int. Ed.* **40,** 362–366 (2001).
24. Rogers, F. M. T. The magnetic anisotropy of coronene, naphthazarin, and other crystals. *J. Am. Chem. Soc.* **69,** 1506–1508 (1947).
25. Aret, E. Growth of Organic Dye Crystals, PhD Thesis, University of Radboud, Nijmegen. http://repository.ubn.ru.nl/handle/2066/112754 (2013).
26. Yamamoto, T. *et al.* Exciton-phonon coupling and pressure-induced structural phase changes in coronene crystals. *Chem. Phys.* **184,** 247–254 (1994).
27. Hummer, K., Puschnig, P. & Ambrosch-Draxl, C. Lowest optical excitations in molecular crystals: bound excitons versus free electron-hole pairs in anthracene. *Phys. Rev. Lett.* **92,** 147402 (2004).
28. Cudazzo, P., Sottile, F., Rubio, A. & Gatti, M. Exciton dispersion in molecular solids. *J. Phys.-Condes. Matter* **27,** 113204 (2015).
29. Nishio, M., Umezawa, Y., Suezawa, H. & Tsuboyama, S. in *The Importance of Pi-Interactions in Crystal Engineering. Frontiers in Crystal Engineering* 2nd edn (eds Tiekink, E. R. T. & Zukerman-Schpector, J.) 1-39 (Wiley (2012).
30. Steiner, T. & Desiraju, G. R. Distinction between the weak hydrogen bond and the van der Waals interaction. *Chem. Comm.* **8,** 891–892 (1998).
31. Allen, F. H. A systematic pairwise comparison of geometric parameters obtained by X-ray and neutron diffraction. *Acta Crystallogr. B* **42,** 515–522 (1986).
32. Clark, S. *et al.* First principles methods using CASTEP. *Z. Kristallogr.* **220,** 567 (2005).
33. Kresse, G. & Furthmüller, J. Efficient iterative schemes for ab initio total-energy calculations using a plane-wave basis set. *Phys. Rev. B.* **54,** 11169 (1996).
34. Monkhorst, H. J. & Pack, J. D. Special points for Brillouin-zone integrations. *Phys. Rev. B* **13,** 5188 (1976).
35. McNellis, E. R., Meyer, J. & Reuter, K. Azobenzene at coinage metal surfaces: role of dispersive van der Waals interactions. *Phys. Rev. B.* **80,** 205414 (2009).
36. Grimme, S. Semi-empirical GGA-type density functional constructed with a long-range dispersion correction. *J. Comput. Chem.* **27,** 1787 (2006).
37. Fedorov, I., Zhuravlev, Y. & Berveno, V. Properties of crystalline coronene: dispersion forces leading to a larger van der Waals radius for carbon. *Phys. Status Solidi B* **249,** 1438–1444 (2012).
38. Blöchl, P. E. Projector augmented-wave method. *Phys. Rev. B.* **50,** 17953 (1994).
39. Tkatchenko, A. & Scheffler, M. Accurate molecular Van der waals interactions from ground-state electron density and free-atom reference data. *Phys. Rev. Lett.* **102,** 073005 (2009).
40. Bučko, T., Lebègue, S., Hafner, J. & Ángyán, J. G. Tkatchenko-Scheffler van der Waals correction method with and without self-consistent screening applied to solids. *Phys. Rev. B.* **87,** 064110 (2013).

Acknowledgements

S.R.H. and J.P. acknowledge the Engineering and Physical Sciences Research Council (EPSRC), UK (grant EP/G036780/1) and the Bristol Centre for Functional Nanomaterials for project funding. The work at Nijmegen was performed as part of the research programme of the 'Stichting voor Fundamenteel Onderzoek der Materie (FOM)', which is financially supported by the 'Nederlandse Organisatie voor Wetenschappelijk Onderzoek (NWO)'. In addition, S.R.H. and J.P. thank Stasja Stanišić for discussions on low-temperature X-ray crystallography and Spartacus Mills for estimations of magnitude. E.D.C. and S.R.H. are grateful to the GW4 for funding through an accelerator grant and L.R.T. is supported in part by a grant from the USAF European Office of Aerospace Research and Development (FA8655-12-1-2078). All authors thank Prof. John Evans and Liana Vella-Zarb of the University of Durham for low-temperature powder XRD, and Dr Chung Choi and Dr George Whittell of the University of Bristol for ICP-AES and DSC, respectively.

Author contributions

S.R.H. initiated and supervised the project. J.P. performed the crystallization experiments at Bristol and characterized the samples optically and with C.B. undertook magnetic characterizations. J.P., A.N.P., P.C.M.C and H.E. performed the crystallization experiments at Nijmegen and undertook ultraviolet–visible characterizations. L.R.T. performed DSC experiments, E.D.C. and C.F. absorbance spectra experiments, G.K.K. single crystal X-ray characterization. S.C and C.F. undertook DFT-D calculations. A.M.C. performed elastic moduli experiments. All authors contributed to the discussion of the results, analysis of the structures and to manuscript preparation.

Additional information

Accession codes: The X-ray crystallographic coordinates for structures reported in this Article have been deposited at the Cambridge Crystallographic Data Centre (CCDC), under deposition number CCDC 1409823. These data can be obtained free of charge from The Cambridge Crystallographic Data Centre via www.ccdc.cam.ac.uk/data_request/cif.

Competing financial interests: The authors declare no competing financial interests.

6

Cold denaturation induces inversion of dipole and spin transfer in chiral peptide monolayers

Meital Eckshtain-Levi[1,*], Eyal Capua[1], Sivan Refaely-Abramson[2,*], Soumyajit Sarkar[2,*], Yulian Gavrilov[3], Shinto P. Mathew[1], Yossi Paltiel[4], Yaakov Levy[3], Leeor Kronik[2] & Ron Naaman[1]

Chirality-induced spin selectivity is a recently-discovered effect, which results in spin selectivity for electrons transmitted through chiral peptide monolayers. Here, we use this spin selectivity to probe the organization of self-assembled α-helix peptide monolayers and examine the relation between structural and spin transfer phenomena. We show that the α-helix structure of oligopeptides based on alanine and aminoisobutyric acid is transformed to a more linear one upon cooling. This process is similar to the known cold denaturation in peptides, but here the self-assembled monolayer plays the role of the solvent. The structural change results in a flip in the direction of the electrical dipole moment of the adsorbed molecules. The dipole flip is accompanied by a concomitant change in the spin that is preferred in electron transfer through the molecules, observed via a new solid-state hybrid organic-inorganic device that is based on the Hall effect, but operates with no external magnetic field or magnetic material.

[1] Department of Chemical Physics, Weizmann Institute of Science, Rehovoth 76100, Israel. [2] Department of Materials and Interfaces, Weizmann Institute of Science, Rehovoth 76100, Israel. [3] Department of Structural Biology, Weizmann Institute of Science, Rehovoth 76100, Israel. [4] Department of Applied Physics and Center for Nano Science and Nanotechnology, The Hebrew University, Jerusalem 91904, Israel. * These authors contributed equally to this work. Correspondence and requests for materials should be addressed to R.N. (email: ron.naaman@weizmann.ac.il).

Self-assembled monolayers (SAM) have been studied intensively due to their technological importance in various fields, from organic electronics to friction and catalysis (see, for example, refs 1–3, and references therein). Commonly it is assumed that the structure of the molecules within the SAM is only weakly dependent on their supra-molecular assembly. However, this is not necessarily the case, especially when dealing with SAM based on molecules well-known to exhibit structural diversity, such as oligopeptides.

Experiments performed more than a decade ago showed that the spin selectivity of photoelectrons transmitted (from a gold substrate) through an oligopeptide monolayer is affected by the direction of the dipole moment of the molecules relative to the electrons' velocity[4] and that both are temperature-dependent[5]. These results are summarized in Fig. 1 for a gold-adsorbed monolayer of oligoalanine consisting of 22 amino acid units. By monitoring the work function of the sample (by means of its contact potential difference (CPD), with a reference gold surface) as a function of temperature, a change in the sign of the CPD is observed (Fig. 1a). Although the null CPD signal does not necessarily coincide with the dipole flipping point, owing to the Pauli Push-back effect that changes the dipole of a monolayer-adsorbed surface even in the absence of a molecular dipole[6,7]. this change does reflect a change in surface dipole[8]. The dipole flip correlates with a change in the sign of the preferred spin of the transmitted photoelectrons (Fig. 1b,c).

While the data of Fig. 1 may hint at a structural effect, the origin of the dipole and spin flipping remained unknown when these results were published. Indeed, determining structural changes in a monolayer at different temperatures, and correlating them with dipole and spin-filtering properties, is not a simple task.

Here, we augment the traditional characterization tools with a thorough computational investigation using molecular dynamics (MD) and density functional theory (DFT), as well as with a new experimental setup—the magnet-less Hall device, based on the recently discovered chirality-induced spin selectivity (CISS) effect[9–12]. The combined computational and experimental results reveal that the oligopeptide structure is highly temperature-dependent. We find that the surrounding chemical environment in the SAM can induce 'denaturation' upon cooling, much as in biological systems 'cold denaturation' can take place due to protein interaction with the surrounding water[13], but here the SAM plays the role of the solvent. This dramatic structural change results in a flip in the direction of the dipole moment of the adsorbed molecules, which, in turn, reverses the polarity of the preferred spin in electron transfer experiments. This phenomenon is of importance not only for the basic understanding of the structure of adsorbed molecules, but also opens new possibilities for controlling spin in organic spintronic devices[14].

Results

Computational results. Our working hypothesis at the outset of this investigation was that the oligopeptide chains may be entropy-driven to fold with increasing temperature. For investigating this initial idea, classical MD simulations were performed on a dense monolayer of 10-mer oligoalanine peptides attached covalently to a surface using the GROMACS package (see Methods section for details). Because the MD force fields are commonly calibrated at room temperature, simulations at very low temperatures are not necessarily reliable as they cannot be expected to represent the system energetics accurately. This difficulty was circumvented by avoiding the need to study the peptide chain explicitly over a wide temperature range. Instead, we tested the effect of folding directly by starting from two distinct organizations of the peptides in the self-assembled monolayer, an α-helical and an extended conformation. Each system was simulated for several nanoseconds at 300 K and was analyzed by focusing on the conformational energetics and dynamics of the central polyalanine peptide in the simulated monolayer. As shown in Fig. 2a, the two configurations of the polyalanine monolayer exhibited different properties, with the monolayer of the helical peptides being more disordered and less packed. We further used the distribution of the inter-peptide angle (Fig. 2c) to estimate the contribution of this degree of freedom for the configurational entropy. Using the simple relation $S = -k\,p\ln p$, where p is the probability of the angle, we found that at room temperature this degree of freedom

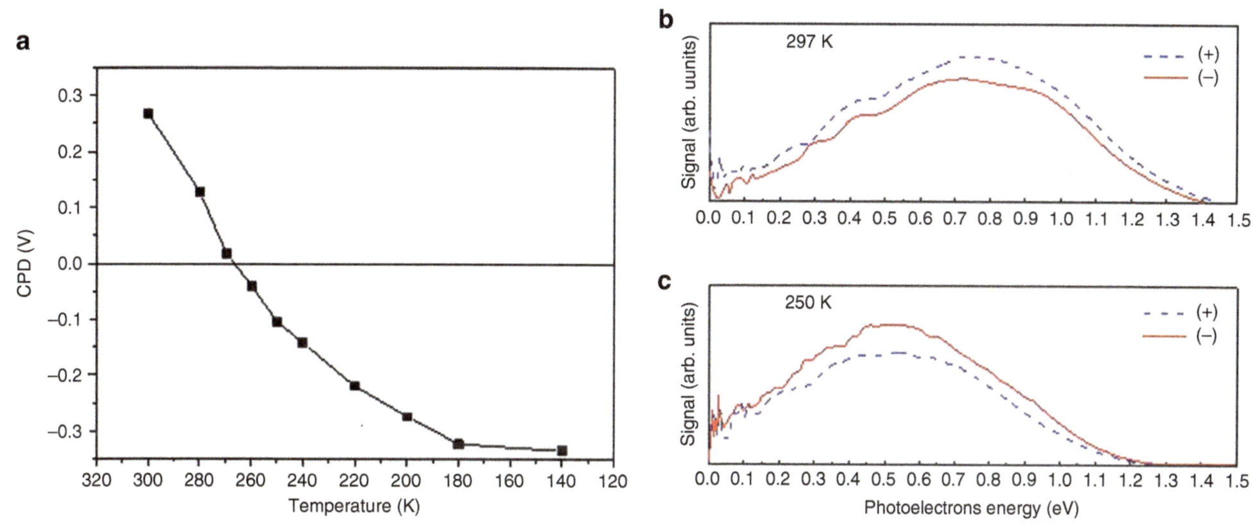

Figure 1 | Temperature-dependent contact potential (surface work function) and spin-polarized photoemission. (**a**) Temperature- dependent contact potential difference (CPD) measurement of gold coated with a self-assembled monolayer of L-polyalanine, bound to the surface through the carbon terminal. Zero voltage indicates the CPD of the bare gold. (**b,c**) Energy distribution of photoelectrons ejected from the gold substrate using a 248 nm laser with clockwise (+, blue) or counter clockwise (−, red) circularly polarized light, at temperatures of 297 K (**b**) or 250 K (**c**). Adapted from ref. 5, used with permission.

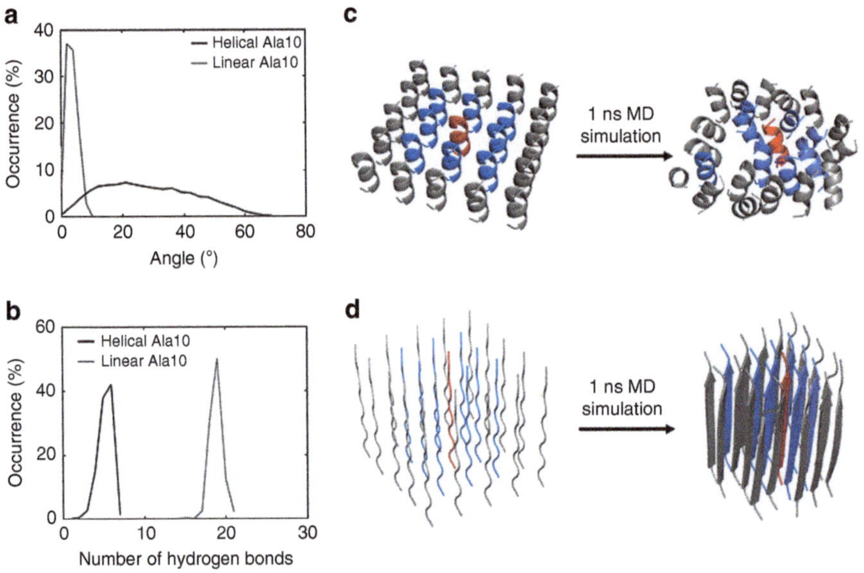

Figure 2 | Molecular dynamics simulations of monolayers of oligoalanine peptides attached covalently to a surface. 5 × 5 monolayer of 10-mer oligoalanine (Ala10), modeled as (**c**) an α-helical or (**d**) a linear conformation. Each model of the monolayer was simulated for 1 ns and further analysis was performed on the central peptide (red) relative to its eight neighbors (blue). Also shown is the distribution of the: (**a**) angle of the central peptide intra-peptide hydrogen bonds; (**b**) number of inter-peptide hydrogen bonds between the central peptide and neighboring ones.

contributes 0.33 kcal mol^{-1} for the helical 10-mer monolayer but only 0.005 kcal mol^{-1} for the linear 10-mer monolayer. Confirming our hypothesis, the stability of the α-helix form at room temperature, thus, is indeed due to its relatively high entropy, while the linear form is more stable at low temperature because of its lower enthalpy. For example, the total number of hydrogen bonds for the central peptide is much larger when the peptide is linear than helical, as shown in Fig. 2b. The detailed calculations suggest that polyalanine adsorbed as a self-assembled monolayer is going through 'cold denaturation' upon cooling and is gradually transformed from the α-helix structure to an extended β-strand shape.

To examine the relation between the cold denaturation and electrical properties, we 'extracted' the central peptide from MD runs with various degrees of stretching and examined the gas-phase dipole and electronic structure obtained from DFT calculations. Importantly, these calculations were performed using the optimally-tuned range-separated hybrid scheme (see Methods section)[15,16]. This approach was shown to provide accurate results for the outer-valence electronic structure of finite systems in general[17,18] and in particular predict their ionization potential and electron affinity from the highest occupied molecular orbital (HOMO) and lowest unoccupied molecular orbital (LUMO), respectively[15,16,19–23], where standard schemes typically fail[24,25]. This is of unique importance for highly polar systems[26], including protein and peptide molecules[27–29], where a strongly underestimated gap may lead, under a large internal electric field, to spurious gap closure and qualitative failure of the calculation.

Representative calculated structures of the model, 10-mer oligoalanine system are shown in Fig. 3a. Importantly, and in agreement with the early experiments of Fig. 1, we find that as the helical structure gradually unfolds and the terminal groups move further apart, the dipole becomes less negative and eventually changes sign. The flip in dipole direction upon structural change is not unique to this specific length of the peptide and also appeared in calculations of longer (20-mer) and shorter (4-mer) oligoalanines, not shown for brevity. The main effect causing the dipole flip is demonstrated schematically in Fig. 3b for a 10-mer oligoalanine. For a linear conformation, the individual contributions of the local dipole moments formed by each of the peptide bonds add up coherently along the molecular axis, in the direction opposite to the dipole formed between the C-terminus and the N-terminus of the peptide. In the α-helical form, the direction of the same peptide-bond related local dipoles is flipped, due to additional hydrogen bonds that set the directionality of the oxygen atoms and change the relative direction of the O–N pairs in the peptide bond. The contribution of these local dipoles to the net dipole is then in the same direction as the contribution of the C-terminus to N-terminus dipole. This results in a larger net dipole moment in the case of the α-helix structure, which has an opposite sign to that of the linear structure. Accordingly, two different structures of similar length can lead to different dipole moments with different signs, as demonstrated by structures 3 and 4 in Fig. 3a. Both structures are ~25 Å long, but while in the somewhat shorter structure 3 many peptide bonds (which are closer to the C-terminus) are still within a helical form, in the somewhat longer structure 4 almost all peptide bonds are linearly aligned. This causes the dipole to flip sign along the molecular axis. Furthermore, the change of dipole sign is accompanied by a related flip in the position of the occupied and unoccupied orbitals, as visualized in Fig. 3a by the charge density distributions of the HOMO and LUMO. This can be easily rationalized in terms of quasi-electron and quasi-hole states responding oppositely to the electric field[30].

Experimental results. Encouraged by the above theoretical findings, we sought direct experimental evidence for chain elongation with decreasing temperature, through a combination of ellipsometry and FTIR (Fourier transform infrared spectroscopy see Methods section). Note that here and throughout this article, experiments were performed on {Ala-Aib} oligopeptides. These were chosen because they possess the same structure as the pure alanine-based ones, discussed so far, but exhibit better solubility. As shown below, they exhibit the exact same phenomena as the oligoalanine peptides. The IR spectroscopy showed consistently an increase in the absorption signal upon cooling to 14 K, with no

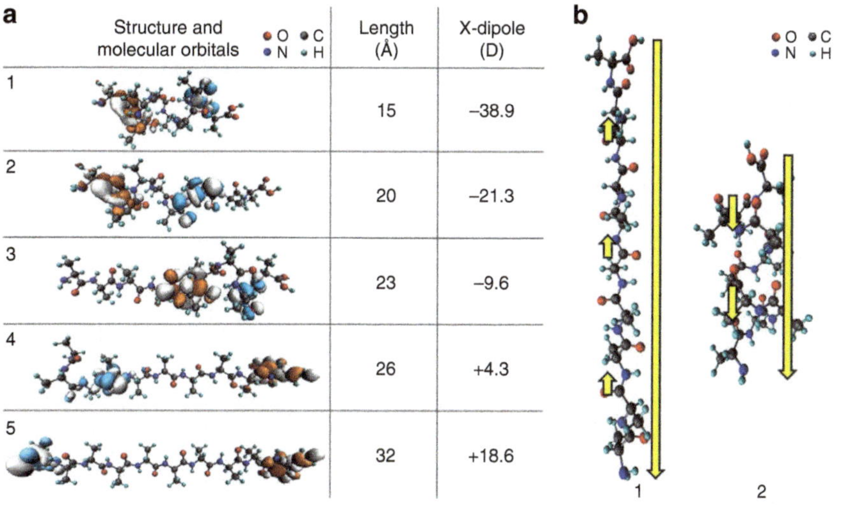

Figure 3 | Calculated structures and dipole moments of oligoalanine molecules. (**a**) Several structures of 10-mer oligoalanine molecules, gradually stretched from their α-helix conformation to a linear one. For each structure, the molecular length (determined by the distance along the molecular axis between the two farthest C atoms), as well as the DFT-computed dipole moment along the molecular axis and the highest occupied (light blue) and lowest unoccupied (orange) molecular orbitals, are given. (**b**) The change in dipole direction upon structural conformation of 10-mer oligoalanines, demonstrated schematically for the most linear and most helical conformations of the examined peptides. Long arrows indicate the dipole formed between the C-terminus and the N-terminus of the peptide. Short arrows indicate local dipoles formed by peptide bonds.

Table 1 | Monolayer thickness as a function of cooling time at 70K extracted from ellipsometry for a monolayer made from NH_2-{Ala-Aib}$_8$-$COCHNH_2CH_2PO_4H_2$ and adsorbed on Si/SiOx.

Cooling time (min)	Thickness of molecule before cooling (Å)	Thickness of molecule after cooling (Å)	Difference in thickness (Å)
30	16 ± 1	20 ± 1	4
60	18 ± 1	25 ± 1	7
120	16 ± 1	29 ± 1	13

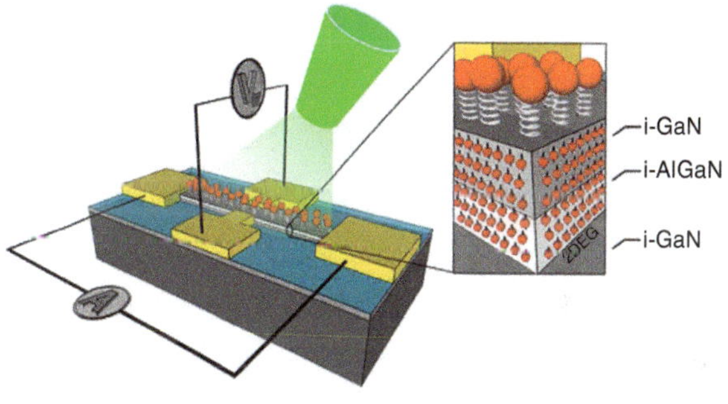

Figure 4 | Schematic structure of a GaN-based magnetless Hall device. A two-dimensional electron gas is formed between the AlGaN layer and the GaN layer underneath it. Current flows between two electrodes and the Hall voltage is measured between the transverse set of electrodes. Nanoparticles are attached to the substrate through chiral molecules (in this case, oligopeptides) adsorbed on top of the conductive channel.

significant changes observed below 100 K. However, while these data are consistent with the denaturation process, they do not constitute a clear indication for it. Ellipsometry data, given in Table 1, did provide clear evidence for chain elongation upon cooling. Experimentally, a length change that depends on the cooling time was observed. For very long cooling times the length is approximately doubled, in agreement with the theoretical prediction.

The remaining question, then, is the origin of the spin-transfer selectivity (Fig. 1b) and whether cold denaturation can also explain its polarity inversion. Generally, spin-selectivity in electron transmission or tunneling through chiral layers is explained by the CISS effect[9,10,31], which has also been used to explain spin-specific electron transfer through oligopeptides[32] and even suggested to play a role in electron transfer through biological systems[33]. Briefly, as an electron moves along a chiral molecule it experiences the electrostatic potential of the molecule, which is chiral. In the electron's rest frame, the charge that defines the electrostatic potential generates a magnetic field which splits the degeneracy of the electron spin states. The sign of the spin being transferred more efficiently through the chiral molecules depends on the sign of the electric field acting on the electron. This is because spin-selectivity is controlled by spin–orbit coupling, which has the well-known form, $H_{so} \propto \vec{v} \times \vec{E}$, where $\vec{v}$ is the electron velocity and $\vec{E}$ is the electric field. Because the dipole flip induced by the cold denaturation necessarily implies a change in sign of the overall electric field across the oligopeptide, reversal in spin-filtering polarity emerges naturally.

While the above explanation is plausible, proving experimentally that indeed the preferred spin inversion is inexorably linked to structural changes is more difficult. To pursue this, we developed a new type of Hall-effect device, shown in Fig. 4 (see Supplementary Fig. 1). The Hall effect is a long-known phenomenon[34]. Briefly, when current is flowing in a substrate between two electrodes, a magnetic field applied perpendicular to the current flow induces an electric potential perpendicular both to the current and the magnetic field. This effect is used as a standard tool for the study of properties of semiconductors, notably charge carrier concentration and mobility[35]. Here, we demonstrate that the spin-filtering capabilities of the chiral monolayer allow for the observation of the Hall effect without either an externally applied magnetic field or a permanent magnet. Instead, as explained below, the hall voltage arises from the accumulation of spin-transfer. This allows us to probe the relation between structural changes and spin inversion directly.

The device, shown schematically in Fig. 4, is based on a two-dimensional electron gas structure, onto which two sets of Hall probe electrodes were evaporated. We have chosen to implement the device on a GaN substrate due to the long spin-lifetime of this material[36]. The width and length of the conducting channel were 40 and 700 μm, respectively. On the conduction channel, a SAM of NH_2-{Ala-Aib}$_8$-$COCHNH_2CH_2PO_4H_2$ or $SHCH_2CH_2CO$-{Ala-Aib}$_5$-COOH oligopeptides were adsorbed, that is, either a phosphonate or a carboxylate group was used to bind the oligopeptide to the GaN surface. The IR spectra of the adsorbed layer are shown in Supplementary Fig. 2. The oligopeptides chosen possess the same structure as the pure alanine-based ones discussed so far and were chosen because of their better solubility. CdSe nano particles (NPs) of ~6 or ~2 nm diameter were then attached to the amine or thiol tail. The SEM image of the layer with the nanoparticles is shown in Supplementary Fig. 3 and additional information is provided in the Supplementary Methods. A constant current of 10 μA was injected along the structure, with the Hall voltage measured in the transverse direction (see Methods section).

The device was calibrated using an external magnetic field as shown in Supplementary Fig. 4. As expected from standard theory, no Hall voltage is observed in the absence of a magnetic field under dark conditions. However, as shown in Fig. 5, a Hall voltage appears - with either binding group, with two different sizes of NPs, and at both high and low temperature when the CdSe NPs are illuminated (with a 2.3 mW, 514 nm laser). Strikingly, the Hall voltage is of opposite sign at low and high temperatures.

Discussion

How can these results be explained and how are they related to the spin-polarization results of Fig. 1? A model explaining these results is given in Fig. 6. Upon excitation of the NPs (Fig. 6a), an electron is transferred from the substrate to the hole state on the NP (Fig. 6b). Because the oligopeptides are chiral, it is expected that the electron transfer process is spin selective, according to the above-explained CISS effect. An electron with the opposite spin is then accumulated on the device surface, acting to magnetize the substrate and create an effective magnetic field for the Hall measurement. If spin–orbit coupling in the NP is large (which is the case for CdSe), the spin of the photo-excited electron on the NP is randomized such that it is spin-selected for transfer to the substrate, still within the lifetime of the net spin formed on the substrate side (Fig. 6c). Both substrate and NP are now neutral again, but the net spin on the substrate is further enhanced (Fig. 6d). If the spin-lifetime in the substrate is long relative to that in the NPs (which is indeed the case for a GaN substrate and CdSe NPs), a specific spin is accumulated on the device surface, creating net magnetization[37]. A Hall voltage can then be measured. Its magnitude must be proportional to the spin accumulation field and its sign must reflect the type of spin accumulated. Under continuous illumination, the net effect will naturally depend critically on the lifetime of the spin state at the substrate surface, the number of spins being transferred between the substrate and the NP during this time, and the rate of spin-randomization. The device presented here is unique because it allows, in principle, a high spin-polarization without a net charge polarization. Therefore, high sensitivity is achieved here due to spins 'loaded' onto the device without encountering the difficulty of Coulomb repulsions that one may face when trying to spin-polarize a medium by injecting spin-polarized electrons. In this sense, the situation is similar to the known case of spin–torque transfer used to spin-polarized ferromagnetic materials[38].

With the model of Fig. 6 in mind, the results of Fig. 5 can now be rationalized. First and foremost, the spin transferred at room temperature is aligned parallel to the electron velocity, while at low temperature it is antiparallel to the velocity. This is in full agreement with an inversion of the CISS effect owing to the electrical dipole flip predicted theoretically in Fig. 3. Other insights can also be drawn from the figure. Clearly, the Hall signal is larger in the case of the phosphonate linker than in the case of the carboxylic one. This result is consistent with reports on the stronger binding of the phosphonates to GaAs[39] and we assume, in light of these results, that the same applies to GaN. In addition, while at room temperature spin randomization is efficient both in the large and small NPs, due to phonon–spin interactions, at low temperature the signal is larger for small NPs. This can be explained by the quenching of the phonon–spin interaction and by the known faster spin-randomization in smaller NPs[40,41], by as much as an order of magnitude, due to spin interaction with the surface[41]. Therefore, when the spin randomization in the NPs is fast, the NPs can be excited many times during the lifetime of the spin polarization on the GaN (see Fig. 5) and many spin-pairs can be injected into the GaN per NPs.

Using a calibration curve (see Supplementary Fig. 4), we can estimate the effective magnetic field that would be required

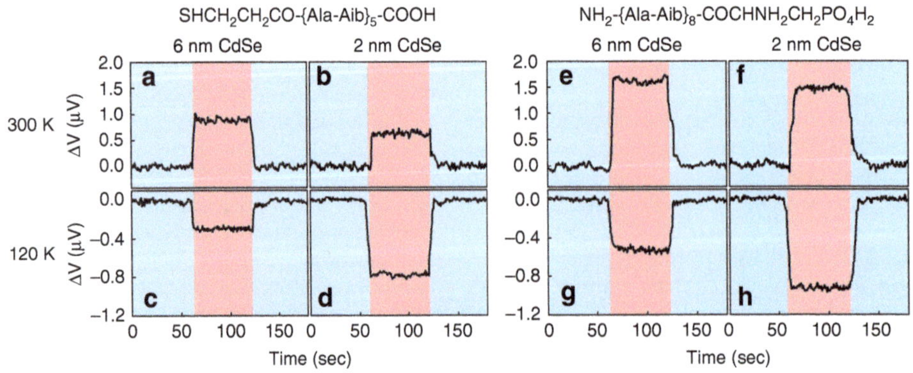

Figure 5 | Hall voltage measurements at different temperatures and for different sizes of nanoparticles. All measurements were performed at 300 K and at 120 K for devices of the type shown in Fig. 4, coated with 6 and 2 nm diameter CdSe nanoparticles that are attached to the GaN surface via NH_2-{Ala-Aib}$_8$-$COCHNH_2CH_2PO_4H_2$ or $SHCH_2CH_2CO$-{Ala-Aib}$_5$-COOH molecules, at a current of 10 μA. The devices were illuminated with a 2.3 mW, 514 nm laser. Three sets of devices were measured and the variation in the signal among them is about 10%. The signal observed corresponds to effective magnetic fields of 20, 16, -5, -18, 73, 36, -8 and -22 G for panels **a,b,c,d,e,f,g** and **h**, respectively. All plots are presented after subtracting a background signal, measured for the same devices but covered with the non-chiral 11-mercaptoundecylphosphoric acid (see Supplementary Fig. 8). Pink and light-blue areas correspond to 'light on' and 'light off' regimes, respectively.

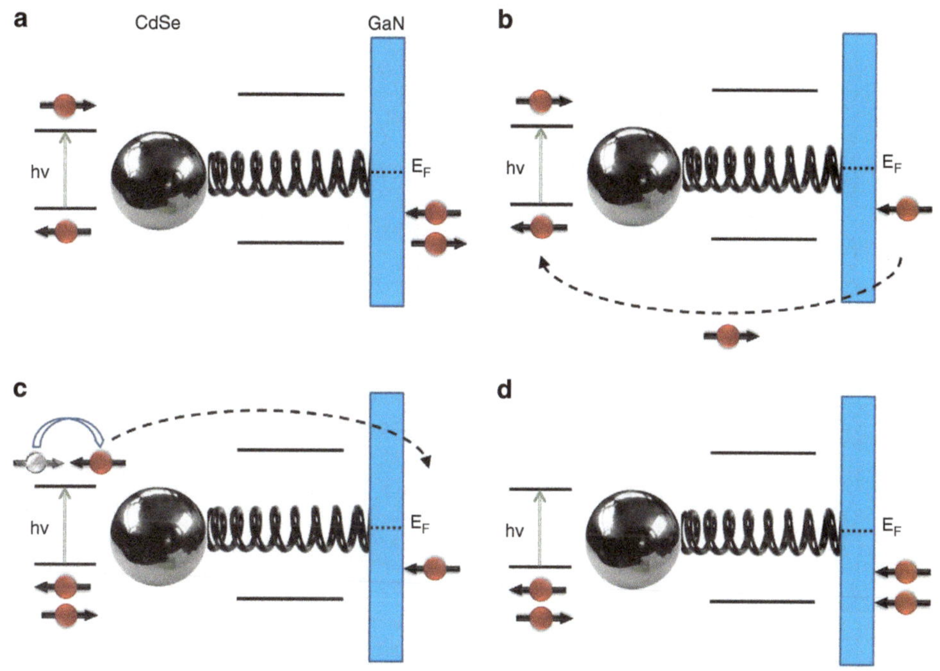

Figure 6 | A schematic explanation of light-induced spin polarization at a semiconductor substrate adsorbed by nanoparticle (NP) bearing chiral molecules. (**a**) Photoexcitation of the NPs creates an electron–hole pair. (**b**) A spin-selected electron is transferred from the substrate to the hole on the NP. An unpaired electron is left on the substrate side. (**c**) Following spin randomization of the excited electron on the NP, this electron is transferred to the substrate. Both NP and substrate are again neutral. (**d**) A net spin polarization at the substrate is formed. This process may occur many times during the substrate spin-polarization lifetime, resulting in very high polarization. Depending on rate constants involved, step (**c**) may precede (**b**).

for obtaining the same Hall voltage from an ordinary Hall measurement. The fields are found to vary between several Gauss to almost 100 G. Because the density of atoms in GaN is $\sim 1 \times 10^{23}$ atoms per cm^3 and the volume of the dielectric ($40\,\mu m \times 700\,\mu m \times 22\,nm$) is $\sim 6 \times 10^{-10}\,cm^3$, the total number of atoms in this volume is about 6×10^{13}. A simple organization of 10^{12} spins can create such a magnetic field if they are equally spaced in a box of the size of the dielectric above the conduction channel. Because the number of spins in this volume is significantly smaller than the number of atoms (by more than an order of magnitude), we can safely assume that the exchange interaction between them, which may limit spin-polarization, is not important. This number of spins implies that $\sim 10^3$–10^4 spins must be transferred from each adsorbed NPs during the lifetime of spin-polarization in the GaN, which is between tens to hundreds of nanoseconds[36]. The spin lifetime on the NPs is of the order of a few psec[40,41]. This ratio of lifetimes is about an order of magnitude more than that required to produce the above net spin polarization, showing that the system can indeed produce the required effective magnetic field by spin polarization that does not involve charge polarization.

Last but not at all least: Beyond the computational evidence, can we establish a direct link between the above-discussed structural changes and the spin-transfer inversion? To probe this, the magnet-less Hall experiment was repeated with a *mixed* monolayer, consisting of 40% NH_2-{Ala-Aib}$_8$-$COCHNH_2CH_2$-PO_4H_2 and 60% 11-mercaptoundecylphosphoric acid [$HSCH_2(CH_2)_9CH_2PO_4H_2$]. The phosphoric acid comprises alkane chains and therefore cannot form hydrogen bonds with the oligopeptide. Therefore the enthalpy gain expected from cold denaturation is much smaller and the effect should be much smaller, if not completely inhibited. The results are shown in Fig. 7. Indeed, for the mixed monolayers, the Hall voltage does not change sign upon cooling. Some signal reduction is observed

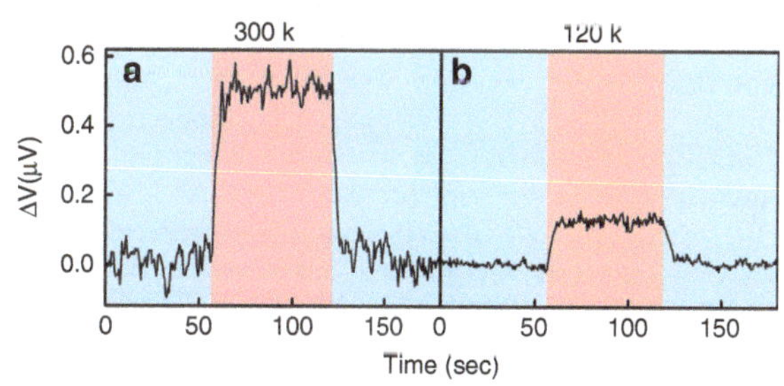

Figure 7 | Hall potential measured for a mixed monolayer. Measurements were conducted at (**a**) 300 K and at (**b**) 120 K, at a current of 10 μA, for devices as in Figs 4 and 5, but with a mixed monolayer consisting of 40% NH_2-{Ala-Aib}$_8$-$COCHNH_2CH_2PO_4H_2$, 60% 11-mercaptoundecylphosphoric acid [$HSCH_2(CH_2)_9CH_2PO_4H_2$] molecules. The signal observed corresponds to effective magnetic fields of 11 and 3 G for the high and low temperature, respectively.

at low temperature, which we attribute to the fact that the two molecules do not mix ideally and there may well be small domains that include the oligopeptides alone and exhibit denaturation as above. Still, the results of Fig. 7 lend very strong support to the theoretical prediction and close the circle between denaturation, dipole inversion, and spin-transfer inversion.

To summarize, using MD and DFT calculations we have been able to explain hitherto baffling experimental results that indicated inversion of the dipole and spin-filtering properties of chiral oligopeptide monolayers upon cooling. The calculations showed that the α-helix structure of the adsorbed oligopeptides 'stretches' upon cooling, in a manner reminiscent of 'cold denaturation' in biological systems, but where the monolayer plays the role of the solvent. Furthermore, this structural change induces an inversion of the peptide dipole, owing primarily to

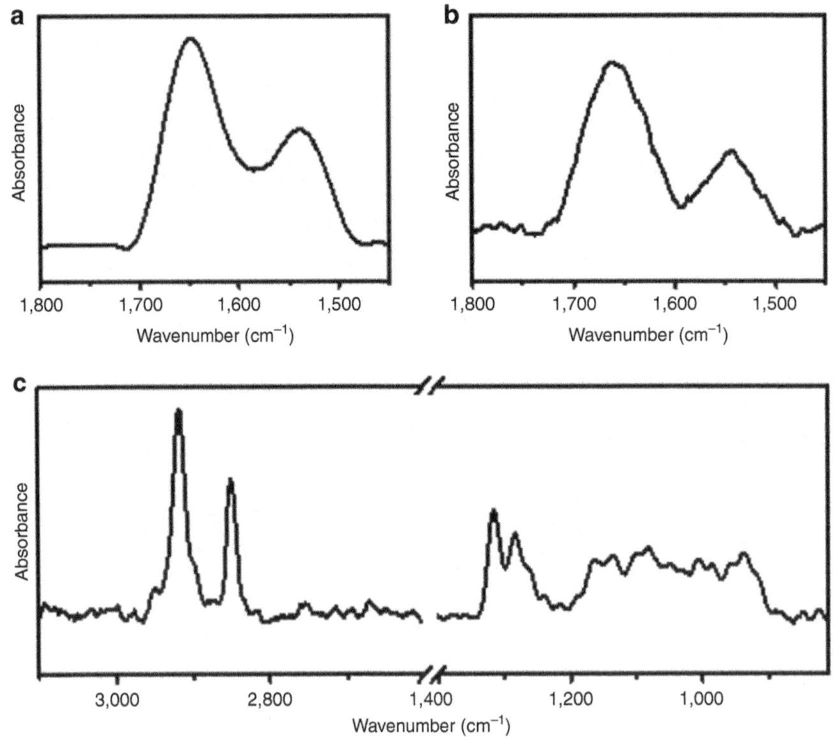

Figure 8 | GATR-FTIR spectra of the adsorbed monolayers. (**a**) NH_2-{Ala-Aib}$_8$-$COCHNH_2CH_2PO_4H_2$, (**b**) $SHCH_2CH_2CO$-{Ala-Aib}$_5$-COOH, and (**c**) 11-mercaptoundecylphosphoric acid monolayers adsorbed on gold.

rearrangement of hydrogen bonds. Phenomenological theory of the CISS effect then explains that this dipole flip must be accompanied by a concomitant change in the spin that is preferred in electron transfer experiments. We have then supported the structural prediction via ellipsometry measurements and confirmed the link between the structural changes and the spin-filtering properties using a novel magnet-less, solid-state hybrid organic–inorganic Hall device, employed with oligopeptide monolayers based on alanine and aminoisobutyric acid. The results were further supported by control experiments on mixed monolayers. The results presented indicate that inter-molecular interactions may be of importance in defining the structure of adsorbed molecules within a monolayer and its temperature-dependence may be strongly related to monolayer packing. In addition, we introduced here a new device that can monitor structural changes in adsorbed chiral molecules by their spin filtering properties. In the future, this may serve to probe the role of spin in electron transfer through bio-systems and its dependence on system structure.

Methods

Device fabrication. A schematic representation of the AlGaN/GaN Hall devices and the setup is given in Fig. 4. The AlGaN/GaN Hall devices were fabricated by standard photolithography. The AlGaN/GaN HEMT Epi wafers on sapphire-substrates were purchased from NTT AT. The structure consist of the following layers, from the bottom up (see Fig. 4d): 1800 nm bulk i-GaN, 20 nm i-AlGaN, and a capping layer of 2 nm i-GaN. Ohmic contacts were achieved by e-beam evaporation of a standard Ti/Al/Ni/Au stack, followed by rapid thermal annealing at 900 °C. Device isolation was done by mesa etching using a BCl_3/Cl_2 base ICP-RIE (Plasma Therm).

Formation of monolayers. Three organic monolayers were adsorbed as SAM on the GaN: 11-mercaptoundecylphosphoric acid (PCI Synthesis), NH_2-{Ala-Aib}$_8$-$COCHNH_2CH_2PO_4H_2$ and $SHCH_2CH_2CO$-{Ala-Aib}$_5$-COOH (both from Genemed Synthesis Inc.). Solvents were reagent grade or better (Merck, Baker, or Bio-Lab). The GaN devices were sonicated prior to molecular adsorption in hot acetone and ethanol for 10 s each, then etched for 30 s in 6 M HCl, rinsed in water, and dried under a N_2 stream. The samples were then cleaned and oxidized in UV/ozone oxidation for 30 min and placed in the adsorption solution (1 mM in toluene) immediately. Vials with the absorption solutions were filled with N_2 and placed in a desiccator for 19 h in the case of 1-mercaptoundecylphosphoric acid and NH_2-{Ala-Aib}$_8$-$COCHNH_2CH_2PO_4H_2$, and for 65 h in the case of $SHCH_2CH_2CO$-{Ala-Aib}$_5$-COOH. After adsorption, the samples were rinsed with toluene and dried with nitrogen.

In order to form a monolayer of CdSe NPs on the top of the SAM coated substrates, the devices were immersed for 4 h in a solution of anhydrous toluene (99.8%, Aldrich) containing core-only CdSe with a diameter of 6.2–7.7 nm (MK Nano) or 2.4–2.6 nm (NN-Labs). The samples were than sonicated for 5–10 s and washed with toluene to remove excess NPs. More details are provided in the Supplementary Methods.

Monolayer characterization. Monolayer formation was confirmed by FTIR in grazing-angle attenuated total reflectance mode (GATR-FTIR), using a Thermo-Scientific FTIR instrument (Nicolet 6700) equipped with a VariGATR accessory (Harrick Scientific) and with a single reflection Ge crystal. The spectra of the two oligopeptides, $SHCH_2CH_2CO$-{Ala-Aib}$_5$-COOH and NH_2-{Ala-Aib}$_8$-$COCHNH_2CH_2PO_4H_2$, exhibit the characteristic peaks that include stretching frequencies at 1659 and 1648 cm^{-1}, related to the amide I band, and peaks at 1544 and 1539 cm^{-1} due to the amide II band (Fig. 8a,b). The spectrum of 1-mercaptoundecylphosphoric acid monolayer exhibits two large peaks at 2850 and 2917 cm^{-1}, attributed to the symmetric and asymmetric CH_2 stretching, respectively (Fig. 8c)[42,43]. A broad band between 1050 and 1200 cm^{-1} is related to the symmetric and antisymmetric P–O stretch and additional peaks associated with the P = O stretch appear at 1284 and 1315 cm^{-1} (ref. 44).

Hall measurements. The GaN Hall device was attached to a sample holder and electrically connected to the measuring units. A constant DC current of 10 μA was applied along the device using a Keithley 6221 current source and the Hall voltage (V_H) across the device was measured using a Keithley Nanovoltmeter 2182A (see Fig. 4). The resistance of the devices was in a range of 3–6 Ω. Chiral molecules were absorbed directly to the GaN, with CdSe NPs bound on top of them. The device was placed in between the magnetic poles and on a cold finger that could be cooled down to 14 K. A magnetic field of up to 0.5 T could be applied perpendicular to the sample plane by an electromagnet. The temperature of the sample holder was controlled by a temperature controller, with a temperature stability of 0.1% at 14 K and 0.3% at 300 K. For calibration, the conventional Hall voltage, as a function of external magnetic field, was measured for each device at several temperatures, 300, 120, 60 and 14 K. A representative plot taken at 300 K with a device coated with a monolayer of NH_2-{Ala-Aib}$_8$-$COCHNH_2CH_2PO_4H_2$ is shown in Fig. 9.

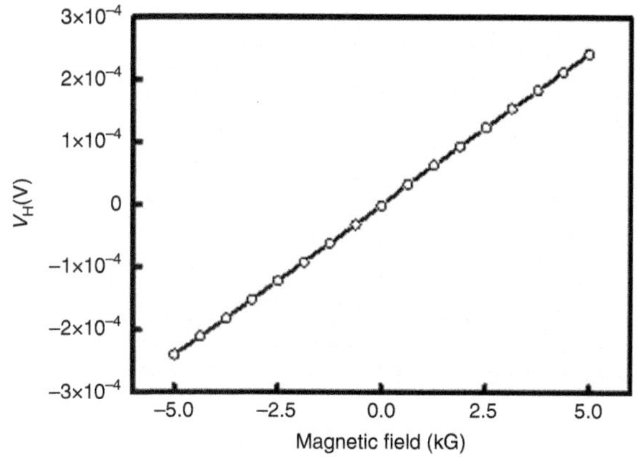

Figure 9 | Hall voltage measured for calibrating the magnet-less Hall devices. The GaN-based device of Fig. 4 was measured, with the Hall voltage shown as a function of applied magnetic field, measured for a device coated with NH_2-{Ala-Aib}$_8$-$COCHNH_2CH_2PO_4H_2$ monolayer.

Several control experiments were performed with circularly polarized light on devices coated with the different monolayers and with large and small nanoparticles, as shown in Supplementary Figures 5,6 and on a bare device (see Supplementary Fig. 7) and with a device coated with 1-mercaptoun-decylphosphoric acid and illuminated with circularly polarized light (see Supplementary Fig. 8).

Molecular dynamics simulations of a polyalanine monolayer. A monolayer composed of 25 Ala10 polyalanines was prepared *in silico* by organizing the peptides initially in 5 × 5 arrays. The longitudinal axes of the peptides were oriented along the Z axis of the simulation box. The peptide molecules were neutralized at the N-terminal end (contained a protonated amino group) and at the C-terminal end (contained a protonated carboxyl group). To imitate the monolayer conditions, the N-terminal nitrogen of the amino group of each peptide was constrained in the Z dimension, which results in the ability of the peptides in the brush to move freely only in the XY plane. Any other peptide atoms were completely free to move. The peptides were attached in either an α-helical or a linear (extended β-strand) conformation.

The initial distance between neighboring peptides was chosen empirically, such that the neighboring chains of ALA10 were placed as closely as possible, but without inter-chain clashes. During the MD the peptides were free to move in the XY plane and change the distance between neighboring peptides. The designed monolayers were studied using all-atom molecular dynamics simulations using the GROMACS package, Version 4.5.4 (ref. 45) and with the CHARMM27 force field in vacuum conditions to mimic the dry environment of the monolayer[46]. In the initial step of the simulation conformations of the studied monolayers, the systems were relaxed using a conjugate gradient method of energy minimization. The LINCS algorithm[47] was used to control bonds during the simulation. The Leapfrog algorithm was employed with a step of 2 fs.

The 5 × 5 monolayers were simulated for 10 ns (analyses of the trajectories showed that longer simulations did not result in additional conformational changes) with a time step of 2 fs (ref. 28). Temperature was controlled at 300 K using the Berendsen thermostat.

The conformations of Ala10 that were studied by DFT were generated by a pulling MD simulation when starting from a helical conformation. In this simulation, a force is applied at each terminal of the peptide at opposite directions. The pulling is done at rate of 0.01 nm/ps and with force constant of 1000 kJ per mol nm^2. Several conformations were sampled along the transition from the α-helical to the β-strand structures, providing conformations with length varying between 15 and 32 Å.

Density functional theory calculations of gas-phase oligopeptides. All gas-phase electronic-structure calculations were performed with Q-Chem[48], version 4.0, using the correlation-consistent polarized valence triple zeta basis set. We used the optimally-tuned range-separated hybrid approach[15], with a functional containing 80% short-range Perdew-Burke-Ernzerhof[49] exchange, 20% short-range Fock exchange, 100% long-range Fock exchange and full Perdew-Burke-Ernzerhof correlation. The range-separation parameter was optimally tuned so as to obey the ionization potential theorem for the peptides in the neutral form in case of an unbound LUMO level, and to obey the same theorem in both neutral and anionic forms for a bound LUMO level.

References

1. Gooding, J. J. & Ciampi, S. The molecular level modification of surfaces: From self-assembled monolayers to complex molecular assemblies. *Chem. Soc. Rev.* **40,** 2704–2718 (2011).
2. Ulman, A *An Introdction to Ulrathin Organic Films: From Langmuir-Blodgett to Self-Assembly* (Academic Press, 1991).
3. Love, C. J., Estroff, L. A., Kriebel, J. K., Nuzzo, R. G. & Whitesides, G. M. Self-assembled monolayers of thiolates on metals as a form of nanotechnology. *Chem. Rev.* **105,** 1103–1169 (2005).
4. Carmeli, I., Skakalova, V., Naaman, R. & Vager, Z. Magnetization of chiral monolayers of polypeptide-A possible source of magnetism in some biological membranes. *Angew. Chem. Int. Ed.* **41,** 761–764 (2002).
5. Carmeli, I., Gefen, Z., Vager, Z. & Naaman, R. Alternation between modes of electron transmission through organized organic layers. *Phys. Rev. B* **68,** 115418 (2003).
6. Bagus, P. S., Staemmler, V. & Wöll, C. Exchange-like effects for closed-shell adsorbates: Interface dipole and work function. *Phys. Rev. Lett.* **89,** 096104 (2002).
7. Vázquez, H., Dappe, Y. J., Ortega, J. & Flores, F. Energy level alignment at metal/organic semiconductor interfaces: 'Pillow' effect, induced density of interface states, and charge neutrality level. *J. Chem. Phys.* **126,** 144703 (2007).
8. Kronik, L. & Shapira, Y. Surface photovoltage phenomena: theory, experiment, and applications. *Surf. Sci. Reports* **37,** 1–206 (1999); section 5.2.3.
9. Naaman, R. & Waldeck, D. H. Spintronics and chirality: Spin selectivity in electron transport through chiral molecules. *Ann. Rev. Phys. Chem.* **66,** 263–281 (2015).
10. Naaman, R. & Waldeck, D. H. The chiral induced spin selectivity effect. *J. Phys. Chem. Lett.* **3,** 2178–2187 (2012).
11. Guo, A.-M. & Sun, Q.-F. Spin-selective transport of electrons in DNA double helix. *Phys. Rev. Lett.* **108,** 218102 (2012).
12. Guo, A.-M. & Sun, Q.-F. Spin-dependent electron transport in protein-like single-helical molecules. *Proc. Natl Acad. Sci. US A.* **111,** 11658–11662 (2014).
13. Privalov, P. L. Cold denaturation of proteins. *Crit. Rev. Biochem. Mol. Biol.* **25,** 281–305 (1990).
14. Sun, D., Ehrenfreund, E. & Vardeny, Z. V. The first decade of organic spintronics research. *Chem. Commun.* **50,** 1781–1793 (2014).
15. Stein, T., Eisenberg, H., Kronik, L. & Baer, R. Fundamental gaps in finite systems from eigenvalues of a generalized Kohn-Sham method. *Phys. Rev. Lett.* **105,** 266802 (2010).
16. Kronik, L., Stein, T., Refaely-Abramson, S. & Baer, R. Excitation gaps of finite-sized systems from optimally tuned range-separated hybrid functionals. *J. Chem. Theory Comput.* **8,** 1515–1531 (2012).
17. Refaely-Abramson, S. *et al.* Quasiparticle spectra from a non-empirical optimally tuned range-separated hybrid density functional. *Phys. Rev. Lett.* **109,** 226405 (2012).
18. Egger, D. A. *et al.* Outer-valence electron spectra of prototypical aromatic heterocycles from an optimally tuned range-separated hybrid functional. *J. Chem. Theory Comput.* **10,** 1934–1952 (2014).
19. Refaely-Abramson, S., Baer, R. & Kronik, L. Fundamental and excitation gaps in molecules of relevance for organic photovoltaics from an optimally tuned range-separated hybrid functional. *Phys. Rev. B* **84,** 075144 (2011).
20. Foster, M. E. & Wong, B. M. Nonempirically tuned range-separated DFT accurately predicts both fundamental and excitation gaps in DNA and RNA nucleobases. *J. Chem. Theory Comput.* **8,** 2682–2687 (2012).
21. Autschbach, J. & Srebro, M. Delocalization Error and 'Functional Tuning' in Kohn–Sham Calculations of Molecular Properties. *Acc. Chem. Res.* **47,** 2592–2602 (2014).
22. Faber, C., Boulanger, P., Attaccalite, C., Duchemin, I. & Blase, X. Excited states properties of organic molecules: From density functional theory to the GW and Bethe–Salpeter Green's function formalisms. *Philos. Trans. R. Soc. London A Math. Phys. Eng. Sci.* **372,** 20130271 (2014).
23. Körzdörfer, T. & Brédas, J.-L. Organic electronic materials: Recent advances in the DFT description of the ground and excited states using tuned range-separated hybrid functionals. *Acc. Chem. Res.* **47,** 3284–3291 (2014).
24. Onida, G., Reining, L. & Rubio, A. Electronic excitations: Density-functional versus many-body Green's-function approaches. *Rev. Mod. Phys.* **74,** 601–659 (2002).
25. Kümmel, S. & Kronik, L. Orbital-dependent density functionals: Theory and applications. *Rev. Mod. Phys.* **80,** 3–60 (2008).
26. Rissner, F. *et al.* Dimensionality effects in the electronic structure of organic semiconductors consisting of polar repeat units. *Org. Electron.* **13,** 3165–3176 (2012).
27. Rudberg, E. Difficulties in applying pure Kohn–Sham density functional theory electronic structure methods to protein molecules. *J. Phys. Condens. Matter* **24,** 72202 (2012).
28. Lever, G., Cole, D. J., Hine, N. D. M., Haynes, P. D. & Payne, M. C. Electrostatic considerations affecting the calculated HOMO–LUMO gap in protein molecules. *J. Phys. Condens. Matter* **25,** 152101 (2013).

29. Sepunaru, L. *et al.* Electronic transport via Homo-peptides: The role of side chains and secondary structure. *J. Am. Chem. Soc.* **137,** 9617–9626 (2015).
30. Rissner, F. *et al.* Collectively induced quantum-confined stark effect in monolayers of molecules consisting of polar repeating units. *J. Am. Chem. Soc.* **133,** 18634–18645 (2011).
31. Gôhler, B. *et al.* Spin selectivity in electron transmission through self-assembled monolayers of dsDNA. *Science* **331,** 894–897 (2011).
32. Kettner, M. *et al.* Spin filtering in electron transport through chiral oligopeptides. *J. Phys. Chem. C* **119,** 14542–14547 (2015).
33. Carmeli, I. *et al.* Spin selectivity in electron transfer in photosystem I. *Angew. Chemie* **53,** 8953–8958 (2014).
34. Hall, E. On a new action of the magnet on electric current. *Amer. J. Math.* **2,** 287–292 (1879).
35. Van der Pauw, L. J. A method of measuring specific resistivity and Hall effect of discs of arbitrary shape. *Philips Res. Rep.* **13,** 1 (1958).
36. Beschoten, B. *et al.* Spin coherence and dephasing in GaN. *Phys. Rev. B* **63,** 121202 (2001).
37. Ben Dor, O., Morali, N., Yochelis, S., Baczewski, L. T. & Paltiel, Y. Local light-induced magnetization using nanodots and chiral molecules. *Nano Lett.* **14,** 6042–6049 (2014).
38. Ralph, D. C. & Stiles, M. D. Spin transfer torques. *J. Magn. Magn. Mater.* **320,** 1190–1216 (2008).
39. Artzi, R., Daube, S. S., Cohen, H. & Naaman, R. Adsorption of organic phosphate as a means to bind biological molecules to GaAs surfaces. *Langmuir* **19,** 7392–7398 (2003).
40. Ma, H., Jin, Z., Zhang, Z., Li, G. & Ma, G. Exciton spin relaxation in colloidal CdSe quantum dots at room temperature. *J. Phys. Chem. A* **116,** 2018–2023 (2012).
41. Kim, J., Wong, C. Y. & Scholes, G. D. Exciton fine structure and spin relaxation in semiconductor colloidal quantum dots. *Acc. Chem. Res.* **42,** 1037–1046 (2009).
42. Kima, H. *et al.* Grafting of molecular layers to oxidized gallium nitride surfaces via phosphonic acid linkages. *Surf. Sci.* **602,** 2382–2388 (2008).
43. Shpaisman, H. *et al.* Electrical transport and photoemission experiments of alkylphosphonate monolayers on GaAs. *J. Phys. Chem. C* **113,** 3313–3321 (2009).
44. Longo, R. C., Cho, K., Schmidt, W. G., Chabal, Y. J. & Thissen, P. Monolayer doping via phosphonic acid grafting on silicon: Microscopic insight from infrared spectroscopy and density functional theory calculations. *Adv. Funct. Mater.* **23,** 3471–3477 (2013).
45. Hess, B. *et al.* GROMACS 4: Algorithms for highly efficient, load-balanced, and scalable molecular simulation. *Chem. Theory Comput.* **4,** 435–447 (2008).
46. Foloppe, N. & MacKerell, Jr. A. D. Intrinsic conformational properties of deoxyribonucleosides: Implicated role for cytosine in the equilibrium among the A, B, and Z forms of DNA. *Biophys. J.* **76,** 3206–3218 (1999).
47. Hess, B. *et al.* LINCS: A linear constraint solver for molecular simulations. *J. Comput. Chem.* **18,** 1463–1472 (1997).
48. Shao, Y. *et al.* Advances in methods and algorithms in a modern quantum chemistry program package. *Phys. Chem. Chem. Phys.* **8,** 3172–3191 (2006).
49. Perdew, J. P., Burke, K. & Ernzerhof, M. Generalized gradient approximation made simple. *Phys. Rev. Lett.* **77,** 3865–3868 (1996).

Acknowledgements

We acknowledge the help of Dr Anup Kumar in performing the ellipsometry measurements. ME, EC, SM and RN acknowledge the support by the ERC-Adv grant and the Isreal Science Foundation. RN and YP acknowledge the support of the VW Foundation and the Israel Ministry of Science. SRA, SS and LK acknowledge support by the European Research Council, the Israel Science Foundation and the Lise Meitner Center for Computational Chemistry. SRA acknowledges an Adams Fellowship of the Israel Academy of Sciences and Humanities.

Author contributions

M.E.-L and S.P.M. performed the experiments. E.C. produced the GaN devices and contributed to the data analysis. S.R.-A and S.S performed the DFT calculations. Y.G. performed the MD calculations. Y. P., Y. L., L. K. and R. N. conceived the project, supervised the experimental and computational work, assisted the analysis of the data and wrote the manuscript.

Additional information

Competing financial interests: The authors declare no competing financial interests.

7

Hetero-site-specific X-ray pump-probe spectroscopy for femtosecond intramolecular dynamics

A. Picón[1], C.S. Lehmann[1,†], C. Bostedt[1,2], A. Rudenko[3], A. Marinelli[4], T. Osipov[4], D. Rolles[3,5], N. Berrah[6], C. Bomme[5], M. Bucher[1,4], G. Doumy[1], B. Erk[5], K.R. Ferguson[4], T. Gorkhover[4], P.J. Ho[1], E.P. Kanter[1], B. Krässig[1], J. Krzywinski[4], A.A. Lutman[4], A.M. March[1], D. Moonshiram[1], D. Ray[4,7], L. Young[1], S.T. Pratt[1] & S.H. Southworth[1]

New capabilities at X-ray free-electron laser facilities allow the generation of two-colour femtosecond X-ray pulses, opening the possibility of performing ultrafast studies of X-ray-induced phenomena. Particularly, the experimental realization of hetero-site-specific X-ray-pump/X-ray-probe spectroscopy is of special interest, in which an X-ray pump pulse is absorbed at one site within a molecule and an X-ray probe pulse follows the X-ray-induced dynamics at another site within the same molecule. Here we show experimental evidence of a hetero-site pump-probe signal. By using two-colour 10-fs X-ray pulses, we are able to observe the femtosecond time dependence for the formation of F ions during the fragmentation of XeF_2 molecules following X-ray absorption at the Xe site.

[1] Argonne National Laboratory, Argonne, Illinois 60439, USA. [2] Department of Physics and Astronomy, Northwestern University, Evanston, Illinois 60208, USA. [3] J.R. Macdonald Laboratory, Department of Physics, Kansas State University, Manhattan, Kansas 66506, USA. [4] SLAC National Accelerator Laboratory, Menlo Park, California 94025, USA. [5] Deutsches Elektronen-Synchrotron (DESY), Hamburg 22607, Germany. [6] Department of Physics, University of Connecticut, Storrs, Connecticut 06269, USA. [7] Lawrence Berkeley National Laboratory, Berkeley, California 94720, USA. † Present address: Fachbereich Chemie, Philipps-Universität Marburg, Marburg, Germany. Correspondence and requests for materials should be addressed to A.P. (email: Antonio.Picon.Alvarez@gmail.com).

The intramolecular flow of charge, energy and nuclear motion is widely encountered in molecular dynamics and underlies chemical reactivity and biological processes[1–3]. The intramolecular flow may start in one site and propagate to another site and involve transient states with a high degree of electron delocalization. In fact, there are numerous processes in nature that involve intramolecular-flow dynamics on the femtosecond timescale, for example, in photosynthesis[4], DNA-mediated electron transfer in biomolecules[5], charge transport in proteins[6,7], relaxation processes in dimers such as interatomic Coulombic decay[8], and changes in chemical bonding and molecular geometry during dissociation[9].

The response of molecules, nanoparticles and condensed matter systems to the absorption of an X-ray photon is another excellent example of intramolecular dynamics[10–13]. X-ray absorption by an inner-shell electron creates a hole that is localized around a specific atomic site of the system. The system relaxes by radiative and nonradiative transitions connecting manifolds of transient states of different charge, electronic configuration and nuclear geometry. In molecules, the potential energy surfaces of the evolving ionic states induce substantial nuclear motion, including dissociation to atomic ions. In polyatomic molecules, the dissociation or the molecular fragmentation is not necessarily localized where the inner-shell hole was produced, and some fragments may be very energetic ions. The dynamics triggered by the inner-shell hole decay plays an important role in X-ray damage and is the major limiting factor in macromolecular crystallography using synchrotron X-rays and X-ray free-electron lasers (XFELs)[14–17].

Because X-rays can provide site-specific excitation, two-colour X-rays enable 'hetero-site-specific' pump-probe experiments in which the pump pulse creates the inner-shell hole in one atomic site whereas the probe pulse, delayed in time, tracks the dynamics at another atomic site (see Fig. 1). Such experiments will yield novel information on X-ray-induced dynamics, providing access to intermediate transient states as well as final products. For an ideal experiment, the XFEL would operate at high repetition rate, generate two or more pulses of controlled colours and time delays, and have narrow bandwidths. New capabilities at Linac Coherent Light Source (LCLS) already allow the generation of two X-ray pulses with some control of the two photon energies, bandwidths, pulse durations and time delay[18].

By using these capabilities, we now demonstrate an experimental study of X-ray-induced intramolecular dynamics. In particular, we study a triatomic molecule, xenon difluoride XeF_2. The inner-shell absorption spectrum of the molecule is ideal for the allowed photon-energy range of the experimental scheme, allowing the excitation of either the 3*d* electron in the Xe site or the 1*s* electron in the F site. By choosing the pump pulse to excite the Xe site and the probe pulse to excite the F site, we observe transient states leading to the break-up of the system to separated atomic ions with femtosecond time resolution.

Results

The two-colour X-ray pump-probe scheme. Our pump-probe scheme is illustrated in Fig. 2. Two X-ray photon energies were used to excite the molecule at the Xe site and probe the dynamics

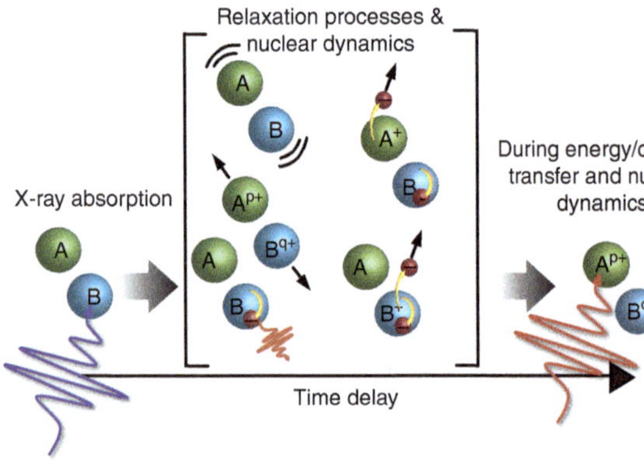

Figure 1 | Hetero-site X-ray pump-probe scheme. Sketch of a two-colour X-ray-pump/X-ray-probe experiment. The X-ray absorbed by B induces inner-shell electronic excitation, ionization, fluorescence, Auger-electron emission and charge redistribution to site A. Nuclear dynamics is also induced—dissociation, isomerization and vibrational excitation. The site-to-site intramolecular dynamics is explored by absorbing a second X-ray at site A after a controlled delay. With this approach, it is possible to characterize the response at one site of the molecule when an X-ray photon has been absorbed in another site of the same molecule.

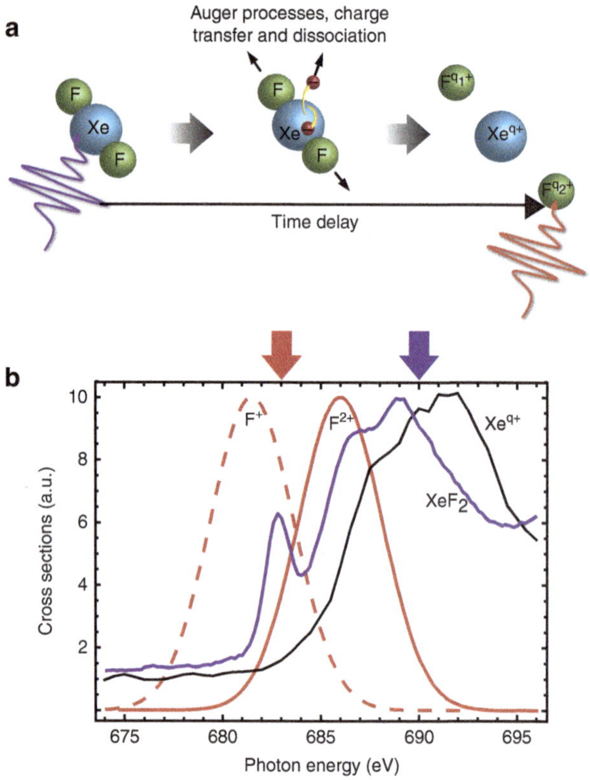

Figure 2 | Hetero-site X-ray pump-probe scheme for xenon difluoride. (**a**) Sketch of the two-colour X-ray-pump/X-ray-probe scheme used for studying the molecular fragmentation of XeF_2. A 690-eV pump pulse excites the Xe $3d \rightarrow \varepsilon_F$ shape resonance and triggers core-hole decay. In the core-hole decay, several nonradiative electronic processes such as Auger decay participate, emitting valence electrons and leaving a highly charged molecule. The induced dynamics leads to fragmentation of the molecule into three separated ions, $F^{q_1+} - Xe^{q+} - F^{q_2+}$. In our experiment, after time delays of 4, 29 and 54 fs, a 683-eV probe pulse excites the $1s \rightarrow 2p$ resonances of F^+ or F^{2+} as the molecule dissociates. (**b**) Photoabsorption cross-sections for the pump (purple arrow) and probe (red arrow) pulses. The cross-sections are arbitrarily scaled. Neutral XeF_2 cross-sections (purple line) are taken from ref. 22, xenon ions Xe^{q+} cross-section (black line) are taken from ref. 30 and averaged over q^+ using the ion-branching ratios of ref. 22, and atomic F $1s \rightarrow 2p$ resonances of F^+ or F^{2+} (red lines) are calculated with Cowan's code[31,32]. Cross-sections have been broadened to account for the ~5-eV widths of the ~10-fs pulses. The two-colour pulses excite the molecule at the Xe site and probe the F ions formed during dissociation.

at the F site. A 690-eV pump pulse excites the Xe $3d \rightarrow \varepsilon_F$ shape resonance and triggers core-hole decay and nuclear dynamics. In an isolated Xe atom, the creation of a $3d$ inner-shell hole induces a multi-step cascade of Auger processes, that is, a sequential emission of Auger electrons. Those processes produce a range of charge states, from Xe^{+} up to Xe^{8+} (refs 19,20). The production of highly charged states, in response to the creation of an inner-shell hole, is the standard behaviour found in heavy elements with many electrons in the outer shells, such as Xe. In a XeF_2 molecule, the dynamics triggered by the $3d$ inner-shell hole relaxation is more interesting and rich. The charge is redistributed from the Xe site to the F sites, and the highly charged states of the molecule end up dissociating into three atomic ions[21,22]. In XeF_2 molecule, this results in a wide range of dissociating channels $F^{q_1+}-Xe^{q+}-F^{q_2+}$. Here, using the absorption of a second X-ray, we investigate the evolution of the system and the formation of the dissociating channels. In our experiment, after time delays of 4, 29 and 54 fs, a 683-eV probe pulse excites the $1s \rightarrow 2p$ resonances of F^{+} and F^{2+} as the molecule dissociates. This excitation increases the charge state of the F ion after probe-induced Auger decay. By capturing the three atomic ions in coincidence, we determine the kinetic energy release (KER) spectrum, that is, the combined kinetic energy of the three ions, and their momentum distribution[23]. These measurements are very sensitive to the charge states and the separation of the ions, and we have observed time-dependent effects in two break-up channels, F^{2+}-Xe^{q+}-F^{3+} and F^{+}-Xe^{q+}-F^{3+}. Thereby we observe X-ray-induced intramolecular dynamics in progress. See Supplementary Notes 1–3 for details on the X-ray pulse generation, data acquistion and analysis, and the cross-sections and transition rates.

Pump-probe pathways for $F^{2+}Xe^{q+}F^{3+}$. In the following we describe the pathways that produce time-dependent features in the KER and the ion momentum distribution of the F^{2+}-Xe^{q+}-F^{3+} break-up channel shown in Figs 3a and 4. We identify two pathways contributing to this channel. The first pathway, after the absorption of one photon from the pump ($h\nu_1$) and one photon from the probe ($h\nu_2$), follows

$$\begin{aligned} F-Xe-F+h\nu_1 &\rightarrow F^{2+}-Xe^{q+}-F^{2+} \\ F^{2+}-Xe^{q+}-F^{2+}+h\nu_2 &\rightarrow F^{2+}-Xe^{q+}-F^{3+}, \end{aligned} \quad (1)$$

while the second pathway follows

$$\begin{aligned} F-Xe-F+h\nu_1 &\rightarrow F^{2+}-Xe^{q+}-F^{+} \\ F^{2+}-Xe^{q+}-F^{+}+h\nu_2 &\rightarrow F^{2+}-Xe^{q+}-F^{3+}. \end{aligned} \quad (2)$$

Note that there are also contributions to the measured KER that do not display a time dependence. These events come from absorption of two photons in the pump pulse or the probe pulse, or from transient states that are bound or quasi-bound with lifetimes longer than 50 fs (see Supplementary Note 4 for more information). The hetero-site pump-probe events are clearly distinct in the lower-energy part of the KER distribution of the F^{2+}-Xe^{q+}-F^{3+} channel. This observation can be understood by classical considerations of the Coulomb explosion of the molecule. The pump excites the molecule at the Xe site at the equilibrium distance of the neutral molecule and populates the intermediate channels F^{2+}-Xe^{q+}-F^{2+} and F^{2+}-Xe^{q+}-F^{+}. At a certain time delay, the probe excites the transient states of the molecule at an F site and transfers population from the intermediate channels to the F^{2+}-Xe^{q+}-F^{3+} channel. Because the intermediate channels have already begun to dissociate, the probe excitation occurs at larger internuclear distances, where the residual Coulomb energy of the F^{2+}-Xe^{q+}-F^{3+} channel is smaller than at short range. This results in a KER feature that shifts to lower energies with increasing time delay.

Pump-probe pathways for $F^{+}Xe^{q+}F^{3+}$. Time-dependent features also appear in the F^{+}-Xe^{q+}-F^{3+} break-up channel as shown in Fig. 3d. Among the measured break up channels, only the F^{2+}-Xe^{q+}-F^{3+} and F^{+}-Xe^{q+}-F^{3+} channels clearly show time-dependent features. To explain this observation, we first note that a $3d$ hole in atomic Xe decays via several pathways[19,20] and produces a range of charge states. The decay of a Xe $3d$ hole in the molecule is also complex and includes charge redistribution to the F ligands[22]. Second, two photons are absorbed in pump-probe events and therefore favour higher total charge states.

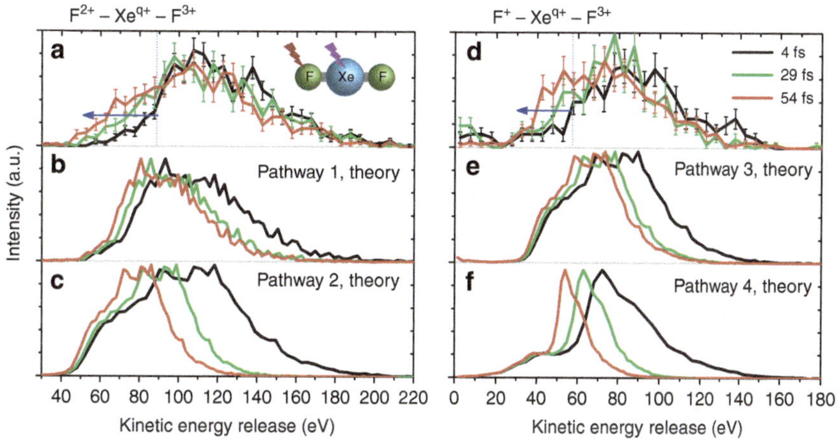

Figure 3 | KER distributions. X-ray pump/X-ray probe KERs of the F^{2+}-Xe^{q+}-F^{3+} and F^{+}-Xe^{q+}-F^{3+} break-up channels, averaging over all Xe charge states (q) observed in the experiment, $\langle q \rangle = +3.9$. The time delays are 4 fs (black line), 29 fs (green line) and 54 fs (red line). A time dependence is observed for a pump pulse that excites the Xe site at 690 eV and a probe pulse that excites strong $1s \rightarrow 2p$ resonances near 683 eV of the transient states leading to F^{+} and F^{2+} ions. (**a**,**d**) Experimental results: the time dependence reveals hetero-site pump-probe events. Pump and probe pulses have ~10-fs FWHM with ~5-eV energy widths. (**b**,**c**,**e**,**f**) Theoretical results based on a classical break-up model for the three time delays, following pathways (1), (2), (3) and (4), respectively. The theoretical results only account for pump-probe events and are consistent with the experimental observations in the KER ranges left from the vertical lines. The error bars represent the standard deviation of the mean.

Third, observations of pump-probe events in the KER and momentum distributions depend on the particular intermediate and final ion charge states and their separations at each time delay. See Supplementary Note 5 for more information.

We identify two pathways contributing time-dependent features to the KER distributions of the F^+-Xe^{q+}-F^{3+} channel. One pathway follows

$$\begin{aligned} F - Xe - F + h\nu_1 &\rightarrow F^+ - Xe^{q+} - F^{2+} \\ F^+ - Xe^{q+} - F^{2+} + h\nu_2 &\rightarrow F^+ - Xe^{q+} - F^{3+}, \end{aligned} \tag{3}$$

while the other pathway follows

$$\begin{aligned} F - Xe - F + h\nu_1 &\rightarrow F^+ - Xe^{q+} - F^+ \\ F^+ - Xe^{q+} - F^+ + h\nu_2 &\rightarrow F^+ - Xe^{q+} - F^{3+}. \end{aligned} \tag{4}$$

Interestingly, note that while in both pathways (2) and (3) the same intermediate state contributes to different final channels, the difference is that the probe X-ray pulse is absorbed in a different fluorine ion.

Classical break-up model. A classical break-up model was used to simulate the time-dependent KERs from pathways (1), (2), (3) and (4), providing an excellent agreement with the observed KER measurements (see Fig. 3b,c,e,f, respectively). In the following we provide a basic explanation of the model, but more detailed information can be found in Supplementary Note 5.

Through a combination of Auger and shakeoff processes, $1s \rightarrow 2p$ excitation of the emerging F ions increases their charge states by one or two. From an initial distribution of internuclear distances for the F^{2+}-Xe^{q+}-F^+, F^{2+}-Xe^{q+}-F^{2+}, F^+-Xe^{q+}-F^{2+} and F^+-Xe^{q+}-F^+ intermediate channels, extracted from the measured KERs when only the pump is present, the time propagation of the charges is calculated by classical-motion equations for the three time delays. The probe is modelled by increasing the charge state of the excited F ion instantly. Finally, by evolving the charges to long distances and calculating their kinetic energy, the simulated F^{2+}-Xe^{q+}-F^{3+} and F^+-Xe^{q+}-F^{3+} KER distributions are obtained. The calculated KERs are consistent with the measured time-dependent features and explain the tendency of the pump-probe events to shift towards lower KERs with increasing time delay.

We cannot describe the parts of the KERs that show no time dependence by using the classical break-up model, because those events come from nonlinear excitations at short internuclear distances, that is, two-photon absorption from the same pulse, either the pump or the probe pulse. However, in the experiment, we measured both pump-only and probe-only KER distributions, shown in Supplementary Fig. 5. With this information and the calculated pump-probe events, we can qualitatively recover the time-dependent experimental distributions, shown in Supplementary Fig. 6. Our model is illustrated for pathway (2) of the F^{2+}-Xe^{q+}-F^{3+} break-up channel in Supplementary Fig. 7.

Time-dependent ion momentum distributions. Figure 4 shows ion momentum distributions for the F^{2+}-Xe^{q+}-F^{3+} break-up channel. We use Dalitz plots[24,25], which are functional ways to visualize momentum distributions of a three-body break-up. The coordinates of the Dalitz plot are given by $x = (\varepsilon_{F^{q_1+}} - \varepsilon_{F^{q_2+}})/\sqrt{3}$ and $y = \varepsilon_{Xe^{q+}} - 1/3$ using $\varepsilon_i = p_i^2/\sum_i p_i^2$, where p_i is the momentum of ion fragment $i = Xe^{q+}$, F^{q_1+}, F^{q_2+}. That the ion momentum distribution is mainly confined in the bottom

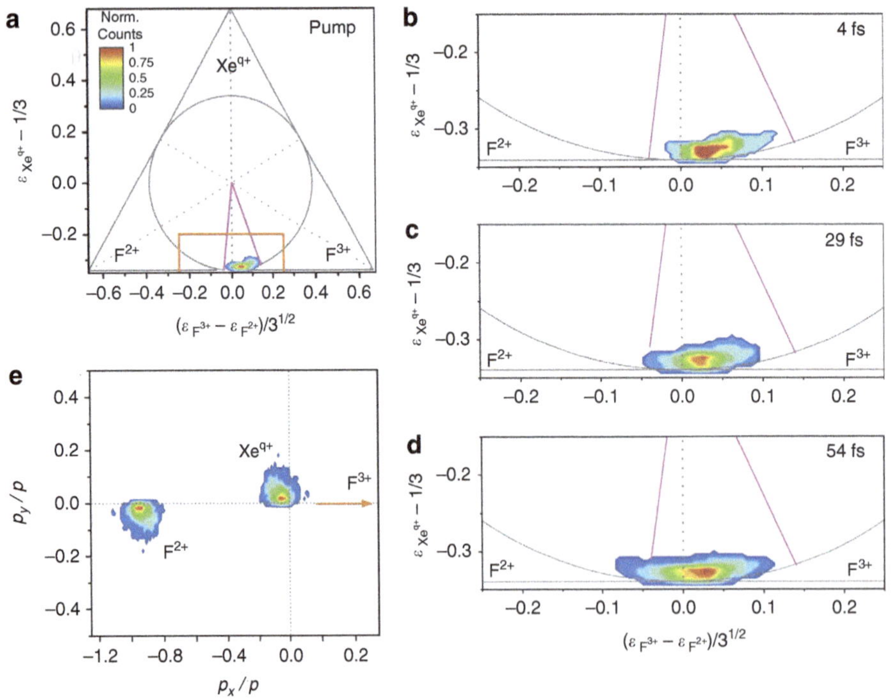

Figure 4 | Time-dependent ion momentum distribution. Dalitz plots of the F^{2+}-Xe^{q+}-F^{3+} break-up channel using 690-eV pump and 683-eV probe pulses. (**a**) The full Dalitz plot is shown for the pump-only data. (**b,c,d**) Pump-probe data for 4-, 29- and 54-fs time delays, respectively, are shown on an expanded scale. Only events with KER below the vertical line near 90 eV in Fig. 3a are plotted. The magenta lines are markers of the pump-only data. The colour label has been normalized to the maximum of counts. Movement of the distributions towards the F^{2+} ion with increasing time delay is consistent with the identified pathways (1) and (2). (**e**) Newton diagram for the break-up channel F^{2+}-Xe^{q+}-F^{3+}: the linear momenta distribution is shown for the Xe^{q+} and F^{2+} ions, always rotating the reference framework in order to fix the direction of the linear momentum of F^{3+} in the (positive) p_x-direction. The distribution indicates linear break-up of the molecular ion.

part of the triangle indicates that the Xe ion has almost zero momentum after the break-up, while the F ions gain the majority of linear momenta.

The momentum distributions of the events below 90-eV KER in Fig. 3a are plotted on an expanded scale in Fig. 4b–d. Those events represent ~20% of the total. In general, the three distributions are weighted towards the F^{3+} ion. This observation is expected from a simultaneous classical break-up of a linear triatomic molecule with asymmetric charges (and fluorines have equal mass). However, the distributions shift toward the F^{2+} ion as the time delay increases. This results from the hetero-site pump-probe events, transferred from pathways (1) and (2), as shown in Supplementary Fig. 9. At long time delays, the ions are sufficiently separated that the momentum distributions of the intermediate channels are preserved in the pump-probe events recorded in the final break-up channel.

We have also analysed the momentum distributions using Newton diagrams (Fig. 4e), which are useful for determining if the break-up geometry is bent or linear. We find that the break-up of XeF_2 is basically linear, that is, there is no bending contribution as in the case of the CS_2 molecule[24].

Discussion

In conclusion, we have used a hetero-site pump-probe technique to observe X-ray-induced intramolecular dynamics in xenon difluoride. The pump pulse creates an inner-shell hole in the Xe site and triggers complex dynamics involving charge redistribution and nuclear motion. By tuning the probe photon energy to excite the F site, we observe transient states leading to the formation of F^+ and F^{2+} ions in the range of 4-54 fs. The time-dependent effects are observed in the measured KER and ion momentum distributions, in agreement with our classical break-up simulations.

The capabilities of XFELs are continuously being improved and several new XFELs will begin operation in the next few years. Tunability, stability, high repetition rate, pulse duration, time delay, bandwidth and coherence are expected to be significantly improved. Much more detailed information on intramolecular dynamics will be obtained using these advanced sources. Furthermore, one may envision using those sources to perform probe-photon-energy scans for every time delay, obtaining X-ray transient absorption spectra. With high repetition rate, one can select specific ion break-up channels by measuring ions in coincidence with electrons. This approach will allow the detailed tracking of the system, including electronic configuration changes. Ultimately these techniques should be applicable to larger molecules of biochemical interest and provide relevant information for X-ray structural imaging techniques at XFELs.

Methods

Experiment. Measurements were made at the LCLS XFEL facility. Two-colour 10-fs (full-width at half-maximum (FWHM)) X-ray pulses separated by 4, 29 and 54 fs were generated by the method explained in ref 18. The X-ray energies were 690 eV for the pump and 683 eV for the probe. The pulse energies for both were roughly balanced, and the combined pulse energies were ~33 μJ. The transport optics have a stable efficiency of ~20%. The nominal focal spot area of the LCLS AMO beamline branch is 5 μm^2. With 10 fs pulses, the estimated peak intensity for the combined pump and probe pulses is $\sim 1.3 \times 10^{16}$ W cm^{-2}. The X-rays interacted with a molecular beam of XeF_2 molecules in the AMO end station[26] and the ion products were measured in coincidence with a reaction microscope and delay-line position-sensitive detector[23–26] (T. Osipov *et al.*, manuscript in preparation). The multi-hit ion data were reduced to determine coincidence events between the two F ions and the Xe ion. The KER and ion momentum distributions of each product channel were determined as a function of pump-probe time delay. Details are given in Supplementary Note 1 and 2.

X-ray pulse analysis. The two-colour X-ray pulses were sampled with the soft X-ray spectrometer[27] to select the pump and probe X-ray energies and to balance their intensities. A gas detector measured the total pulse energy. While the separation between pump and probe X-ray energies remained ~7–8 eV, the absolute energies varied by ~2–3 eV due to the SASE bandwidth and another ~2–3 eV due to jitter of the electron beam energy. The combined X-ray energy spreads of the individual pump and probe pulses were ~5 eV FWHM, on average. The sampled data shows that the pump and probe pulse intensities were roughly balanced, but with significant fluctuations of ~50%. Based on the settings of the slotted spoiler[28] and measurements with the XTCAV analyzer[29], the pulse durations were ~10 fs. However, the XTCAV analyzer is not senstive to the individual pulses, because the same electrons are used to lase twice in the adopted scheme.

Break-up model. By using a model based on classical interactions (Newton equations for charged particles), we are able to qualitatively describe the time dependence observed in the F^{2+}–Xe^{q+}–F^{3+} and F^{+}–Xe^{q+}–F^{3+} break-up channels, as shown in Fig. 3. The probe can induce population transfer either from the F^{2+}–Xe^{q+}–F^{+} or from the F^{2+}–Xe^{q+}–F^{2+} channels into the F^{2+}–Xe^{q+}–F^{3+} channel, as represented in the pathways given by equations (1) and (2). Similarly, the probe can induce population transfer either from the F^{+}–Xe^{q+}–F^{+} or from the F^{+}–Xe^{q+}–F^{2+} channels into the F^{+}–Xe^{q+}–F^{3+} channel, as represented in the pathways given by equations (3) and (4). The probe pulse can excite either the F^+ or the F^{2+} ion. Resonant excitation from the probe pulse will result in the charge states increasing by 1 or 2 by a combination of Auger and shake-off processes. By using classical-motion equations, each KER of the intermediate channels can be mapped onto an initial distribution of internuclear distances (the total kinetic energy is converted to the Coulomb energy of the potential). Here we make the approximation that the initial internuclear distances between Xe and the two F sites are equal. The KER distribution induced by the probe pulse is calculated by evolving the charged particles for every internuclear distance in the initial distribution during the three time delays. The differential equation given by three point-like charges separated by a certain internuclear distance and repelling each other by the Coulomb force is solved by the Runge-Kutta method. Pure Coulomb potential curves are assumed. After a specific time delay, we assume that the probe instantly increases the charge of a F particle, plus one or two, and we obtain a F^{2+}–Xe^{q+}–F^{3+} KER distribution, as shown in Fig. 3b,c, in good agreement with the time-dependent feature observed in the experimental data. Similarly, we obtain a F^{+}–Xe^{q+}–F^{3+} KER distribution for that channel, as shown in Fig. 3e,f. See Supplementary Note 5 for details.

Over-the-barrier model. The Coulomb potential created by two separated classical ions is quite dependent on the internuclear distance. If one electron is assumed to be in one of the ions, and we need an energy I_p to remove the electron from that ion, one can calculate which is the required internuclear distance that induces a Coulomb barrier between the two ions to be equal to the energy required to remove the electron. For shorter internuclear distances, the electron will be energetically allowed to be transferred to the neighbouring ion. This classical model, the so-called 'over-the-barrier' model, has been successfully applied to ion-scattering experiments, and it has been recently used to describe changes due to nuclear motion in the charge redistribution during inner-shell excitation[10]. By using the 'over-the-barrier' model we can estimate the internuclear distances in which charge transfer is not possible for several break-up channels, and in combination with a classical break-up model as described above we can obtain an estimate of the expected timescales for charge redistribution. We concluded that for 29-fs and 54-fs time delays, charge redistribution is unlikely. Hence, the time-dependent features observed in Fig. 3a,d cannot depend on charge transfer processes initiated during the absorption of the probe photon. See Supplementary Note 6 for details.

References

1. Lépine, F., Ivanov, M. Y. & Vrakking, M. J. J. Attosecond molecular dynamics: fact or fiction? *Nature Photon* **8**, 195–204 (2014).
2. Zewail, A. H. Femtochemistry: atomic-scale dynamics of the chemical bond. *J. Phys. Chem. A* **104**, 5660–5694 (2000).
3. Sundström, V. Femtobiology. *Annu. Rev. Phys. Chem.* **59**, 53–77 (2008).
4. Cheng, Y.-C. & Fleming, G. R. Dynamics of light harvesting in photosynthesis. *Ann. Rev. Phys. Chem.* **60**, 241–262 (2009).
5. Wan, C. *et al.* Femtosecond dynamics of DNA-mediated electron transfer. *Proc. Natl Acad. Sci. USA* **96**, 6014–6019 (1999).
6. Schlag, E. W., Sheu, S.-Y., Yang, D.-Y., Selzle, H. L. & Lin, S. H. Distal charge transport in peptides. *Angew. Chem. Int. Ed.* **46**, 3196–3210 (2007).
7. Vyalikh, D. V. *et al.* Charge transport in proteins probed by resonant photoemission. *Phys. Rev. Lett.* **102**, 098101 (2009).
8. Sisourat, N. *et al.* Ultralong-range energy transfer by interatomic Coulombic decay in an extreme quantum system. *Nature Phys.* **6**, 508–511 (2010).
9. Li, W. *et al.* Visualizing electron rearrangement in space and time during the transition from a molecule to atoms. *Proc. Natl Acad. Sci. USA* **107**, 20219–20222 (2010).
10. Erk, B. *et al.* Imaging charge transfer in iodomethane upon x-ray photoabsorption. *Science* **345**, 288–291 (2014).

11. Liekhus-Schmaltz, C. E. *et al.* Ultrafast isomerization initiated by x-ray core ionization. *Nat. Commun.* **6,** 8199 (2015).
12. Ferguson, K. R. *et al.* Transient lattice contraction in the solid-to-plasma transition. *Sci. Adv.* **2,** e1500837 (2016).
13. Inoue, I. *et al.* Observation of femtosecond X-ray interactions with matter using an X-ray-X-ray pump-probe scheme. *Proc. Natl Acad. Sci. USA* **6,** 1492–1497 (2016).
14. Carugo, O. & Carugo, K. D. When x-rays modify the protein structure: radiation damage at work. *Trends Biochem. Sci.* **30,** 213–219 (2005).
15. Neutze, R., Wouts, R., van der Spoel, D., Weckert, E. & Hajdu, J. Potential for biomolecular imaging with femtosecond x-ray pulses. *Nature* **406,** 752–757 (2000).
16. Quiney, H. M. & Nugent, K. A. Biomolecular imaging and electronic damage using x-ray free-electron lasers. *Nature Phys.* **7,** 142–146 (2011).
17. Nass, K. *et al.* Indications of radiation damage in ferredoxin microcrystals using high-intensity X-FEL beams. *J. Synchrotron Radiat.* **22,** 225–238 (2015).
18. Lutman, A. A. *et al.* Experimental demonstration of femtosecond two-color x-ray free-electron lasers. *Phys. Rev. Lett.* **110,** 134801 (2013).
19. Jonauskas, V. *et al.* Auger cascade satellites following 3d ionization in xenon. *J. Phys. B At. Mol. Opt. Phys.* **36,** 4403–4416 (2003).
20. Partanen, L. *et al.* Multiple ionization of Xe—comparison of de-excitation pathways following $3d_{5/2}$ ionization and $3d_{5/2} \rightarrow 6p$ resonance excitation. *J. Phys. B At. Mol. Opt. Phys.* **38,** 1881–1893 (2005).
21. Dunford, R. W. *et al.* Evidence for interatomic Coulombic decay in Xe *K*-shell-vacancy decay of XeF_2. *Phys. Rev. A* **86,** 033401 (2012).
22. Southworth, S. H. *et al.* Inner-shell photoionization and core-hole decay of Xe and XeF_2. *J. Chem. Phys.* **142,** 224302 (2015).
23. Ullrich, J. *et al.* Recoil-ion electron momentum spectroscopy: reaction microscopes. *Rep. Prog. Phys.* **66,** 1463–1545 (2003).
24. Guillemin, R. *et al.* Selecting core-hole localization or delocalization in CS_2 by photofragmentation dynamics. *Nat. Commun.* **6,** 6166 (2015).
25. Dalitz, R. H. Decay of τ mesons of known charge. *Phys. Rev.* **94,** 1046–1051 (1954).
26. Ferguson, K. R. *et al.* The atomic, molecular and optical science instrument at the Linac Coherent Light Source. *J. Synchrotron Radiat.* **22,** 492–497 (2015).
27. Heimann, P. *et al.* Linac Coherent Light Source soft x-ray materials science instrument optical design and monochromator commissioning. *Rev. Sci. Instrum.* **82,** 093104 (2011).
28. Emma, P. *et al.* Femtosecond and subfemtosecond x-ray pulses from a self-amplified spontaneous-emission–based free-electron laser. *Phys. Rev. Lett.* **92,** 074801 (2004).
29. Behrens, C. *et al.* Few-femtosecond time-resolved measurements of X-ray free-electron lasers. *Nat. Commun.* **5,** 3762 (2014).
30. Schippers, S. *et al.* Absolute cross sections for photoionization of Xe^{q+} ions ($1 \leq q \leq 5$) at the 3d ionization threshold. *J. Phys. B At. Mol. Opt. Phys.* **47,** 115602 (2014).
31. Robert, D. Cowan. in *Los Alamos Series in Basic and Applied Sciences* (ed. Robert D. Cowan.) (University of California Press, Berkely, 1981).
32. Los Alamos National Laboratory. Atomic Physics Codes. http://aphysics2.lanl.gov/tempweb/lanl/.

Acknowledgements

A.P. and C.S.L. acknowledge Lutz Foucar and Achim Czasch for fruitful discussions and their help with the data analysis. We thank Jeff Hammond, Lan Cheng and John Stanton for discussions of quantum chemical methods. The US Department of Energy, Office of Science, Basic Energy Sciences, Chemical Sciences, Geosciences, and Biosciences Division supported the Argonne group under contract no. DE-AC02-06CH11357, A.R. and D.R. under contract no. DE-FG02-86ER13491, and N.B. under contract no. DE-SC0012376. D.R. also acknowledges support from the Helmholtz Gemeinschaft through the Young Investigator Program. T.G. acknowledges the Peter-Ewald-Fellowship from the Volkswagen foundation. Use of the Linac Coherent Light Source (LCLS), SLAC National Accelerator Laboratory, is supported by the U.S. Department of Energy, Office of Science, Office of Basic Energy Sciences under Contract No. DE-AC02-76SF00515.

Author contributions

A.P. and S.H.S. conceived the idea with contributions from C.B., A.R., S.T.P., G.D. and P.J.H. A.P. and P.J.H. performed theoretical calculations of inner-shell dynamics and ionization cross-sections. The experiments were performed by all authors, and the data analysis and scientific interpretation was done by C.S.L. and A.P. with input from S.H.S., T.O., A.R., D.R. and C.B. A.P., C.S.L. and S.H.S. wrote the manuscript with contributions from all authors.

Additional information

Competing financial interests: The authors declare no competing financial interests.

8

Structure and transformation of tactoids in cellulose nanocrystal suspensions

Pei-Xi Wang[1], Wadood Y. Hamad[2] & Mark J. MacLachlan[1]

Cellulose nanocrystals obtained from natural sources are of great interest for many applications. In water, cellulose nanocrystals form a liquid crystalline phase whose hierarchical structure is retained in solid films after drying. Although tactoids, one of the most primitive components of liquid crystals, are thought to have a significant role in the evolution of this phase, they have evaded structural study of their internal organization. Here we report the capture of cellulose nanocrystal tactoids in a polymer matrix. This method allows us to visualize, for the first time, the arrangement of cellulose nanocrystals within individual tactoids by electron microscopy. Furthermore, we can follow the structural evolution of the liquid crystalline phase from tactoids to iridescent-layered films. Our insights into the early nucleation events of cellulose nanocrystals give important information about the growth of cholesteric liquid crystalline phases, especially for cellulose nanocrystals, and are crucial for preparing photonics-quality films.

[1] Department of Chemistry, University of British Columbia, 2036 Main Mall, Vancouver, British Columbia, Canada V6T 1Z1. [2] FPInnovations, 2665 East Mall, Vancouver, British Columbia, Canada V6T 1Z4. Correspondence and requests for materials should be addressed to M.J.M. (email: mmaclach@chem.ubc.ca).

Cellulose nanocrystals (CNCs) have impressive structural and mechanical properties that are being exploited for diverse applications[1–5]. In 1959, Marchessault *et al.*[6] reported the formation of permanently birefringent gels from colloidal dispersions of cellulose crystallites after acid hydrolysis. Interestingly, liquid crystalline droplets (called 'tactoids') spontaneously nucleate from aqueous CNC suspensions above ~3 wt% CNCs[7]. Drying the suspensions affords iridescent films[8,9] with a helicoidal-layered structure resembling the Bouligand structure of chitin observed in crabs, beetles and other arthropods[10,11]. As the reflected colour of the films can be changed by varying the pitch of the helical structure[12,13], these iridescent materials are of great interest for coatings[14], security features[15], and sensors[16].

Despite growing interest in the solution and solid-state properties of CNCs, there is still a large gap in our understanding of the liquid crystallinity and structure of these materials. One hypothesis is that the chirality arises from a helical twist in the crystals[17,18], and several researchers have observed twisting of cellulose microfibrils[19,20] and CNCs[21] by electron microscopy and atomic force microscopy, but the twisting is quite subtle. Bergström and coworkers[22] showed that the chiral nematic phase is already established when the CNCs are ~50 nm apart in water. It is not clear how chirality is mediated between the CNCs over these distances. Some researchers have postulated that there is a chiral charge distribution of the sulfate ester groups on the surface emanating from the H_2SO_4 hydrolysis process[7]. The transformation from isotropic to fully lyotropic liquid crystal in the case of CNCs has been difficult to study, but critical understanding may come from the structure of tactoids, which are key components in the evolution of liquid crystallinity in CNCs.

Tactoids are spherical, ellipsoidal or spindle-shaped anisotropic droplets nucleated with sharp edges in isotropic dispersions[23]. They have been observed in many liquid crystalline substances, including polypeptides[24,25]. As these tactoids with short-range-ordered structures are the intermediate state that bridges the isotropic phase and the macroscopic liquid crystalline phase with long-range anisotropy[26,27], great efforts have been made in the past 50 years to explain their origin, shape and structure; alas, these studies were nearly all based only on polarized optical microscopy (POM) or X-ray diffraction[28–30]. The significance of tactoids in improving the properties of liquid crystal-derived materials has been noticed, since Lagerwall and coworkers[31] showed that applying shear forces to suspensions of CNCs as they dried improved the alignment of the tactoids and, consequently, improved the optical properties of the films. While hypothetical models have been built, the structure of these tactoids and the arrangement of the dispersed substances within them have not been directly observed with electron microscopy due to the difficulty of capturing these droplets in a solid-state matrix while maintaining their ordered structures.

Here we demonstrate a way to capture CNC tactoids from suspensions and solidify them in polymer matrices as 'fossil records'. This new method enables us to directly observe, for the first time, the arrangement of CNCs inside the tactoids by electron microscopy, and study the morphological transformation process from distinct tactoids to long-range lamellar structures with strong iridescence.

Results

Solidification of CNC tactoids in PAAm matrices. When the lyotropic liquid crystal of CNCs first forms, one can observe tactoids by POM under crossed polarizers (Supplementary Movie 1). These tactoids appear as anisotropic droplets with parallel birefringent bands that are ~10 µm apart. Newborn tactoids (those having the fewest bands) usually have an ellipsoidal shape, possibly due to the interfacial tension between the isotropic and anisotropic phases[32]. The spacing between adjacent periodic bands is much longer (more than 10 times) than the length of the CNCs, which is consistent with a chiral nematic structure rather than smectic order, as in smectic phases the spacing of periodic bands should correspond to the length of the mesogens[7,33]. Our initial efforts to capture these tactoids by rapidly freezing the suspensions and then freeze-drying failed to show anything sensible due to collapse of the structure when water is removed. We postulated that capturing the tactoid in a polymer matrix may prevent the collapse. As it was previously demonstrated that the precursors of polyacrylamide (PAAm) hydrogels are compatible with the self-assembly of CNCs in aqueous suspensions[16], we wondered if a similar polymerization process could be used to capture the microscopic structure and transformation of tactoids.

Suspensions of CNCs in water were combined with PAAm precursors and a photoinitiator. After sonication, the mixtures were allowed to stand in 60 mm Petri dishes for a period ('evaporation time') of 1–12 h under ambient conditions. The mass of these CNC–PAAm precursor mixtures decreases linearly with time, as shown in Supplementary Table 1, causing the concentration of CNCs to vary from ~4 to 5 wt% CNCs over this time. After a period of evaporation, tactoids were clearly visible by POM and photopolymerization was initiated with UV light. Figure 1a shows a photograph of the freshly prepared hydrogel that was polymerized after an evaporation time of 6 h (~4.4 wt% CNCs), and Fig. 1b shows the POM image of a cross-section of this hydrogel, clearly revealing tactoids with periodically spaced birefringent bands. After solvent removal, the hydrogels turned into hard and brittle plastic blocks (Fig. 1c) with the microstructures of the tactoids stabilized inside the matrix (Fig. 1d). Thus a 'fossil record' that represents a transition stage of the CNC liquid crystalline phase morphological transformation was obtained, giving us an unprecedented chance to reveal the microscopic origins of tactoids, fingerprints and other optical phenomena of CNC suspensions by electron microscopy.

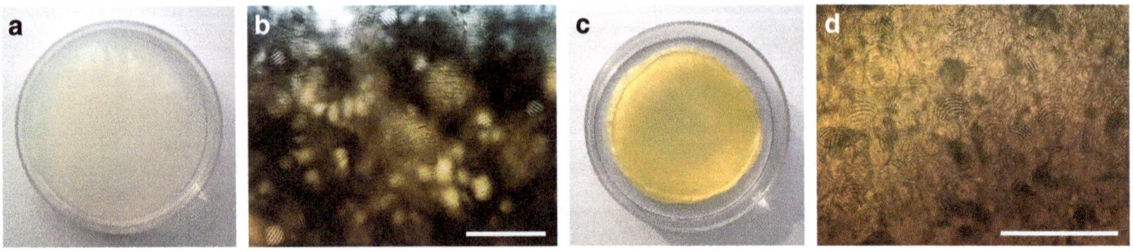

Figure 1 | Photos and POM images of polymer matrices with CNC tactoids. (**a**) Freshly prepared CNC-PAAm composite hydrogel. (**b**) POM image of a fresh hydrogel section. (**c**) Dried gel. (**d**) POM image of a dried hydrogel section. Evaporation time and estimated CNC concentration for these images are 6 h and 4.4 wt%. Diameter of the Petri dishes in **a** and **c** is 60 mm. Scale bars, 200 µm (**b** and **d**).

Chiral nematic organization of CNCs in tactoids. Scanning electron microscopy (SEM) of dried hydrogel cross sections revealed tactoids of various sizes. Figure 2a shows the SEM image of a 'baby' tactoid formed, a tiny droplet without birefringent bands, as verified by POM (Supplementary Fig. 1). SEM clearly reveals that these 'baby' tactoids are ellipsoidal, anisotropic microdomains with a lamellar-like structure. CNCs in these 'baby' tactoids are arranged parallel into thin nematic layers, as shown in Fig. 2b,c. No periodic bands (that is, birefringent lines) could be observed in these tiny tactoids as their diameters are $<5\,\mu m$, which is not enough to contain a half-helical pitch of the CNC chiral nematic structure at that stage (usually in the range of 5–10 μm). Although individual CNC rods can been seen, their helical twisting could not be resolved at this stage.

As a liquid crystalline domain, the tactoid is separated from the isotropic region (where CNCs show no preferential alignment) by an arc-shaped sharp edge (Fig. 2d,e and also Fig. 3d). CNCs inside the tactoid are arranged in nematic layers (Fig. 2c), while outside this domain they are substantially more disordered, consistent with an isotropic phase (Fig. 2f). Additional images showing the details of this tactoid can be found in Supplementary Figs 2 and 3. The diameter of the nanocrystals in these tactoids is about 20–30 nm, while the CNCs used in these studies have widths of ~10–30 nm and lengths of ~100–400 nm (Supplementary Fig. 4). This difference could be caused by the polymer matrix, sputter coating (8 nm of 80/20 Pt–Pd alloy) or the possible aggregation of CNCs.

Tactoids with a diameter of 10–15 μm usually have only one periodic band, as shown in Fig. 3a, which was captured after an evaporation time of 1 h (~4.1 wt% CNCs). Here it should be noticed that the tactoid looks flattened in the vertical direction, possibly due to the anisotropic shrinkage of the hydrogel when it was being dried (*vide infra*). The first periodic band in the newly formed tactoid looks like a wave crest (Fig. 3b). At this early stage, CNCs in a tactoid are twisted about 180° from one end to the other, which means a half-helical pitch. Another tactoid with only one band is shown in Supplementary Fig. 5.

After further water evaporation, larger tactoids with more birefringent bands are generated. Figure 3c shows a 30-micron-sized tactoid with four periodic bands. An unexpected

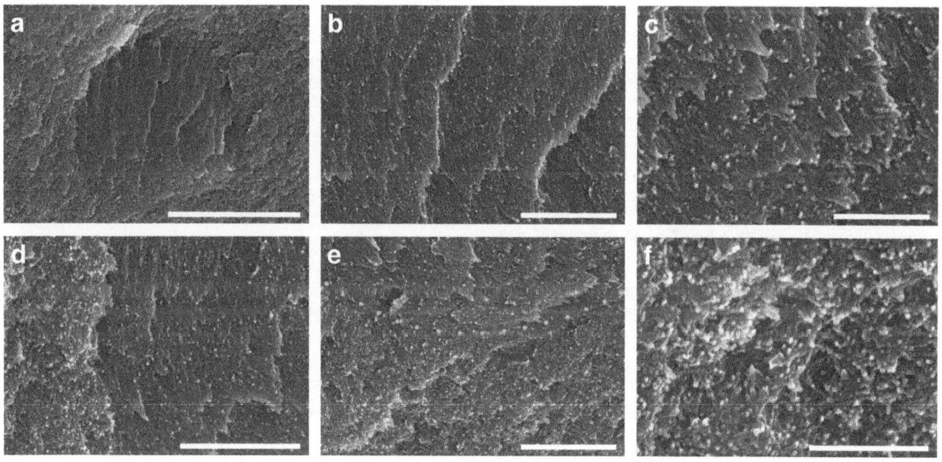

Figure 2 | SEM images of a baby tactoid captured in a polymer matrix. (**a**) Cross-sectional SEM image of a newly formed CNC tactoid (1 h evaporation time, about 4.1 wt% CNCs). The ordered arrangement of CNCs inside this tactoid is shown in **b** and **c**. (**d**) The left edge of this tactoid. (**e**) The bottom edge of the same tactoid. (**f**) The randomly arranged CNCs in the nearby isotropic region outside this tactoid. Scale bars, (**a**) 5 μm, (**b**) 1 μm, (**c**) 500 nm, (**d**,**e**) 1 μm and (**f**) 500 nm.

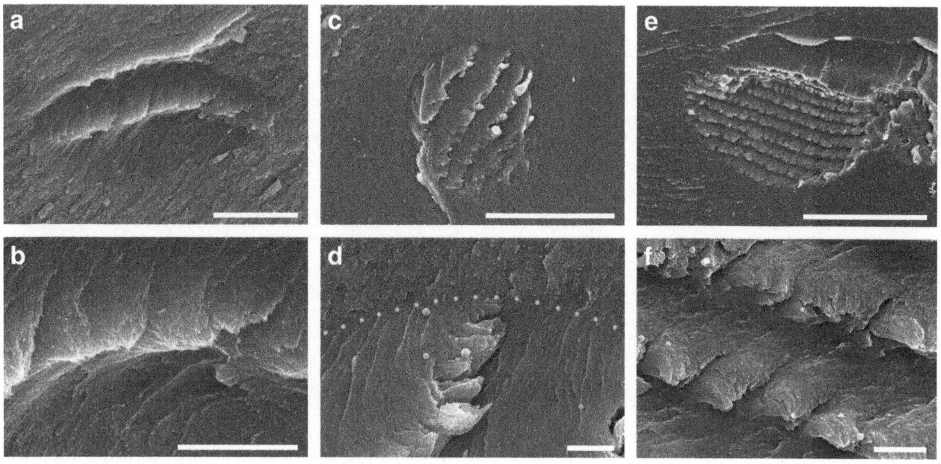

Figure 3 | SEM images of tactoids with different numbers of periodic bands. (**a**) Cross-sectional SEM image of a tactoid with only one band. This band is magnified in **b**. (**c**) A tactoid with four periodic bands. Its top edge is indicated in **d** by blue dots. (**e**) A tactoid with nine periodic bands. A half helical pitch of this tactoid is shown in **f**. Evaporation time and estimated CNC concentration for these images are as follows: (**a**,**b**) 1 h, 4.1 wt%; (**c**,**d**) 3 h, 4.2 wt%; and (**e**,**f**) 6 h, 4.4 wt%. Scale bars, (**a**) 5 μm, (**b**) 2 μm, (**c**) 30 μm, (**d**) 2 μm, (**e**) 40 μm and (**f**) 2 μm.

aggregation of the chiral nematic layers into thick sheets is observed as a stepped geometry; a cross-section of such a sheet is shown in Supplementary Fig. 6a, which contains a half-helical pitch. This phenomenon may be caused by shear forces during the drying process of the hydrogels, cleaving the layers at a particular orientation of the CNCs in the layered structure. More SEM images pertaining to this phenomenon can be found in Supplementary Figs 6 and 7.

Figure 3e shows a tactoid with nine periodic bands, where CNCs are well organized into a chiral nematic structure with a periodic spacing of about 5 μm (Fig. 3f and Supplementary Fig. 8). We even observed individual tactoids with up to ~30 bands forming in suspensions after a long evaporation time of 12 h (~4.9 wt% CNCs). Supporting SEM images showing different growth stages of tactoids can be found in Supplementary Fig. 9.

We were fortunate to successfully capture tactoids at the intersection of two fracture surfaces, which allowed us the opportunity to view them from multiple angles. Figure 4 shows a tactoid with a right-angled edge (6 h evaporation time, ~4.4 wt% CNCs), which was imaged from both the left (Fig. 5a–c), the front (Fig. 5d–f) and the right (Fig. 5g–i) sides. CNCs projecting outward perpendicular to the left side cross-section surface are clearly shown in Fig. 5c, which are indicated by a yellow dotted line. CNCs projecting out on the right side of the tactoid are indicated by a red dotted line (Fig. 5i). The region between two adjacent lines represents a half-helical pitch as shown in Fig. 5b,h, while the long-range spacing of these lines is depicted in Fig. 5a (left side) and Fig. 5 g (right side). By comparing all of these images, we are able to locate both the left- and right-oriented CNCs as shown by the yellow and red lines, respectively, in the front viewings of Fig. 5d–f. The relationship between the CNC orientations on the two sides is strong evidence for the chiral nematic ordering in the CNC tactoids. Additional SEM images of tactoids with a right-angled edge are shown in Supplementary Fig. 10.

Although the chiral nematic ordering of CNC tactoids was successfully captured by the PAAm hydrogels, the shrinkage of these hydrogels on drying inevitably decreased the size of the tactoids inside, and sometimes flattened them in the vertical direction. We found that a typical hydrogel that evaporated for 6 h before photopolymerization (that is, 4.4 wt% CNCs) shrank ~66% in thickness (from 3.53 ± 0.03 to 1.20 ± 0.03 mm), and

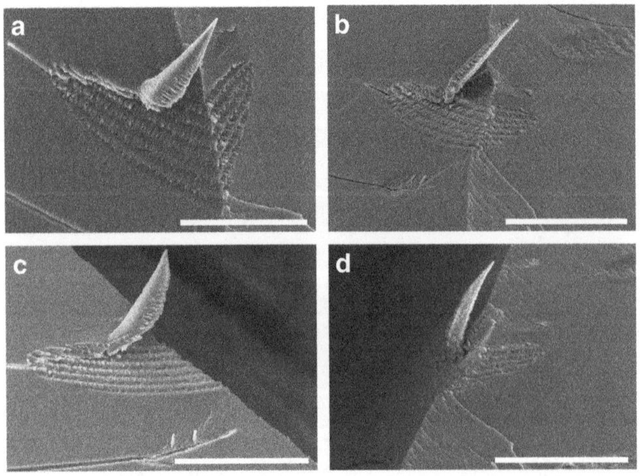

Figure 4 | SEM images of a tactoid sitting at a right-angled edge. This tactoid has seven periodic bands (6 h evaporation time, about 4.4 wt% CNCs). Its top, front, left and right views are shown in **a**,**b**,**c** and **d**, respectively. Scale bars, (**a**) 40 μm and (**b–d**) 50 μm.

~43% in diameter (from 60 to 34 ± 1 mm) after complete drying. If the hydrogel is polymerized with a lower volume fraction of water, it shrinks less on drying. For example, a hydrogel prepared after an evaporation time of 12 h shrank ~61% in thickness (from 3.20 ± 0.03 to 1.24 ± 0.01 mm), and ~43% in diameter. Thus the anisotropic shrinkage of hydrogels would vertically flatten the tactoids inside. In addition, the cross sections of tactoids are not always parallel to their helical axes, prohibiting accurate measurements of the helical pitch based on SEM images. This problem has been thoroughly discussed in the situation of chiral nematic CNC films by Majoinen *et al.*[34] Nevertheless, we measured the pitch of the chiral nematic structure in 29 tactoids as accurately as possible (Supplementary Table 2). The pitch ranges from 4 to 12 μm, consistent with the ranges observed by POM. If the lateral growth of layers in a tactoid is faster than the generation of new layers, its helical axis would be directed along the short axis of the ellipsoid; otherwise it would be parallel to the long axis. Both of these two cases have been observed during the drying process of CNC suspensions (Supplementary Movie 1).

Fusion of tactoids and formation of defects. As the concentration of CNC suspensions is increased, bigger tactoids with more periodic bands appear. Watching the tactoids in a drying suspension by POM for hours, we observed that in many situations the tactoids can grow by a coalescence mechanism, where smaller tactoids fuse to form larger tactoids (Supplementary Movies 2–5). This is a similar mechanism to the growth of oil droplets in emulsions. We were fortunate to capture these fusions in the PAAm hydrogels. Figure 6a shows a collision between two tactoids and Fig. 6b shows the orientations of the periodic bands of these tactoids at the contact point. We cannot preclude the possibility that other growth mechanisms, such as Ostwald ripening, are simultaneously active.

In some cases (especially when two tactoids come together with a right angle between their helical axes), the fusion between the tactoids leads to the formation of defects as their periodic bands bend, fold, elongate or dislocate in the process. Interestingly, the layers are soft and flexible enough to deform while accommodating the union, but often the two tactoids never completely fuse into a single uniform tactoid, possibly due to the slow relaxation rate of CNCs as their size is much larger than molecular cholesteric liquid crystalline substances. It was previously noted that healing of the defects into a uniform domain must take a long time under ambient conditions, particularly when evaporation is continuing[31]. Two hybrid tactoids formed by fusion processes are shown in Fig. 6c,d, where newly formed ones usually have irregular shapes (Fig. 6c), but gradually appear to become more regular ellipsoids (Fig. 6d, Supplementary Movie 4).

Sedimentation of tactoids. We hypothesized that CNCs would be more closely packed inside tactoids as they adopt a similar orientation in each layer, enabling them to occupy the space more efficiently. This would lead to a higher density of CNCs inside the tactoids than in the isotropic regions. Quantifying the density of the tactoids has proven to be difficult since they are predominantly water. However, we could clearly observe aggregation of tactoids in the lower regions of the hydrogel (Supplementary Fig. 12d), and even observe sedimentation of birefringent layers at the bottom of the suspension from vertically sliced cross sections of fresh hydrogels, as shown in Fig. 6e,f. This phenomenon is consistent with the isotropic and ordered phase separation observed in CNC suspensions[26,35]. In our work, a suspension of 4.08 wt% CNCs was allowed to stand in a sealed

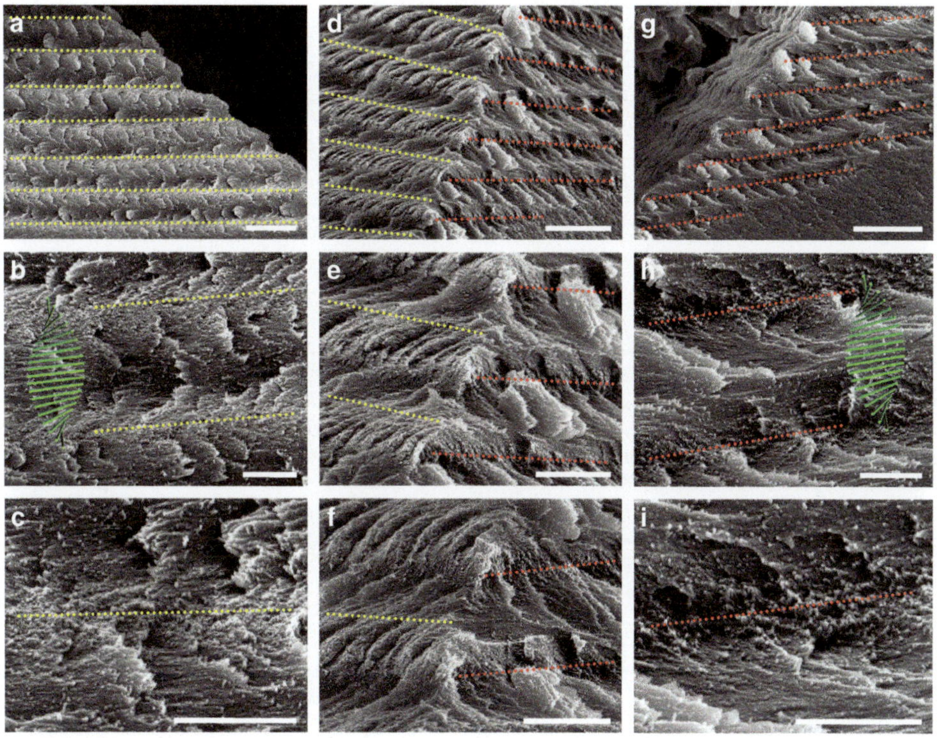

Figure 5 | The three dimensional spatial structure of a tactoid. The tactoid shown in Fig. 4 is examined by SEM from different directions. Left, front and right views are placed in the left, middle, and right columns, respectively. On the left side, CNCs perpendicular to the cross-section (the CNCs projecting out) are indicated by yellow lines (**c**), while at the right side they are indicated by red lines (**i**). The region between two adjacent lines is a half helical pitch as shown in **b** and **h**, while the long-range spacing is depicted in **a** and **g**. From **d–f** we can see that the CNCs projecting out on the left and right sides of the tactoid are alternately arranged, which is a solid confirmation of the chiral nematic ordering in CNC tactoids. Scale bars, (**a**) 5 μm, (**b,c**) 1 μm, (**d**) 5 μm, (**e,f**) 2 μm, (**g**) 5 μm and (**h,i**) 1 μm.

separatory funnel for 1 week to equilibrate, and the concentration of the CNCs in the ordered phase was measured to be 4.46 wt%, while the CNC concentration in the isotropic phase was 3.75 wt% (densities of 1.0165 and 1.0150 $g\,cm^{-3}$, respectively). Now it is apparent that the so called 'ordered phase' in CNC suspensions has a long-range lamellar structure composed of fused tactoids, while discrete tactoids still exist in the upper phase. Fusions also happen between tactoids and continuous layers when discrete tactoids reach the bottom of the suspension and join the long-range lamellar structure, leading to the formation of fusion defects. More images showing the fusions of CNC tactoids can be found in Supplementary Fig. 11.

Eventually, the evaporation of water from the CNC suspension will press and flatten all the microstructures inside. The periodic bands will be mostly horizontally aligned and tightly packed together with the spacing among them being greatly decreased in comparison to the tactoids. However, the fusion defects of tactoids will persist in the continuous lamellar structure as shown in Fig. 6g–i. These defects generally have a much longer spacing (usually $>10\,\mu m$) compared with other regions. We believe that these fusion defects are one of the possible origins of the fingerprint textures observed from the top of dried CNC films (pure or composite) with POM (Supplementary Fig. 12f). Recently Gray and Mu[36] provided another explanation for the fingerprint textures based on the POM observation of concentrated (6–8 wt%) CNC suspensions. The predominant part of these systems is the liquid crystalline phase, and the long-range lamellar structures are already formed throughout. Flow of the suspension causes distortions and gives rise to an oblique viewing angle for the chiral nematic bands, which increases the apparent spacing of the fingerprint lines. But drying these concentrated suspensions can also lead to the formation of folded layers, and the periodic spacing of the vertexes of these folded layers will eventually be much larger than the half helical pitch.

Discussion

Our electron microscopy results show that the tactoids themselves have chiral nematic structure and they are already well organized. When tactoids grow by a coalescence mechanism (which is one possibility for the growth of tactoids; other mechanisms may also occur), the periodic bands of a large tactoid are gathered by fusions of small ones (Supplementary Fig. 13). Thus, each band represents a baby tactoid at the very early state, resulting in defects between every two adjacent bands. Although CNCs are reoriented to minimize these defects, they still undermine the continuity of the chiral nematic ordering in the whole tactoid. When these organized tactoids fuse and sediment, they form the layered, iridescent chiral nematic structure that is familiar for films of CNCs. Figure 7a–c are cartoon depictions of the tactoids and their transformations as the solvent evaporates, yielding long-range chiral nematic layers of CNCs. Two simplified situations for the fusion of tactoids are depicted. In one case, two tactoids come together with their helical axes perpendicular to each other (a 'head-to-side' mode), afterwards a Y-shaped fusion defect is formed, as the bands of one tactoid are folded and inserted between two adjacent bands of the other tactoid. In this case, the fusion defect would be eventually retained in the continuous lamellar structure and then the dried CNC film. In another case, two tactoids fuse with their helical axes parallel to each other (a 'head-to-tail' mode), thus a larger tactoid

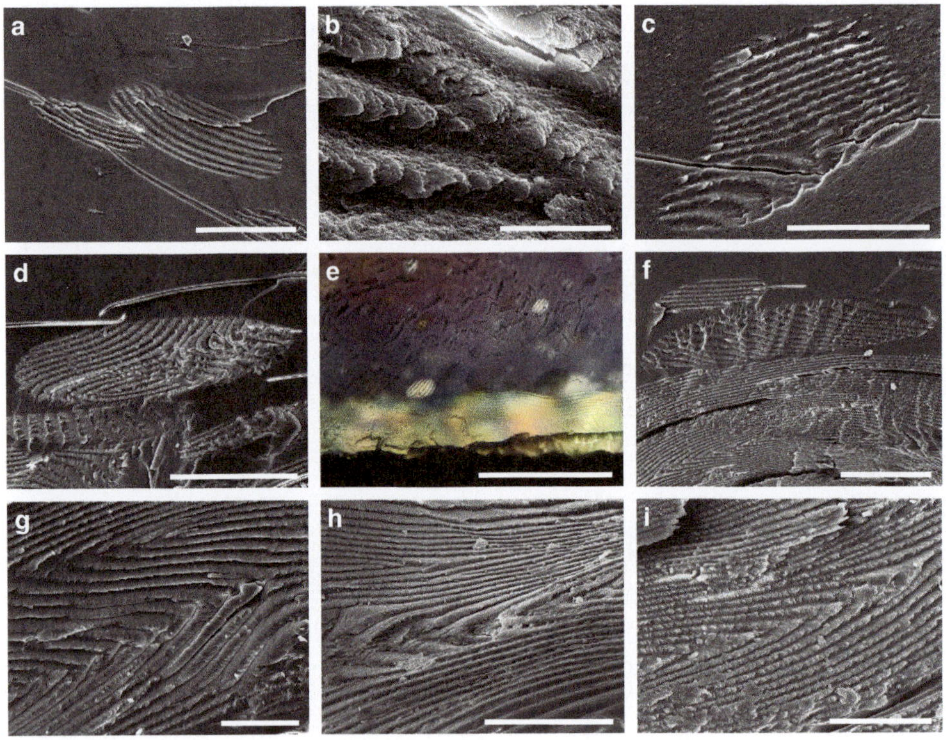

Figure 6 | Fusion of tactoids and formation of defects. (**a**) SEM image depicting the initiation of fusion between two tactoids. (**b**) SEM image of the fusion point shown in **a** at a higher magnification. (**c,d**) 'Hybrid tactoids' formed by the fusion of several individual ones. (**e,f**) POM and SEM images illustrating the sedimentation of CNC tactoids to the lower part and the formation of continuous layers at the bottom of the suspension. Tactoid fusion defects that persist in continuous layers are shown in SEM images (**g,h** and **i**). Evaporation time and estimated CNC concentration for these images are as follows: (**a,b**) 9 h, 4.6 wt%; (**c-i**) 12 h, 4.9 wt%. Scale bars, (**a**) 50 μm, (**b**) 5 μm, (**c,d**) 50 μm, (**e**) 200 μm, (**f**) 50 μm and (**g-i**) 20 μm.

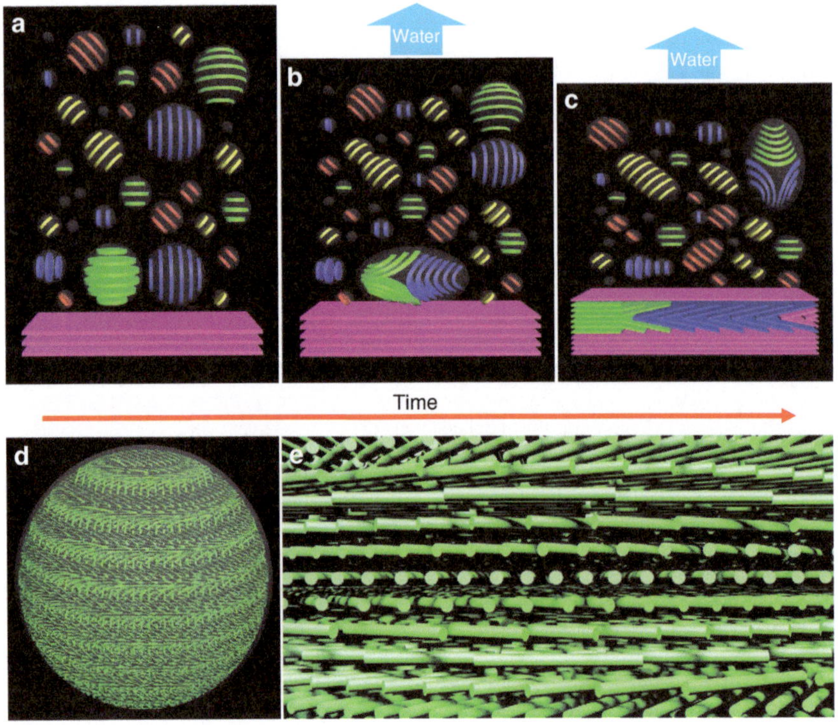

Figure 7 | Models of the structure and transformation of tactoids. (**a-c**) Cartoon depictions of the tactoids and their transformations as the solvent evaporates from the CNC suspension. A typical fusion mode which leads to the defects of folded layers is shown. **d** and **e** are three-dimensional models showing the ellipsoidal shape of CNC tactoids and the chiral nematic ordering of CNCs inside them.

with more periodic bands is formed, and ideally no bands would be folded in this process. It should be noted that tactoids gradually flatten from spheres into oblate spheroids or ellipsoids during this process, while the orientation of the helical axis can change among tactoids and is not always aligned along the tactoid's short or long axis. Figure 7d depicts the chiral nematic arrangement of idealized CNCs in a typical tactoid; Fig. 7e is a magnified image of the helicoidal organization of the rods within the tactoid.

In summary, we report an effective method to capture the tactoids formed in aqueous suspensions of CNCs, and stabilize them in solid matrices. This method provides us with an unprecedented opportunity to directly observe the organization of CNCs in the tactoids at the nanoscale using electron microscopy. These studies reveal that the tactoids formed from the isotropic phase have chiral nematic order with a hierarchical-layered structure. Also, the morphological transformation from tactoids to long-range-ordered structures is studied, and the fusion defects of tactoids are observed. These defects become part of the fingerprint texture in dried CNC films. We hope that these results will help to guide the construction of CNC-based materials with improved order, and that our methods may be applied to understand the arrangements of mesogens in other liquid crystalline substances.

Methods

Capture of tactoids in polyacrylamide hydrogels. In a typical procedure, 1.0 g acrylamide (monomer), 100 mg N,N′-methylenebisacrylamide (cross-linker) and 5 mg 2-hydroxy-4′-(2-hydroxyethoxy)-2-methylpropiophenone (or 50 mg 2,2-diethoxyacetophenone, photoinitiator) were mixed with 10 ml CNC suspension (about 400 mg CNCs). After sonication for 1 h, the homogeneous mixture (total volume ~11.2 ml) was poured into a 60-mm polystyrene Petri dish to evaporate until a desired microstructure was formed, which could be monitored in real-time by POM. This delay time is termed the 'evaporation time' in the manuscript. A 300-nm ultraviolet-B light source (8 W) was used in the photo-polymerization process. Ultraviolet irradiation was applied for 3 h to give CNC–PAAm composite hydrogels (in general, robust hydrogels could be obtained in 20 min).

In the morphological transformation experiments, a series of CNC–PAAm precursor suspensions was prepared as above and dried under ambient conditions in parallel. Ultraviolet light was applied to different batches of these samples after 1, 3, 6, 9, 12 and 24 h to capture the microstructures at different evaporation stages.

Measurements. The cross sections used for POM imaging were vertically sliced from freshly prepared hydrogels with a razor blade, and then horizontally placed on glass slides for imaging. Regions at different depths in the suspensions could be observed when the cross sections were moved horizontally on the POM stage.

Samples for SEM imaging were obtained by heating the hydrogels in air at 60 °C for 12 h, then breaking the resulting hard and brittle plastic blocks into small pieces with a hammer.

In the phase separation experiment, a CNC suspension (~200 ml) was placed in a sealed separatory funnel to stand for 7 days at 8 °C. The two phases were then separated and CNC concentrations were determined as follows: a 10 ml CNC suspension in a pre-weighed vial was weighed using an analytical balance, which was then completely dried at 75 °C for 24 h, cooled and weighed again. Densities were determined by measuring the mass of a suspension in a 10-ml volumetric flask and comparing with the mass of deionized water of the same volume.

References

1. Habibi, Y., Lucia, L. A. & Rojas, O. J. Cellulose nanocrystals: chemistry, self-assembly, and applications. *Chem. Rev.* **110,** 3479–3500 (2010).
2. Hamad, W. Y. in *ACS Symposium Series 1067* (eds Zhu, J. Y., Zhang, X. & Pan, X. J.) 301-321 (American Chemical Society, 2011).
3. Moon, R. J., Martini, A., Nairn, J., Simonsen, J. & Youngblood, J. Cellulose nanomaterials review: structure, properties and nanocomposites. *Chem. Soc. Rev.* **40,** 3941–3994 (2011).
4. Klemm, D. *et al.* Nanocelluloses: a new family of nature-based materials. *Angew. Chem. Int. Ed.* **50,** 5438–5466 (2011).
5. Lagerwall, J. P. F. *et al.* Cellulose nanocrystal-based materials: from liquid crystal self-assembly and glass formation to multifunctional thin films. *NPG Asia Mater.* **6,** e80 (2014).
6. Marchessault, R. H., Morehead, F. F. & Walter, N. M. Liquid crystal systems from fibrillar polysaccharides. *Nature* **184,** 632–633 (1959).
7. Revol, J.-F., Bradford, H., Giasson, J., Marchessault, R. H. & Gray, D. G. Helicoidal self-ordering of cellulose microfibrils in aqueous suspension. *Int. J. Biol. Macromol.* **14,** 170–172 (1992).
8. Revol, J.-F., Godbout, L. & Gray, D. G. Solid self-assembled films of cellulose with chiral nematic order and optically variable properties. *J. Pulp Pap. Sci.* **24,** 146–149 (1998).
9. Roman, M. & Gray, D. G. Parabolic focal conics in self-assembled solid films of cellulose nanocrystals. *Langmuir* **21,** 5555–5561 (2005).
10. Bouligand, Y. Twisted fibrous arrangements in biological materials and cholesteric mesophases. *Tissue Cell* **4,** 189–217 (1972).
11. Sharma, V., Crne, M., Park, J. O. & Srinivasarao, M. Structural origin of circularly polarized iridescence in jeweled beetles. *Science* **325,** 449–451 (2009).
12. Dong, X. M., Kimura, T., Revol, J.-F. & Gray, D. G. Effects of ionic strength on the isotropic-chiral nematic phase transition of suspensions of cellulose crystallites. *Langmuir* **12,** 2076–2082 (1996).
13. Shopsowitz, K. E., Qi, H., Hamad, W. Y. & MacLachlan, M. J. Free-standing mesoporous silica films with tunable chiral nematic structures. *Nature* **468,** 422–425 (2010).
14. Hoeger, I., Rojas, O. J., Efimenko, K., Velev, O. D. & Kelley, S. S. Ultrathin film coatings of aligned cellulose nanocrystals from a convective-shear assembly system and their surface mechanical properties. *Soft Matter* **7,** 1957–1967 (2011).
15. Zhang, Y. P. *et al.* in. *Proceedings of SPIE 9172, Nanostructured Thin Films VII* 91720R (2014).
16. Kelly, J. A. *et al.* Responsive photonic hydrogels based on nanocrystalline cellulose. *Angew. Chem. Int. Ed.* **52,** 8912–8916 (2013).
17. Orts, W. J., Godbout, L., Marchessault, R. H. & Revol, J.-F. Enhanced ordering of liquid crystalline suspensions of cellulose microfibrils: A Small Angle Neutron Scattering Study. *Macromolecules* **31,** 5717–5725 (1998).
18. Araki, J. & Kuga, S. Effect of trace electrolyte on liquid crystal type of cellulose microcrystals. *Langmuir* **17,** 4493–4496 (2001).
19. Hanley, S. J., Revol, J.-F., Godbout, L. & Gray, D. G. Atomic force microscopy and transmission electron microscopy of cellulose from *Micrasterias denticulata*; evidence for a chiral helical microfibril twist. *Cellulose* **4,** 209–220 (1997).
20. Usov, I. *et al.* Understanding nanocellulose chirality and structure-properties relationship at the single fibril level. *Nat. Commun.* **6,** 7564 (2015).
21. Elazzouzi-Hafraoui, S. *et al.* The shape and size distribution of crystalline nanoparticles prepared by acid hydrolysis of native cellulose. *Biomacromolecules* **9,** 57–65 (2008).
22. Schütz, C. *et al.* Rod packing in chiral nematic cellulose nanocrystal dispersions studied by small-angle x-ray scattering and laser diffraction. *Langmuir* **31,** 6507–6513 (2015).
23. Sonin, A. S. Inorganic lyotropic liquid crystals. *J. Mater. Chem.* **8,** 2557–2574 (1998).
24. Bawden, F. C., Pirie, N. W., Bernal, J. D. & Fankuchen, I. Liquid crystalline substances from virus-infected plants. *Nature* **138,** 1051–1052 (1936).
25. Robinson, C. Liquid-crystalline structures in solutions of a polypeptide. *Trans. Faraday Soc.* **52,** 571–592 (1956).
26. Revol, J.-F. *et al.* Chiral nematic suspensions of cellulose crystallites; phase separation and magnetic field orientation. *Liq. Cryst.* **16,** 127–134 (1994).
27. Dumanli, A. G. *et al.* Controlled, bio-inspired self-assembly of cellulose-based chiral reflectors. *Adv. Opt. Mater.* **2,** 646–650 (2014).
28. Watson, J. H. L., Heller, W. & Wojtowicz, W. Comparative electron and light microscopic investigations of tactoid structures in V_2O_5-sols. *Science* **109,** 274–278 (1949).
29. Kaznacheev, A. V., Bogdanov, M. M. & Taraskin, S. A. The nature of prolate shape of tactoids in lyotropic inorganic liquid crystals. *J. Exp. Theor. Phys.* **95,** 57–63 (2002).
30. Prinsen, P. & van der Schoot, P. Shape and director-field transformation of tactoids. *Phys. Rev. E* **68,** 021701 (2003).
31. Park, J. H. *et al.* Macroscopic control of helix orientation in films dried from cholesteric liquid-crystalline cellulose nanocrystal suspensions. *ChemPhysChem* **15,** 1477–1484 (2014).
32. Chen, W. & Gray, D. G. Interfacial tension between isotropic and anisotropic phases of a suspension of rodlike particles. *Langmuir* **18,** 633–637 (2002).
33. Gray, D. G. Iridescent films from cellulose nanocrystals: chiral nematic or smectic multi-lamellar structure? *J-FOR* **3,** 6–8 (2013).
34. Majoinen, J., Kontturi, E., Ikkala, O. & Gray, D. G. SEM imaging of chiral nematic films cast from cellulose nanocrystal suspensions. *Cellulose* **19,** 1599–1605 (2012).
35. Hirai, A., Inui, O., Horii, F. & Tsuji, M. Phase separation behavior in aqueous suspensions of bacterial cellulose nanocrystals prepared by sulfuric acid treatment. *Langmuir* **25,** 497–502 (2009).
36. Gray, D. G. & Mu, X. Chiral nematic structure of cellulose nanocrystal suspensions and films; polarized light and atomic force microscopy. *Materials* **8,** 7873–7888 (2015).

Acknowledgements

This work was supported by the Natural Sciences and Engineering Research Council (NSERC) of Canada. CNCs were kindly provided by CelluForce. P.-X.W. is grateful to UBC for a graduate fellowship.

Author contributions

P.-X.W. conducted all of the synthesis and characterization of the materials. W.Y.H. supplied the CNCs and contributed expertise on CNCs. M.J.M. initiated and guided this work. All three authors participated in the writing and editing of the manuscript.

Additional information

Competing financial interests: The authors declare no competing financial interests.

9

A stable room-temperature sodium-sulfur battery

Shuya Wei[1], Shaomao Xu[1], Akanksha Agrawral[1], Snehashis Choudhury[1], Yingying Lu[2], Zhengyuan Tu[3], Lin Ma[3] & Lynden A. Archer[1]

High-energy rechargeable batteries based on earth-abundant materials are important for mobile and stationary storage technologies. Rechargeable sodium-sulfur batteries able to operate stably at room temperature are among the most sought-after platforms because such cells take advantage of a two-electron-redox process to achieve high storage capacity from inexpensive electrode materials. Here we report a room-temperature sodium-sulfur battery that uses a microporous carbon-sulfur composite cathode, and a liquid carbonate electrolyte containing the ionic liquid 1-methyl-3-propylimidazolium-chlorate tethered to SiO_2 nanoparticles. We show that these cells can cycle stably at a rate of 0.5 C (1 C = 1675 mAh g^{-1}) with 600 mAh g^{-1} reversible capacity and nearly 100% Coulombic efficiency. By means of spectroscopic and electrochemical analysis, we find that the particles form a sodium-ion conductive film on the anode, which stabilizes deposition of sodium. We also find that sulfur remains interred in the carbon pores and undergo solid-state electrochemical reactions with sodium ions.

[1] School of Chemical and Biomolecular Engineering, Cornell University, Ithaca, New York 14853, USA. [2] College of Chemical and Biological Engineering, Zhejiang University, Hangzhou 310027, China. [3] Department of Material Science and Engineering, Cornell University, Ithaca, New York 14853, USA. Correspondence and requests for materials should be addressed to Y.L. (email: yingyinglu@zju.edu.cn) or to L.A.A. (email: laa25@cornell.edu).

The importance of rechargeable lithium batteries in portable electronics and their potential for electrifying transportation have been well described in several reviews[1–4]. Various recent efforts have focused on the lithium–sulfur (Li–S) chemistry due to the high theoretical-specific energy (2,500 W h kg^{-1}), high natural abundance and environmental benignity of the sulfur cathode, with great progress being achieved during the past decade[5–11]. Although many technical challenges remain, the cost and feasibility of batteries that use metallic lithium as the anode and sulfur as the cathode appear good for applications in transportation, but less so for grid-related applications, where scale and cost are as important as performance[12,13]. Sodium, the second lightest and smallest alkali metal is a low-cost alternative to lithium as anode and is available in regions all over the world, it is therefore unsurprising that interest in Na-based batteries predate those in Li-based ones[14,15].

High-temperature sodium–sulfur (Na–S) batteries operated at >300 °C with molten electrodes and a solid β-alumina electrolyte have been commercialized for stationary-energy-storage systems, confirming that this cell chemistry can meet the scale and cost requirements for feasibility in grid-scale applications[16,17]. A stable room-temperature analogue of the rechargeable Na–S battery with a higher theoretical-specific energy of 1,274 W h kg^{-1} (refs 18,19) has to date proven remarkably elusive, despite its superficial analogies to room-temperature Li–S batteries under active study. The large difference in size between Na atom and Na^+ ion defines one aspect of the challenge, as it is thought to make sodium more prone than lithium to form unstable electrodeposits and dendrites[15]. Sodium is also more reactive with aprotic liquid electrolyte solvents and forms a less-stable protective solid electrolyte interface (SEI) in aprotic liquids[18,20], which leads to lower electrochemical conversion efficiency. Na^+ ions are larger and less reducing than Li^+ ions[15], which implies that transport and kinetics of electrochemical processes in the cathode are more sluggish. Finally, Na reduction products with sulfur are more soluble than the analogous ones for lithium[14]. Taken together, these traits mean that a successful Na–S cell must overcome multiple new challenges, in addition to the already well-known ones facing Li–S batteries: the insulating nature of sulfur and its solid-state discharge product; the solubility of intermediate lithium polysulfides (LiPS) species and their associated shuttling between the electrodes, which lowers the Coulombic efficiency of the cell; and volume expansion of the cathode on cell discharge[6,7,21,22]. It is significant that some of these problems remain even when a solid-state electrolyte is employed in high-temperature Na–S cells in which the Na metal anode is a liquid.

To construct a room-temperature rechargeable Na–S battery, a conductive cathode substrate able to overcome the electronically insulating nature of both the fully charged and discharged products (S and Na_2S) is required for high active material utilization. To maintain stable cell performance, the substrate must also be able to prevent loss of the intermediate sodium polysulfides (NaPS) species[23] to the electrolyte. Sulfur infused into microporous carbon materials with small pore sizes ($d_p < 1.8$ nm) and high surface areas ($S_A \geq 843\ m^2\ g^{-1}$) have been reported previously[19,24–28]. When employed as cathodes in Li–S cells, the materials have been reported to display only one of the two discharge plateaus observed in traditional Li–S batteries, which has been argued to lend support to the hypothesis that in microporous carbons sulfur undergoes a solid-state electrochemical reaction with Li^+ to form solid sulfide product species in the cathode—that is, without forming soluble LiPS[25,28]. Xin *et al.*[19,26] have presented an alternative argument that supports formation of smaller sulfur (S_{2-4}) species in microporous carbon substrates that on reduction with Li^+ cannot form soluble high-order LiPS. Although this argument is a reasonable interpretation of the electrochemistry data, support from thermodynamic analysis of the electrode has been lacking so far. On the anode side, fundamentally based strategies for preventing dendrite formation in lithium metal batteries should necessarily be applicable for the sodium anode. Among the most fundamental of these approaches, are the efforts reported by Schaefer *et al.*[29,30] and Lu *et al.*[31–34] to reduce the magnitude of destabilizing electric fields near the anode by tethering anions to slow-moving or immobile supports[35,36]. Other successful methods, such as introduction of LiF[37] or fluoroethylene carbonate[20,38] in the electrolyte, or coating a protective hollow carbon sphere layer[39] on lithium metal, to allow stable Li deposition and prevent dendrite formation could also potentially work in Na-metal battery systems.

We herein report a stable room-temperature rechargeable Na–S battery (Fig. 1a) that overcomes all of the aforementioned challenges. The battery utilizes a Na metal anode, a metal-organic framework (MOF)-derived[27] microporous carbon polyhedron-sulfur composite (MCPS) cathode, and a liquid electrolyte comprised of a 1:1 mixture of ethylene carbonate (EC) and propylene carbonate (PC) containing 1 M $NaClO_4$ salt and 1-methyl-3-propylimidazolium-chlorate ionic liquid tethered silica nanoparticle (SiO_2–IL–ClO_4) additives as an agent for stabilizing electrodeposition. Na–S cells with this design are shown to achieve excellent cycling performance with nearly 100 % Coulombic efficiency at higher current density and with relatively high sulfur loadings in the cathode. Reversible storage capacities of over 860 mAh g^{-1} at 0.1 C (1 C = 1,675 mA g^{-1}) and 600 mAh g^{-1} at 0.5 C based on active sulfur mass are reported. Even at the higher current density (0.5 C) the batteries are able to cycle stably for over 100 cycles with 0.31% capacity decay per cycle. The fundamental origins of the superior performance of the constructed Na–S cells are studied using spectroscopic tools and analysis to understand the electrochemistry at the cathode on sodiation and desodiation processes. Notably, we find that no soluble NaPS species are formed and that the diffusivity of Na^+ into the composite cathode is consistent with expectations for solid-state transport. Altogether, these results indicate that the Na–S cells follow a different electrochemical reaction mechanism compared to traditional metal–sulfur batteries, which likely contributes to the stability and high capacity retention on cycling.

Results

Electrolyte stability. The SiO_2–IL–ClO_4 particle additives in the electrolyte play a significant role in ensuring stable cell performance during the recharge cycle. To understand the role played by the particle additive, an electrochemical floating test was performed in the potential range from 3.0 to 5.0 V. As shown in Fig. 1b, electrolytes without particles exhibit an increase of current as the potential increases and display an unstable time-dependent current response when the potential reaches 4 V. In contrast, electrolytes containing SiO_2–IL–ClO_4 particle additives exhibit much lower leakage current and are stable at least up to 4.5 V. We believe that these effects stem from the same source—immobilization of a fraction of anions near the anode as a supporting electrolyte during cell recharge—as reported in lithium electrodeposition studies, where similar SiO_2–IL–TFSI (bis(trifluoromethane)sulfonimide) particles were shown to provide orders of magnitude enhancements in the stability of lithium deposition in a PC-1M LiTFSI electrolyte[31,33]. Figure 1c reports the ionic conductivity as a function of temperature for EC/PC 1 M $NaClO_4$ electrolytes containing different concentrations of SiO_2–IL–ClO_4 particles. The measurements were

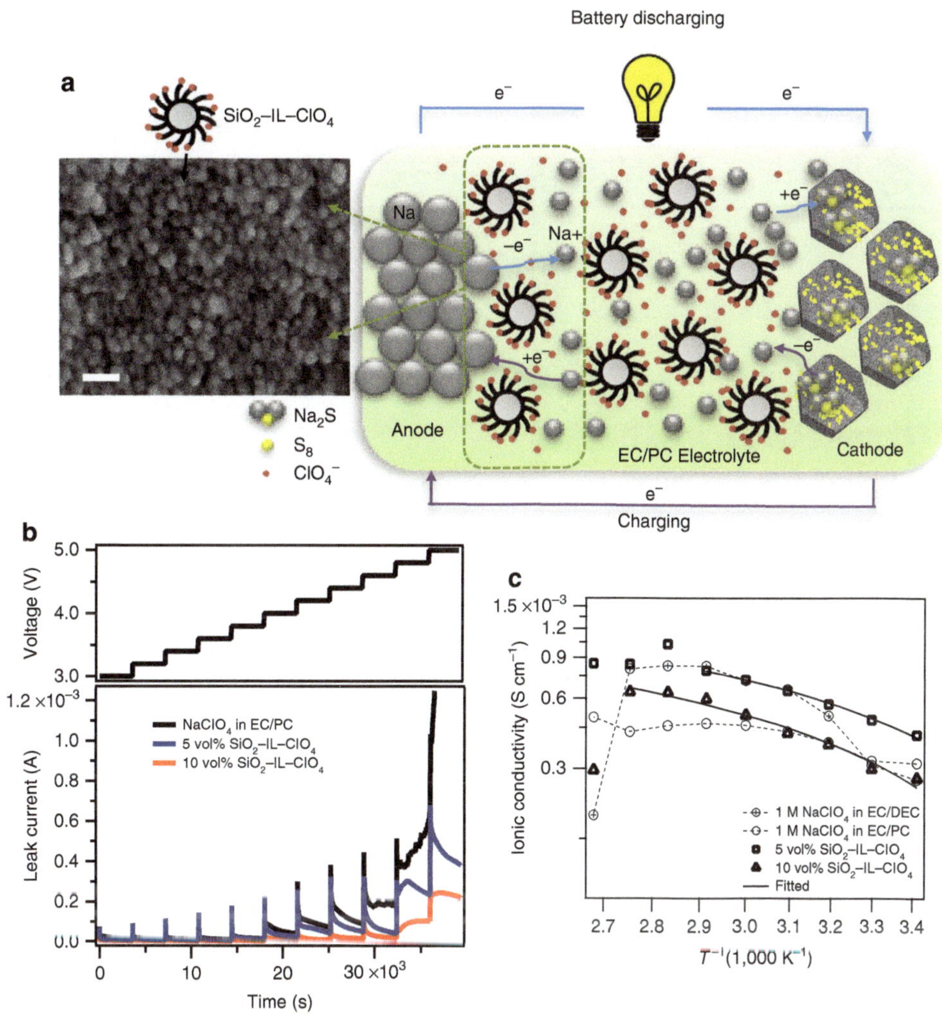

Figure 1 | A stable sodium-sulfur (Na-S) cell. (**a**) Schematic drawing of the Na-S cell during galvanostatic cycling, using 1-methyl-3-propylimidazolium-chlorate ionic liquid tethered silica nanoparticle (SiO_2-IL-ClO_4) as additive in 1 M $NaClO_4$ in a mixture of ethylene carbonate and propylene carbonate (EC/PC) (v:v = 1:1). On the anode side, sodium atom loses electron to form sodium ion during discharge. Sodium ion diffuses inside the microporous carbon-sulfur composite and reacts with sulfur to form sodium sulfide (Na_2S) on the cathode side, and the reverse reaction takes place during charging, where SiO_2-IL-ClO_4 helps stabilize sodium anode. The SEM image of the sodium metal surface cycled in a cell with 10 vol% of SiO_2-IL-ClO_4 in the electrolyte show for the first time that the particles form a conformal layer on the anode surface. Scale bar, 30 nm. (**b**) Constant voltage-charge profile of the Na-S cells with different volume fraction of SiO_2-IL-ClO_4 in the electrolyte mentioned in **a** maintained at 3.0, 3.2, 3.4...5.0 V for 1 h at room temperature. (**c**) Ionic conductivity of the Na-S cells with different volume fraction of SiO_2-IL-ClO_4 in the electrolyte as a function of temperature. EC/DEC represents a mixture of ethylene carbonate and diethyl carbonate (v:v = 1:1). The solid lines are linear Vogel-Fulcher-Tammann (VFT) fits of the temperature dependent ionic conductivity.

performed using coin cells with a Na metal electrode. It can be seen that the particles stabilize the ionic conductivities of the electrolytes, particularly at intermediate temperatures. Cells with 10 vol% particles in the electrolyte exhibit stable bulk ion transport until just below the melting point of Na metal (97.72 °C). In contrast, control cells with no particles present in the electrolyte exhibit irregular changes in conductivity with temperature, suggesting less stability. A scanning electron micrograph is provided in Fig. 1a of the sodium metal surface harvested from a cell in which 10 nm SiO_2–IL–ClO_4 particles were present in the electrolyte. It is obvious from the figure that the particles form a dense monolayer on the Na-metal surface, which we hypothesize is the fundamental source of the enhanced electrochemical and thermal stability. On the basis of these observations, we propose that SiO_2–IL–ClO_4 additive could play two roles in stabilizing the cell: (i) the tethered ionic liquid forms a mechanically robust and chemically stable SEI layer on the surface of sodium metal, which limits contact and parasitic thermal and electrochemical side reactions with the electrolyte[33]; (ii) the silica particles serve as anchor points for ClO_4^- anions, which function as supporting electrolyte and reduce the electric field through the tethered anion effect discussed previously[35,36] (Supplementary Fig. 1).

Cathode characteristics. In metal–sulfur batteries, cathode design is now understood to play an important role in improving cycling performance. Due to the highly reactive nature of sodium, reaction between sodium and dissolved NaPS species is anticipated to be more vigorous, compared to lithium and LiPS. Cathode configurations that allow stronger anchoring of PS species are therefore required for Na-S cells[19,40]. Microporous carbon materials are thought to provide the strongest physical confinement/immobilization for sulfur and its reduction

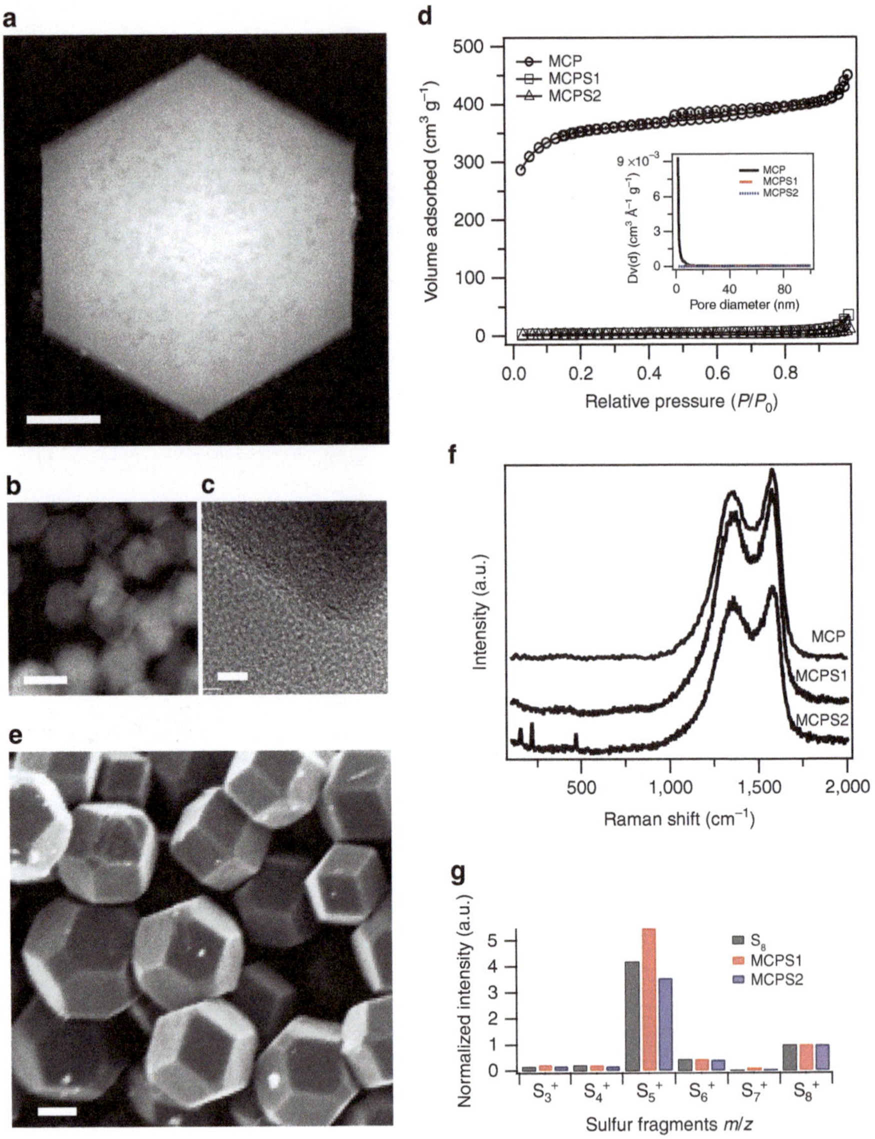

Figure 2 | Physical characterization of the microporous carbon polyhedrons (MCP) and microporous carbon polyhedron-sulfur composites. (**a**,**b**) Scanning transmission electron microscopy (STEM) and (**c**) transmission electron microscopy (TEM) images of MCP. The TEM image shows the edge of the MCP, indicating uniform porous structure of the MCP. Scale bar, (**a**) 200 nm; (**b**) 1 μm; (**c**) 5 nm. (**d**) N_2 adsorption-desorption isotherm and the corresponding Barrett-Joyner-Halenda (BJH) pore size distribution (inset) of MCP and MCPS composites. (**e**) Scanning electron microscopy (SEM) image of MCPS1. Scale bar, 200 nm. (**f**) Raman spectra of the MCP and MCPS composites. Three peaks in MCPS2 between 100 and 500 cm^{-1} are the signature peaks of crystalline sulfur. (**g**) Normalized positive sulfur fragmentation intensities with respect to S_8^+ (intensity is 1) for elemental sulfur, MCPS1 and MCPS2. m/z for S_3^+, S_4^+, S_5^+, S_6^+, S_7^+, S_8^+ are 95.916, 127.888, 159,860, 191.832, 223.804 and 255.776, respectively.

products due to their extremely small pore sizes, high surface area, and good affinity of carbon for sulfur. This material has been successfully applied to create Na–S cathodes with low sulfur loadings (32 %)[19]. Substantial increases in the sulfur loading in microporous carbon are needed to create Na–S cells that live up to the potential of this chemistry outlined in the introduction.

To achieve this goal, designs of the porous carbon host with homogeneous pore size distribution, high pore volume and increased surface area[41] are needed. We employed a facile synthesis route to create well-patterned microporous carbon polyhedrons (MCP) using zeolite-type MOF (ZIF-8) rhombic dodecahedra as both the template and precursor[27,42]. Figure 2a–c reports scanning transmission electron microscopy and transmission electron microscopy images of the MCP, indicating a uniform microporous sponge-like texture. The abundant micropores give rise to a high Brunauer–Emmett–Teller surface area of 833 $m^2\,g^{-1}$ calculated from the N_2 adsorption–desorption isotherm (Fig. 2d; Supplementary Table 1), which also give rise to a high (708.5 $m^2\,g^{-1}$) micropore surface area and pore size distribution ranging from 0.6 to 1.8 nm. To create cathodes, different amounts of sulfur were infused into the MCP and the resultant composites denoted as MCPS1 and MCPS2 with 47% and 65% sulfur loading, respectively (verified by TGA curve in Supplementary Fig. 2). The weight loss for MCPS1 due to the evaporation of sulfur occurs in a wide temperature range up to 450 °C, indicating strong nonpolar interaction between sulfur and the carbon matrix, while MCPS2 shows a two-step weight loss, representing sulfur species outside and inside the MCP, respectively. A scanning electron microscopy (SEM) image (Fig. 2e) of the as-synthesized MCPS1 suggests that the rhombic dodecahedra morphology from ZIF-8 (Supplementary Fig. 3) is well maintained after the carbonization and sulfur

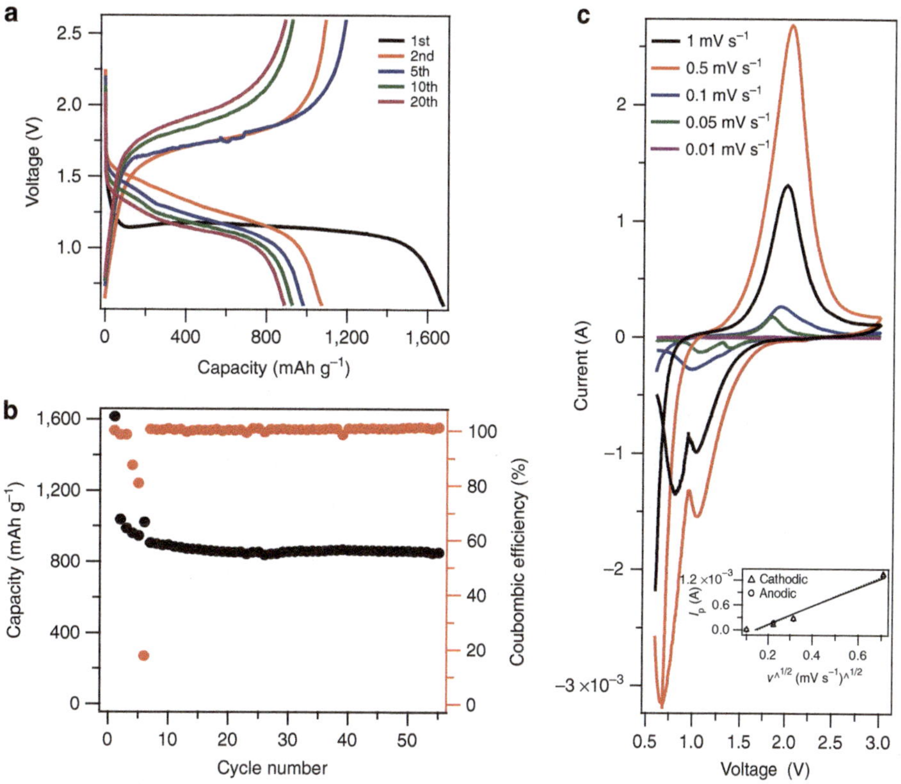

Figure 3 | Electrochemical characterization of the Na-S cell in a carbonate electrolyte. (**a**) Electrochemical discharge and charge curves of the cell at various cycles. The tests were performed at 0.1 C for both charge and discharge in the potential range of 0.6-2.6 V versus Na/Na^+. (**b**) Capacity and Coulombic efficiencies versus cycle number for the cell. (**c**) Cyclic voltammograms (CV) of the Na-S cells at various scan rates; inset: relation between peak cathodic and anodic currents verse square root of scan rate derived from CV.

infusion processes, and most of the sulfur is trapped inside the micropores.

The state of sulfur inside micropores can play a significant role in the electrochemical stability of the cathode, but remains unclear[43]. Raman spectra (Fig. 2f) of MCP and MCPS composites indicate carbonization of ZIF-8 and good dispersion of sulfur in the micropores in MCPS1 as no crystalline sulfur peaks can be observed. Sulfur fragments obtained from different sulfur-containing species were evaluated by direct analysis in real-time mass spectra (DART-MS). Normalized intensities for positively charged sulfur fragments are summarized in Fig. 2g. S_8^+ was identified both from elemental sulfur and MCPS composites, and S_5^+ was the dominant positively charged sulfur species in all samples. When a negative ion source was applied in DART-MS to the MCPS composites, S_3^- is found to be the dominant sulfur species (Supplementary Fig. 4). Combining the dominant sulfur fragments both from positive- and negative-ion-source measurements, yields S_8 as the main species in MCPS. This observation indicates that sulfur inside microporous carbon used in the present work is not the smaller sulfur S_{2-4} species seen by Xin *et al.*[26] in their studies, but rather still exists as S_8. We also note that even the results in ref. 26 may be doubtful for while Xin *et al.* used DFT simulation of different sulfur allotropes to argue that S_8 cannot fit inside microporous carbon, considering the ring diameter, 0.69 nm, and crown-like ring structure[44], S_8 is actually able to be accommodated inside extremely small micropore since the pore size distribution is an average value and the shape of the micropores in carbon is typically slit-like[24,28]. Galvanostatic discharge experiments by Moon *et al.*[45] show that introduction of β-monoclinic S_8 inside a vertically aligned carbon nanotube with a diameter of 3 nm eliminates the upper (2.4 V) discharge plateau associated with formation of soluble polysulfide species in Li–S cells. This suggests that S_8 can probably show different electrochemistry when under confinement in porous carbon.

Electrochemical properties. Galvanostatic cycling experiments were performed to assess the electrochemical properties of Na–S cells in which MCPS1 is used as cathode. Results reported in Fig. 3a show that the cell exhibits a high initial discharge capacity of $1614\,mAh\,g^{-1}$ at a current density of 0.1 C (1 C = 1,675 $mAh\,g^{-1}$). The dimple and the lower-voltage plateau at the beginning of discharge, compared with the following cycles, indicates that Na^+ ions need to go through a barrier, which is probably related to desolvation or solvation shell distortion[46,47] to accommodate the extremely small pore size to diffuse inside the micropores. The higher irreversible capacity is partially attributed to initial SEI formation and electrolyte decomposition. The reversible discharge plateau in the following cycles ranges from 1.6 to 1 V. The lower voltage is consistent with direct formation of Na_2S/Na_2S_2, without creation of intermediate soluble NaPS species. A reversible discharge capacity of $800\,mAh\,g^{-1}$ is stably achieved for 50 cycles (Fig. 3b).

In contrast, when TEGDME, which has high solubility for NaPS, is used as the electrolyte solvent, shuttling is observed in the voltage profile (see Supplementary Fig. 5) and corrosion of the sodium anode is readily seen from post-mortem studies (Supplementary Fig. 6). This has been known for some time to be the cause of cell failure in sodium–sulfur batteries using ether-based electrolytes[14,18,48–53]. Those batteries can neither bear high current densities (1/64C is employed in ref. 48)[48,49] nor exhibit satisfactory cycle life[50–53]. Even though $NaNO_3$ is used as an electrolyte additive to passivate sodium metal—with the

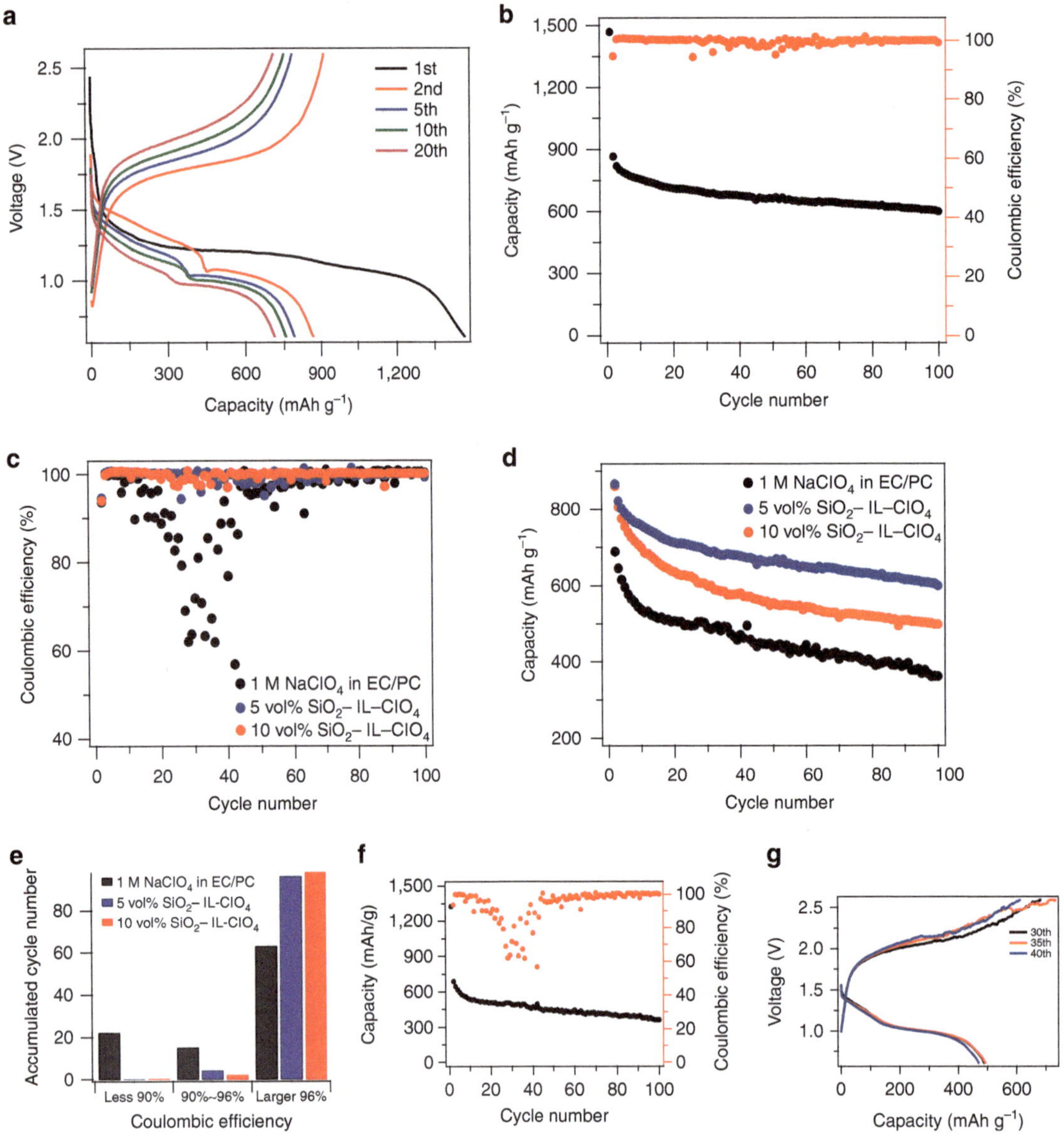

Figure 4 | Galvanostatic cycling performance of the Na-S cell in a carbonate electrolyte with different amounts of SiO_2-IL-ClO_4. (**a**) Electrochemical discharge and charge curves of the cell at various cycles with 5 vol% of SiO_2-IL-ClO_4 in the electrolyte. The tests were performed at 0.1 C for the first discharge and 0.5 C for the following cycles in the potential range of 0.6-2.6 V versus Na/Na^+. (**b**) Capacity and Coulombic efficiencies versus cycle number for the cell with 5 vol% of SiO_2-IL-ClO_4 in the electrolyte. (**c**) Coulombic efficiency and (**d**) capacity versus cycle number for the cell with different amounts of SiO_2-IL-ClO_4 in the electrolytes, respectively, at a current density of 0.5 C. (**e**) Electrolyte stability analysis for the three cases in **c** in term of Coulombic efficiency for the first 100 cycles. (**f**) cycling performance and (**g**) voltage profile of the cell without SiO_2-IL-ClO_4 in the electrolyte at 0.5 C.

expectation that it is as effective as $LiNO_3$ on passivating lithium anode in Li–S batteries[54]—the cells do not achieve outstanding performance in any of the studied configurations. A similar result is found when MCPS2 is used as cathode with the EC/PC based electrolyte, where data reported in Supplementary Fig. 7 indicate that higher-order NaPSs is formed and react with the carbonate electrolyte. Thus, carbonate-based electrolytes are shown to be incompatible with metal-sulfur batteries if insufficient care is taken to sequester NaPS in the cathode[43,55]. These results therefore clearly show that the electrochemistry of sulfur in micporours carbon is affected by subtle features related the solubility of PS in electrolyte; competition between the affinity of sulfur for the microporous carbon and electrolyte determines whether and what intermediate sulfide specie is observed.

Low self-discharge is another required feature of a stable electrochemical energy-storage device[56,57]. Metal–sulfur batteries, unfortunately, have strong self-discharge behaviour in ether-based electrolytes due to the formation and dissolution of metal polysulfides. Comparison of the initial discharge voltage profile (Supplementary Fig. 5c,d) between fully charged Na–S cells based on carbonate electrolytes and the MCPS1 composite cathodes show little difference after 10-min and 2-week resting times, indicating relatively small decrease in capacity (<22%). Na–S cells containing TEGDME electrolyte and MCPS1 composite cathodes exhibit a 42% capacity decay after 2 weeks. In contrast Na–S cells based on a physical sulfur–carbon blend cathode exhibit an immediate voltage drop and short circuit. This observation confirms the importance of confinement of sulfur in MCP for minimizing self-discharge and parasitic internal chemical reactions in the cell.

In the carbonate electrolyte, interaction between sulfur and the microporous carbon appears to be strong enough to completely prevent sulfur loss to the electrolyte. This raises the possibility that sulfur undergoes a solid-state electrochemical reaction in the microporous carbon. To investigate this possibility, cyclic voltammetry measurements were performed at various scan rates. Results reported in Fig. 3c clearly shows that a two-electron transfer process occurs in the discharge cycle. The reduction peaks are seen to shift towards more negative values and oxidation peak towards more positive values with increasing scan rate, indicative of an electrochemical process in which mixed kinetics of charge transfer process and diffusion of electroactive species control[58]. The cathodic and anodic peak currents are also seen to be similar in magnitude, which indicates good reversibility and similar reaction mechanisms occurring during charge and discharge. Finally, the peak current is found to increase linearly with the square root of scan rate (Fig. 3c), which is a classical characteristic of a diffusion-limited process[59].

A superficial assessment of the cycling results in Fig. 3a would conclude that the Na–S cells cycle well. More careful scrutiny of the large drop in Columbic efficiency seen in Fig. 3b at around the 6th cycle and the slight ripple in the charge profile at the 10th cycle shown in Fig. 3a reveal serious stability problems with the Na–S cells (see Supplementary Fig. 8). Measurements at higher current density show that the cells become progressively more unstable and the effects seen in Fig. 3a,b become more severe. Because the unstable operation is only evident during cell recharge, we suspected that it originated from unstable Na deposition and/or side reactions of the freshly created Na surface area with the electrolyte[60,61]. Motivated by the earlier work by Lu *et al.*[31–33], which showed that SiO_2–IL–TFSI particles can stabilize electrodeposition of Li by both the tethered anion mechanism and by protecting Li metal against parasitic side reactions with liquid electrolytes, we investigated the effect of SiO_2–IL–ClO_4 nanoparticles as electrolyte additives for Na–S batteries. Previous studies have reported that at 10% SiO_2–IL–TFSI particle additives in 1 M NaTFSI in PC electrolytes stabilized the charging process in Na-CO_2/O_2 cells[62]; leading to rechargeable batteries based on even that novel chemistry. In the present studies, either 5 vol% or 10 vol% of the SiO_2–IL–ClO_4 was added to the electrolyte as a stabilizer and the cell response in galvanostatic cycling experiments compared with those obtained from control experiments in which the IL-tethered particles were not present. Consistent with the previously reported results, it is seen that as little as 5 vol% of the SiO_2–IL–ClO_4 additive could stabilize charging to a large extent. Figure 4a,b report the voltage profile and cycling stability of these cells. The first discharge is performed at 0.1 C to fully activate the electrode. A discharge capacity of around 866 mAh g^{-1} is achieved initially and maintained to 600 mAh g^{-1} at the 100th cycle, indicating a small capacity decay of 0.31% per cycle, which is comparable to current Li–S

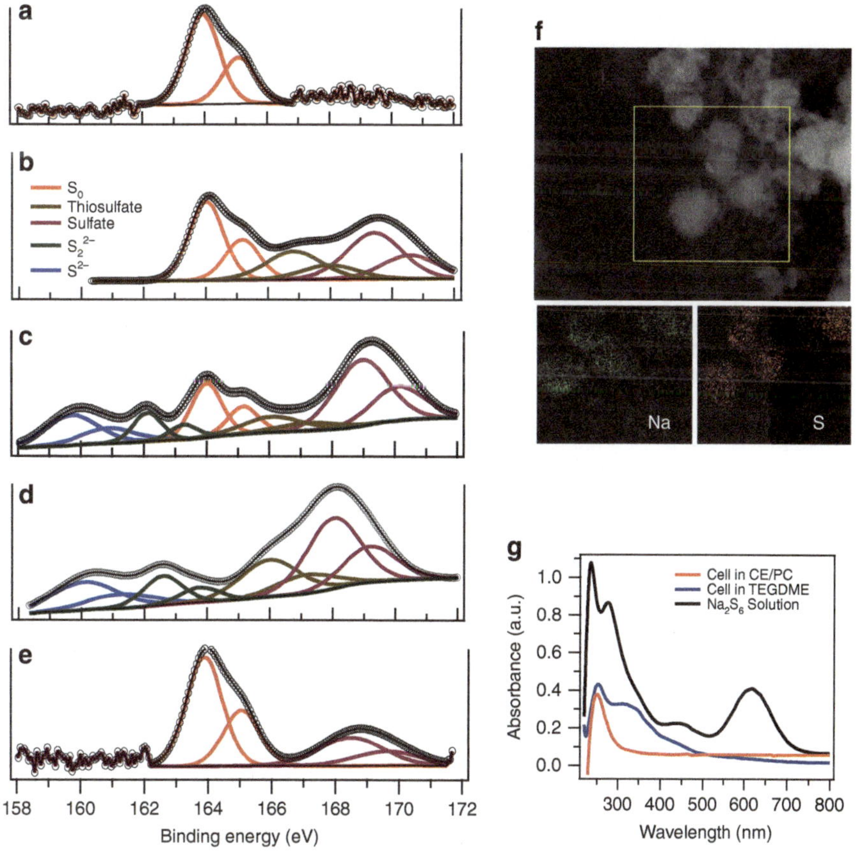

Figure 5 | Post-mortem characterization of the MCPS1 cathodes in carbonate electrolytes. (**a**–**e**) *Ex situ* XPS spectra of S 2p in MCPS1 cathodes at pristine and different cycling states in a carbonate electrolyte. (**a**), pristine cathode. Cell was discharged to (**b**) 1.4 V, (**c**) 1 V and (**d**) 0.6 V and was recharged to (**e**) 2.6 V at the first cycle, respectively. Cathodes were disassembled in an argon-filled grove-box and washed with electrolyte solvent before characterization. (**f**) STEM image and EDX maps of the MCPS1 cathode after first discharge. The cathode was washed with electrolyte solvent and sonicated to form a homogeneous suspension in a sealed vial. EDX, energy dispersive X-ray. (**g**) Ultraviolet-vis spectra of the cathodes solutions cycled in different electrolytes after 10 cycles at 0.1 C. The MCPS1 cathodes cycled in different electrolytes were soaked in 2 ml TEGDME for 4 days to extract PS species. For the Na_2S_6 solution, 1 M Na_2S_6, which is synthesized by mixing Na_2S and sulfur in a stoichiometric ratio of 2:6 in TEGDME. It was diluted 200 times and subjected to test. Peak assignment: S^{2-} and S_2^{2-}, ~260 nm; S_6^{2-}, 340 nm and 450 nm; S_3^{2-}, 330 nm; $S_3^{\cdot -}$: 610 nm (ref. 65).

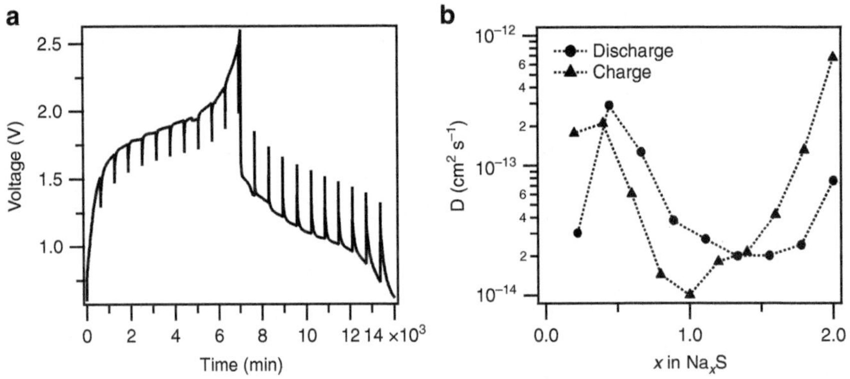

Figure 6 | Diffusivity analysis of the Na-S cell. (**a**) Galvanostatic intermittent titration technique (GITT) curves of MCPS1 in EC/PC electrolyte and (**b**) Diffusion coefficients derived from **a**.

batteries at the same C rate. The Coulombic efficiencies for the batteries with and without SiO_2–IL–ClO_4 are compared in Fig. 4c–g to evaluate their stability. It is apparent that cells without SiO_2–IL–ClO_4 show diverging Coulombic efficiency between the 10th to 60th cycle, while cells with only small amounts of SiO_2–IL–ClO_4 exhibit improved Coulombic efficiency to over 90% each cycle, which is enhanced with increasing SiO_2–IL–ClO_4 amount. A benefit of the improved charging stability of the cells is that their cycling performance is enhanced over multiple discharge cycles as shown in Fig. 4d. We tentatively attribute this effect to a reduction in electrolyte loss as a result of side reactions with the anode during cell recharge.

Reaction mechanism. To investigate the discharge reaction mechanism, X-ray photoelectron spectroscopy (XPS) was applied on the cathode side after galvanostatic cycling at different stages of discharge/charge to study the species formed at each stage (Fig. 5a–e). The pristine cathode exhibits an elemental-state sulfur doublet with S2p 2/3 at 164 eV. When the cell is discharged below 1 V, S2p 2/3 peaks at 162.1 and 160 eV representing Na_2S_2 and Na_2S rise. The peak beyond 166 eV is probably due to thiosulfate/sulfate complex species originating from oxidized sulfide species[63]. On full discharge, the elemental sulfur peak disappeared; only the sulfide peaks remain, suggesting the final discharge product is Na_2S. This can explain why a higher capacity is achieved compared with high-temperature Na–S batteries, where the final discharge product is Na_2S_x ($x \geq 3$) because of the phase limitation[16]. Energy dispersive X-ray analysis of the cathode after full discharge reveal the atomic ratio of Na and S is about 2.1, consistent with an almost full reduction from S to Na_2S, and sulfur intensities concentrated inside the micropores and suggesting the solid-state reaction (Fig. 5f; Supplementary Fig. 9). To further understand the reaction mechanism, an organic conversation technique was utilized to characterize the reaction species during initial cycling in both EC/PC and TEGDME electrolytes. In this approach, highly reactive sulfide species are first converted to their stable analogue benzyl sulfides (BzS_x, $x = 1$–5)[43,64], and nuclear magnetic resonance (NMR) spectroscopy of the analogues applied to analyse the organic molecules (Supplementary Fig. 10 and Supplementary Note 1). Remarkably, chemical shifts representing Bz_2S were observed in the whole process for the cell utilizing EC/PC carbonate electrolyte and in the first discharge process for the cell based on the TEGDME electrolyte. In contrast, high-order BzS_x are clearly observed in recharged cathodes in TEGDME. Ultraviolet-visible (ultraviolet–vis) spectra of a dilute Na_2S_6 solution, battery cathodes after 10 cycles in TEGDME, and carbonate electrolyte soaked in TEGDME are shown in Fig. 5g. In the carbonate electrolyte, there are only insoluble S_2^{2-} or S^{2-} formed, while the high-order soluble PS are formed when the cells are cycled in TEGDME[65]. This again confirms our hypothesis that the MCPS cathode coupled with carbonate electrolyte undergoes solid-state reaction with no soluble intermediate polysulfide formed in Na–S batteries.

On the basis of the spectroscopic study above, one may more firmly hypothesize that the electrochemistry in the cathode reaction occurs completely inside the MCPS, which means that both the transport of Na^+ into the cathode and the electrochemical reaction with sulfur in the cathode progress as solid-state processes. To verify this hypothesis, we first extract an approximate value for the Na^+ diffusivity in the cathode based on the electrochemical data. Measurements utilizing a galvanostatic intermittent titration technique (GITT) was performed by discharging the cell for 30 min at 0.1 C followed by a 10-h relaxation (Fig. 6a). The diffusion coefficient (Fig. 6b) at different stages during reversible charging and discharging can be calculated from the transient voltage response using an expression developed by Weppner and Huggins[66] for solid-state diffusion processes in batteries. The Na^+ diffusivity deduced from this analysis is found to be the lowest in the region where the discharge profile exhibits a clear plateau, consistent with the idea that a kinetics-controlled mechanism is overlayed on the discharge. In addition, the equilibrium potential determined at the end of each titration step changes very slightly and all below 2 V, suggesting the formation of Na_2S_x ($x \leq 2$). The reason why the solid-state reaction occurs only in microporous carbons with small pore sizes, but not in other carbon materials is open to argument and will likely be a subject for many future studies. On the basis of the empirical evidence in the present study, we attribute the difference to the stronger interactions between sulfur/sulfur species in the carbon relative to the strength of sulfur/sulfur species solvation due to their poor solubility in carbonate electrolytes[67] and to the short electronic transport lengths, which permit good active material utilization during normal battery operation even under slow solid-state transport kinetics.

Discussion

We report an example of a room-temperature, rechargeable Na–S battery that can be cycled stably with high Coulombic efficiency at low and moderate current densities. The battery utilizes a micoporous carbon/sulfur composite in the cathode and an EC/PC-1M $NaClO_4$ electrolyte. The combination of cathode substrate and electrolyte are shown to provide sufficiently strong association of sulfur in the cathode to confine the electrochemical reactions in the cathode to an all-solid-state process in which Na_2S appears to be the only product. We use both spectroscopic and analytical tools to show that for the carbons used in the present work, sulfur remains as S_8 and that the cathode reaction occurs inside the microporous carbon composite. An additional problem associated with instability during the recharge process of Na–S cells operated at moderate and high current density was identified and resolved using SiO_2–IL–ClO_4 particles as additives in the electrolytes. Electron microscopy and electrochemical analysis indicate that the particles form a dense protective coating on the Na anode and stabilize deposition of sodium by at least two mechanisms. First, they form a particle-rich, mechanically strong SEI layer that protects sodium metal from parasitic side reactions with the liquid carbonate electrolyte. Second, they appear to utilize a previously reported tethered anion effect to stabilize deposition of Na. Our finding underscores the benefits of micorporous carbon–sulfur composite and nanoparticles for guiding new material designs for inexpensive rechargeable metal–sulfur batteries. Further investigations are needed to fully understand the interaction of microporous carbon and sulfur species as well as the specific role SiO_2–IL–ClO_4 plays on metal anode protection.

Methods

Materials synthesis. MCPS and SiO_2–IL–ClO_4 electrolyte were synthesized according to the previous methods with modifications[27,32]. Briefly, the synthesis of MCPS is the same except the final sulfur infusion step. A sealed Pyrex tube was used to hold samples and a ramp rate of 1 °C min^{-1} was used for both heating and cooling. The final mass fraction of sulfur in the composites was determined by TGA (Q5000 IR Thermogravimetric Analyzer). The synthesis of SiO_2–IL–ClO_4 was the same as well except the anion exchange step. In this work, $NaClO_4$ was used to as anion exchange source.

Material characterization. The morphology and elemental mappings of the materials were studied using a FEI Tecnai F20 Transmission Electron Microscope and A LEO 1550 high-resolution SEM. The nitrogen adsorption–desorption isotherms of the MCP and MCPS were obtained with a Brunauer–Emmett–Teller (Micromeritics ASAP2020). AccuTOF DART was used to get mass spectra for sulfur and MCPS composites. Raman spectra were collected using a Renishaw InVia Confocal Raman Microscope ($\lambda = 488$ nm). ^{1}H NMR spectra were taken by Inova-400 Spectrometer. Ultraviolet-vis spectra were collected by Shimadzu UV-Vis-NIR Spectrometer. XPS measurements were performed with a Surface Science SSX-100 spectrometer using a monochromatic Al Kα source (1486.6 eV). Non-linear least squares curve fitting was applied to high-resolution spectra, using CasaXPS software.

Electrochemical measurements. The cathodes were prepared with MCPS1 or MCPS2, carbon black (Super-P, TIMCAL), and polymer binder (poly(vinylidene difluoride), PVDF, Aldrich) in a weight ratio of 8:1:1. A carbon-coated aluminium foil (0.018 mm in thick, 1.27 cm in diameter, MTI Corp.) was used as the current collector. The typical thickness of the active material film is ~20 µm and sulfur loading is around 0.73 to 1 mg. Sodium foil (Alfa Aesar) was used as the counter and reference electrode. A glass fibre filtre paper (Watchman 934-AH) was used as separator. 80 µl 1 M sodium perchlorate ($NaClO_4$) in a mixture ethylene carbonate (EC) and diethyl carbonate (DEC; v:v = 1:1) or in tetraethylene glycol dimethyl ether (TEGDME) or in a mixture of EC and propylene carbonate (PC; v:v = 1:1) with different amount of SiO_2–IL–ClO_4 were used as electrolyte for the cells. Cell assembly was carried out in an argon-filled glove-box (MBraun Labmaster) by using coin cell 2032 type. The room-temperature cycling characteristics of the cells were evaluated under galvanostatic conditions using Neware CT-3008 battery testers and electrochemical processes in the cells were studied by cyclic voltammetry using a CHI600D potentiostat. Electrochemical impedance and floating tests were conducted by using a Solartron Cell Test System model 1470E potentiostat/galvanostat. Ionic conductivities were measured using a Novocontrol N40 broadband dielectric spectrometer.

For post-mortem studies, cells were disassembled in an argon-filled glove-box and the electrodes were harvested and rinsed thoroughly with the electrolyte solvent before analysis.

References

1. Armand, M. & Tarascon, J. M. Building better batteries. *Nature* **451,** 652–657 (2008).
2. Dunn, B., Kamath, H. & Tarascon, J.-M. Electrical energy storage for the grid: a battery of choices. *Science* **334,** 928–935 (2011).
3. Tarascon, J. M. & Armand, M. Issues and challenges facing rechargeable lithium batteries. *Nature* **414,** 359–367 (2001).
4. Whittingham, M. S. Electrical Energy Storage and Intercalation Chemistry. *Science* **192,** 1126–1127 (1976).
5. Bruce, P. G., Freunberger, S. A., Hardwick, L. J. & Tarascon, J.-M. Li-O2 and Li-S batteries with high energy storage. *Nat. Mater.* **11,** 19–29 (2012).
6. Ma, L., Hendrickson, K. E., Wei, S. & Archer, L. A. Nanomaterials: science and applications in the lithium-sulfur battery. *Nano Today* **10,** 315–338 (2015).
7. Manthiram, A., Chung, S.-H. & Zu, C. Lithium-Sulfur Batteries: Progress and Prospects. *Adv. Mater.* **27,** 1980–2006 (2015).
8. Wang, Z. *et al.* Enhancing lithium-sulphur battery performance by strongly binding the discharge products on amino-functionalized reduced graphene oxide. *Nat. Commun.* **5,** 50052 (2014).
9. Ji, X., Lee, K. T. & Nazar, L. F. A highly ordered nanostructured carbon-sulphur cathode for lithium-sulphur batteries. *Nat. Mater.* **8,** 500–506 (2009).
10. Zhao, Q. *et al.* Sulfur nanodots electrodeposited on Ni foam as high-performance cathode for Li-S batteries. *Nano Lett.* **15,** 721–726 (2015).
11. Guo, J., Yang, Z., Yu, Y., Abruña, H. D. & Archer, L. A. Lithium-sulfur battery cathode enabled by lithium-nitrile interaction. *J. Am. Chem. Soc.* **135,** 763–767 (2013).
12. Yabuuchi, N. *et al.* P2-type Nax[Fe1/2Mn1/2]O2 made from earth-abundant elements for rechargeable Na batteries. *Nat. Mater.* **11,** 512–517 (2012).
13. Lu, Y., Wang, L., Cheng, J. & Goodenough, J. B. Prussian blue: a new framework of electrode materials for sodium batteries. *Chem. Commun.* **48,** 6544–6546 (2012).
14. Manthiram, A. & Yu, X. Ambient temperature sodium-sulfur batteries. *Small* **11,** 2108–2114 (2015).
15. Adelhelm, P. *et al.* From lithium to sodium: cell chemistry of room temperature sodium-air and sodium-sulfur batteries. *Beilstein J. Nanotechnol.* **6,** 1016–1055 (2015).
16. Sudworth, J. L. & Tilley, A. R. *The Sodium Sulfur Battery* Chapter 2 Chapman & Hall, 1985).
17. Hueso, K. B., Armand, M. & Rojo, T. High temperature sodium batteries: status, challenges and future trends. *Energy Environ. Sci.* **6,** 734–749 (2013).
18. Seh, Z. W., Sun, J., Sun, Y. & Cui, Y. A highly reversible room-temperature sodium metal anode. *ACS Cent. Sci.* **1,** 449–455 (2015).
19. Xin, S., Yin, Y.-X., Guo, Y.-G. & Wan, L.-J. A high-energy room-temperature sodium-sulfur battery. *Adv. Mater.* **26,** 1261–1265 (2014).
20. Komaba, S. *et al.* Fluorinated ethylene carbonate as electrolyte additive for rechargeable Na batteries. *ACS Appl. Mater. Interfaces* **3,** 4165–4168 (2011).
21. Yin, Y.-X., Xin, S., Guo, Y.-G. & Wan, L.-J. Lithium-sulfur batteries: electrochemistry, materials, and prospects. *Angew. Chem. Int. Ed.* **52,** 13186–13200 (2013).
22. Su, Y.-S., Fu, Y., Cochell, T. & Manthiram, A. A strategic approach to recharging lithium-sulphur batteries for long cycle life. *Nat. Commun.* **4,** 2985 (2013).
23. Peng, H.-J. & Zhang, Q. Designing host materials for sulfur cathodes: from physical confinement to surface chemistry. *Angew. Chem. Int. Ed.* **54,** 11018–11020 (2015).
24. Zhang, B., Qin, X., Li, G. R. & Gao, X. P. Enhancement of long stability of sulfur cathode by encapsulating sulfur into micropores of carbon spheres. *Energy Environ. Sci.* **3,** 1531–1537 (2010).
25. Li, Z. *et al.* Insight into the electrode mechanism in lithium-sulfur batteries with ordered microporous carbon confined sulfur as the cathode. *Adv. Energy Mater.* **4,** 1301473–1301480 (2014).
26. Xin, S. *et al.* Smaller sulfur molecules promise better lithium-sulfur batteries. *J. Am. Chem. Soc.* **134,** 18510–18513 (2012).
27. Wu, H. B. *et al.* Embedding sulfur in MOF-derived microporous carbon polyhedrons for lithium-sulfur batteries. *Chemistry* **19,** 10804–10808 (2013).
28. Fu, C., Wong, B. M., Bozhilov, K. N. & Guo, J. Solid state lithiation-delithiation of sulphur in sub-nano confinement: a new concept for designing lithium-sulphur batteries. *Chem. Sci.* **7,** 1224–1232 (2016).
29. Schaefer, J. L., Moganty, S. S., Yanga, D. A. & Archer, L. A. Nanoporous hybrid electrolytes. *J. Mater. Chem.* **21,** 10094–10101 (2011).
30. Schaefer, J. *et al.* Electrolytes for high-energy lithium batteries. *Appl. Nanosci.* **2,** 91–109 (2012).
31. Lu, Y., Das, S. K., Moganty, S. S. & Archer, L. A. Ionic liquid-nanoparticle hybrid electrolytes and their application in secondary lithium-metal batteries. *Adv. Mater.* **24,** 4430–4435 (2012).
32. Lu, Y., Moganty, S. S., Schaefer, J. L. & Archer, L. A. Ionic liquid-nanoparticle hybrid electrolytes. *J. Mater. Chem.* **22,** 4066–4072 (2012).
33. Lu, Y., Korf, K., Kambe, Y., Tu, Z. & Archer, L. A. Ionic-liquid-nanoparticle hybrid electrolytes: applications in lithium metal batteries. *Angew. Chem. Int. Ed.* **53,** 488–492 (2014).
34. Lu, Y. *et al.* Stable cycling of lithium metal batteries using high transference number electrolytes. *Adv. Energy Mater.* **5,** 1402073–1402079 (2015).

35. Tikekar, M. D., Archer, L. A. & Koch, D. L. Stability analysis of electrodeposition across a structured electrolyte with immobilized anions. *J. Electrochem. Soc.* **161,** A847–A855 (2014).
36. Tu, Z., Nath, P., Lu, Y., Tikekar, M. D. & Archer, L. A. Nanostructured electrolytes for stable lithium electrodeposition in secondary batteries. *Acc. Chem. Res.* **48,** 2947–2956 (2015).
37. Lu, Y., Tu, Z. & Archer, L. A. Stable lithium electrodeposition in liquid and nanoporous solid electrolytes. *Nat. Mater.* **13,** 961–969 (2014).
38. McMillan, R. S., Worsfold, D. J., Murray, J. J., Davidson, I. & Shu, Z. X. Electrolyte comprising fluoro-ethylene carbonate and propylene carbonate, for alkali metal-ion secondary battery, US patent 6506524 B1 (2003).
39. Zheng, G. *et al.* Interconnected hollow carbon nanospheres for stable lithium metal anodes. *Nat. Nano* **9,** 618–623 (2014).
40. Hwang, T. H., Jung, D. S., Kim, J.-S., Kim, B. G. & Choi, J. W. One-dimensional carbon-sulfur composite fibers for Na-S rechargeable batteries operating at room temperature. *Nano Lett.* **13,** 4532–4538 (2013).
41. Slater, A. G. & Cooper, A. I. Function-led design of new porous materials. *Science* **348,** 988–998 (2015).
42. Cravillon, J. *et al.* Controlling zeolitic imidazolate framework nano- and microcrystal formation: insight into crystal growth by time-resolved in situ static light scattering. *Chem. Mater.* **23,** 2130–2141 (2011).
43. Wei, S., Ma, L., Hendrickson, K. E., Tu, Z. & Archer, L. A. Metal-sulfur battery cathodes based on PAN-sulfur composites. *J. Am. Chem. Soc.* **137,** 12143–12152 (2015).
44. Meyer, B. Elemental sulfur. *Chem. Rev.* **76,** 367–388 (1976).
45. Moon, S. *et al.* Encapsulated monoclinic sulfur for stable cycling of Li-S rechargeable batteries. *Adv. Mater.* **25,** 6547–6553 (2013).
46. Erlich, R. H. & Popov, A. I. Spectroscopic studies of ionic solvation. X. Study of the solvation of sodium ions in nonaqueous solvents by sodium-23 nuclear magnetic resonance. *J. Am. Chem. Soc.* **93,** 5620–5623 (1971).
47. Ong, S. P. *et al.* Voltage, stability and diffusion barrier differences between sodium-ion and lithium-ion intercalation materials. *Energy Environ. Sci.* **4,** 3680–3688 (2011).
48. Kim, I. *et al.* A room temperature Na/S battery using a β'' alumina solid electrolyte separator, tetraethylene glycol dimethyl ether electrolyte, and a S/C composite cathode. *J. Power Sources* **301,** 332–337 (2016).
49. Yu, X. & Manthiram, A. Ambient-temperature sodium-sulfur batteries with a sodiated nafion membrane and a carbon nanofiber-activated carbon composite electrode. *Adv. Energy Mater.* **5,** 1500350–1500355 (2015).
50. Yu, X. & Manthiram, A. Capacity enhancement and discharge mechanisms of room-temperature sodium-sulfur batteries. *ChemElectroChem* **1,** 1275–1280 (2014).
51. Yu, X. & Manthiram, A. Room-temperature sodium-sulfur batteries with liquid-phase sodium polysulfide catholytes and binder-free multiwall carbon nanotube fabric electrodes. *J. Phys. Chem. C* **118,** 22952–22959 (2014).
52. Yu, X. & Manthiram, A. Highly reversible room-temperature sulfur/long-chain sodium polysulfide batteries. *J. Phys. Chem. Lett.* **5,** 1943–1947 (2014).
53. Bauer, I., Kohl, M., Althues, H. & Kaskel, S. Shuttle suppression in room temperature sodium-sulfur batteries using ion selective polymer membranes. *Chem. Commun.* **50,** 3208–3210 (2014).
54. Li, W. *et al.* The synergetic effect of lithium polysulfide and lithium nitrate to prevent lithium dendrite growth. *Nat. Commun.* **6,** 7436 (2015).
55. Yim, T. *et al.* Effect of chemical reactivity of polysulfide toward carbonate-based electrolyte on the electrochemical performance of Li-S batteries. *Electrochim. Acta* **107,** 454–460 (2013).
56. Xu, W.-T. *et al.* Towards stable lithium-sulfur batteries with a low self-discharge rate: ion diffusion modulation and anode protection. *ChemSusChem* **8,** 2892–2901 (2015).
57. Manthiram, A., Fu, Y., Chung, S.-H., Zu, C. & Su, Y.-S. Rechargeable lithium-sulfur batteries. *Chem. Rev.* **114,** 11751–11787 (2014).
58. Cohn, G., Ma, L. & Archer, L. A. A novel non-aqueous aluminum sulfur battery. *J. Power Sources* **283,** 416–422 (2015).
59. Mirkin, M. V., Bulhoes, L. O. S. & Bard, A. J. Determination of the kinetic parameters for the electroreduction of fullerene C60 by scanning electrochemical microscopy and fast scan cyclic voltammetry. *J. Am. Chem. Soc.* **115,** 201–204 (1993).
60. Kundu, D., Talaie, E., Duffort, V. & Nazar, L. F. The emerging chemistry of sodium ion batteries for electrochemical energy storage. *Angew. Chem. Int. Ed.* **54,** 3431–3448 (2015).
61. Wang, L. *et al.* A Superior low-cost cathode for a Na-ion battery. *Angew. Chem. Int. Ed.* **52,** 1964–1967 (2013).
62. Xu, S., Lu, Y., Wang, H., Abruna, H. D. & Archer, L. A. A rechargeable Na-CO2/O2 battery enabled by stable nanoparticle hybrid electrolytes. *J. Mater. Chem. A* **2,** 17723–17729 (2014).
63. Liang, X. *et al.* A highly efficient polysulfide mediator for lithium-sulfur batteries. *Nat. Commun.* **6,** 5682 (2015).
64. Kawase, A., Shirai, S., Yamoto, Y., Arakawa, R. & Takata, T. Electrochemical reactions of lithium-sulfur batteries: an analytical study using the organic conversion technique. *Phys. Chem. Chem. Phys.* **16,** 9344–9350 (2014).
65. Manan, N. S. A. *et al.* Electrochemistry of Sulfur and Polysulfides in Ionic Liquids. *The Journal of Physical Chemistry B* **115,** 13873–13879 (2011).
66. Weppner, W. & Huggins, R. A. Determination of the kinetic parameters of mixed-conducting electrodes and application to the system Li3Sb. *J. Electrochem. Soc.* **124,** 1569–1578 (1977).
67. Kamlet, M. J., Carr, P. W., Taft, R. W. & Abraham, M. H. Linear solvation energy relationships. 13. Relationship between the Hildebrand solubility parameter, .delta.H, and the solvatochromic parameter, .pi.*. *J. Am. Chem. Soc.* **103,** 6062–6066 (1981).

Acknowledgements

This work was supported by the National Science Foundation, Award No. DMR-1609125. Electron microscopy, X-ray diffractometry, X-ray spectroscopy facilities and optical spectrometers available through the Cornell Center for Materials Research (CCMR) were used for this work (NSF Grant DMR-1120296).

Author contributions

S.W., Y.L. and L.A.A. conceived the idea. S.W. carried out the synthesis, characterizations and electrochemical measurements. S.W., S.X. and Y.L. synthesized ionic liquid tethered nanoparticles and assembled the coin cells. S.W., A.A., S.C. Z.T. and L.M. performed the sodium surface characterization experiments in the paper. Everyone participated in the data analysis and discussion of the results reported in the paper. S.W. and L.A.A. wrote the paper.

Additional information

Competing financial interests: L.A.A. is a founder and Scientific Advisor to NOHMS Technologies, a technology concern seeking to commercialize functional electrolytes for rechargeable lithium ion and lithium/sulfur batteries. The remaining authors declare no competing financial interest.

10

Thermoelectric effect and its dependence on molecular length and sequence in single DNA molecules

Yueqi Li[1,2], Limin Xiang[1,2], Julio L. Palma[1,2], Yoshihiro Asai[3] & Nongjian Tao[1,4]

Studying the thermoelectric effect in DNA is important for unravelling charge transport mechanisms and for developing relevant applications of DNA molecules. Here we report a study of the thermoelectric effect in single DNA molecules. By varying the molecular length and sequence, we tune the charge transport in DNA to either a hopping- or tunnelling-dominated regimes. The thermoelectric effect is small and insensitive to the molecular length in the hopping regime. In contrast, the thermoelectric effect is large and sensitive to the length in the tunnelling regime. These findings indicate that one may control the thermoelectric effect in DNA by varying its sequence and length. We describe the experimental results in terms of hopping and tunnelling charge transport models.

[1]Center for Bioelectronics and Biosensors, Biodesign Institute, Arizona State University, Tempe, Arizona 85287-5801, USA. [2]School of Molecular Sciences, Arizona State University, Tempe, Arizona 85287-5801, USA. [3]Research Center for Computational Design of Advanced Functional Materials (CD-FMat), National Institute of Advanced Industrial Science and Technology (AIST), Central 2, Umezono 1-1-1, Tsukuba, Ibaraki 305-8568, Japan. [4]School of Chemistry and Chemical Engineering, Nanjing University, Nanjing, Jiangsu 210093 China. Correspondence and requests for materials should be addressed to N.T. (email: Nongjian.Tao@asu.edu).

The thermoelectric effect is a basic property of materials and has found applications in energy conversion, temperature sensing and regulation. This important effect in single molecules is expected to be distinctly different from that in bulk materials[1]. In addition to potential applications, studying the thermoelectric effect in single molecules helps assess the molecular orbital level alignment, understand the energy conversion mechanism associated with charge transport and determine whether electrons or holes are responsible for the charge transport. The work by Reddy *et al.*[2] has stimulated many efforts in investigating the thermoelectric effect in molecules[3–6]. However, to date, research has been limited to the regime where the Fermi levels of the electrodes are in the highest occupied molecular orbital (HOMO)–lowest unoccupied molecular orbital (LUMO) gaps of the molecules, and electrons or holes transport through molecules via coherent tunnelling, a quantum mechanical process with a characteristic exponential dependence of the molecular resistance with length.

Here we report a study of the thermoelectric effect in double-stranded DNA (dsDNA) molecules. By selecting different DNA sequences, we study the thermoelectric effect not only in the tunnelling regime but also in the hopping regime, in which electrons or holes hop along a molecule from one end to another sequentially via multiple steps, leading to a linear dependence of the molecular resistance with length. The work demonstrates that DNA thermoelectricity can be tuned by its sequences and length, which provides new insights into the thermoelectric effect in single molecules, and basic knowledge for potential applications with programmable DNA nanostructures, a field that has been advancing rapidly in recent years[7].

Results

dsDNA sequences. To study the thermoelectric effect in the hopping regime, we selected DNA molecules with sequences of 5′-A(CG)$_n$T-3′ ($n = 3$, 4, 5, 6 and 7; Fig. 1b), where A, C, G and T denote the four DNA bases, adenine, cytosine, guanine and thymine, respectively. Note that this sequence notation shows only one strand of the dsDNA, and the second strand has the complementary sequence. The complete sequences of all the dsDNA studied in this work are listed in Table 1. Previous studies have shown that charge transport in GC sequences is dominated by hopping of holes along the molecules with G as hopping sites[8–13]. To investigate the thermoelectric effect in the tunnelling regime, we studied DNA sequences, 5′-ACGC(**AT**)$_m$ GCGT-3′, where $m = 1$, 2, 3 and 4, and 5′-ACGC(**AT**)$_{m-1}$ **A**GCGT-3′, where $m = 1$, 2 and 3 (Fig. 1b; Table 1). The insertion of a block of AT bases in the middle of the GC sequences introduces a tunnelling barrier into the DNA molecules, as shown in literature[14–19], allowing us to investigate the thermoelectric effect in the tunnelling regime. To measure the DNA conductance and thermal electric effect, we modified T base at the 3′ end with an amino group, which binds to gold electrodes to establish electrical contact between the electrodes and the molecule[20].

Conductance. We measured the DNA conductance and thermoelectric effect with a temperature controlled scanning tunnelling microscopy (STM) set-up shown in Fig. 1a, following a procedure described in detail elsewhere[21] (see also the Methods section). The experiment generated conductance versus STM tip–substrate distance traces, showing plateaus associated with the formation and breakdown of single-molecule junctions (marked with arrows in the red traces in Fig. 1c). As a control experiment, we performed the experiment without DNA and observed no characteristic plateaus in the conductance traces (black traces). By collecting thousands of conductance traces, we constructed a conductance histogram (Fig. 1d), which shows a broad peak, where the peak position measures the average conductance of a single dsDNA molecule. The width of the histogram peak is not due to experimental error, but rather it reflects the intrinsic variability and distribution in the molecule–electrode binding geometry[22]. To determine the experimental error, we repeated the above experiment more than three times, each using a new sample to generate thousands of conductance traces, for every DNA sequence, and used the s.d. in the conductance values as the error bar.

We characterized the DNA samples by performing non-denaturing polyacrylamide gel electrophoresis, circular dichroism spectroscopy and melting temperature experiments in phosphate buffer (Supplementary Figs 1, 2 and 3; Supplementary Methods 1, 2 and 3). These experiments confirmed that each DNA sequence studied here was in B-form double-helical structure in phosphate buffer. To ensure the DNA molecule retained its double-helical structure in the experiments ($\sim$20% relative humidity), we compared the conductance of DNA measured in humidified air with that measured in phosphate buffer and found similar values within the experimental uncertainty (Supplementary Figs 4 and 5). Another parameter that characterizes molecular junction stability is the stretching length, which is the average distance that one can stretch the junction before it breaks. We also found similar stretching lengths measured in phosphate buffer and in humidified air (Supplementary Fig. 6; Supplementary Discussion 1). These facts led us to conclude that the DNA molecules in humidified air retained its double-helical structure, possibly due to a thin absorbed water layer on the substrate surface. This conclusion is further supported by atomic force microscopy[23], STM[24], mechanical break junction[25] and polarization modulation infrared[26] measurements of dsDNA on a surface in the air.

The conductance (or resistance) of DNA depends on both the molecular length and sequence. For A(CG)$_n$T ($n = 3$–7), the resistance increases linearly with the molecular length (Fig. 2a), which is expected for hopping transport[20,27,28]. According to the hopping model[29,30], holes hop along a DNA molecule from one end to another, where the individual G bases act as hopping sites such as stepping stones. Consequently, the total resistance is proportional to the number of hopping sites (G bases), and thus the length of the molecule (see more discussions in Supplementary Fig. 7 and Supplementary Discussion 2). When DNA is connected to two electrodes (tip and substrate), an additional contribution to the resistance arises from the two contacts. The total resistance of the DNA is given by[29,30]

$$R = R_c + R_h = R_c + (N-1)R_{G-G}, \qquad (1)$$

where R_c is contact resistance, R_h is intrinsic DNA resistance, N is number of base pairs and R_{G-G} is resistance of a hopping step (one GC base pair). Table 2 summarizes R_c and R_h of all the A(CG)$_n$T sequences. By fitting the experimental data with equation (1), we found that the contact resistance, $R_c = 0.48 \pm 0.16\,\mathrm{M\Omega}$, and the resistance of one step hopping, $R_{G-G} = 0.28 \pm 0.02\,\mathrm{M\Omega}$ per GC base pair. Note that the contact resistance is small compared with the total resistance for A(CG)$_n$T (see Table 2), indicating efficient electronic coupling between the electrodes and the molecule, and the DNA resistance is dominated by the molecule itself, rather by the contact.

In contrast to A(CG)$_n$T described above, the resistance length dependence of ACGC(**AT**)$_m$GCGT/ACGC(**AT**)$_{m-1}$**A**GCGT is significantly different (Fig. 2b). For $m = 0$, 1 and 2, the resistance increases more rapidly with length (L) and can be best fitted with an exponential function, rather than with a shallow linear dependence found for A(CG)$_n$T. For $m > 2$, however, the resistance becomes weakly length dependent, indicating a

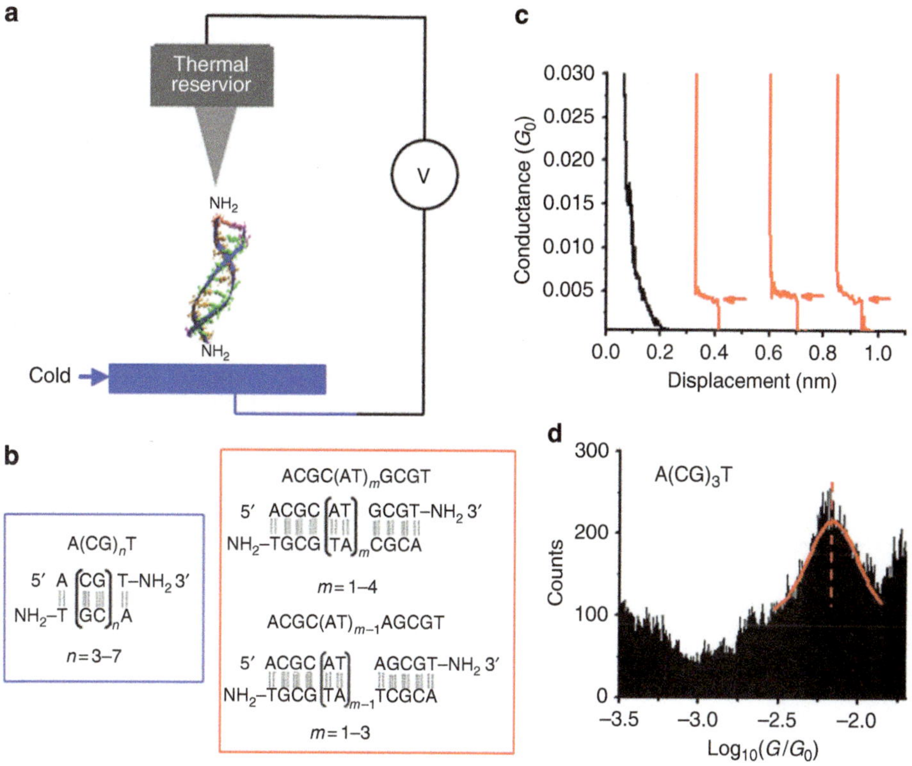

Figure 1 | Thermoelectric and conductance measurements of DNA. (**a**) Conductance and thermoelectric effect of a DNA molecule bridged between STM tip (kept at 295 K) and substrate (cold). (**b**) Two families of DNA sequences are studied in this work, denoted as (1) $A(CG)_nT$, (2) $ACGC(\mathbf{AT})_mGCGT$ and $ACGC(\mathbf{AT})_{m-1}\mathbf{A}GCGT$ (see Table 1 for a full list). (**c**) Conductance traces without (black) and with (red) dsDNA $A(CG)_3T$, where plateaus are marked with red arrows. (**d**) Conductance histogram of $A(CG)_3T$, where the solid red curve is Gaussian fit to the conductance peak. Note: $G_0 = 2e^2/h = 77.48\,\mu S$.

Table 1 | Sequences of dsDNA studied in this work.

Name	Duplex	Abbreviation
$A(CG)_3T$	5′-ACGCGCGT-3′; 3′-TGCGCGCA-5′	$A(CG)_nT$
$A(CG)_4T$	5′-ACGCGCGCGT-3′; 3′-TGCGCGCGCA-5′	$A(CG)_nT$
$A(CG)_5T$	5′-ACGCGCGCGCGT-3′; 3′-TGCGCGCGCGCA-5′	$A(CG)_nT$
$A(CG)_6T$	5′-ACGCGCGCGCGCGT-3′; 3′-TGCGCGCGCGCGCA-5′	$A(CG)_nT$
$A(CG)_7T$	5′-ACGCGCGCGCGCGCGT-3′; 3′-TCGGCGCGCGCGCGCA-5′	$A(CG)_nT$
ACGC**A**GCGT	5′-ACGCAGCGT-3′; 3′-TGCGTCGCA-5′	$ACGC(AT)_{m-1}AGCGT$
ACGC**AT**GCGT	5′-ACGCATGCGT-3′; 3′-TGCGTACGCA-5′	$ACGC(AT)_mGCGT$
ACGC**ATA**GCGT	5′-ACGCATAGCGT-3′; 3′-TGCGTATCGCA-5′	$ACGC(AT)_{m-1}AGCGT$
ACGC(**AT**)$_2$GCGT	5′-ACGCATATGCGT-3′; 3′-TGCGTATACGCA-5′	$ACGC(AT)_mGCGT$
ACGC(**AT**)$_2$**A**GCGT	5′-ACGCATATAGCGT-3′; 3′-TGCGTATATCGCA-5′	$ACGC(AT)_{m-1}AGCGT$
ACGC(**AT**)$_3$GCGT	5′-ACGCATATATGCGT-3′; 3′-TGCGTATATACGCA-5′	$ACGC(AT)_mGCGT$
ACGC(**AT**)$_4$GCGT	5′-ACGCATATATATGCGT-3′; 3′-TGCGTATATATACGCA-5′	$ACGC(AT)_mGCGT$

dsDNA, double-stranded DNA.

transition in the charge transport mechanism at $m = 2$ (or 4 A-T base pairs). This observation agrees with previous photochemical and direct conductance measurements of DNA, which have shown that inserting a short AT block in the middle of a GC sequence introduces a tunnelling barrier, leading to an exponential increase in the DNA resistance with the AT block length[14–19], and inserting a longer AT block results in the transition in the charge transport mechanism from tunnelling to hopping[14,16]. The tunnelling-hopping transition has also been predicted by theoretical calculations[14–19]. See Supplementary Discussion 2 for further discussion.

On the basis of the above results, we modelled the $ACGC(\mathbf{AT})_mGCGT$ and $ACGC(\mathbf{AT})_{m-1}\mathbf{A}GCGT$ resistance as a sum of three contributions, electrode–DNA contact, hopping via GC sequences and tunnelling via the AT block, given by $R = R_c + R_h + R_t$, where R_c is the contact resistance, R_h is the GC hopping resistance and R_t is the AT tunnelling resistance. Since both R_c and R_h are known from the $A(CG)_nT$ measurement, the above relation allowed us to determine the resistance of the inserted AT block, which is listed in Table 2. Semi-logarithmic plot of the AT block resistance (R_t) versus the AT block length shows an exponential dependence, following $R_t = R_0\exp(\beta L)$, where R_0 is the contact resistance between the GC and AT block, and β is the decay constant and L is the AT block length (Fig. 2c). Fitting the experimental data with the above exponential relation leads to a β of $2.03 \pm 0.12\,nm^{-1}$ (Fig. 2c). This decay constant is smaller than $\sim 4\,nm^{-1}$ reported by Xu *et al.*[15] for thiolated DNA molecules, but is within the

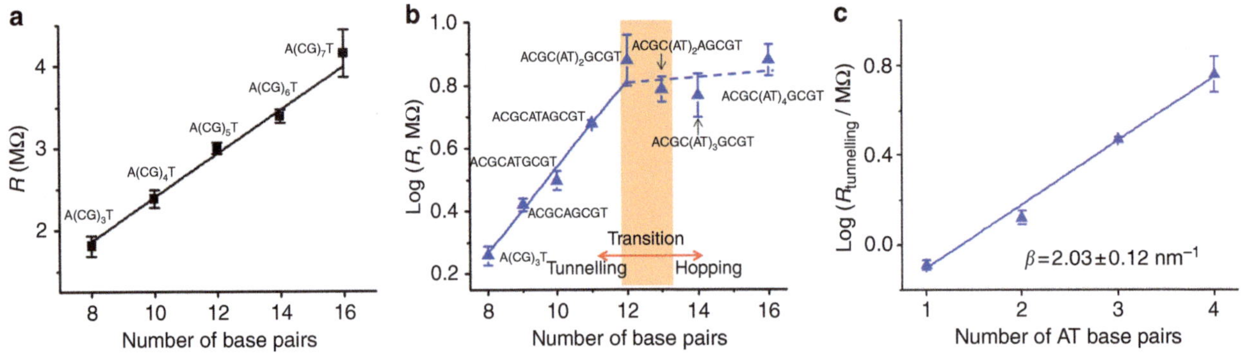

Figure 2 | DNA resistance versus molecular length. (**a**) A(CG)$_n$T (n=3-7), (**b**) ACGC(**AT**)$_m$GCGT (m=0-4) and ACGC(**AT**)$_{m-1}$**A**GCGT (m=1-3) sequences, and (**c**) the AT blocks ((AT)$_m$ and (AT)$_{m-1}$A) in ACGC(AT)$_m$GCGT and ACGC(AT)$_{m-1}$AGCGT sequences. The solid lines in **a**,**b** and **c** are linear fits to the data, and the dashed line in **b** is a guide to eye. The transition from tunnelling to hopping occurs the AT block is longer than 4 base pairs, which is markers with a shaded orange region.

Table 2 | Conductance and Seebeck coefficients of DNA with different sequences.

Sequence	Conductance peak position (log(G/G$_0$))*	Conductance peak width	Resistance (MΩ)	R_c (MΩ)†	R_h (MΩ)†	R_t (MΩ)†	Seebeck coefficient (μV K^{-1})
A(CG)$_3$T	−2.17 ± 0.03	0.20	1.9 ± 0.1	0.48	1.40	NA	1.4 ± 0.2
A(CG)$_4$T	−2.27 ± 0.02	0.18	2.4 ± 0.1	0.48	1.96	NA	0.8 ± 0.2
A(CG)$_5$T	−2.37 ± 0.01	0.17	3.0 ± 0.1	0.48	2.52	NA	1.0 ± 0.2
A(CG)$_6$T	−2.42 ± 0.01	0.18	3.4 ± 0.1	0.48	3.08	NA	0.6 ± 0.2
A(CG)$_7$T	−2.51 ± 0.03	0.20	4.2 ± 0.3	0.48	3.64	NA	0.8 ± 0.2
ACGC**A**GCGT	−2.31 ± 0.02	0.25	2.6 ± 0.1	0.48	1.12	1.0	5.0 ± 0.3
ACGC**AT**GCGT	−2.36 ± 0.03	0.35	3.0 ± 0.2	0.48	1.12	1.4	5.9 ± 0.4
ACGC**ATA**GCGT	−2.57 ± 0.01	0.30	4.8 ± 0.1	0.48	1.12	3.2	6.5 ± 0.8
ACGC(**AT**)$_2$GCGT	−2.76 ± 0.08	0.30	7.4 ± 1.4	0.48	1.12	5.8	7.9 ± 0.5
ACGC(**AT**)$_2$**A**GCGT	−2.68 ± 0.04	0.50	6.1 ± 0.6	0.48	1.12	4.5	4.9 ± 0.2
ACGC(**AT**)$_3$GCGT	−2.65 ± 0.07	0.52	5.8 ± 0.9	0.48	1.12	4.2	2.0 ± 0.2
ACGC(**AT**)$_4$GCGT	−2.77 ± 0.05	0.40	7.6 ± 0.9	0.48	1.12	6.0	2.0 ± 0.3

NA, not applicable.
*G_0 (=$2e^2/h$=77.48 μS) is the conductance quantum.
†R_c, R_h and R_t are the contact resistance, hopping part resistance (CG sequences) and tunnelling part (AT sequences) resistance, respectively.

range reported by various other works[8-13] and consistent with the calculations by Torisi *et al.*[19] and Voityuk *et al.*[31] for dsDNA with alternating AT sequences.

Thermoelectric effect. The thermoelectric effect of a single DNA-electrode junction can be expressed by

$$S_{junction} = S_{Au} - \frac{\Delta U_{TE,m}}{\Delta T} = S_{Au} - \frac{-\Delta\left(I_{TE,m}/G_m\right)}{\Delta T}, \quad (2)$$

where S_{Au} is the Seebeck coefficient of gold (~2 μV K^{-1} (ref. 32)), $U_{TE,m}$ is the open-circuit voltage, $I_{TE,m}$ is the short-circuit current, G_m is the conductance of the molecular junction and ΔT is the temperature difference between the STM tip and substrate electrodes. We measured the thermoelectric effect by holding the tip at 295 K while cooling the substrate from 295 to 275 K (ref. 33). The selection of this temperature range ensures that all the dsDNA sequences are structurally stable, and their resistance does not change significantly with temperature[34].

The thermoelectric effect measurement started by detecting plateaus in the conductance traces during the retraction of the STM tip from the substrate (Fig. 3a). Once a plateau was detected, the tip retraction was halted and the bias voltage between the tip and substrate electrodes was swept over ±10 mV to obtain a current (I)–voltage (V) characteristic curve. After recording an I–V curve, the tip was further retracted to a new position, and the I–V measurement was repeated until the molecular junction broke down. Figure 3b shows several I–V curves for ACGC(**AT**)$_2$GCGT measured at different locations of a conductance plateau, where the colours of curves match the colours of the dots marked on the conductance plateau shown in Fig. 3a. These I–V curves are linear within the bias range.

From the I–V characteristics, we obtained the average Seebeck coefficient of a DNA molecule with the following procedure. First, each I–V curve was fitted with a linear function to obtain G, and then a corresponding I/G–V curve was obtained by dividing I by G. From thousands of I/G–V curves, a two-dimensional I/G–V histogram was constructed for each temperature difference (Fig. 3c,d). When the substrate is held at the same temperature as the tip (room temperature), the centre of the histogram passes through the origin of I/G–V plot (Fig. 3c). However, when the substrate is cooled below the tip temperature, there is an offset in the histogram in the I/G–V plot, due to the thermoelectric effect (Fig. 3d). By extracting the voltage offsets (at zero current) from the I/G–V histograms, we constructed the thermoelectric voltage histograms at different tip–substrate temperature differences (Fig. 3e; Supplementary Figs 8 and 9; Supplementary Table 1; Supplementary Discussion 3). The peak in each thermoelectric voltage histogram was fitted with a Gaussian distribution, from which $U_{TE,m}$ was determined. Figure 3f shows that $U_{TE,m}$ depends linearly on ΔT, and the slope was used to determine $S_{junction}$, the Seebeck coefficient of the dsDNA molecule according to equation (2).

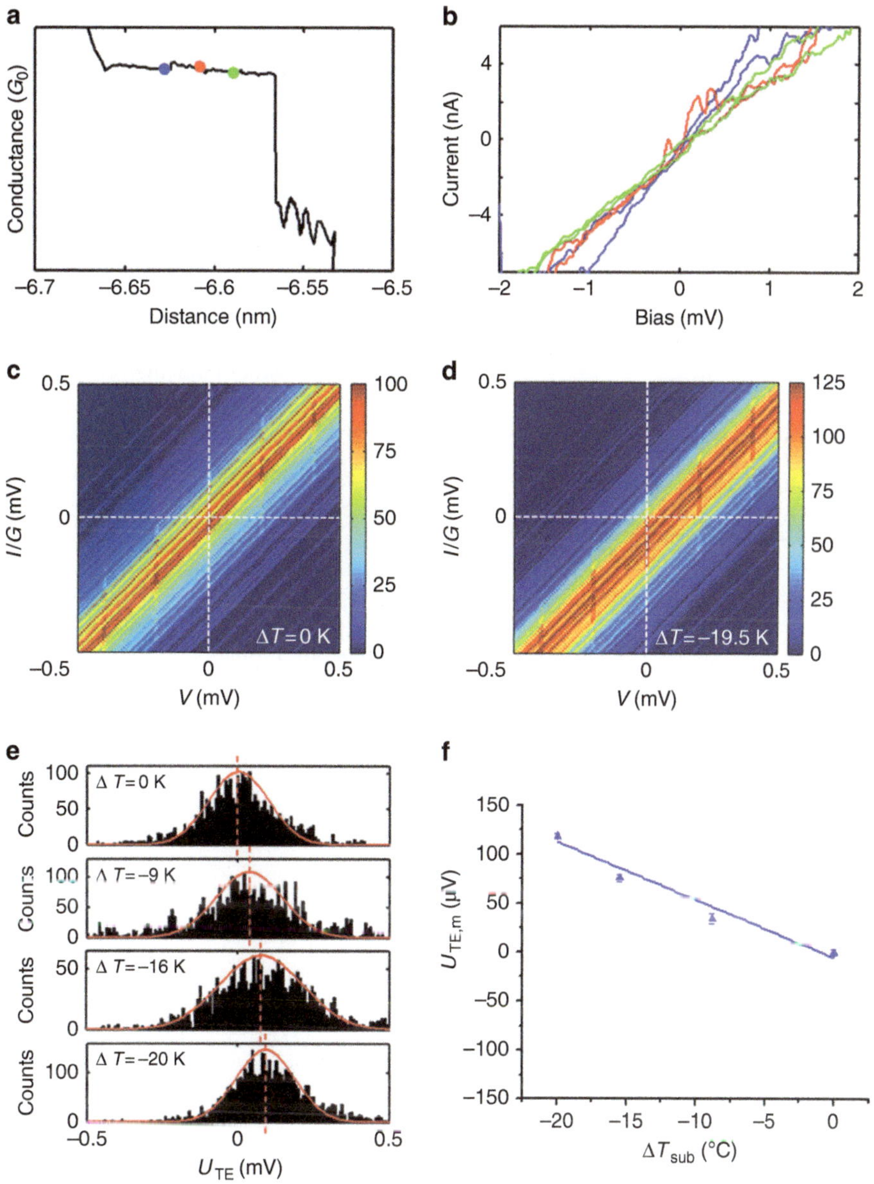

Figure 3 | Thermoelectric measurement of DNA. (**a**) Semi-logarithmic plot of conductance-distance trace (the coloured dots mark, where *I*-*V* curves are measured). (**b**) Individual *I*-*V* curves measured at the locations marked with corresponding colour in **a**. (**c**) *I*/*G*-*V* histogram at $\Delta T = 0$ K. (**d**) *I*/*G*-*V* histogram at $\Delta T = 19.5$ K, showing an offset due to the thermoelectric effect. (**e**) Thermal voltage histogram, where the red curves show Gaussian fits. (**f**) Thermoelectric voltage versus ΔT, where the straight line is linear fit to the data, and error bars are fitting errors. DNA sequence: ACGC(**AT**)$_2$GCGT.

Figure 4 summarizes the Seebeck coefficients of DNA with different lengths and sequences, from which we can draw several important conclusions. First, the Seebeck coefficients are positive for all the DNA sequences in both the tunnelling and hopping regimes studied here. It has been predicted that a positive Seebeck coefficient corresponds to hole-dominated charge transport and a negative Seebeck coefficient signals electron-dominated charge transport[2]. The observation of positive Seebeck coefficients here indicates that holes dominate charge transport in DNA. This is expected because the DNA HOMO level is close to the electrode Fermi level compared with its LUMO level, and is also consistent with the previous experiments[14,16,18,19,35]. However, the prediction of the Seebeck coefficient sign is based on a coherent tunnelling model, which is not necessarily applicable to hopping transport. The data shown in this indicate that this prediction appears to be valid also in the hopping regime. Second, the Seebeck coefficients of DNA in the hopping regime (in A(CG)$_n$T) are small, and weakly depend on the molecular length compared with other organic molecules[36,37]. Last, inserting a short AT block (shorter than 5 AT base pairs) into the middle of A(CG)$_n$T leads to a much greater Seebeck coefficient, and it increases with the AT block length. However, when the AT block is longer than 5 AT base pairs, the Seebeck coefficient drops to the level of A(CG)$_n$T and become insensitive the AT length. This transition coincides with the observed tunnelling-hopping transition near 4–5 AT base pairs, which strongly suggests that the thermoelectric effect is large in the tunnelling regime and small in the hopping regime. We discuss these observations below.

For bulk materials, the Seebeck coefficient can be expressed by the Mott formula[38],

$$S = \frac{\pi^2 k_B^2 T}{3e} \cdot \left(\frac{1}{R} \cdot \frac{\partial R}{\partial E}\right)_{E=E_F}, \tag{3}$$

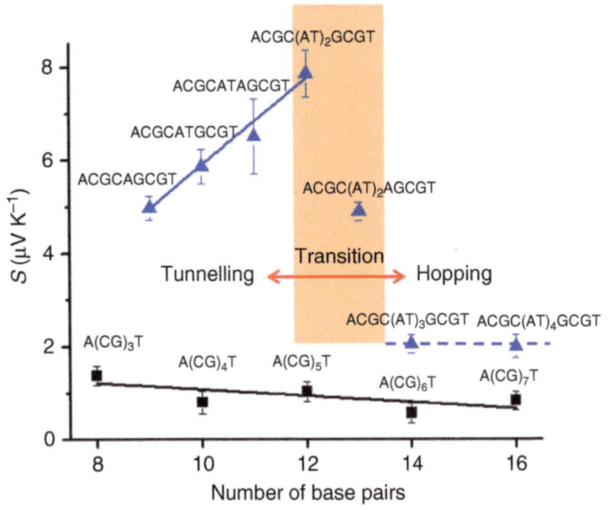

Figure 4 | Seebeck coefficients of DNA with different molecular lengths and sequences. The black squares are the Seebeck coefficients of A(CG)$_n$T ($n=1, 2, 3$). The blue triangles are the Seebeck coefficients of ACGC(**AT**)$_m$GCGT ($m=1$-4) and ACGC(**AT**)$_{m-1}$**A**GCGT ($m=1-3$) sequences, showing a transition when the AT block length is longer than 4 base pairs. The solid and dashed lines are guides to eye.

where R is the energy-dependent resistance function and E is the energy of the carriers. k_B is the Boltzmann constant, T is temperature and e is elementary charge. The Mott formula assumes incoherent transport, which is applicable to hopping transport in DNA. For coherent tunnelling transport, the Seebeck coefficient can be obtained with Landauer's formula, which leads to a similar expression as equation (3) except that R is replaced by the energy-dependent transmission probability function[5,39].

We analyse the experimental thermoelectric effect in the hopping regime with equation (3) by assuming that the resistance function, $R(E)$, can be expressed as a sum of the contact and hopping resistance functions (equation (1)), which is modelled as Fig. 5a. Consequently, the overall Seebeck coefficient (S) consists of contributions from the molecule-electrode contact (S_c) and hopping along the dsDNA sequences (S_h), given by

$$S\left[\mathrm{A(CG)}_n\mathrm{T}\right] = \frac{R_c}{R}S_c + \frac{R_h}{R}S_h, \quad (4)$$

where R, R_c and R_h in equation (4) are the total resistance, contact resistance and DNA resistance, respectively. We discuss the validity of the assumption and derivation of equation (4) in Supplementary Discussion 4. Both R_c and R_h are known from the linear fitting to the experimental data shown in Fig. 2a (Table 2).

In the case of ACGC(**AT**)$_m$GCGT and ACGC(**AT**)$_{m-1}$ **A**GCGT sequences, the total resistance R[AT] must include a contribution, R_t, arising from the inserted AT block in the middle of the GC sequence, as shown in Fig. 5b. The overall Seebeck coefficient of the DNA junction must also include the contribution from the inserted AT block, expressed by,

$$S[\mathrm{AT}] = \frac{R_c}{R}S_c + \frac{R_h}{R}S_h + \frac{R_i}{R}S_i, \quad (5)$$

where S_t is the Seebeck coefficient of the inserted AT part.

In equation (4), although R_h is proportional to the molecular length, S_h is independent of the length because it is proportional to the relative change in R_h (a normalized quantity). S_c is also independent of the molecular length. So, we conclude that the molecular length dependence of the overall Seebeck coefficient, S, arises from R in equation (4), which as shown in Fig. 2a is weakly dependent on molecular length. By fitting the measured Seebeck coefficients times R with R_h at different lengths (Fig. 5c), we obtained the slope, $S_h = 0.1 \pm 0.3\,\mu\mathrm{V\,K^{-1}}$, and from the intercept we found that $S_c = 4.8 \pm 1.7\,\mu\mathrm{V\,K^{-1}}$, showing that the contact plays a dominant role in the thermoelectric effect of A(CG)$_n$T. The above model explains the weak length dependence of the observed thermoelectric effect.

For equation (5), we plotted $S \times R$ versus R_t (Table 2) in Fig. 5d for short AT blocks ($m = 1$–2), which follows a linear function. According to equation (5), the slope of the linear relation is S_t, from which we found that $S_t = 8.9 \pm 1.4\,\mu\mathrm{V\,K^{-1}}$. For longer AT blocks ($m > 2$), we determined S_t from equation (5) (Table 3). Compared with the short AT blocks, these sequences have smaller Seebeck coefficients ($\sim 2\,\mu\mathrm{V\,K^{-1}}$).

To further understand the tunnelling Seebeck coefficient, S_t, from the AT block, we express the transmission function of the tunnelling barrier with a Lorentzian distribution by[33,36,37,40]

$$\tau(E) = \frac{4\Gamma^2}{4\Gamma^2 + \Delta E_H^2}, \quad (6)$$

where Γ is the width of the HOMO level of the AT block and ΔE_H is the energy difference between the AT block and GC HOMO levels. Using the above transmission function, the contributions of the AT block to the resistance and Seebeck coefficient are expressed as

$$\frac{1}{R_t} = G_0\frac{4\Gamma^2}{4\Gamma^2 + \Delta E_H^2}, \quad (7)$$

$$S_t = \frac{\pi^2 k_B^2 T}{3e} \cdot \frac{2\Delta E_H}{4\Gamma^2 + \Delta E_H^2}. \quad (8)$$

By fitting the experimental data with equations (7) and (8), we obtained Γ and ΔE_H (Table 3), which shows that $\Gamma \ll \Delta E_H$, indicating a weak coupling between the AT block and the GC sequences and electrodes. In this weak-coupling limit, S_t is proportional to $1/\Delta E_H$ (refs 33,41). Because the electronic coupling between alternating AT base pairs in the AT block is weak[31], ΔE_H is expected to be weakly dependent on the AT block length, which explains the constant slope (S_t) in Fig. 5d.

The above hopping and tunnelling models help understand the length dependence of the Seebeck coefficients in DNA. However, to explain that the insertion of a short AT block increases the Seebeck coefficient, a model including sufficient chemical and structural details of dsDNA is needed. We used an extended Su–Schrieffer–Heeger model with the Lennard–Jones potential to describe the dsDNA (see Supplementary Fig. 10 and Supplementary Discussion 5 for details). The theory reproduces sequence dependence, and also the observed large increase in the Seebeck coefficient associated with the insertion of an AT block in the middle of CG sequences (Supplementary Fig. 11). The increase can be attributed to the difference in the site energies between AT and GC base pairs. However, we emphasized that the calculation was performed at 8 K. As such, the theory cannot provide quantitative agreement with the experimental Seebeck coefficient nor it predicts the transition from tunnelling to hopping. Further investigation is thus still needed to quantitatively describe the measured Seebeck coefficients and the tunnelling–hopping transition.

In this paper we, have studied the charge transport and thermoelectric effect in single DNA molecules with different sequences and lengths. The transport mechanism in CG sequences (5′-A(CG)$_n$T-3′, with $n = 3$–7) is hole hopping, as reflected by linear dependence of the resistance on the molecular length. The Seebeck coefficient that describes the thermoelectric effect is small and weakly dependent on the molecular length. Inserting a block of AT base pairs in the middle of a CG sequence

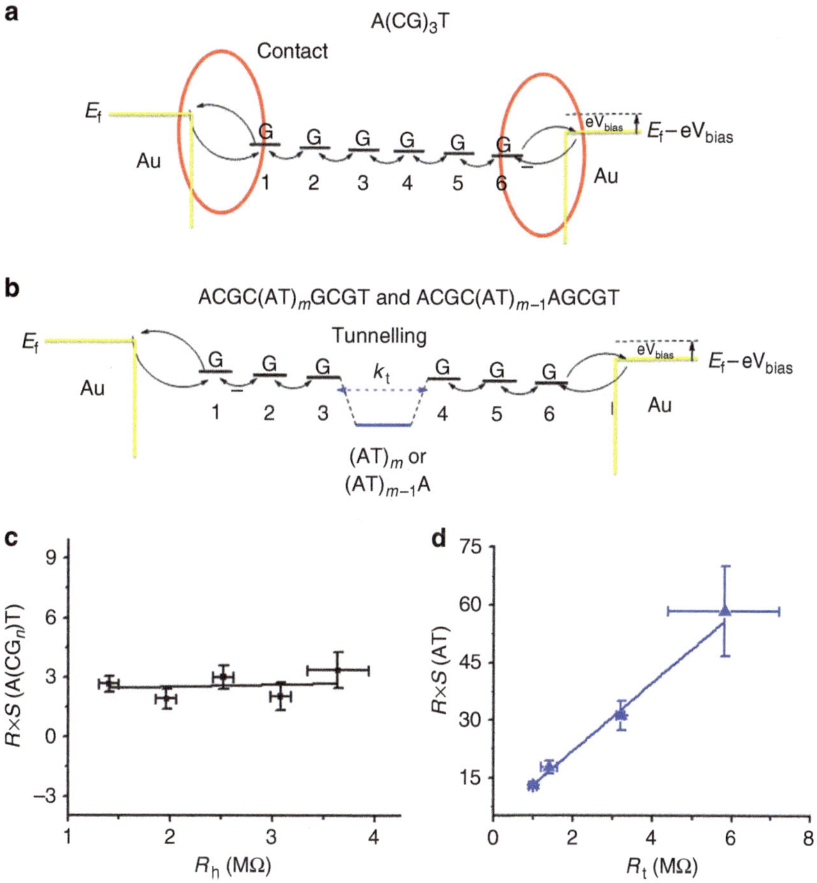

Figure 5 | Thermoelectric effect in DNA in terms of tunnelling and hopping models. (**a**) Charge transport in $A(CG)_nT$, where a hole is injected from left electrode into the first G, then hops along the molecule with each G as a hopping site, and eventually reaches the right electrode. The charge transfer rates at the contacts (red part) are energy dependent, which is a dominant contribution to the Seebeck coefficient. (**b**) Charge transport in ACGC(**AT**)$_m$GCGT and ACGC(**AT**)$_{m-1}$AGCGT, where a tunnelling barrier (marked blue) arises from the AT block. (**c**) $R \times S$ versus R_h for $A(CG)_nT$ sequences. (**d**) $R \times S$ versus R_t for ACGC(**AT**)$_m$GCGT and ACGC(**AT**)$_{m-1}$**A**GCGT ($m=1$ and 2). Solid lines in **c** and **d** are linear fits to the data.

Table 3 | The resistance (R_t) and the Seebeck coefficients (S_t) of the AT blocks, from which the broadening of the HOMO levels (Γ), and the energy difference between the HOMO levels of AT and GC blocks were determined.

Sequences	R_t (MΩ)	S_t (μV K^{-1})	Γ (eV)	ΔE_H (eV)
ACGC**A**GCGT	1.0 ± 0.3	8.9 ± 1.4	0.07 ± 0.02	1.3 ± 0.4
ACGC**AT**GCGT	1.4 ± 0.4	8.9 ± 1.4	0.06 ± 0.02	1.3 ± 0.4
ACGC**ATA**GCGT	3.2 ± 0.3	8.9 ± 1.4	0.05 ± 0.01	1.6 ± 0.3
ACGC(**AT**)$_2$GCGT	5.8 ± 1.6	8.9 ± 1.4	0.03 ± 0.01	1.4 ± 0.5
ACGC(**AT**)$_2$**A**GCGT	4.5 ± 0.8	6.1 ± 1.3	NA	NA
ACGC(**AT**)$_3$GCGT	4.2 ± 1.1	2.2 ± 0.8	NA	NA
ACGC(**AT**)$_4$GCGT	6.0 ± 1.1	2.1 ± 0.7	NA	NA

HOMO, highest occupied molecular orbital; NA, not applicable.

(ACGC(**AT**)$_m$GCGT-3′, $m=1$–4; and 5′-ACGC(**AT**)$_{m-1}$AGCGT-3′, $m=1$–3) changes both the charge transport and thermoelectric effect substantially. When the inserted AT block is shorter than 4 base pairs, it acts as a tunnelling barrier with resistance increases exponentially with the block length, and its Seebeck coefficient is large compared with the CG sequences. However, when the inserted AT block is longer than 4 base pairs, the charge transport mechanism changes from tunnelling to hopping, the resistance becomes weakly length dependent and the Seebeck coefficient drops to smaller values. The experimental results have been analysed in terms on tunnelling and hopping models, and compared with an extended Su–Schrieffer–Heeger model with the Lennard–Jones potential. The work demonstrates that the DNA thermoelectricity may be tuned by its length and sequence, and studying the thermoelectric effect provides new insights into charge transport in DNA.

Methods

Preparation of dsDNA SAM. DNA molecules with different sequences and lengths were purchased from Bio-Synthesis Inc. The T terminal of each DNA molecule was modified with an amine group[20], allowing binding to gold electrodes for electrical measurement. The name and structure of all the sequences are listed in Table 1. The DNA molecules were annealed using a thermal cycler (TC-050-18, Labnet Inc) to form the B-form double-helical structure in phosphate buffer, which was verified by gel electrophoresis, circular dichroism and melting curve (Supplementary Figs 1, 2 and 3; Supplementary Methods 1, 2 and 3). The phosphate buffer was prepared with $Na_2HPO_4.2H_2O$ (for high-performance liquid chromatography (HPLC) ≥98.5%) and NaH_2PO_4 (for HPLC ≥99%) purchased from Fluka. The DNA sample was immobilized on a gold substrate (gold with thickness of ~160 nm on mica) prepared in a vapour deposition system under ultrahigh vacuum. Before each experiment, the gold substrate was annealed with hydrogen flame briefly, and immediately immersed in phosphate buffer (5 mM, pH = 7) containing 5 μM double-helical DNA, incubated overnight. The substrate electrode was then rinsed with phosphate buffer and dried with nitrogen gas. The DNA-coated gold substrate was attached to the sample holder of the STM, which was placed in a glove box under nitrogen atmosphere (relative humidity ~20%).

STM set-up. The charge transport and thermoelectric measurements were carried out with the STM consisting of a controller (Nanoscope IIIA, Digital Instruments Inc.) and a STM scanner (Molecular Imaging). The STM tip was freshly cut from a gold wire (0.25 mm diameter, 99.95%, Alfa Aesar). A piece of electrically shielded copper strip (0.7 × 0.4 × 2 cm) was thermally connected to the STM tip holder to serve as a thermal reservoir. The STM sample holder was mounted on a semiconductor cooler with a thermistor to measure the temperature (controlled by Lake Shore 331 temperature controller).

Measurement of thermoelectricity. The STM break junction technique was used to measure the conductance of DNA with a procedure described in detail elsewhere[21] (Fig. 1a). Briefly, the STM tip was first brought into contact with the DNA molecules on the gold substrate, allowing the amine-terminated DNA to bind with the tip and substrate electrodes. The tip was then retracted from the substrate, during which a conductance–distance trace was recorded. The last plateau in the conductance trace corresponded to a single DNA molecule attached to the tip and substrate electrodes to form a molecular junction. The process was repeated thousands of times. A conductance histogram was constructed from the individual conductance traces, and the peak in the conductance histogram gave the average conductance of the DNA molecule.

To measure the thermoelectric effect of DNA molecules, a temperature gradient was created across the molecule by cooling the sample holder while maintaining the STM tip at the room temperature[33]. The temperature of the sample was monitored with thermistors during the experiment. When a molecular junction was formed in the STM break junction measurement as indicated by the formation of a plateau feature in the conductance–distance trace, the tip was held in position and an I–V curve was recorded. The tip was then retracted further away from the substrate by an additional distance. If the current did not drop abruptly, signalling that the molecular junction was still intact, another I–V curve was recorded. Otherwise, the measurement was started over again.

Thousands of I–V curves were collected for each temperature gradient. These I–V curves were normalized by conductance (G) and presented as a two-dimensional I/G versus V histogram. The voltage offsets in the I/G versus V histogram were used to construct a histogram of the thermoelectric effect, and Gaussian fitting of the thermoelectric effect histogram provided the Seebeck coefficient of DNA. Controlling the STM tip movement, recording and processing the experimental data described above were achieved with Labview 8.5 and data acquisition card.

References

1. Mahan, G. D. & Sofo, J. O. The best thermoelectric. *Proc. Natl Acad. Sci. USA* **93,** 7436–7439 (1996).
2. Reddy, P., Jang, S.-Y., Segalman, R. A. & Majumdar, A. Thermoelectricity in molecular junctions. *Science* **315,** 1568–1571 (2007).
3. Dubi, Y. & Di Ventra, M. Thermoelectric effects in nanoscale junctions. *Nano Lett.* **9,** 97–101 (2009).
4. Bergfield, J. P. & Stafford, C. A. Thermoelectric signatures of coherent transport in single-molecule heterojunctions. *Nano Lett.* **9,** 3072–3076 (2009).
5. Ke, S. H., Yang, M., Curtarolo, S. & Baranger, H. U. Thermopower of molecular junctions: an ab initio study. *Nano Lett.* **9,** 1011–1014 (2009).
6. Liu, Y., Chen, Y. & Chen, Y. Thermoelectric efficiency in nanojunctions: a comparison between atomic junctions and molecular junctions. *ACS Nano* **3,** 3497–3504 (2009).
7. Seeman, N. C. Nanomaterials Based on DNA. *Annu. Rev. Biochem.* **79,** 65–87 (2010).
8. Fukui, K. & Tanaka, K. Distance dependence of photoinduced electron transfer in DNA. *Angew. Chem. Int. Ed.* **37,** 158–161 (1998).
9. Kelley, S. O. & Barton, J. K. Electron transfer between bases in double helical DNA. *Science* **283,** 375–381 (1999).
10. Lewis, F. D. *et al.* Distance-dependent electron transfer in DNA hairpins. *Science* **277,** 673–676 (1997).
11. Meade, T. J. & Katten, J. F. Electron-transfer through DNA-site-specific modification of duplex DNA with Ruthenium donors and accepters. *Angew. Chem. Int. Ed.* **34,** 352–354 (1995).
12. Meggers, E., Kusch, D., Spichty, M. & Wille, U. Electron transfer through DNA in the course of radical-induced strand cleavage. *Angew. Chem. Int. Ed.* **37,** 460–462 (1998).
13. Brun, A. M. & Harriman, A. Dynamics of electron transfer between intercalated polycyclic molecules: effect of interspersed bases. *J. Am. Chem. Soc.* **114,** 3656–3660 (1992).
14. Giese, B., Amaudrut, J., Koehler, A.-K., Spormann, M. & Wessely, S. Direct observation of hole transfer through DNA by hopping between adenine bases and by tunnelling. *Nature* **412,** 318–320 (2001).
15. Zhang, X. u. & Tao, L. i. Direct conductance measurement of sngle DNA molecules in aqueous solution. *Nano Lett.* **4,** 1105–1108 (2004).
16. Lewis, F. D., Zhu, H., Daublain, P., Cohen, B. & Wasielewski, M. R. Hole mobility in DNA a tracts. *Angew. Chem. Int. Ed.* **45,** 7982–7985 (2006).
17. Jortner, J., Bixon, M., Voityuk, A. A. & Rösch, N. Superexchange mediated charge hopping in DNA. *J. Phys. Chem. A* **106,** 7599–7606 (2002).
18. Renaud, N., Berlin, Y. A., Lewis, F. D. & Ratner, M. A. Between superexchange and hopping: an intermediate charge-transfer mechanism in Poly(A)-Poly(T) DNA hairpins. *J. Am. Chem. Soc.* **135,** 3953–3963 (2013).
19. Troisi, A. & Orlandi, G. The hole transfer in DNA: calculation of electron coupling between close bases. *Chem. Phys. Lett.* **344,** 509–518 (2001).
20. Xiang, L. *et al.* Intermediate tunnelling–hopping regime in DNA charge transport. *Nat. Chem.* **7,** 221–226 (2015).
21. Xu, B. & Tao, N. J. Measurement of single-molecule resistance by repeated formation of molecular junctions. *Science* **301,** 1221–1223 (2003).
22. Williams, P. D. & Reuter, M. G. Level alignments and coupling strengths in conductance histograms: the information content of a single channel peak. *J. Phys. Chem. C* **117,** 5937–5942 (2013).
23. Uchihashi, T. *et al.* Identification of B-form dna in an ultrahigh vacuum by noncontact-mode atomic force microscopy. *Langmuir* **16,** 1349–1353 (1999).
24. Tanaka, H., Tanaka, H. & Kawai, T. Partial sequencing of a single DNA molecule with a scanning tunnelling microscope. *Nat. Nanotechnol.* **4,** 518–522 (2009).
25. Kang, N. Electrical characterization of DNA in mechanically controlled break-junctions. *New J. Phys.* **10,** 023030 (2008).
26. Brewer, S. H., Anthireya, S. J., Lappi, S. E., Drapcho, D. L. & Franzen, S. Detection of DNA Hybridization on Gold Surfaces by Polarization Modulation Infrared Reflection Absorption Spectroscopy. *Langmuir* **18,** 4460–4464 (2002).
27. McCreery, R. L. Molecular electronic junctions. *Chem. Mater.* **16,** 4477–4496 (2004).
28. Luo, L., Choi, S. H. & Frisbie, C. D. Probing hopping conduction in conjugated molecular wires connected to metal electrodes. *Chem. Mater.* **23,** 631–645 (2010).
29. Segal, D., Nitzan, A., Ratner, M. & Davis, W. B. Activated conduction in microscopic molecular junctions. *J. Phys. Chem. B* **104,** 2790–2793 (2000).
30. Nitzan, A. The relationship between electron transfer rate and molecular conduction 2. The sequential hopping case. *Isr. J. Chem.* **42,** 163–166 (2002).
31. Voityuk, A. A., Jortner, J., Bixon, M. & Rösch, N. Electronic coupling between Watson–Crick pairs for hole transfer and transport in desoxyribonucleic acid. *J. Chem. Phys.* **114,** 5614–5620 (2001).
32. Blatt, F. J., Schroeder, P. A., Foiles, C. L. & Greig, D. *Thermoelectric Power of Metals* (Plenum Press, 1976).
33. Guo, S., Zhou, G. & Tao, N. Single molecule conductance, Thermopower, and Transition Voltage. *Nano Lett.* **13,** 4326–4332 (2013).
34. Joshua, H., Fang, C., Peiming, Z. & Nongjian, T. Thermal and electrochemical gate effects on DNA conductance. *J. Phys. Condens. Matter* **19,** 215202 (2007).
35. Di Felice, R., Calzolari, A., Molinari, E. & Garbesi, A. *Ab initio* study of model guanine assemblies: The role of pi-pi coupling and band transport. *Phys. Rev. B* **65,** 045104 (2001).
36. Malen, J. A. *et al.* Identifying the length dependence of orbital alignment and contact coupling in molecular heterojunctions. *Nano Lett.* **9,** 1164–1169 (2009).
37. Widawsky, J. R. *et al.* Length-dependent thermopower of highly conducting Au–C bonded single molecule junctions. *Nano Lett.* **13,** 2889–2894 (2013).
38. Mott, N. F. & Davis, E. A. *Electronic Processes in Non-Crystalline Materials* (Clarendon Press, 1979).
39. Paulsson, M. & Datta, S. Thermoelectric effect in molecular electronics. *Phys. Rev. B* **67,** 241403 (2003).
40. Kim, Y., Jeong, W., Kim, K., Lee, W. & Reddy, P. Electrostatic control of thermoelectricity in molecular junctions. *Nat. Nanotechnol.* **9,** 881–885 (2014).
41. Bâldea, I. Interpretation of stochastic events in single-molecule measurements of conductance and transition voltage spectroscopy. *J. Am. Chem. Soc.* **134,** 7958–7962 (2012).

Acknowledgements

We thank Hao Yan for the help on the gel electrophoresis experiments, Mark A. Ratner, Yuri A. Berlin and Vladimiro Mujica for the support and discussions. Financial support from Office of Naval Research (ONR N00014-11-1-0729) is gratefully acknowledged.

Author contributions

N.T. conceived the project. Y.L. designed, conducted and analysed the experiments. N.T., L.X., J.L.P. and Y.L. made the theoretical explanation with hopping model. Y.A. completed the SSH calculation. N.T., L.X. and Y.L. wrote the paper.

Additional information

Competing financial interests: The authors declare no competing financial interests.

11

Graphene-modified nanostructured vanadium pentoxide hybrids with extraordinary electrochemical performance for Li-ion batteries

Qi Liu[1], Zhe-Fei Li[1], Yadong Liu[1], Hangyu Zhang[2], Yang Ren[3], Cheng-Jun Sun[3], Wenquan Lu[4], Yun Zhou[2], Lia Stanciu[2,5], Eric A. Stach[6] & Jian Xie[1]

The long-standing issues of low intrinsic electronic conductivity, slow lithium-ion diffusion and irreversible phase transitions on deep discharge prevent the high specific capacity/energy (443 mAh g^{-1} and 1,550 Wh kg^{-1}) vanadium pentoxide from being used as the cathode material in practical battery applications. Here we develop a method to incorporate graphene sheets into vanadium pentoxide nanoribbons via the sol-gel process. The resulting graphene-modified nanostructured vanadium pentoxide hybrids contain only 2 wt. % graphene, yet exhibits extraordinary electrochemical performance: a specific capacity of 438 mAh g^{-1}, approaching the theoretical value (443 mAh g^{-1}), a long cyclability and significantly enhanced rate capability. Such performance is the result of the combined effects of the graphene on structural stability, electronic conduction, vanadium redox reaction and lithium-ion diffusion supported by various experimental studies. This method provides a new avenue to create nanostructured metal oxide/graphene materials for advanced battery applications.

[1] Department of Mechanical Engineering, Purdue School of Engineering and Technology, Indiana University-Purdue University, Indianapolis, Indiana 46202, USA. [2] School of Materials Engineering, Purdue University, West Lafayette, Indiana 47907, USA. [3] X-ray Science Division, Advanced Photon Source, Argonne National Laboratory, 9700 South Cass Avenue, Lemont, Illinois 60439, USA. [4] Chemical Science and Engineering Division, Argonne National Laboratory, 9700 South Cass Avenue, Lemont, Illinois 60439, USA. [5] Weldon School of Biomedical Engineering, Purdue University, West Lafayette, Indiana 47907, USA. [6] Center for Functional Nanomaterials, Brookhaven National Laboratory, Upton, New York 11973, USA. Correspondence and requests for materials should be addressed to J.X. (email: jianxie@iupui.edu).

For lithium-ion batteries (LIBs), the cathode is typically a metal oxide and serves as the host for Li^+ ion intercalation during the charge/discharge process. Different metal oxides have been explored as cathode materials[1]. Among the commonly used cathode materials such as $LiCoO_2$ (274 mAh g^{-1})[2] and $LiFePO_4$ (170 mAh g^{-1})[3], vanadium pentoxide (V_2O_5) has a theoretical capacity of 443 mAh g^{-1} (assuming intercalation of three lithium ions) and a specific energy of 1,218 mWh g^{-1} (assuming a nominal 2.75 V discharge voltage). In addition, vanadium is abundant in the crust of the earth, which makes it inexpensive. The combination of the high specific capacity/energy and its abundance make V_2O_5 a very attractive candidate for LIB applications. However, the high specific capacity/energy has not been realized in practical LIB applications[4–10] because of these three issues: (1) low electrical conduction, (2) slow lithium-ion diffusion and (3) irreversible phase transitions upon deep discharge. Like most metal oxides, V_2O_5 has very low electronic conductivity due to its low *d*-band mobility[11]. The lithiation/delithiation processes in crystalline V_2O_5 are accompanied by structural phase transitions[12]. Such a phase transition process induces lattice strain due to phase coexistence within the same electrode particle. It is known that the lattice mismatch-induced mechanical strain can cause irreversible structural damage, which not only limits battery life, but also causes irreversible capacity loss[13]. Consequently, V_2O_5 currently has very poor rate capability and very limited long-term cyclability.

Compared with crystalline V_2O_5, the amorphous V_2O_5 xerogel offers a considerable advantage by virtue of its morphology. The vanadium oxide xerogel, $V_2O_5 \cdot nH_2O$ has a ribbon-like structure, providing more access for Li^+ ion insertion than its crystalline counterpart, leading to a more versatile host for Li^+ ion intercalation and exhibiting improved capacity for the lithium[14]. The basic units of the V_2O_5 xerogel are sheets comprised of two vanadium oxide layers[1]. When the distance of the adjacent layers of V_2O_5 increases, the insertion capacity will increase. Hence, in hydrated V_2O_5 xerogel, $V_2O_5 \cdot nH_2O$, the water molecules function as pillars between the V_2O_5 layers and support a basal spacing of ~11.5 Å[15], leading to a Li intercalation capacity about 1.4 times larger than that of orthorhombic V_2O_5,(~4.56 Å). In addition, the larger interlayer spacing also facilitates the Li^+ diffusion within the V_2O_5 matrix[16]. In addition, the interlayered water molecules result in a distorted square-pyramidal ligand field that favours electron transfer and Li^+ intercalation within the V_2O_5 matrix[16]. However, even for V_2O_5 xerogel, the same challenges of low electrical conduction (both intraparticle, within a V_2O_5 particle, and interparticle, between V_2O_5 particles), slow lithium diffusion and structure stability/reversibility still remain. Efforts have been made to improve the conductivity by coating the V_2O_5 xerogel with conductive materials (that is, single-wall carbon nanotubes)[17], doping metals[18] and organic polymers[19–23]. These measures can improve the V_2O_5 xerogel only to a certain degree, and none of them significantly improve the structural stability and reversibility. Hence, a comprehensive approach that can simultaneously deal with all three of these issues is needed.

Graphene is a single-atomic layer of sp^2-bonded carbon atoms arranged in a honeycomb crystal structure[24], which provides extraordinary electrical (that is, extremely high electric conductivity—6.29*10^7 S cm^{-1}), mechanical (that is, fracture strength ~130 GPa) and thermal properties (that is, 3,000 W/m-K in plane)[25]. Graphene can be prepared using either the chemical or thermal reduction of graphene oxide (GO), which is a layered stack of oxidized graphene sheets with different functional groups. Thus, GO can be easily dispersed in the form of a single sheet in water at low concentrations. Our comprehensive approach for dealing with these three issues of the V_2O_5 xerogel is to insert a single graphene sheet to, at most, a few layers of graphene sheets between the V_2O_5 nanoribbons to modify the V_2O_5 xerogel structure via the sol–gel process. Such a structure can be expected to improve both the intra-particle electronic conduction and the lithium-ion diffusion.

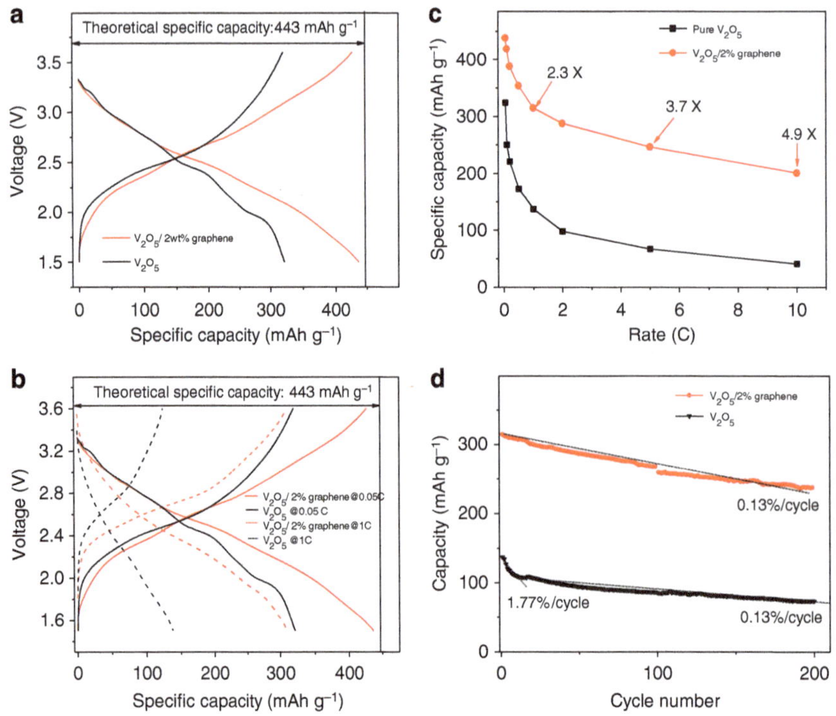

Figure 1 | Electrochemical characterization. Charge/discharge curves of pure V_2O_5 and V_2O_5-G/2% graphene cells at (**a**) 0.05 and (**b**) 0.05 and 1.0 C; (**c**) rate performance of pure V_2O_5 and V_2O_5-G/2% graphene cells based on C-rate and (**d**) cycle life of pure V_2O_5 and V_2O_5-G with 2% graphene cells at a 1 C rate.

Furthermore, the resulting smaller V_2O_5 grain size in the V_2O_5 xerogel can be expected to reduce the stress within the particles, leading to better structural stability and cycle life. Here we report a simple and unique method for synthesizing graphene-modified, nanostructured V_2O_5 via the sol–gel process, leading to the creation of a class of V_2O_5 hybrids with intercalated graphene at low concentrations (that is, 2 wt. %). We demonstrate that these V_2O_5/graphene hybrid materials have overcome the above-mentioned three major issues and exhibit extraordinary electrochemical performance as cathode materials for LIBs. In operando synchrotron high-energy X-ray diffraction (HEXRD), X-ray absorption near-edge spectroscopy (XANES), cryo-transmission electron microscopy (cryo-TEM) and low-voltage, aberration-corrected high-resolution TEM are used to characterize their structure and to elucidate the formation mechanism of these materials, providing a foundation for understanding their extraordinary performance.

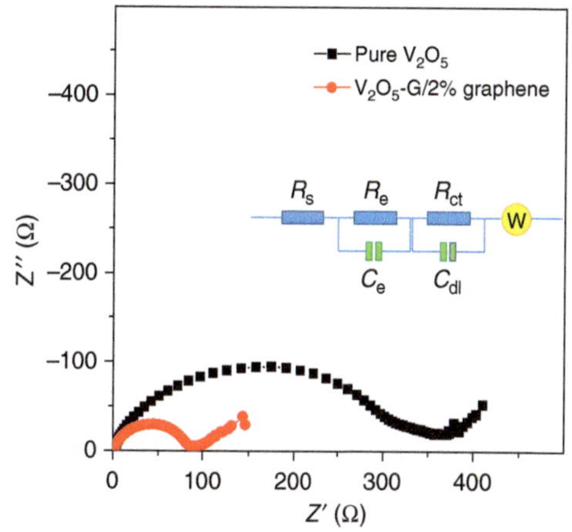

Figure 2 | Electrochemical impedance spectroscopy (Nyquist plot). Electrochemical impedance spectroscopy (Nyquist plot) of the pure V_2O_5 and the V_2O_5-G/2% graphene cells. Amplitude: 5 mV, frequency range: 1 MHz–0.01 Hz

Results:

Electrochemical performance. The introduction of a tiny amount of graphene (that is, 2 wt. %) into the V_2O_5 has an extraordinary effect on its electrochemical performance. In the following discussion, we define the graphene-modified nanostructure–V_2O_5 hybrid as 'V_2O_5–G' and V_2O_5 without graphene as 'pure V_2O_5'. A specific capacity of 438 mAh g^{-1} (corresponding to 1,034 Wh kg^{-1} and 3,118 Wh l^{-1}) has been achieved at 0.05 C (Fig. 1a) for the V_2O_5–G with 2% graphene, which is almost to the theoretical specific capacity, 99% of 443 mAh g^{-1} (theoretical value), while the pure V_2O_5 delivered only 324 mAh g^{-1} (corresponding to 777 Wh kg^{-1}). The achieved 438 mAh g^{-1} also suggests that the three Li^+ ions have been reversibly inserted/de-inserted into each V_2O_5 molecule in the V_2O_5–G hybrid. In addition to the high specific capacity, this V_2O_5–G hybrid shows the excellent reversibility (of capacity) from the voltage profile, which indicates that most of the inserted lithium ions can be removed during the discharge/charge process. Even at a higher rate, 1 C, such V_2O_5–G still delivered 315 mAh g^{-1} (corresponding to 768 Wh kg^{-1}; Fig. 1b), which is 2.23 times that of the pure V_2O_5, 137 mAh g^{-1} (corresponding to 299 Wh kg^{-1}). The pure V_2O_5 discharge profile shows three distinct voltage stages: 3.3-2.5 V, 2.5-2.0 V and 2.0-1.75 V, corresponding to the first, second and third Li^+ ion intercalations into the V_2O_5 (Fig. 1b). However, no such voltage plateaus were seen for the V_2O_5–G, which is typical for a disordered layer or single-phase material due to the absence of the crystalline phase transitions[26]. It is expected that the lower crystallinity should lead to less stress on the material and better cycle performance.

The introduction of graphene also has a significant effect on the rate performance, which is a major issue for pure V_2O_5. At fairly higher current densities, the V_2O_5–G still retains a high capacity: 419 mAh g^{-1} at 0.1 C, 354 mAh g^{-1} at 0.5 C, 315 mAh g^{-1} at 1 C, 247 mAh g^{-1} at 5 C and 201 mAh g^{-1} at 10 C, as compared with those of pure V_2O_5: 250 mAh g^{-1} at 0.1 C, 173 mAh g^{-1} at 0.5 C, 137 mAh g^{-1} at 1 C, 67 mAh g^{-1} at 5 C and 41 mAh g^{-1} at 10 C (Fig. 1c). This corresponds to a 67.6, 104.6, 130.0, 268.7 and 390.2% specific capacity increase at the different rates, respectively. Such a huge improvement on rate performance, particularly at high rates (that is, 10 C), suggests that it not only has the electric conductivity of the V_2O_5–G tremendously increased (both interparticle and intraparticle conductivity), but also that the ionic conductivities have increased. This is clearly seen from the alternating current (AC) impedance data, shown in Fig. 2 and described in detail below.

The synthesized V_2O_5–G also exhibits improved cycling stability. The V_2O_5–G achieved 150 cycles with 80% initial capacity at a 1 C rate, while the pure V_2O_5 achieved only 11 cycles (Fig. 1d). (The death criteria of a battery for electric vehicle and hybrid electric vehicle is defined as 80% of its initial capacity; hence, we compare the cycle life at 80% initial capacity.) The capacity decay rate is relatively stable for V_2O_5–G, at 0.13%/cycle, while the pure V_2O_5 shows two distinctly different decay rates: it initially decays sharply at 1.77%/cycle to 80% initial capacity after 11 cycles; after 12 cycles, it decays at a much slower rate of 0.13%/cycle, indicating that the pure V_2O_5 might experience a structure change during the initial cycles and that then the structure becomes stabilized. The content of graphene in the V_2O_5–G plays a critical role. Clearly, 2% graphene results in the highest specific capacity, 438 mAh g^{-1} (Supplementary Fig. 1), while 10% graphene leads to a lower specific capacity, 278 mAh g^{-1}, but much improved cycle stability. It is speculated that the lower graphene content may result in a better dispersion of the graphene sheets in the hydrogel of the V_2O_5, while the higher graphene content may lead to the restacking of the graphene sheets, as we observed in our previous work[27]. The thicker stacks of graphene (that is, 3–5 graphene sheets per stack) may hold the V_2O_5 nanoribbons tighter and, thus, help maintain the structural integrity, consequently leading to a much better cycle life. Another possibility is that the higher graphene content may lead to more complete coverage over the V_2O_5 nanoribbons, which could help to hold the ribbons together and prevent them from collapsing. A detailed understanding of the effects of the graphene content in the V_2O_5–G is another topic of study and will be reported separately.

AC impedance spectra of both pure V_2O_5 and V_2O_5–G were measured to validate that our approach of inserting the graphene sheets between the V_2O_5 nanoribbons via the sol–gel process leads to improved electric conductivity, Li^+ ion diffusion and structural stability. The results were fitted using the model[28] shown in Fig. 2, and the fitting results are listed in Supplementary Table 1. Clearly, the small amount of graphene in the V_2O_5 caused a huge change in the electric conductivity, R_e, from 309.48 Ω in pure V_2O_5 to 86.55 Ω in V_2O_5–G, an order of magnitude change. Both the pure V_2O_5 and the V_2O_5–G electrodes had the exact same composition and were made using the same procedures under the same conditions; therefore, the change in the R_e must be due to the conductivity of V_2O_5. This is similar to the doping of Cu into the V_2O_5 xerogel by Symerl *et al.*[18] The redox reaction of vanadium in the V_2O_5 also increased significantly. The R_{ct} changed from 46.88 Ω in the pure

V_2O_5 to 10.94 Ω in the V_2O_5–G. This is a 4.28X change that also explains the increased rate performance. Finally, the Li^+ ion diffusion within the V_2O_5 has improved. The W_d changed from 0.451 to 0.396, corresponding to a Li^+ diffusion coefficient in the V_2O_5 that changed from 1.21 E − 12 to 1.57 E − 12 $cm^2 s^{-1}$, a 12% increase. Thus, the AC impedance results show that our approach, indeed, works as we designed.

From the data, it is clear that the introduction of a small amount (that is, 2%) of graphene has a profound effect on the electrochemical performance of V_2O_5, and the V_2O_5–G shows the best electrochemical performance of any reported V_2O_5 hybrids in the coin cell configuration, as summarized in Supplementary Table 2 (refs 5,7,29–43). However, all performance changes are rooted in the material structure. Hence, in operando XANES, in operando HEXRD, cryo-TEM and aberration-corrected TEM were carried out to understand the structure and the formation mechanisms of the V_2O_5–G. The results are presented below.

Formation mechanism and morphology of pure V_2O_5 and V_2O_5–G. As demonstrated previously, the introduction of a small amount of graphene can cause a huge performance change for the formed V_2O_5–G. Naturally, a question arises: how could such a small amount of graphene (∼2% weight) change the structure of the V_2O_5 xerogel? Aiming to understand the effects of graphene on the structure of the V_2O_5 xerogel, the synthesis process of V_2O_5–G was studied using cryo-TEM. As described below, sodium metavanadate ($NaVO_3$) becomes yellow coloured decavanadic acid (HVO_3) after passing through an ion exchange column. This diluted HVO_3 starts to slowly form a V_2O_5 hydrogel via the protonation of HVO_3 (usually within several minutes)[23]; the solution gradually changes colour from yellow to dark brown and, eventually (usually after 1–2 weeks), to dark red, indicating the completion of the formation of a three-dimensional network of V_2O_5 hydrogel. The 3.5 μl aliquots of HVO_3 solution were taken at 0, 30, 45, 60, 90, 120, 360 min and 1, 2 and 3 weeks to monitor the process of initializing, nucleating and growing the ribbons for the V_2O_5 xerogel (the time at 0 min refers to the time when about 5 ml of the HVO_3 solution came out from the ion exchange column). The advantage of cryo-TEM is that it can directly observe the microgeometry and the morphology of particles within a liquid without disturbance by fast freezing the liquid sample using liquid N_2, which preserves the morphology and microgeometry of the particles in the original liquid. We have successfully used cryo-TEM in our previous work to perform similar characterizations[44].

The nucleation immediately (0 min) occurred once the HVO_3 solution was formed (right after $NaVO_3$ passed through the ion exchange column) (Fig. 3a). Then, small V_2O_5 ribbons started to grow into 100 nm long ribbons with diameters of a few nanometres, as shown in the 60-min image (Fig. 3b). The V_2O_5 ribbons grew more rapidly in one direction, with the shorter dimension never growing larger than those observed after 90 min (Fig. 3c). Finally, after 2 weeks, the V_2O_5 hydrogel network was formed with a similar length but much larger width than the V_2O_5 ribbons (Fig. 3d). It is very interesting to see that when the GO was added into the HVO_3 solution, the V_2O_5 hydrogel formation took place on the surface of the GO sheets. First, the solution nucleated, small ribbons formed, the ribbons grew and finally, the V_2O_5 hydrogel network formed. On adding the GO solution into the HVO_3 solution, the nuclei that formed in the beginning tended to adsorb on the GO sheets due to the Coulombic interaction and van der Waals force between the nuclei and the GO.

In contrast to pure V_2O_5, the nucleation of the V_2O_5 hydrogel in the presence of GO sheets took a much longer time: after

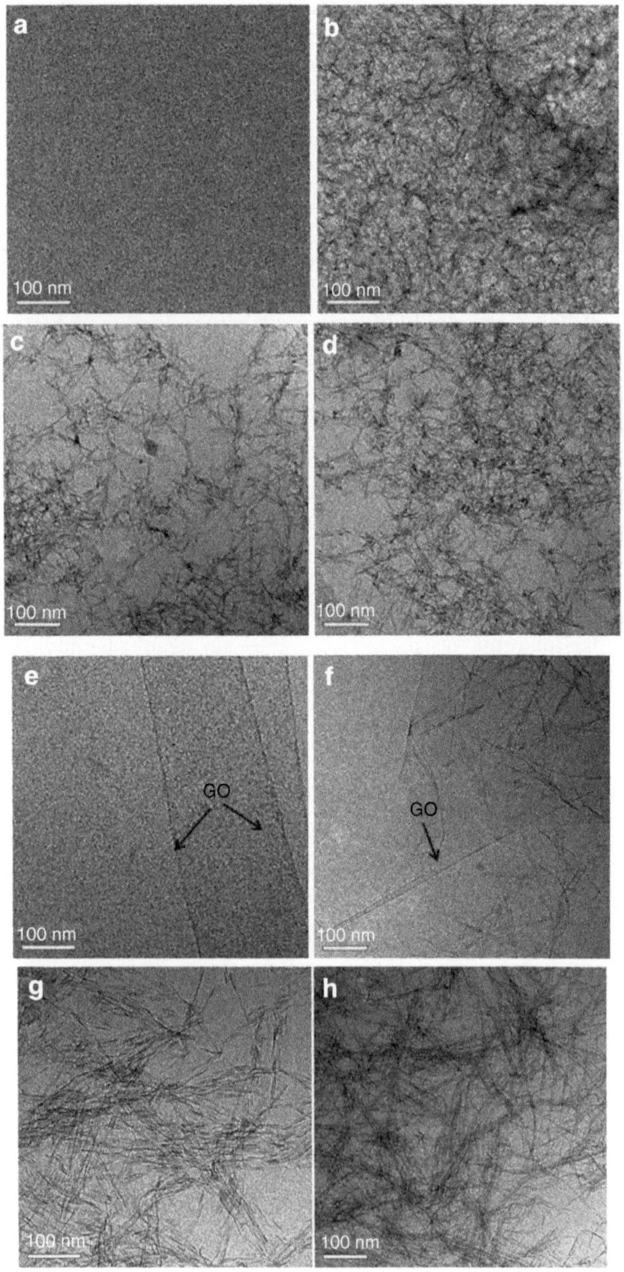

Figure 3 | Cyro-TEM images of the frozen solution during different aging processes. For pure HVO3: (**a**) nucleation (0 min), (**b**) vanadium oxide ribbon growth (1 h), (**c**) continuous growth of V_2O_5 ribbons (1 h 30 min.) and (**d**) fully grown vanadium oxide ribbons (2 weeks). For the mixture of HVO_3 and the GO solution: (**e**) nucleation (30 min), (**f**) vanadium oxide ribbon growth (2 h), (**g**) continuous growth of V_2O_5 ribbons (6 h) and (**h**) fully grown vanadium oxide ribbons (3 weeks)

30 min, the nucleation started (Fig. 3e), which probably was due to the repulsive effect from some regions of the GO surface having negative charges for the different functional groups (that is, phenol, carbonyl, ketone, etc.)[45]. After 120 min, very few pieces of the V_2O_5 ribbons could be seen (Fig. 3f). Even after 6 h, the V_2O_5 ribbons continuously grew, but with much lower density than pure V_2O_5 (Fig. 3g). Finally, it took 3 weeks to form the fully grown V_2O_5 hydrogel network (Fig. 3h). However, the V_2O_5 ribbons in the fully grown V_2O_5 hydrogel on the GO surface looked more uniform, had a much smaller range of width, and were much less densely arranged than the pure V_2O_5 hydrogel

(compared in Fig. 3d,h). This must be attributed to the existence of the GO sheets, which provides a substrate for the V_2O_5 hydrogel formation but at a much slower rate. It appears that this facilitates crystal growth rather than nucleation due to the negative charge repulsion. Thus, the V_2O_5 ribbons are formed with a smaller width and are much less densely arranged over the GO surface. This leads to the smaller grain size of the V_2O_5 ribbons, which correlates to improved Li^+ diffusion as indicated by AC impedance results and to reduced lattice stress as evidenced by the increased cycle life. The less densely arranged V_2O_5 ribbons over the GO surface also result in more gaps between the V_2O_5 ribbons, providing more surface area for Li^+ diffusion into the ribbons. Clearly, the graphene sheets serve as a substrate for the V_2O_5 ribbons and lead to the tremendous increase in electric conductivity. On the other hand, the V_2O_5 formation over the GO surface requires that the V_2O_5 anchor on the GO first, which makes the overall V_2O_5 ribbon formation take a much longer time than it does in the liquid. This GO-aiding synthesis led to a uniform distribution of the GO sheets within the V_2O_5 nanoribbons, as evidenced by the scanning electron microscopy images and energy-dispersive X-ray spectroscopy mapping (Supplementary Fig. 2). The proposed synthesis scheme of a V_2O_5/GO hybrid is shown in Fig. 4, and, for clarity, only the V_2O_5 on one side of the GO sheet was drawn. As the GO is a single layer, the GO sheets act as spacers to create gaps between the formed V_2O_5 nanoribbons once the water is removed from the V_2O_5 hydrogel by heating. In other words, the V_2O_5 nanoribbons may be sandwiched between the layers of graphene after annealing. Such V_2O_5 nanoribbons with a 5–10 nm diameter were anchored or sandwiched between the graphene sheets, as clearly seen in Fig. 5a (indicated by the yellow dashed circle region). In addition, a low-voltage, aberration-corrected high-resolution TEM image of the synthesized V_2O_5–G in Fig. 5b clearly shows the V_2O_5 nanoribbons sandwiched in between the graphene sheets. The electron energy loss spectroscopy (EELS) data (Supplementary Fig. 3) shows that the oxidation state of vanadium in the V_2O_5–G hybrid remained unchanged, indicating that the intercalation of the GO/graphene sheets into the V_2O_5 layers did not cause the change in the oxidation state of the vanadium.

In operando HEXRD characterization of V_2O_5–G. To further understand the structure of V_2O_5–G, in operando HEXRD was measured for both V_2O_5–G and pure V_2O_5 during the heating process. The results for pure V_2O_5 are shown in Fig. 6a. Initially, the sample showed the layered hydrated V_2O_5 *(00l)* reflections, a typical disordered layer structure. The layered structure was maintained until about 200 °C; then, the *(00l)* reflection shifted slowly to a higher 2-theta angle. The shift was primarily caused by the loss of water molecules between the V_2O_5 layers[16], resulting in the shortening of the interlayer spacing between the V_2O_5 layers. The transition from the disordered layer phase to the crystalline phase started around 200 °C, (Fig. 6a; inset): the newly emerged peak, around 1.42°, is attributed to the orthorhombic V_2O_5 (JCPDS No, 41-1426). The loss of water from in between the V_2O_5 layers correlates with the observed phase transformation. As the temperature continued to increase, the peak intensity of the crystalline V_2O_5 rapidly increased at the expense of the relative peak intensity of the layer hydrated

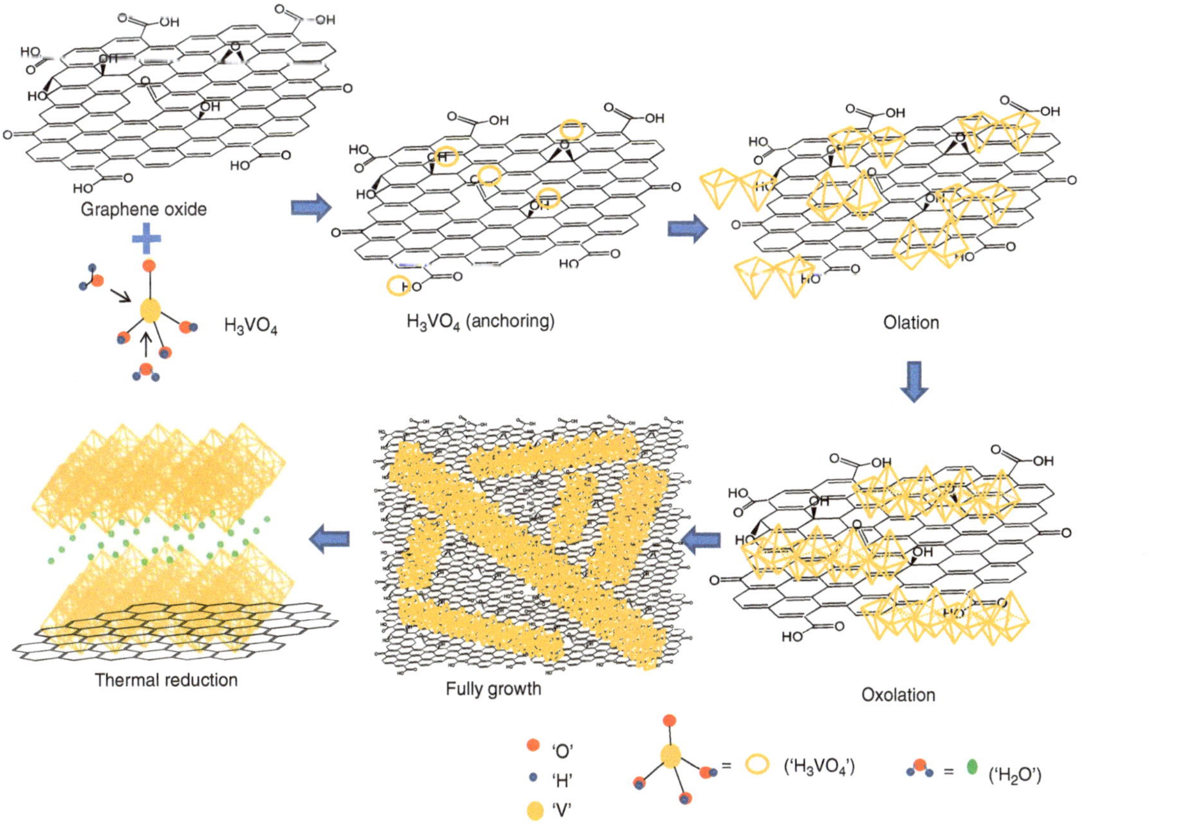

Figure 4 | A proposed synthesis scheme of V_2O_5/graphene hybrids. Olation: each $VO(OH)_3(OH)_2$ molecule is bridged by OH group to form a polymer chain; oxolation: slower oxolation reactions convert unstable 2(OH)1 (2(OH)1 repents a OH group bridges two $VO(OH)_3(OH2)_2$ molecules) into stable 3(O)1 bridges (an oxygen connects three $VO(OH)_3(OH2)_2$ molecules, forming double chains). Further oxolation involving the last OH groups link these double chains together so that V_2O_5 nanofibres formed

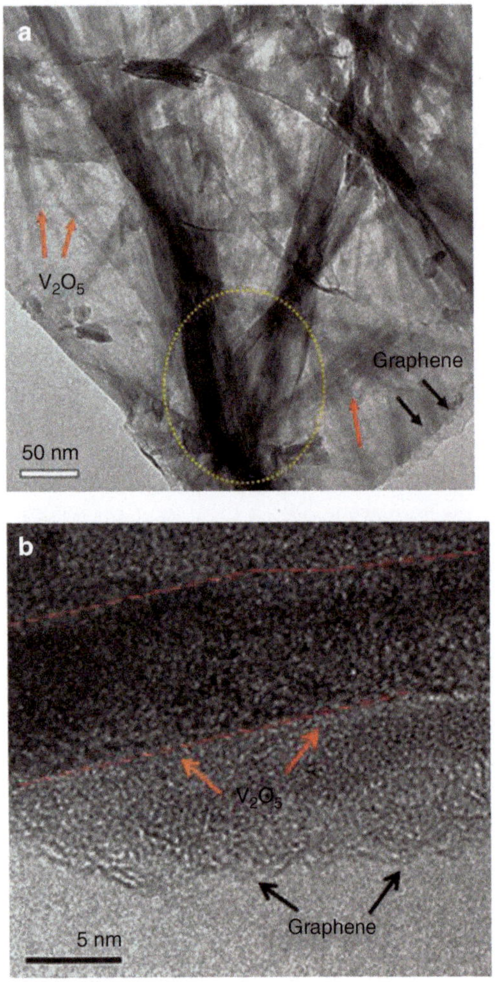

Figure 5 | TEM characterization of the V_2O_5-G hybrids. (**a**) TEM of as-synthesized V_2O_5-G hybrid after calcination. (Note: in panels, the black arrows point to the graphene sheets, the red arrows point to the V_2O_5 nanoribbons and the yellow dashed circle points out the area where the graphene and nanoribbons twist together.). (**b**) HRTEM of as-synthesized V_2O_5-G hybrid after calcination. (Note: the red dashed line points out the area where a V_2O_5 nanoribbon is located).

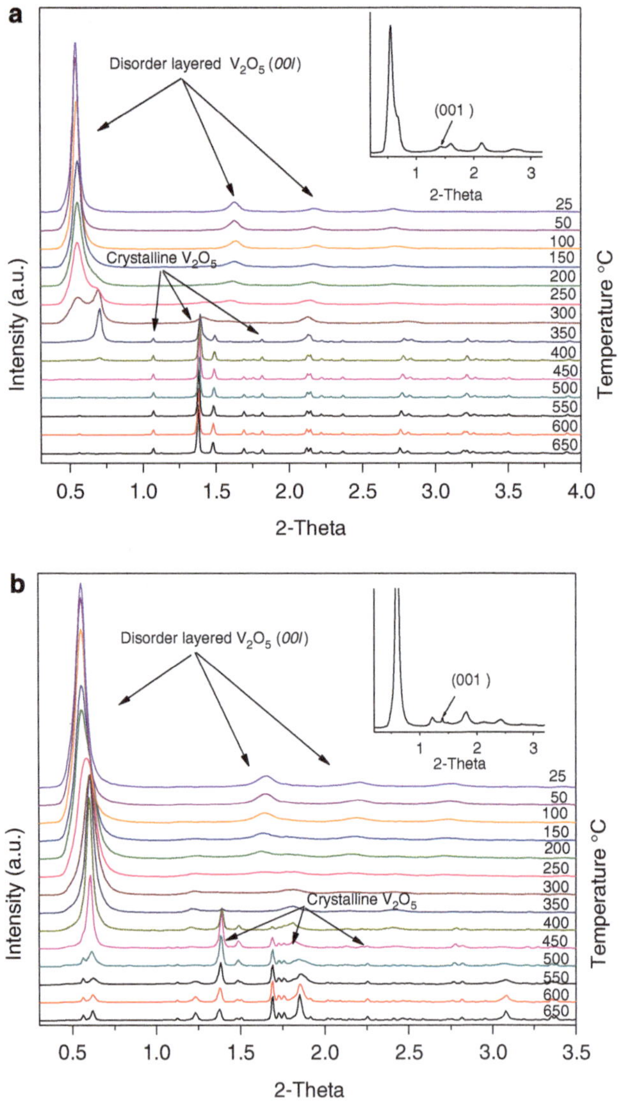

Figure 6 | *In operando* HEXRD characterization of pure V_2O_5 and V_2O_5-G. (**a**) XRD patterns of the V_2O_5 xerogel during the heat treatment process between room temperature and 600 °C, showing that the bi-pyramid structure completely collapses at around 400 °C and (**b**) XRD patterns of V2O5-G with graphene during the heat treatment process between room temperature and 600 °C, showing that the bi-pyramid structure persists until 450 °C

structure (amorphous). Finally, at around 400 °C, the amorphous phase almost completely diminished and the phase transformation was completed. The V_2O_5 xerogel structure collapsed due to the complete removal of crystal water from the V_2O_5 on heating to 400 °C; then, the V_2O_5 xerogel completely transformed into a V_2O_5 nanocrystal, which shows three distinct discharge stages (Fig. 1a,b).

It is interesting to note that the graphene had a significant impact on the structure of the V_2O_5 hybrid. Initially, the V_2O_5–G showed a layered hydrated structure similar to that of pure V_2O_5 but with larger interlayer spacing (*d*-spacing), indicating that the intercalation of the GO sheets leads to increased interlayer spacing between the V_2O_5 layers. At room temperature, the *d*-spacing of the V_2O_5–G is about 13.65 Å, which is much higher than that of the pure V_2O_5, 11.40 Å. Similar XRD results on the intercalation of polymers between V_2O_5 layers in xerogel have been observed, suggesting that the interlayer spacing increased to between ~12.0 and ~15.0 Å, depending on the type of the polymer[46–49]. As the temperature increased, the (*00l*) reflection shifted to a higher 2-theta angle in a manner similar to that of the pure V_2O_5 sample, but at a much slower rate. Unlike the pure V_2O_5, which transitioned from the disordered layer phase to the crystal phase starting around 200 °C, the phase transition for V_2O_5–G started around 400 °C as indicated by the emerging peak at 1.42° (Fig. 6b). Clearly, with the presence of graphene, the disordered-to-crystalline phase transition was delayed. Obviously, in analogy to the pure V_2O_5, for the hybrid, the thermal stability will be greatly enhanced when the V_2O_5 layer is affixed to the graphene sheets.

For the electrochemical performance testing, the obtained V_2O_5–G was annealed at 400 °C in N_2 before being used as the cathode. The HEXRD data in Fig. 6b show that the calculated V_2O_5 interlayer spacing, *d*-spacing, was 10.2 Å, corresponding to 0.3 water per V_2O_5, $V_2O_5 \cdot 0.3H_2O$ (refs 8,16). However, for the pure V_2O_5, there was almost no water left at 400 °C and the structure completely changed to a nanocrystal instead of the layered V_2O_5 xerogel. Crystallized water inside the V_2O_5 xerogel functions as 'pillars' to keep the interlayer spacing fixed between the two V_2O_5 sheets. For the V_2O_5 xerogel, the large amount of water may react with the lithium to form Li_2O, which deteriorates intercalation performance[8]. Without water molecules to act as the pillars, the V_2O_5 network in the disordered layer V_2O_5 xerogel will collapse during annealing and become crystalline V_2O_5, which has much lower electrochemical performance. Hence, a minimum amount of crystal water is needed to keep the disordered layer phase. Thus, the graphene-sheet-modified V_2O_5 layers do increase the thermal stability and preserve the disordered layer phase even at 400 °C with 0.3 water per V_2O_5.

Thermogravimetric analysis (TGA) was also carried out for both pure V_2O_5 and V_2O_5–G to study the structure change during annealing, and the results are shown in Fig. 7. The pure V_2O_5 xerogel had a rapid weight loss (0.15%/°C) until 80 °C, followed by a gradual weight loss at a much slower rate (0.018%/°C) up to 300 °C, which corresponds to the loss of weakly bonded water molecules in the V_2O_5 xerogel[23]. As temperatures went to beyond 200 °C, the tightly bonded water molecules—crystallized water molecules—were removed, and the phase conversion from the disordered layer phase to the orthorhombic phase started[16]. This is consistent with our HEXRD results. Compared with the pure V_2O_5, the V_2O_5–G showed a completely different profile; it had a gradual weight loss at a rate of 0.024%/°C until 250 °C, which is characteristic of the loss of weakly bonded water from the V_2O_5–G. This was followed by another gradual weight loss with a slightly faster slope (0.05%/°C) until 450 °C. In terms of the crystal structure, the bi-pyramid structure persisted below 450 °C, although there were indications of the nucleation of the orthorhombic V_2O_5 phase at around 400 °C. The TGA results further verified that the thermal stability of the V_2O_5 was greatly enhanced by the presence of graphene.

In operando XANES analysis. In operando XANES characterization of V_2O_5–G during the electrochemical process was carried out at 20-BM-B, Advanced Photon Source. The two-dimensional (2D) contour plots of in operando XANES spectra for the V_2O_5–G are shown in Fig. 8a. In addition, the voltage profile is displayed on the side of the XANES data for comparison. Obviously, the electrochemical process correlates well with the evolution of the XANES data. Typical vanadium XANES during the first discharge process is shown in Fig. 8b and illustrates the insertion of lithium ions into the V_2O_5–G. On the insertion of the lithium ions, the K-edge of the vanadium continuously shifted towards the lower binding energy. This shift indicates a decrease in the average valence state of vanadium, which corresponds to the increase in the number of inserted Li^+ ions in each V_2O_5.

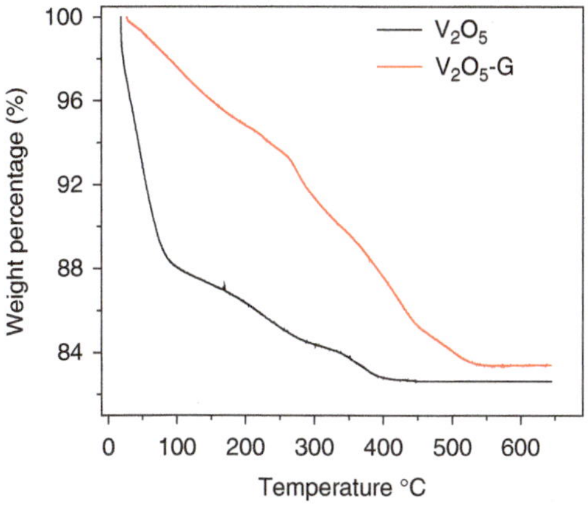

Figure 7 | Thermogravimetric analysis. Comparison of TGA curves between pure V_2O_5 xerogel and V_2O_5-G.

At the fully discharged state, the emerging shoulder (labelled by the an arrow) was found, which is consistent with the reduction of V(IV) to V(III)[50]. During the charge process after the first discharge, the K-edge of the vanadium shifted towards the higher binding energy (Supplementary Fig. 4).

Besides the K-edge, there are two other important features in the XANES region: the pre-edge (labelled 1 in Fig. 8b) and the edge resonance (labelled 2 in Fig. 8b), which are also sensitive to the valence state of the vanadium and the chemical environment surrounding the vanadium site[51]. As the intensity of the pre-edge peak was consistent with the deviation from the octahedral symmetry of the vanadium site[52], the intensity of the pre-edge decreased and the peak '1' position shifted towards the lower energy during the discharge process (Fig. 8b). The edge resonance, which is related to the energy absorption by core electrons, showed major changes in both intensity and shape[53]. During the discharge process, the shape changed from two well-defined peaks to a single peak (peak '2' in Fig. 8b) and the intensity increased, while during the charge process, the intensity decreased and the shape changed from a single peak back to two well-defined peaks (Supplementary Fig. 4). A similar reversible behaviour for the pre-edge, the edge and the edge resonance was also observed by Smyrl *et al.*[54] on V_2O_5 xerogel cathodes.

To further establish the intensity change and energy shift of the pre-edge, the edge and the edge resonance, a combined Gaussian-arctangent fit was performed[55], and the results are shown in Fig. 8c. As discussed earlier, the V_2O_5 gel-based materials consist of distorted, square-based pyramids with four basal oxygen atoms and one apical oxygen atom[1]. The intensity of the pre-edge directly correlates to the distortion of the V_2O_5 square pyramid, while the intensity of the edge resonance is associated with the symmetry of the basal oxygen configuration around the vanadium site[53]. Obviously, the intensity of the pre-edge peak and the edge resonance (Fig. 8c, inset) peak showed major changes but exhibited a linear decrease. Importantly, the intensities during the charging process also changed linearly, which confirms the structural reversibility of the hybrid. The XANES K-edge and pre-edge shift as a function of Li composition in the V_2O_5–G is shown in Fig. 8d. Obviously, both the K-edge and pre-edge energy shifts (Fig. 8d, inset) varied linearly with the valence of the vanadium atom. Importantly, the energy shifts during the charge process also showed a linear change, which confirms the electronic reversibility of the hybrid. In summary, our in operando XANES measurements clearly reveal the electronic and structural reversibility of the V_2O_5–G during the lithium insertion and de-insertion process. On the other hand, the excellent performance of the V_2O_5–G can be ascribed to the reversibility of the intercalation process and preserved single phase of the composite, as can be seen in the in operando HEXRD results (Fig. 6b). Compared with the crystalline V_2O_5, the stabilized layered structure will undoubtedly allow easier lattice expansion and contraction during the discharge/charge process without any major distortion of the V_2O_5 hybrid structure.

Discussion

To date, graphene has been considered an ideal conducting material to improve the electrical conductivity and enhance the structure of V_2O_5[10,43,56,57]. However, most composites have been created by simply mixing graphene with the V_2O_5. Such simple physical mixing usually requires high graphene loadings (for example, 30% graphene[56]), which leads to a significant improvement on the cycle life and rate performance, but with a heavy penalty on the specific capacity. In this work, a method of creating V_2O_5–G via the sol–gel process has been developed, and

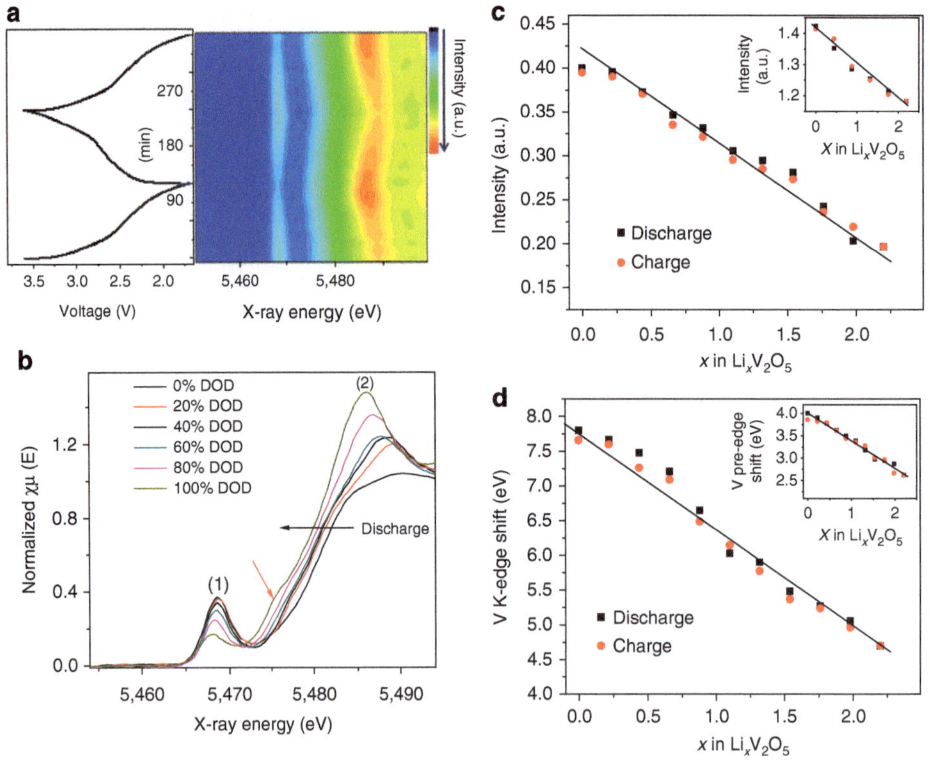

Figure 8 | In operando XANES characterization of the Li/V_2O_5-G coin cells during discharge/charge cycling. (**a**) Normalized XANES spectra (2D contour plot) and voltage profiles of Li/V_2O_5-G coin cells during cycling. (**b**) Typical vanadium K-edge XANES spectra during the first discharge process. (**c**) XANES pre-edge peak and main peak (inset) intensities as a function of lithium composition during the first charge/discharge process. (**d**) V K-edge energy (at half weight) shift and pre-edge shift (inset) as a function of lithium composition during the first charge/discharge process. The Li content for the lithiated samples (that is, x in $Li_xV_2O_5$) was calculated from the electrochemical results. The energy shift uses vanadium as the reference

we have demonstrated that although a small amount of graphene is used, there is a profound effect on the structure. This structural change consequently results in extraordinary electrochemical performance without a heavy penalty on specific capacity; in fact, the material nearly achieves the theoretical specific capacity. We clearly demonstrated through our in operando HEXRD that the graphene sheets help to preserve the V_2O_5 layered structure and keep the xerogel from collapsing to form the orthorhombic crystalline phase during the annealing process. In addition, the AC impedance proves that the electric conductivity, the vanadium redox reaction and the Li^+ diffusion have been improved due to the introduction of such a small amount of graphene.

The performance of electrode materials is always rooted in the materials structure[58,59]. The graphene sheets sandwiched in between the V_2O_5 layers not only enhance the electron conduction between the V_2O_5 layers (intraparticle conduction, supported by TEM and XRD results) and between the V_2O_5 ribbons (inter-particles conduction, supported by Cryo-TEM results), but also preserve the water molecules between the two layers of V_2O_5, which facilitates the Li^+ diffusion (as can be seen from the AC impedance results). The combined effect on electronic conduction and Li^+ diffusion significantly improves the electrochemical performance. Furthermore, the presence of the graphene sheets between the V_2O_5 layers enhances the structural integrity of the disordered V_2O_5-G during cycling, resulting in the improved reversibility of the intercalation process and in the preserved single phase of the hybrid composite, which in turn, leads to a significantly improved cycle life.

In summary, we have demonstrated a novel and simple method to incorporate graphene sheets into the nanostructure of V_2O_5 xerogel via a sol–gel process, forming a graphene-modified V_2O_5 hybrid with a stabilized layered structure. The reversible charge/discharge of 438 mAh g^{-1} capacity indicated that the three Li^+ ions have been successfully inserted/desinserted to the V_2O_5/graphene hybrid. The introduction of a small amount of graphene into the V_2O_5 can effectively alter the structure of the nanocomposite, resulting in significant improvements in electric conductivity, structural stability and ion diffusion. This, in turn, results in extraordinary electrochemical performance of the graphene-modified V_2O_5 hybrid. The hybrid nearly reached the theoretical specific capacity and demonstrated excellent rate performance and greatly enhanced cycle life. This method provides a new avenue to create nanostructured materials with improved properties for metal oxides as long as they can be synthesized via the sol–gel process or by a reaction in a solution. The sol–gel process and the solution method are both easy for scale-up, and the starting materials are abundant, which makes widespread industrial application of these new materials feasible.

Methods

Synthesis of V_2O_5 xerogel. The V_2O_5 xerogel was prepared by a simple, modified ion exchange method[23,48,60,61]. A 0.1 M solution of $NaVO_3$ (Sigma, >99.5%) was eluted through a column loaded with a proton-exchange resin (Dowex-50-WX2, 50–100 mesh). The obtained yellow solution of HVO_3 was aged in a glass container for 2 weeks to obtain a mature homogeneous vanadium oxide hydrogel. Dried xerogel was obtained by freeze drying the hydrogel under vacuum.

Synthesis of GO. GO was prepared using two methods. First, 2 g of graphite flakes were mixed with 10 ml of concentrated H_2SO_4, 2 g of $K_2S_2O_8$ and 2 g of P_2O_5. The obtained mixture was heated at 80 °C for 4 h with constant stirring. Then the mixture was filtered and washed thoroughly with deionized (DI) water. After drying in an oven at 80 °C overnight, this pre-oxidized graphite was then subjected to oxidation using Hummer's method[45]. Next, 2 g of pre-oxidized graphite, 1 g of sodium nitrate and 46 ml of sulfuric acid were mixed and stirred for 15 min in an ice bath. Then, 6 g of potassium permanganate was slowly added to the obtained suspension solution for another 15 min. After that, 92 ml DI water was slowly added to the suspension, while the temperature was kept constant at about 98 °C for 15 min. After the suspension was diluted by 280 ml DI water, 10 ml of 30% H_2O_2 was added to reduce the unreacted permanganate. Finally, the resulting suspension was centrifuged several times to remove the unreacted acids and salts. The purified GO was dispersed in DI water by sonication for 1 h to form a 0.2 mg ml^{-1} solution. Then the GO dispersion was subjected to another centrifugation to remove the un-exfoliated GO. The resulting GO diluted solution could remain a very stable suspension without any precipitation for a few months[27].

Synthesis of a graphene-modified nanostructured V_2O_5 hybrid. The hybrid was prepared by mixing the prepared GO suspension and the yellow solution of HVO_3 at the desired ratio. The obtained dark yellow solution was aged in a glass container for 3 weeks to obtain a completely cured, homogeneous V_2O_5/GO hydrogel. The dried V_2O_5/GO xerogel was obtained by freeze drying the V_2O_5/GO hydrogel in a vacuum. The formed V_2O_5/GO xerogel was heated and annealed under N_2 at a rate of 5 °C min^{-1} up to 400 °C and kept constant at 400 °C for 6 h, during which the GO was reduced to graphene and the graphene-modified V_2O_5 hybrid was formed.

Electrochemical characterization. The electrodes were prepared by casting a slurry of 80% V_2O_5–G, 10% polyvinylidence difluoride, and 10% carbon black onto a 10-μm-thick sheet of Al foil. For comparison, the pure $V_2O_5 \cdot nH_2O$ was synthesized using the same conditions except for the addition of the GO. The corresponding electrodes were prepared using the same procedure. The prepared electrodes were placed in a vacuum oven and allowed to dry at 90 °C for 24 h. The electrolyte consisted of a solution of 1.2 M $LiPF_6$ (Novolyte, Ohio, USA) in a mixture of solvent from ethylene carbonate and ethyl methyl carbonate (EMC; 3:7, by weight; Novolyte, Ohio, USA).

The prepared V_2O_5–G and V_2O_5 were assembled into 2016-type coin cells using Li metal anodes (Aldrich, USA) and Celgard 3501 separators (Celgard, Ohio, USA) for characterizing their electrochemical performance. These cells were tested with an Arbin Battery cycler (BT-2000, Arbin, TX, USA) using different C-rates between 1.5 and 3.6 V. A Solartron 1287A/1260A Potentiostat/Impedance System (Solartron Analytical, England, UK) was used to measure the AC impedance of these cells in the frequency range of 0.01 Hz–1 MHz with an amplitude of 5 mV. The results were fitted using the model[28], where R_0 is the contact resistance, R_e and C_e refer to the resistance and capacitance of the V_2O_5 electrode, R_{ct} and C_{dl} stand for the charge-transfer resistance of the redox reaction of vanadium in V_2O_5 and the double-layer capacitance in the electrode, respectively, and W_d refers to the Warburg diffusion impedance, which could reflect the diffusion of Li ions in the V_2O_5.

Materials characterization. Scanning electron microscopy and energy-dispersive X-ray spectroscopy characterization were performed with a JEOL 7800 (Japan) operated at 5 kV. Standard high-resolution TEM characterization was performed with a JEOL 2100F (Japan) operated at 200 kV. Low-voltage (80 kV), aberration-corrected high-resolution TEM and EELS were conducted using a Titan 80/300 TEM. Cryogenic temperature TEM analysis was carried out for the synthesized V_2O_5 hydrogel solutions with and without GO aged at different times to elucidate the formation mechanism. For the cryo-TEM, a 3.5-μl aliquot of the aged solution samples was placed on a copper grid (400 mesh) coated with a perforated/porous carbon film. The excess solution was blotted off with filter paper. The grid was then immediately plunged into liquid ethane cooled by liquid N_2. After that, the sample grid was loaded into the microscope with a Gatan side-entry cryogenic holder. Low-dose images were collected using a CM 200 and CM 300 cryomicroscope with a field emission gun operating at 200 and 300 kV, respectively. The TGA was performed for both pure V_2O_5 and V_2O_5–G using a TA 2000 thermoanalyzer.

Temperature-dependent in operando HEXRD measurements during the annealing processes were performed on the beam line 11-ID-C at Advanced Photon Source, Argonne National Laboratory. A monochromator with a Si (311) single crystal was used to provide an X-ray beam with 115 keV of energy. A high-energy X-ray with a beam size of 0.2 × 0.2 mm and wavelength of 0.108 Å was used to obtain 2D diffraction patterns in the transmission geometry. X-rays were collected with a Perkin-Elmer large-area detector placed at 1,800 mm from the sample. The synthesized pure V_2O_5 and V_2O_5–G were dried at 80 °C overnight and were pressed into pellets about 1-mm thick. The pellet was placed between a ceramic can and a platinum cover with a hole (D = 1 mm) in the centre of both the can and the cover. After that, the ceramic can was placed vertically in a programmable furnace with glass windows, and nitrogen was used as the protective gas. The sample was heated up to 600 °C with a heating rate of 2 °C per minute. The diffraction data from the sample was collected every 34 s. The obtained 2D diffraction patterns were calibrated using a standard CeO_2 sample and converted to one-dimensional patterns using Fit2D software[62].

Synchrotron radiation-based in operando X-ray absorption spectroscopy can be very informative when applied to complex battery systems[63]. Particularly, the in operando XANES technique is very powerful for determining the average oxidation state and local structure of the elements in all phases. With these considerations in mind, we carried out in operando XANES measurements during the cycling of coin cells that contained the V_2O_5–G electrode on the beam line 20-BM-B at Advanced Photon Source, Argonne National Laboratory. Before the in operando synchrotron experiment, the XANES spectrum of the reference (commercial V_2O_5 powder) was taken for comparison. As can be seen in Supplementary Fig. 5, the XANES spectrum comparison between the commercial V_2O_5 and the V_2O_5–G overlapped very well, suggesting that the intercalation did not cause a significant change in the valence state of the vanadium, which is consistent with the EELS results. The Li/V_2O_5 coin cells with holes ($D = 2$ mm) at the centre were assembled for the in operando XANES study. The holes were sealed using Kapton tapes, which allowed the X-rays through but prevented air from entering the cell. In operando XANES was performed at the vanadium K-edge to monitor the change of the valence state of the vanadium in the cathode material. The XANES measurements were carried out in transmission mode at the beamline 20-BM-B of the APS using a Si (111) monochromator. Energy calibration was performed by using the first derivative point of the XANES spectrum for the V (K-edge = 5463.76 eV). In the meantime, the reference spectra were collected for each in operando spectrum where vanadium metal was used in the reference channel. The coin cells with the exact same electrodes were also used because of the better signal-to-noise ratio. All cells were charged/discharged with a constant current of about 0.5 C between 1.7 and 3.6 V, while the XANES spectra data were collected every 262 s. The XANES data reduction and analysis followed standard methods using the ATHENA software package.

References

1. Whittingham, M. S. Lithium batteries and cathode materials. *Chem. Rev.* **104**, 4271–4302 (2004).
2. Alcantara, R., Lavela, P., Tirado, J., Stoyanova, R. & Zhecheva, E. Structure and electrochemical properties of boron-doped $LiCoO_2$. *J. Solid State Chem.* **134**, 265–273 (1997).
3. Huang, H., Yin, S.-C. & Nazar, L. F. Approaching theoretical capacity of LiFePO4 at room temperature at high rates. *Electrochem. Solid-State Lett* **4**, A170–A172 (2001).
4. Cao, A. M., Hu, J. S., Liang, H. P. & Wan, L. J. Self- assembled vanadium pentoxide (V2O5) hollow microspheres from nanorods and their application in lithium-ion batteries. *Angew. Chem. Int. Ed.* **44**, 4391–4395 (2005).
5. Mai, L. *et al.* Electrospun ultralong hierarchical vanadium oxide nanowires with high performance for lithium ion batteries. *Nano Lett.* **10**, 4750–4755 (2010).
6. Pan, A. *et al.* Facile synthesized nanorod structured vanadium pentoxide for high-rate lithium batteries. *J. Mater. Chem.* **20**, 9193–9199 (2010).
7. Mai, Y. J. *et al.* CuO/graphene composite as anode materials for lithium-ion batteries. *Electrochim. Acta* **56**, 2306–2311 (2011).
8. Wang, Y., Takahashi, K., Lee, K. H. & Cao, G. Nanostructured vanadium oxide electrodes for enhanced lithium-ion intercalation. *Adv. Funct. Mater.* **16**, 1133–1144 (2006).
9. Gu, L. *et al.* Direct observation of lithium staging in partially delithiated lifepo4 at atomic resolution. *J. Am. Chem. Soc.* **133**, 4661–4663 (2011).
10. Li, Z.-F. *et al.* Hierarchical nanocomposites of vanadium oxide thin film anchored on graphene as high-performance cathodes in li-ion batteries. *ACS Appl. Mater. Interfaces* **6**, 18894–18900 (2014).
11. Cox, P. A. *Transition Metal Oxides: An Introduction to Their Electronic Structure and Properties* (Clarendon Press, 1992).
12. Liu, Y. *et al.* Facile synthesis of nanostructured vanadium oxide as cathode materials for efficient Li-ion batteries. *J. Mater. Chem.* **22**, 24439–24445 (2012).
13. Park, M.-S. *et al.* Preparation and electrochemical properties of sno2 nanowires for application in lithium-ion batteries. *Angew. Chem. Int. Ed.* **119**, 764–767 (2007).
14. Rolison, D. R. & Dunn, B. Electrically conductive oxide aerogels: new materials in electrochemistry. *J. Mater. Chem.* **11**, 963–980 (2001).
15. Petkov, V. *et al.* Structure of V2O5 n H2O xerogel solved by the atomic pair distribution function technique. *J. Am. Chem. Soc.* **124**, 10157–10162 (2002).
16. Wang, J., Curtis, C. J., Schulz, D. L. & Zhang, J.-G. Influences of treatment temperature and water content on capacity and rechargeability of V_2O_5 xerogel films. *J. Electrochem. Soc.* **151**, A1–A7 (2004).
17. Sakamoto, J. S. & Dunn, B. Vanadium oxide-carbon nanotube composite electrodes for use in secondary lithium batteries. *J. Electrochem. Soc.* **149**, A26–A30 (2002).
18. Coustier, F., Jarero, G., Passerini, S. & Smyrl, W. H. Performance of copper-doped V_2O_5 xerogel in coin cell assembly. *J. Power Sources* **83**, 9–14 (1999).
19. Harreld, J. H., Dunn, B. & Nazar, L. F. Design and synthesis of inorganic–organic hybrid microstructures. *Int. J. Inorg. Mater.* **1**, 135–146 (1999).
20. Huguenin, F., do Prado Gambardella, M. T., Torresi, R. M., de Torresi, S. I. C. & Buttry, D. A. Chemical and electrochemical characterization of a novel nanocomposite formed from V_2O_5 and poly (N-propane sulfonic acid aniline), a self-doped polyaniline. *J. Electrochem. Soc.* **147**, 2437–2444 (2000).

21. Leroux, F., Goward, G., Power, W. & Nazar, L. Electrochemical Li insertion into conductive polymer/V_2O_5 nanocomposites. *J. Electrochem. Soc.* **144,** 3886–3895 (1997).
22. Lira-Cantú, M. & Gómez-Romero, P. The organic-inorganic polyaniline/V_2O_5 system. application as a high-capacity hybrid cathode for rechargeable lithium batteries. *J. Electrochem. Soc.* **146,** 2029–2033 (1999).
23. Livage, J. Synthesis of polyoxovanadates via "chimie douce". *Coord. Chem. Rev.* **178,** 999–1018 (1998).
24. Lui, C. H., Liu, L., Mak, K. F., Flynn, G. W. & Heinz, T. F. Ultraflat graphene. *Nature* **462,** 339–341 (2009).
25. Allen, M. J., Tung, V. C. & Kaner, R. B. Honeycomb carbon: a review of graphene. *Chem. Rev.* **110,** 132–145 (2009).
26. Cocciantelli, J. *et al.* On the $\delta\rightarrow\gamma$ irreversible transformation in Li//V_2O_5 secondary batteries. *Solid State Ionics* **78,** 143–150 (1995).
27. Li, Z.-F. *et al.* Fabrication of high-surface-area graphene/polyaniline nanocomposites and their application in supercapacitors. *ACS Appl. Mater. Interfaces* **5,** 2685–2691 (2013).
28. Liu, Y. *et al.* Failure study of commercial LiFePO4 cells in over-discharge conditions using electrochemical impedance spectroscopy. *J. Electrochem. Soc.* **161,** A620–A632 (2014).
29. Almeida, E., Abbate, M. & Rosolen, J. Enhancement of the electrochemical performance of a Li-intercalated V_2O_5 xerogel doped with Eu. *Solid State Ionics* **160,** 61–67 (2003).
30. Boyano, I. *et al.* Improvement in the Ppy/V_2O_5 hybrid as a cathode material for Li ion batteries using PSA as an organic additive. *J. Power Sources* **166,** 471–477 (2007).
31. Cheah, Y. L. *et al.* Morphology, structure and electrochemical properties of single phase electrospun vanadium pentoxide nanofibers for lithium ion batteries. *J. Power Sources* **196,** 6465–6472 (2011).
32. Coustier, F., Jarero, G., Passerini, S. & Smyrl, W. H. Performance of copper-doped V_2O_5 xerogel in coin cell assembly. *J. Power Sources* **83,** 9–14 (1999).
33. Coustier, F., Passerini, S. & Smyrl, W. Dip-coated silver-doped V_2O_5 xerogels as host materials for lithium intercalation. *Solid State Ionics* **100,** 247–258 (1997).
34. Du, G. *et al.* Graphene–V2O5 · nH2O xerogel composite cathodes for lithium ion batteries. *RSC Adv.* **1,** 690–697 (2011).
35. Yu, D. *et al.* Mesoporous vanadium pentoxide nanofibers with significantly enhanced Li-ion storage properties by electrospinning. *Energy Environ. Sci.* **4,** 858–861 (2011).
36. Ng, S. H. *et al.* Synthesis and electrochemical properties of V_2O_5 nanostructures prepared via a precipitation process for lithium-ion battery cathodes. *J. Power Sources* **174,** 1032–1035 (2007).
37. Pan, A. *et al.* Facile synthesized nanorod structured vanadium pentoxide for high-rate lithium batteries. *J. Mater. Chem.* **20,** 9193–9199 (2010).
38. Rui, X. *et al.* Facile preparation of hydrated vanadium pentoxide nanobelts based bulky paper as flexible binder-free cathodes for high-performance lithium ion batteries. *RSC Adv.* **1,** 117–122 (2011).
39. Tang, Y. *et al.* Vanadium pentoxide cathode materials for high-performance lithium-ion batteries enabled by a hierarchical nanoflower structure via an electrochemical process. *J. Mater. Chem. A* **1,** 82–88 (2013).
40. Tipton, A. L., Passerini, S., Owens, B. B. & Smyrl, W. H. Performance of lithium/V_2O_5 xerogel coin cells. *J. Electrochem. Soc.* **143,** 3473–3477 (1996).
41. Wang, B., Wang, Y., Sun, B., Munroe, P. & Wang, G. Coral-like V_2O_5 nanowhiskers as high-capacity cathode materials for lithium-ion batteries. *RSC Adv.* **3,** 5069–5075 (2013).
42. Wang, S. *et al.* Porous monodisperse V_2O_5 microspheres as cathode materials for lithium-ion batteries. *J. Mater. Chem.* **21,** 6365–6369 (2011).
43. Liu, H. & Yang, W. Ultralong single crystalline V2O5 nanowire/graphene composite fabricated by a facile green approach and its lithium storage behavior. *Energy Environ. Sci.* **4,** 4000–4008 (2011).
44. Xu, F. *et al.* Investigation of a catalyst ink dispersion using both ultra-small-angle X-ray scattering and cryogenic TEM. *Langmuir.* **26,** 19199–19208 (2010).
45. Wang, M.-x. *et al.* Structural modification of graphene sheets to create a dense network of defect sites. *J. Phys. Chem. Lett.* **4,** 1484–1488 (2013).
46. Kanatzidis, M. G. & Huang, S. P. Synthesis of polymeric tetraphenylphosphonium (tetraselenido)argentate(1-) [(Ph4P)AgSe4]n. A novel one-dimensional inorganic polymer. *J. Am. Chem. Soc.* **111,** 760–761 (1989).
47. Kanatzidis, M. G., Wu, C. G., Marcy, H. O., DeGroot, D. C. & Kannewurf, C. R. Conductive polymer/oxide bronze nanocomposites. Intercalated polythiophene in vanadium pentoxide (V_2O_5) xerogels. *Chem. Mater.* **2,** 222–224 (1990).
48. Livage, J. Vanadium pentoxide gels. *Chem. Mater.* **3,** 578–593 (1991).
49. Wu, C. G. *et al.* Redox intercalative polymerization of aniline in V_2O_5 xerogel. The postintercalative intralamellar polymer growth in polyaniline/metal oxide nanocomposites is facilitated by molecular oxygen. *Chem. Mater.* **8,** 1992–2004 (1996).
50. Mansour, A., Smith, P., Baker, W., Balasubramanian, M. & McBreen, J. A comparative in situ X-ray absorption spectroscopy study of nanophase V_2O_5 aerogel and ambigel cathodes. *J. Electrochem. Soc.* **150,** A403–A413 (2003).
51. Liu, Q. *et al.* The structural evolution of V_2O_5 nanocystals during electrochemical cycling studied using in operando synchrotron techniques. *Electrochim. Acta* **136,** 318–322 (2014).
52. Passerini, S. *et al.* XAS and electrochemical characterization of lithium intercalated V_2O_5 xerogels. *Solid State Ionics* **90,** 5–14 (1996).
53. Wong, J., Lytle, F. W., Messmer, R. P. & Maylotte, D. H. K-edge absorption spectra of selected vanadium compounds. *Phys. Rev. B* **30,** 5596–5610 (1984).
54. Giorgetti, M. *et al. In situ* X-ray absorption spectroscopy characterization of V_2O_5 xerogel cathodes upon lithium intercalation. *J. Electrochem. Soc.* **146,** 2387–2392 (1999).
55. Mansour, A. N., Smith, P. H., Baker, W. M., Balasubramanian, M. & McBreen, J. *In situ* XAS investigation of the oxidation state and local structure of vanadium in discharged and charged V_2O_5 aerogel cathodes. *Electrochim. Acta* **47,** 3151–3161 (2002).
56. Du, G. *et al.* Graphene-V_2O_5[middle dot]nH2O xerogel composite cathodes for lithium ion batteries. *RSC Adv* **1,** 690–697 (2011).
57. Qian, Y., Vu, A., Smyrl, W. & Stein, A. Facile preparation and electrochemical properties of V_2O_5-graphene composite films as free-standing cathodes for rechargeable lithium batteries. *J. Electrochem. Soc.* **159,** A1135–A1140 (2012).
58. Wang, J. *et al.* Accurate control of multishelled Co3O4 hollow microspheres as high-performance anode materials in lithium-ion batteries. *Angew. Chem. Int. Ed.* **52,** 6417–6420 (2013).
59. Lai, X., Halpert, J. E. & Wang, D. Recent advances in micro-/nano-structured hollow spheres for energy applications: From simple to complex systems. *Energy Environ. Sci.* **5,** 5604–5618 (2012).
60. Le, D. B., Passerini, S., Tipton, A. L., Owens, B. B. & Smyrl, W. H. Aerogels and xerogels of V_2O_5 as intercalation hosts. *J. Electrochem. Soc.* **142,** L102–L103 (1995).
61. Lee, C.-Y., Marschilok, A. C., Subramanian, A., Takeuchi, K. J. & Takeuchi, E. S. Synthesis and characterization of sodium vanadium oxide gels: the effects of water (n) and sodium (x) content on the electrochemistry of NaxV2O5[middle dot]nH2O. *Phys. Chem. Chem. Phys.* **13,** 18047–18054 (2011).
62. Liu, Q. *et al.* Rate-dependent, Li-ion insertion/deinsertion behavior of LiFePO4 cathodes in commercial 18650 LiFePO4 cells. *ACS Appl. Mater. Interfaces* **6,** 3282–3289 (2014).
63. Cui, Y. *et al.* (De)Lithiation mechanism of Li/SeSx (x = 0–7) batteries determined by *in situ* synchrotron X-ray diffraction and X-ray absorption spectroscopy. *J. Am. Chem. Soc.* **135,** 8047–8056 (2013).

Acknowledgements

We like to express our appreciation for Dr Corinne Renguette's help with editing and Dr Qingliu Wu's help with partially drawing Fig. 4. Use of the Advanced Photon Source was supported by the U.S. Department of Energy, Office of Science, Office of Basic Energy Science, under Contract No. DE-AC02-06CH11357. This research was also carried out in part at the Center for Functional Nanomaterials at Brookhaven National Laboratory (U.S. DOE contract DE-AC02-98CH10886). A patent has been filed for this novel synthesis method and the corresponding V_2O_5–G hybrid materials.

Author contributions

J.X. designed the work and conceived the mechanism. Q.L., Y.L. and Z.-F.L. synthesized the materials. Q.L., Y.L. and Y.Z. tested the materials and analysed the electrochemical data. W.L. helped to analyse the electrochemical data. H.Z., Z.-F.L. and L.S. performed the cryo-TEM work. E.A.S. performed the TEM work. Y.R. and Q.L. performed the in operando HEXRD work; C.-J.S., Q.L. and Z.-F.L. performed the in operando XANES work. J.X. and Y.R. supervised the project. Q.L. and J.X. wrote the manuscript. All of the authors discussed the results and reviewed the manuscript.

Additional information

Competing financial interests: The authors declare no competing financial interests.

12

Separating hydrogen and oxygen evolution in alkaline water electrolysis using nickel hydroxide

Long Chen[1], Xiaoli Dong[1], Yonggang Wang[1] & Yongyao Xia[1]

Low-cost alkaline water electrolysis has been considered a sustainable approach to producing hydrogen using renewable energy inputs, but preventing hydrogen/oxygen mixing and efficiently using the instable renewable energy are challenging. Here, using nickel hydroxide as a redox mediator, we decouple the hydrogen and oxygen production in alkaline water electrolysis, which overcomes the gas-mixing issue and may increase the use of renewable energy. In this architecture, the hydrogen production occurs at the cathode by water reduction, and the anodic $Ni(OH)_2$ is simultaneously oxidized into NiOOH. The subsequent oxygen production involves a cathodic NiOOH reduction ($NiOOH \rightarrow Ni(OH)_2$) and an anodic OH^- oxidization. Alternatively, the NiOOH formed during hydrogen production can be coupled with a zinc anode to form a NiOOH-Zn battery, and its discharge product (that is, $Ni(OH)_2$) can be used to produce hydrogen again. This architecture brings a potential solution to facilitate renewables-to-hydrogen conversion.

[1] Department of Chemistry, Shanghai Key Laboratory of Molecular Catalysis and Innovative Materials, Institute of New Energy, iChEM (Collaborative Innovation Center of Chemistry for Energy Materials), Fudan University, Shanghai 200433, China. Correspondence and requests for materials should be addressed to Y.W. (email: ygwang@fudan.edu.cn) or to Y.X. (email: yyxia@fudan.edu.cn).

Hydrogen has long been considered a promising alternative to unsustainable fossil fuels because it is vital for the production of commodity chemicals such as ammonia and has great potential as a clean-burning fuel[1,2]. However, 90% of the world's hydrogen is currently obtained by the reformation of fossil fuels[3,4], which consumes much energy and is accompanied by serious CO_2 emissions. To realize a hydrogen-based economy, hydrogen must be efficiently and sustainably produced[1–8]. Accordingly, the state-of-the-art (photo) electrocatalysts for water splitting are attracting extensive attention[9–13]. Advanced water electrolysis has been considered as one of the most efficient and reliable approaches to producing hydrogen from renewable energy, such as solar, wind and hydropower, for grid-scale energy storage[14–16] because the electrolysis of water at room temperature stands out as a scalable technology, for which the only required inputs are water and energy (in the form of electricity)[17]. Therefore, electrochemical water splitting is attracting extensive attention[18–31]. However, the general application of water electrolysis currently faces a great challenge.

Room-temperature water electrolysis is generally performed under acidic or alkaline conditions. The water electrolysis under acidic condition is performed in an electrolyser with a proton exchange membrane (PEM); thus, it is called PEM water electrolysis[32–35]. Although PEM water electrolysis systems offer several advantages, such as high energy efficiency, a great hydrogen production rate and a compact design, their application remains hampered by the high cost of the catalysts and membranes[36,37]. The acidic environment limits the catalysts for the oxygen-evolving reaction (OER) and hydrogen-evolving reaction (HER) to noble metals. The expensive PEM, which is necessary to prevent the H_2/O_2 mixing, also significantly increases the cost. Furthermore, the short durability of the membrane makes PEM electrolysers too expensive for general applications[37,38]. Cronin's group recently developed a new method to split the conventional PEM water electrolysis process into two steps using the polyoxometalate $H_3PMo_{12}O_{40}$ as a buffer for redox equivalents[39,40]. The separate generation of O_2 and H_2 simplifies gas handling, puts less stringent demands on the membrane of PEM water electrolysis, and potentially reduces cost[41]. Cronin *et al.* undoubtedly provided an interesting and bright idea for future water electrolysis technology. However, their results[39,40] only focus on PEM water electrolysis (acid condition) and cannot be used in alkaline water electrolysis.

Compared with PEM water electrolysis, alkaline water electrolysis exhibits inherent low-cost characteristic because it can use a non-precious catalyst and a porous separator[36,37]. However, alkaline water electrolysers are difficult to shut down/start up, and their output cannot be quickly ramped because the pressures on the anode and cathode sides of the cell must always be equalized to prevent gas crossover through the porous cell separator[36]. Therefore, it is notably difficult to efficiently use the intermittent and fluctuating power from renewable energy for conventional alkaline water electrolysis. As an alternative to the porous separator, alkaline anion exchange membranes are considered suitable candidate materials that can easily prevent the gas mixing in alkaline water electrolysers[36,42]. However, similar to PEM water electrolysis, the cost and short durability of alkaline anion exchange membranes limit the scalable application. In particular, high-pressure gases in the electrolytic cell aggravate the membrane degradation[40]. Thus, to make renewables-to-hydrogen conversion both practically and economically more attractive, new electrolyser systems must be developed to prevent product gases from mixing over a range of current densities and effectively use the low-cost characteristic of alkaline water electrolysis. It should be a promising solution to separate H_2 production and O_2 production. However, the method to separate conventional alkaline water electrolysis into two steps has never been reported.

In the report by Cronin *et al.*[39], a polyoxometalate-based redox mediator was used as an electron-coupled-proton buffer in an acid solution to decouple the H_2 and O_2 evolution in PEM water electrolysis. Similarly, it should be a precondition to find a stable electron-coupled-proton buffer in an alkaline solution to separate the H_2 and O_2 evolution in alkaline water electrolysis. Nickel hydroxide has been widely used in rechargeable batteries using alkaline electrolytes[43]. Its charge and discharge depend on the reversible transformation of $Ni(OH)_2/NiOOH$[44–47], which can be explained as an electron-coupled proton release/storage process (that is, H^+ de-intercalation from $Ni(OH)_2$ or intercalation into NiOOH, Supplementary Fig. 1). Therefore, it should be an interesting topic to use nickel hydroxide as an electron-coupled-proton buffer to separate the H_2 and O_2 evolution in alkaline water electrolysis.

Here, using nickel hydroxide ($Ni(OH)_2/NiOOH$) as a redox mediator, we successfully split the conventional alkaline water electrolysis process into separate H_2 and O_2 production steps, which well overcomes the gas-mixing issue and increases the use of renewable energy. In this architecture, the H_2 production involves the cathodic H_2O reduction and anodic $Ni(OH)_2$ oxidization ($Ni(OH)_2 \rightarrow NiOOH$), and the subsequent O_2 production depends on the cathodic NiOOH reduction and anodic OH^- oxidization. In addition, the formed NiOOH during the H_2 production can be coupled with a zinc anode to form a NiOOH-Zn battery, which provides an interesting rechargeable system that produces H_2 during the electrolysis and delivers energy on the discharge of the NiOOH-Zn battery.

Results

Mechanism of the two-step alkaline water electrolysis. As shown in Fig. 1a, the H_2 production (Step 1) involves a cathodic reduction of H_2O on the HER electrode ($H_2O \rightarrow H_2$) and a simultaneous anodic oxidization of the $Ni(OH)_2$ electrode ($Ni(OH)_2 \rightarrow NiOOH$). The subsequent O_2 production (Step 2) occurs on the OER electrode by an anodic oxidization of OH^- ($OH^- \rightarrow O_2$), whereas the NiOOH cathode is reduced to $Ni(OH)_2$. This approach leads to a device architecture for the alkaline electrolytic cell with several important advantages. First, the separate generation of O_2 and H_2 prevents the product gases from mixing over a range of current densities and simplifies the gas handling, which greatly increase the operation flexibility of alkaline electrolytic cells and make them suitable to be driven by sustainable energy (such as solar energy). Second, this device architecture can produce highly pure H_2 and O_2 with no membrane, which further reduces the cost of the alkaline water electrolysis technology. Third, the separate H_2 and O_2 productions require different driving voltages (or power inputs), which implies that we can flexibly use sustainable energy (such as solar or wind power) for H_2 production or O_2 production based on the output variation in these unstable power sources. Finally, the NiOOH that forms during the H_2 production (that is, Step 1) can be coupled with a zinc anode to form a NiOOH-Zn battery for energy storage, and its discharge depends on the cathodic reduction of the NiOOH electrode ($NiOOH \rightarrow Ni(OH)_2$) and the anodic oxidization of the zinc electrode ($Zn \rightarrow ZnO_2^{2-}$). Herein, it should be noted that the cathodic reduction potential of NiOOH (0.45 V versus Hg/HgO) is significantly higher than the anodic oxidization potential of zinc (− 1.15 V versus Hg/HgO) (ref. 48). Therefore, the NiOOH cathode and Zn anode can be coupled to form the NiOOH-Zn battery system that has been commercialized[49–51].) Its discharge product ($Ni(OH)_2$) can be used to produce H_2 again, which provides an interesting

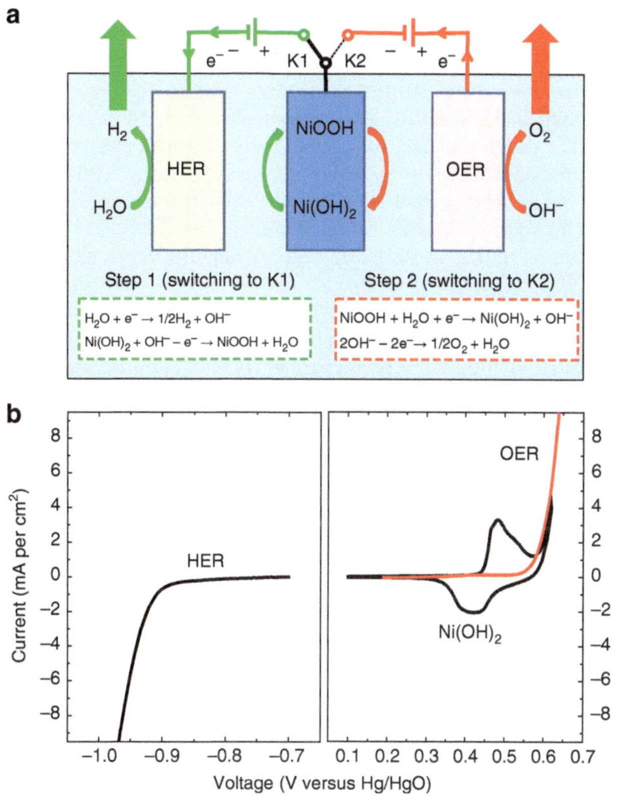

Figure 1 | Mechanism of the two-step alkaline water electrolysis. (**a**) A schematic of the operation mechanism of the cell, where Step 1 (H_2 production; switching to K1) involves the cathodic reduction of H_2O on the HER electrode ($H_2O + e^- \rightarrow 1/2H_2 + OH^-$) and anodic oxidization in the $Ni(OH)_2$ electrode ($Ni(OH)_2 + OH^- - e^- \rightarrow NiOOH + H_2O$). Step 2 ($O_2$ production; switching to K2) includes the cathodic reduction of the NiOOH electrode ($NiOOH + H_2O + e^- \rightarrow Ni(OH)_2 + OH^-$) and anodic OH^- oxidization on the OER electrode ($2OH^- - 2e^- \rightarrow 1/2O_2 + H_2O$). (**b**) CV curve of the $Ni(OH)_2$ film electrode at a scan rate of $5\,mV\,s^{-1}$ in 1 M KOH (black line), the linear sweep voltammetric (LSV) data of the commercial RuO_2/IrO_2-coated Ti-mesh electrode for the OER at a scan rate of $5\,mV\,s^{-1}$ in 1 M KOH (red line) and the LSV data of the commercial Pt-coated Ti-mesh electrode for the HER in 1 M KOH with a sweep rate of $5\,mV\,s^{-1}$ (blue line).

rechargeable cycle that produces H_2 with the charge (that is, electrolysis in Step 1) and delivers energy with the discharge of the NiOOH-Zn battery.

In this work, nickel hydroxide, which is the conventional electrode material for commercial rechargeable Ni-MH or Ni-Cd batteries, was used as a redox mediator to split the conventional alkaline water electrolysis process into two steps. Before the fabrication of this alkaline water electrolytic cell, the electrochemical profile of $Ni(OH)_2$ in an alkaline electrolyte (1 M KOH) was investigated using a cyclic voltammogram (CV) with a typical three-electrode system, which used a Pt plate and a Hg/HgO electrode as the counter and reference electrodes, respectively. Carbon-nanotube-supported $Ni(OH)_2$ particles were used as the active material to prepare the $Ni(OH)_2$-based film electrode (see the Methods and Supplementary Fig. 2 for details) for the CV measurement, where the carbon nanotube support with high electronic conductivity was only used to alleviate the polarization that arose from the electrode impedance. The CV curve of $Ni(OH)_2$ at a scan rate of $5\,mV\,s^{-1}$ is shown in Fig. 1b (black line). The OER and HER potentials of the commercial RuO_2/IrO_2-coated Ti-mesh electrode and Pt-coated Ti-mesh electrode were also investigated using the three-electrode method for comparison (see the red and blue lines in Fig. 1b). As shown in Fig. 1b, a couple of redox peaks are clearly observed at 0.43 and 0.49 V (versus Hg/HgO) in the CV curve of the $Ni(OH)_2$ electrode because of the reversible cycling between $Ni(OH)_2$ and NiOOH. Obviously, the special potential window for the $Ni(OH)_2$/NiOOH redox couple is located between the onset potential for the OER and the onset potential for the HER. The result indicates that $Ni(OH)_2$ can be used as a redox mediator to split the conventional alkaline water electrolysis process into two steps according to Fig. 1a. The galvanostatic charge-discharge curve of the $Ni(OH)_2$ electrode at a current density of $0.2\,A\,g^{-1}$ is shown in Supplementary Fig. 3 to clarify the specific capacity of $Ni(OH)_2$ (see the corresponding discussion about Supplementary Fig. 3).

Performance of the two-step alkaline water electrolysis. To test the hypothesis in Fig. 1a, an alkaline water electrolytic cell was constructed with a commercial Pt-coated Ti-mesh electrode (Supplementary Fig. 4) for the HER, a commercial RuO_2/IrO_2-coated Ti-mesh electrode for the OER (Supplementary Fig. 5) and a commercial $Ni(OH)_2$ electrode of conventional Ni-MH or Ni-Cd batteries (Supplementary Fig. 6). The photo profile of the cell is shown in Supplementary Fig. 7, which shows that the $Ni(OH)_2$ electrode ($2.5 \times 4\,cm^2$) is located between the HER electrode ($2.5 \times 4\,cm^2$) and the OER electrode ($2.5 \times 4\,cm^2$). The water electrolysis of the cell was investigated by chronopotentiometry measurements with different applied currents of 100–500 mA. The chronopotentiometry curve (cell voltage versus time) of the electrolytic cell at a constant applied current of 200 mA is shown in Fig. 2a. The chronopotentiometry data of the anode (anodic potential versus time) and cathode (cathodic potential versus time) were also investigated during the electrolysis process and are provided in Fig. 2a. The electrolysis process includes two steps (Steps 1 and 2) with different cell voltages. As shown in Fig. 2a, Step 1 (that is, the H_2-production process) exhibits a cell voltage of ~ 1.6 V, which arises from the difference between the anodic potential of 0.5 V (versus Hg/HgO) of the $Ni(OH)_2$ oxidation ($Ni(OH)_2 \rightarrow NiOOH$) and the cathodic potential of -1.1 V (versus Hg/HgO) of the H_2O reduction ($H_2O \rightarrow H_2$). In Step 2 (that is, the O_2-production process), the cell voltage is 0.4 V, which is equal to the potential difference (0.7–0.3 V versus Hg/HgO) between the anodic oxidation of OH^- ($OH^- \rightarrow O_2$) and the cathodic reduction of NiOOH ($NiOOH \rightarrow Ni(OH)_2$). In Step 2, the cell voltage sharply increases sharply at the end of electrolysis (Fig. 2a), which indicates that all of the NiOOH has been reduced to $Ni(OH)_2$. In other words, the electrolysis in Step 2 automatically finished after 600 s (= 1,200–600 s), which is equal to the electrolysis time in Step 1 at the identical current of 200 mA. The equal electrolysis time clearly indicates a Coulombic efficiency of $\sim 100\%$. Photo profiles of the H_2 generation in Step 1 and O_2 generation in Step 2 are shown in Fig. 2b,c to further characterize the separated steps. In addition, the video evidence also clearly demonstrates the separate H_2/O_2 generation directly (Supplementary Movies 1 and 2). To clarify the operation flexibility of this electrolyser, the water electrolysis was also investigated at a lower current of 100 mA and a higher current of 500 mA (Supplementary Fig. 8). It should be noted that as a notably mature electrode material, $Ni(OH)_2$ exhibits high efficiency and a long cycle life. These characteristics are also notably important to facilitate the cycle of H_2 generation (Step 1) and O_2 generation (Step 2). To demonstrate this point, the H_2/O_2 generation cycle performance was investigated with an applied current of 200 mA. As shown in Fig. 2d, this alkaline electrolytic cell exhibits stable H_2 and O_2 generation over 20 consecutive cycles. Furthermore,

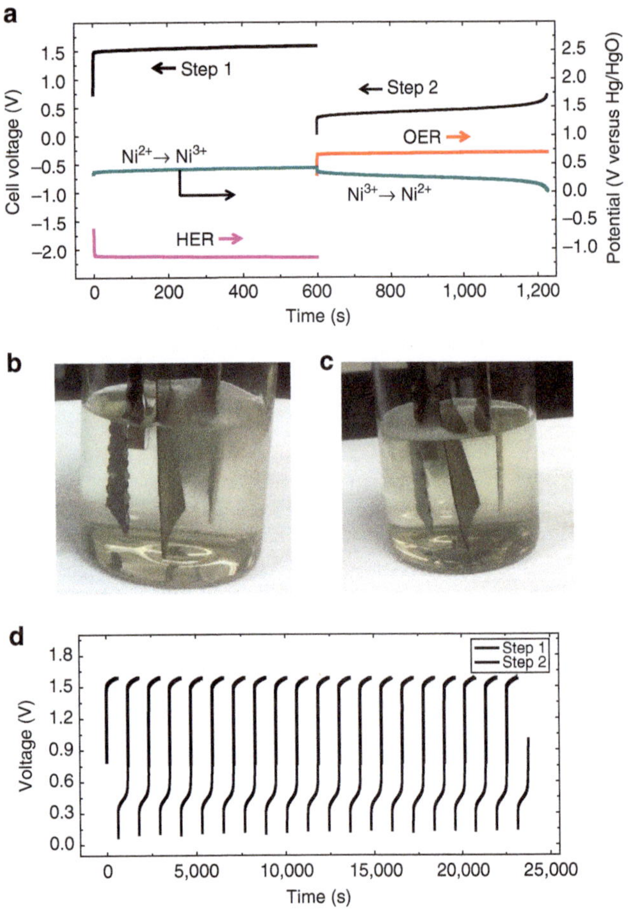

Figure 2 | Performance of the two-step alkaline water electrolysis. (**a**) Chronopotentiometry curve (cell voltage versus time) of the cell at a constant applied current of 200 mA, where the voltages for H_2 production (Step 1) and O_2 production (Step 2) are labelled by the blue and black lines, respectively. Chronopotentiometry data (potential versus time) of the HER electrode (pink line), $Ni(OH)_2$ electrode (green line) and OER electrode (red line) are provided. ((Voltage of Step 1) = (Potential of $Ni^{2+} \rightarrow Ni^{3+}$)—(Potential of HER); (Voltage of Step 2) = (Potential of OER)—(Potential of $Ni^{3+} \rightarrow Ni^{2+}$)). (**b**,**c**) Photo profiles of the H_2/O_2 generation in Steps 1 and 2, where it can be detected that H_2 and O_2 are produced on the HER (**b**) and OER electrodes (**c**), respectively, at separate times (Supplementary Movies 1 and 2 further confirm this point). (**d**) Chronopotentiometry curve (cell voltage versus time) of the H_2/O_2 generation cycle with an applied current of 200 mA and a step time of 600 s, where the chronopotentiometry data of Step 1 (H_2 generation) and Step 2 (O_2 generation) are labelled with the blue and black lines, respectively.

100 consecutive cycles of H_2/O_2 generation are shown in Supplementary Fig. 9 to further demonstrate the stability. As shown in Fig. 2d (or Fig. 2a), the separate H_2 production (Step 1) and O_2 production (Step 2) require different driving voltages (or power inputs), which implies that we can flexibly use renewable energy, such as solar or wind power, to produce H_2 or O_2 based on the output variation in these unstable power sources. For example, we can use solar energy at noon to drive the H_2-production step, which requires a high driving voltage (or power input), and solar energy at dusk to power the O_2-production step, which requires a low driving voltage. The flexibility can increase the use of sustainable energy.

In the above investigation (Fig. 2), a step time of only 10 min (600 s) was used to characterize the separate H_2 and O_2 production. Such a short time was used to emphasize that we can flexibly change the operation of our system even within a notably short time. In fact, the electrolysis time in each step can be easily controlled by the applied current. As shown in Supplementary Fig. 10, the electrolysis time in each step can be increased to 12 h with a low current of 20 mA. In addition, the $Ni(OH)_2$ electrode can be cycled with different charge depths (Supplementary Fig. 11). Thus, we can also control the electrolysis time in each step by adjusting the charge depths of the $Ni(OH)_2$ electrode (Supplementary Fig. 12). However, as mentioned in the introduction section, alkaline water electrolysis can use non-precious electrodes for the H_2/O_2 production[36,37]. Therefore, non-precious electrodes (a Co_3O_4-based OER electrode and a metal-Ni-foam-based HER electrode) were used to further demonstrate the separate H_2 and O_2 production (see Supplementary Fig. 13 and Supplementary Movies 3 and 4 for details). Furthermore, according to the previous report by Cronin *et al.*[39], the efficiency of two-step water electrolysis can be evaluated by comparing its total driving voltage (Step 1 + Step 2) to the driving voltage of the corresponding one-step water electrolysis. Therefore, the efficiency of the two-step alkaline water electrolysis using precious or non-precious HER/OER electrodes was calculated according to the method described by Cronin *et al.* (see Supplementary Fig. 14 for detail). As shown in Supplementary Fig. 14a,b, the efficiency of the two-step water electrolysis using precious electrodes (a RuO_2/IrO_2-coated Ti-mesh electrode for the OER and a Pt-coated Ti-mesh electrode for the HER) is 92% (= 1.829/1.985) compared with its corresponding one-step water electrolysis. According to the data shown in Supplementary Fig. 14c,d, the efficiency of the two-step water electrolysis using non-precious electrodes (a Co_3O_4-based electrode for the OER and a metal Ni-foam electrode for the HER) is also ~92% (= 1.973/2.137) compared with its corresponding one-step water electrolysis. The achieved efficiency is slightly higher than that (79%) of the two-step PEM water electrolysis reported by Cronin's group[39].

Purity of the generated H_2/O_2. To confirm the purity of the H_2/O_2 in the separate steps, *in situ* differential electrochemical mass spectrometry was used to measure the gas evolution of the total water electrolysis process at a constant applied current of 200 mA. In this experiment, a quadrupole mass spectrometer with a leak inlet was connected to the alkaline water electrolytic cell with two tubes as the purge/carrier gas inlet and outlet (see the Methods and Supplementary Fig. 15 for details). A pure Ar gas stream was used as the purge gas before electrolysis and the carrier gas during the electrolysis process. Before the online gas analysis, the system was purged with a pure Ar stream for 1 h. The system was further purged with a pure Ar stream for another 1 h, with an online analysis record (Fig. 3a) showing that both O_2 and H_2 reached a stable background line. Then, the H_2-production step (Step 1) was started, and the H_2 evolution is clearly observed in the online analysis record. The ion current intensity of O_2 obviously remained at the background level in Step 1, which indicates that no O_2 was generated in the H_2 production process of 30 min (Fig. 3a,b). After the H_2 production (Step 1) finished, a rest step of 130 min was performed with a pure Ar stream to eliminate remnant H_2 in the system, and a hysteresis of H_2 could be observed in the online analysis record. Afterward, the O_2-production step (Step 2) was started. As shown in Fig. 3b, the O_2 production automatically finished with a total electrolysis time of ~30 min (1,700 s), which is close to the H_2 production time (1,800 s). The minor difference of 100 s may be because of the slight self-discharge of NiOOH in the rest step. However, this description does not indicate that the self-discharge of the NiOOH electrode will be significantly aggravated with a longer rest time (see Supplementary Fig. 16 for an extended

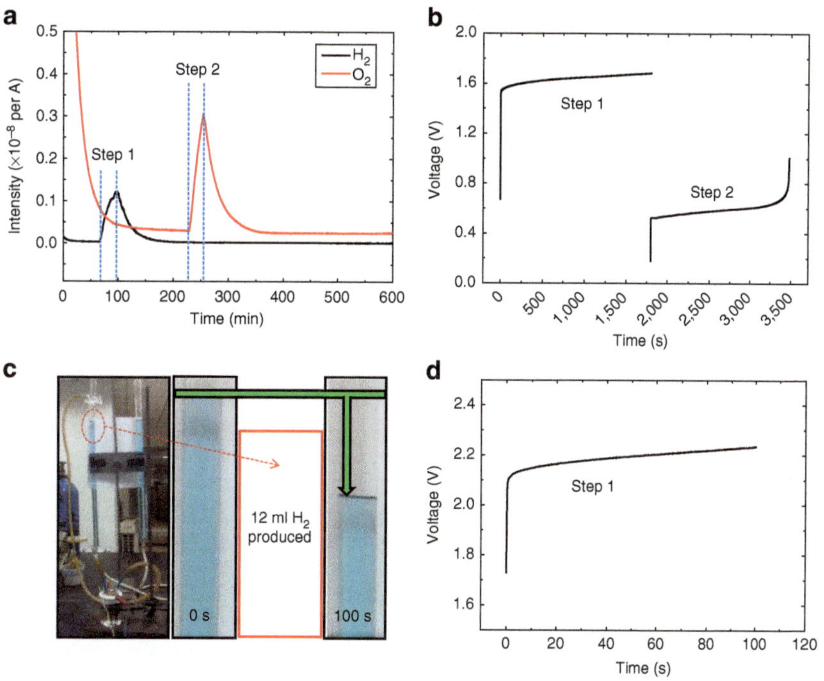

Figure 3 | Purity of the generated H_2/O_2. (**a**) *In situ* differential electrochemical mass spectrometry (DEMS) curves of the H_2 evolution (black line) and O_2 evolution (red line) in the total water electrolysis process at a constant applied current of 200 mA. A pure Ar gas stream was used as a purge gas before electrolysis and the carrier gas in the total electrolysis process. There is a rest time of 130 min between Steps 1 and 2. (**b**) The chronopotentiometry curve (cell voltage versus time) of the electrolytic cell corresponds to the *in situ* DEMS test with two steps: H_2-production process of 30 min (Step 1, the blue line) and O_2-production process (Step 2, black line). (**c**) A typical apparatus configuration to determine the evolution of the H_2 volumes. (The H_2 production rate (ml s^{-1}) was measured with an applied current of 1,000 mA for 100 s.) Approximately 12 ml H_2 was generated after 100 s of electrolysis, as indicated by the green arrow. (**d**) The chronopotentiometry curve (cell voltage versus time) of the electrolytic cell corresponds to the H_2-production volume test with an applied current of 1,000 mA for 100 s.

discussion about the self-discharge of the nickel hydroxide electrode). As shown in Fig. 3a, there is no H_2 evolution in the O_2 production process. Therefore, the results in Fig. 3a,b well demonstrate the purity of the H_2/O_2 in the separate steps. Herein, it should be noted that the online gas analysis in our experiment was only used to characterize the purity of the H_2 and O_2 in separate steps. A typical drainage method (Supplementary Fig. 17) was used to quantify the H_2 generation over a specific time length. In this experiment, the H_2 production rate (ml s^{-1}) was measured with an applied current of 1,000 mA for 100 s (Fig. 3c,d). Figure 3c,d shows that ~12 ml H_2 was generated in the 100 s electrolysis, which is close to the theoretical value (12.67 ml). Therefore, the Faradaic efficiency is 94.7% (12/12.67). In theory, the Faradaic efficiency should be 100%, but the impedance and dissolution of H_2 in the aqueous solution may slightly reduce the efficiency. This method was used to measure the generated O_2 volume in Step 2 at the identical current of 1,000 mA. The obtained result indicates that ~6 ml O_2 was generated in Step 2. Therefore, the H_2-to-O_2 ratio is 2:1 in the consecutive cycle of Steps 1 and 2.

Combination between the H_2-production and NiOOH-Zn battery. Interestingly, the aforementioned O_2-production step (Step 2) can be replaced by the discharge step of the NiOOH-Zn battery (Step 2′), which will enable the coupling of H_2 production with a discharge step of the NiOOH-Zn battery (Fig. 4a). As shown in Fig. 4a, the H_2 production (Step 1) includes the cathodic reduction of H_2O on the HER electrode ($H_2O \rightarrow H_2$) and the anodic oxidization of the $Ni(OH)_2$ electrode ($Ni(OH)_2 \rightarrow$ NiOOH). Next, the NiOOH electrode that is formed in Step 1 is coupled with a zinc anode to form a NiOOH-Zn battery. The subsequent discharge step (Step 2′) of the NiOOH-Zn battery is based on the cathodic reduction of the NiOOH electrode ($NiOOH \rightarrow Ni(OH)_2$) and the anodic oxidization ($Zn \rightarrow ZnO_2^{2-}$) of zinc[48–50]. In other words, the architecture in Fig. 4a provides an interesting rechargeable cycle that produces H_2 with charge (that is, electrolysis in Step 1) and delivers energy with the discharge of the NiOOH-Zn battery (Step 2′). To confirm this hypothesis, the NiOOH electrode, which formed after the electrolysis for H_2 production with an applied current of 200 mA for 600 s, was directly coupled with a zinc-plate electrode in the electrolysis cell to construct a NiOOH-Zn battery. The discharge profile of the NiOOH-Zn battery was investigated with a current of 200 mA. As shown in Fig. 4b, the NiOOH-Zn battery displays a discharge voltage of ~1.6 V with a total discharge time of ~600 s. Furthermore, the consecutive H_2-production step (Step 1) and discharge step (Step 2′) of the NiOOH-Zn battery can be cycled exactly like a rechargeable battery (inset of Fig. 4b). Further cycle data of Step 1 (H_2-production step) and Step 2′ (discharge of the NiOOH-Zn battery) are provided in Supplementary Fig. 18. Therefore, the architecture in Fig. 4a introduces a new energy storage/conversion approach, in which solar energy can be used to drive the electrolysis (charge process) of Step 1 to produce H_2 during the daytime, and the discharge step (Step 2′) of the NiOOH-Zn battery can be used to deliver energy to power electronic devices overnight. To clarify this point for the layperson reader, the NiOOH-Zn battery that formed after the H_2-production step was used to power a 1.6 V electric fan (Supplementary Movie 5). The cycle of Steps 1 and 2′ also increases the $Zn(OH)_2$ concentration in the alkaline electrolyte. The $Zn(OH)_2$ alkaline solution can be used to produce O_2 and

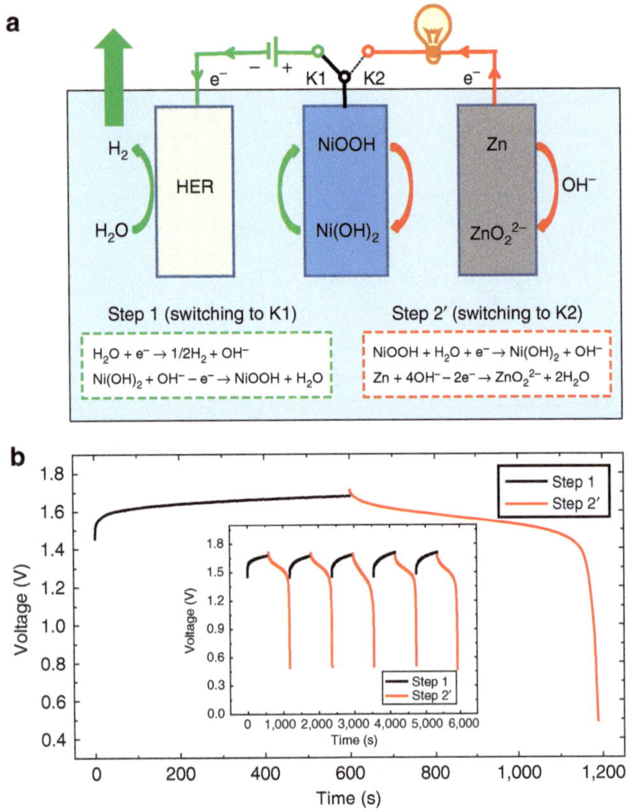

Figure 4 | Combination of the H2-production and the NiOOH-Zn battery. (**a**) A schematic of the operation mechanism of the cell. (Herein, the H_2-production process (Step 1; switching to K1) includes the cathodic reduction of H_2O on the HER electrode ($H_2O + e^- \rightarrow 1/2H_2 + OH^-$) and the anodic oxidization of the $Ni(OH)_2$ electrode ($Ni(OH)_2 + OH^- - e^- \rightarrow NiOOH + H_2O$). Then, by switching to K2, the NiOOH electrode that was formed during Step 1 is coupled with a zinc anode to form a NiOOH-Zn battery, and its discharge (Step 2′) depends on the cathodic reduction of the NiOOH electrode ($NiOOH + H_2O + e^- \rightarrow Ni(OH)_2 + OH^-$) and the anodic oxidization of the zinc electrode ($Zn + 4OH^- - 2e^- \rightarrow ZnO_2^{2-} + 2H_2O$)). (**b**) Chronopotentiometry curve (cell voltage versus time) of the H_2-production process (Step 1, black line) and the discharge curve of the NiOOH-Zn battery (Step 2′, red line). The electrolysis for H_2 production applied a current of 200 mA for 600 s; then, the discharge profile of the NiOOH-Zn battery was also investigated with a current of 200 mA. The inset is the cycle performance of the H_2-production step (black line) and discharge step (red line) of the NiOOH-Zn battery with an applied current of 200 mA. As shown in **a**,**b**, Step 1 requires a power input to produce H_2 gas, whereas Step 2′ (discharge of the NiOOH-Zn battery) can deliver energy to power other devices.

metallic Zn through electrolysis at the proper time with other energy inputs, such as at night-time with wind power, nuclear fission and so on. In addition, the rechargeable system based on the H_2 production (Step 1) and discharge of the NiOOH-Zn battery (Step 2′) exhibits a theoretical energy density of 280 Wh kg^{-1} (see Supplementary Fig. 19 for details), which is close to the theoretical energy density of conventional Ni-MH batteries, Ni-Cd batteries or Ni-Zn batteries and higher than the theoretical energy density of lead-acid batteries and aqueous Li-ion batteries[51].

Discussion

We successfully split the conventional alkaline electrolysis into two steps by using nickel hydroxide as a recyclable redox mediator. The separate H_2 and O_2 production overcomes the challenge of H_2/O_2 mixing and facilitates the operation of alkaline electrolysis even with unstable power inputs. The separate H_2 and O_2 productions require different driving voltages (or input power), which implies that we can flexibly use sustainable energy, such as solar or wind power, with higher efficiency. Finally, the combination of H_2 production and discharge of the NiOOH-Zn battery potentially provides a new energy storage/conversion approach for human life. It should also be noted that the electrochemical redox process of $Ni(OH)_2$/NiOOH is generally limited by the proton diffusion in the crystalline framework of $Ni(OH)_2$ or NiOOH[52], which limits the electrolysis rate. Therefore, solar energy, with the characteristic of low power loads, should be suitable to drive the new type alkaline electrolytic cell. However, much effort has been made to develop high-rate $Ni(OH)_2$ electrodes to increase the power density of $Ni(OH)_2$-based batteries, which may further enhance the electrolysis rate of the new type alkaline electrolytic cell. The data in the result section were obtained using a one-compartment water electrolytic cell, where all three electrodes (HER cathode, OER anode and $Ni(OH)_2$ electrode) were immersed in the same electrolyte. In practical applications, the two-step alkaline water electrolysis can also be operated with two separate rooms, where the nickel hydroxide ($Ni(OH)_2$/NiOOH) electrode is used as 'a solid-state proton buffer' (see the proton buffer mechanism in Supplementary Fig. 1), which can be moved between room 1 for H_2 production (Step 1) and room 2 for O_2 production (Step 2; see the extended discussion in Supplementary Figs 20 and 21 and Supplementary Movies 6 and 7).

Methods

Synthesis and characterization of $Ni(OH)_2$/MWNT composite. First, multi-walled carbon nanotubes (MWNTs) underwent a hydrophilic treatment before use. The MWNTs were sonicated in 30% HNO_3 solution for 30 min, filtered and washed with distilled water, and finally dried at 100°C for 12 h. According to our previous report, $Ni(OH)_2$/MWNT composites were prepared by loading $Ni(OH)_2$ on the treated MWNTs in an alkaline medium: 0.3 g of accurately weighed MWNTs was immersed and dispersed in a bath that contained 2.2 g of 0.1 M $Ni(NO_3)_2 \cdot 6H_2O$ solution. Then, 0.1 M KOH solution was dropped into the bath while stirring until the pH of the aqueous solution was 8.5. The resulted product was filtered and repeatedly washed with distilled water, dried at 80 °C and weighed. The 70 wt % mass load of $Ni(OH)_2$ in the composites was evaluated by calculating the weight difference of MWNTs. The $Ni(OH)_2$/MWNT composite was characterized using S-4800 scanning electron microscopes and a Joel JEM2011 transmission electron microscope.

Preparation and electrochemical test of the electrode. The $Ni(OH)_2$/MWNT composite electrode was prepared according to the following steps: a mixture of 85 wt% $Ni(OH)_2$/MWNT composites, 10 wt% acetylene black and 5 wt% polytetrafluoroethylene was thoroughly mixed to form a film, which was pressed onto a nickel grid (1.2×10^7 Pa) that served as a current collector surface (1 cm^2). The $Ni(OH)_2$/MWNT composite electrode was characterized by CV with a scan rate of 5 mV s^{-1} and a galvanostatic charge-discharge test at a current density of 0.2 A g^{-1}. The electrolyte was 1 M KOH solution. The onset potential of the OER on the commercial RuO_2/IrO_2-coated Ti-mesh electrode and onset potential of the HER on the commercial Pt-coated Ti-mesh electrode in 1 M KOH were investigated by linear sweep voltammetric measurements with a sweep rate of 5 mV s^{-1} in 1 M KOH. The aforementioned experiments were performed with a typical three-electrode method, in which a Pt plate and Hg/HgO (0.098 V versus the standard hydrogen electrode) were used as the counter and reference electrodes, respectively. All electrochemical measurements were performed with a CHI 660D electrochemistry workstation.

Fabrication of the new type electrolytic cell. The cell was constructed with a commercial Pt-coated Ti-mesh electrode (Supplementary Fig. 4) for the HER, a commercial RuO_2/IrO_2-coated Ti-mesh electrode for the OER (Supplementary Fig. 5) and a commercial $Ni(OH)_2$ electrode of conventional Ni-MH or Ni-Cd batteries (Supplementary Fig. 6). The photo profile of the cell is shown in Supplementary Fig. 7, where the $Ni(OH)_2$ electrode (2.5×4 cm^2) is located between the HER electrode (2.5×4 cm^2) and OER electrode (2.5×4 cm^2).

Water electrolysis investigation. The water electrolysis of the new type alkaline water electrolysis cell was investigated using chronopotentiometry measurements with applied currents of 100, 200 and 500 mA. In Step 1, the HER electrode (that is, Pt coated Ti-mesh electrode) and $Ni(OH)_2$ electrode were connected to the cathode and anode, respectively, of the power source for electrolysis. The duration time of Step 1 (that is, the H_2-production step) is 600 s with an applied current of 100, 200 or 500 mA. After Step 1, Step 2 (that is, the O_2-production step) was started. In Step 2, the NiOOH electrode, which formed during Step 1, and the OER electrode (that is, the RuO_2/IrO_2-coated Ti-mesh electrode) were connected to the cathode and anode, respectively, of the power source for the electrolysis. Step 2 automatically stopped when the cell voltage sharply increased at the end of the electrolysis, which indicated that all the NiOOH was converted into $Ni(OH)_2$. The cell voltages (V versus time) of Steps 1 and 2 were recorded to characterize the electrolysis profile. With an additional reference electrode (that is, the Hg/HgO electrode), the chronopotentiometry data (potential versus time) of a single electrode (the HER electrode, $Ni(OH)_2$ electrode or OER electrode) were also recorded during Steps 1 and 2. A schematic illustration of the voltage (or potential) record is provided in Supplementary Fig. 22. As shown in Supplementary Fig. 22, V_1 is the cell voltage in Step 1; V_2 is the cell voltage in Step 2; P_1 is the potential of the HER electrode in Step 1; P_2 is the potential of the $Ni(OH)_2$ electrode in Steps 1 and 2 and P_3 is the potential of the OER in Step 2. A PARSTAT MC multi-channel workstation was used to perform the water electrolysis investigation of Fig. 2.

***In situ* differential electrochemical mass spectrometry.** A quadrupole mass spectrometer (NETZSCH QMS 403 C) with a leak inlet was applied to measure the gas evolution of the total water electrolysis process at a constant applied current of 200 mA. The electrolysis process was performed using a Solarton Instrument Model 1287 electrochemical interface. As shown in Supplementary Fig. 15, the mass spectrometer is connected to the new-type alkaline water electrolysis cell with two tubes as the purge/carrier gas inlet and outlet. A pure Ar gas stream was used as the purge gas before electrolysis and the carrier gas during the electrolysis process. The gas flows were typically 10 ml min^{-1}. Before the online gas analysis, the system was purged with a pure Ar stream for 1 h. The system was further purged with a pure Ar stream for another 1 h before Step 1 began. The duration time of Step 1 was 30 min. After the end of Step 1, a rest step of 130 min was performed with a pure Ar stream to eliminate remnant H_2 in the system. Then, Step 2 was started and continued until the cell voltage sharply increased. The experiment completed when all remaining O_2 in the electrolysis cell was removed.

References

1. Goff, A. L. *et al.* From hydrogenases to noble metal-free catalytic nanomaterials for H_2 production and uptake. *Science* **326,** 1384–1387 (2009).
2. Armaroli, N. & Balzani, V. The hydrogen issue. *ChemSusChem* **4,** 21–36 (2011).
3. Reece, S. Y. *et al.* Wireless solar water splitting using silicon-based semiconductors and earth-abundant catalysts. *Science* **334,** 645–648 (2011).
4. Tachibana, Y., Vayssieres, L. & Durrant, J. R. Artificial photosynthesis for solar water-splitting. *Nat. Photonics* **6,** 511–518 (2012).
5. Lewis, N. S. & Nocera, D. G. Powering the planet: chemical challenges in solar energy utilization. *Proc. Natl Acad. Sci. USA* **103,** 15729–15735 (2006).
6. Gust, D., Moore, T. A. & Moore, A. L. Solar fuels via artificial photosynthesis. *Acc. Chem. Res.* **42,** 1890–1898 (2009).
7. Cook, T. R. *et al.* Solar energy supply and storage for the legacy and nonlegacy worlds. *Chem. Rev.* **110,** 6474–6502 (2010).
8. Swierk, J. R. & Mallouk, T. E. Design and development of photoanodes for water-splitting dye-sensitized photoelectrochemical cells. *Chem. Soc. Rev.* **42,** 2357–2387 (2013).
9. Kanan, M. W. & Nocera, D. G. *In situ* formation of an oxygen-evolving catalyst in neutral water containing phosphate and Co^{2+}. *Science* **321,** 1072–1075 (2008).
10. Luo, J. S. *et al.* Water photolysis at 12.3% efficiency via perovskite photovoltaics and earth-abundant catalysts. *Science* **345,** 1593–1596 (2014).
11. Dasgupta, N. P., Liu, C., Andrews, S., Prinz, F. B. & Yang, P. D. Atomic layer deposition of platinum catalysts on nanowire surfaces for photoelectrochemical water reduction. *J. Am. Chem. Soc.* **135,** 12932–12935 (2013).
12. Swierk, J. R. *et al.* Metal-free organic sensitizers for use in water-splitting dye-sensitized photoelectrochemical cells. *Proc. Natl Acad. Sci. USA* **112,** 1681–1686 (2015).
13. Seabold, J. A. & Choi, K. S. Efficient and stable photo-oxidation of water by a bismuth vanadate photoanode coupled with an iron oxyhydroxide oxygen evolution catalyst. *J. Am. Chem. Soc.* **134,** 2186–2192 (2012).
14. Guerrini, E. & Trasatii, S. in *Catalysis for Sustainable Energy Production* (eds Barbaro, P. & Bianchini, C.) 235–269 (Wiley-VCH Verlag GmbH & Co. KGaA, 2009).
15. Barbir, F. PEM electrolysis for production of hydrogen from renewable energy sources. *Sol. Energy* **78,** 661–669 (2005).
16. Gandía, L. M., Oroz, R., Ursúa, A., Sanchis, P. & Diéguez, P. M. Renewable hydrogen production: performance of an alkaline water electrolyzer working under emulated wind conditions. *Energy Fuels* **21,** 1699–1706 (2007).
17. Holladay, J. D., Hu, J., King, D. L. & Wang, Y. An overview of hydrogen production technologies. *Catal. Today* **139,** 244–260 (2009).
18. McCrory, C. C. L., Jung, S., Peters, J. C. & Jaramillo, T. F. Benchmarking heterogeneous electrocatalysts for the oxygen evolution reaction. *J. Am. Chem. Soc.* **135,** 16977–16987 (2013).
19. Suntivich, J., May, K. J., Gasteiger, H. A., Goodenough, J. B. & Yang, S. H. A perovskite oxide optimized for oxygen evolution catalysis from molecular orbital principles. *Science* **334,** 1383–1385 (2011).
20. Subbaraman, R. *et al.* Enhancing hydrogen evolution activity in water splitting by tailoring Li^+-$Ni(OH)_2$-Pt interfaces. *Science* **334,** 1256–1260 (2011).
21. Jin, H. Y. *et al.* *In situ* cobalt-cobalt oxide/N-doped carbon hybrids as superior bifunctional electrocatalysts for hydrogen and oxygen evolution. *J. Am. Chem. Soc.* **137,** 2688–2694 (2015).
22. Nakagawa, T., Beasley, C. A. & Murray, R. W. Efficient electro-oxidation of water near its reversible potential by a mesoporous IrO_x nanoparticle film. *J. Phys. Chem. C* **113,** 12958–12961 (2009).
23. Gao, M. R. *et al.* Efficient water oxidation using nanostructured α-nickel-hydroxide as an electrocatalyst. *J. Am. Chem. Soc.* **136,** 7077–7084 (2014).
24. Louie, M. W. & Bell, A. T. An investigation of thin-film Ni-Fe oxide catalysts for the electrochemical evolution of oxygen. *J. Am. Chem. Soc.* **135,** 12329–12337 (2013).
25. Kong, D. S., Wang, H. T., Lu, Z. Y. & Cui, Y. $CoSe_2$ nanoparticles grown on carbon fiber paper: An efficient and stable electrocatalyst for hydrogen evolution reaction. *J. Am. Chem. Soc.* **136,** 4897–4900 (2014).
26. Voiry, D. *et al.* Enhanced catalytic activity in strained chemically exfoliated WS_2 nanosheets for hydrogen evolution. *Nat. Mater.* **12,** 850–855 (2013).
27. Esposito, D. V., Hunt, S. T., Kimmel, Y. C. & Chen, J. G. G. A new class of electrocatalysts for hydrogen production from water electrolysis: metal monolayers supported on low-cost transition metal carbides. *J. Am. Chem. Soc.* **134,** 3025–3033 (2012).
28. Zhang, P. L., Wang, M., Yang, Y., Yao, T. Y. & Sun, L. C. A molecular copper catalyst for electrochemical water reduction with a large hydrogen-generation rate constant in aqueous solution. *Angew. Chem. Int. Ed.* **53,** 13803–13807 (2014).
29. Menezes, P. W. *et al.* Cobalt-manganese-based spinels as multifunctional materials that unify catalytic water oxidation and oxygen reduction reactions. *ChemSusChem* **8,** 164–171 (2015).
30. Bloor, L. G., Molina, P. I., Symes, M. D. & Cronin, L. Low pH electrolytic water splitting using earth-abundant metastable catalysts that self-assemble in situ. *J. Am. Chem. Soc.* **136,** 3304–3311 (2014).
31. Trotochaud, L., Ranney, J. K., Williams, K. N. & Boettcher, S. W. Solution-cast metal oxide thin film electrocatalysts for oxygen evolution. *J. Am. Chem. Soc.* **134,** 17253–17261 (2012).
32. Grigoriev, S. A., Porembsky, V. I. & Fateev, V. N. Pure hydrogen production by PEM electrolysis for hydrogen energy. *Int. J. Hydrogen Energy* **31,** 171–175 (2006).
33. Marshall, A. T., Sunde, S., Tsypkin, M. & Tunold, R. Performance of a PEM water electrolysis cell using $Ir_xRu_yTa_zO_2$ electrocatalysts for the oxygen evolution electrode. *Int. J. Hydrogen Energy* **32,** 2320–2324 (2007).
34. Cheng, J. B., Zhang, H. M., Chen, G. B. & Zhang, Y. N. Study of $Ir_xRu_{1-x}O_2$ oxides as anodic electrocatalysts for solid polymer electrolyte water electrolysis. *Electrochim. Acta* **54,** 6250–6256 (2009).
35. Wei, G. Q. *et al.* The stability of MEA in SPE water electrolysis for hydrogen production. *Int. J. Hydrogen Energy* **35,** 3951–3957 (2010).
36. Leng, Y. J. *et al.* Solid-state water electrolysis with an alkaline membrane. *J. Am. Chem. Soc.* **134,** 9054–9057 (2012).
37. Zeng, K. & Zhang, D. K. Recent progress in alkaline water electrolysis for hydrogen production and applications. *Progr. Energy Combust. Sci.* **36,** 307–326 (2010).
38. Lu, P. W. T. & Srinivasan, S. Advances in water electrolysis technology with emphasis on use of the solid polymer electrolyte. *J. Appl. Electrochem.* **9,** 269–283 (1979).
39. Symes, M. D. & Cronin, L. Decoupling hydrogen and oxygen evolution during electrolytic water splitting using an electron-coupled-proton buffer. *Nat. Chem* **5,** 403–409 (2013).
40. Rausch, B., Symes, M. D., Chisholm, G. & Cronin, L. Decoupled catalytic hydrogen evolution from a molecular metal oxide redox mediator in water splitting. *Science* **345,** 1326–1330 (2014).
41. Mallouk, T. E. Divide and conquer. *Nat. Chem.* **5,** 362–363 (2013).
42. Vengatesan, S., Santhi, S., Jeevanantham, S. & Sozhan, G. Quaternized poly (styrene-co-vinylbenzyl chloride) anion exchange memebrances for alkaline water electrolysers. *J. Power Sources* **284,** 361–368 (2015).
43. Dai, J. X., Li, S. F. Y., Xiao, T. D., Wang, D. M. & Reisner, D. E. Structural stability of aluminum stabilized alpha nickel hydroxide as positive electrode for alkaline secondary batteries. *J. Power Sources* **89,** 40–45 (2000).
44. Oliva, P., Leonardi, J., Laurent, J. F., Delmas, C., Braconnier, J. J. & Figlarz, M. Review of the structure and the electrochemistry of nickel hydroxides and oxy-hydroxides. *J. Power Sources* **8,** 229–255 (1982).
45. Ógrady, W. E., Pandya, K. I., Swider, K. E. & Corrigan, D. A. *In situ* X-Ray absorption near-edge structure evidence for quadrivalent nickel in nickel battery electrodes. *J. Electrochem. Soc.* **143,** 1613–1616 (1996).

46. Corrigan, D. A. & Knight, S. L. Electrochemical and spectroscopic evidence on the participation of quadrivalent nickel in the nickel hydroxide redox reaction. *J. Electrochem. Soc.* **136,** 613–619 (1989).
47. Bardé, F., Palacin, M. R., Chabre, Y., Isnard, O. & Tarascon, J.-M. *In situ* neutron powder diffraction of nickel hydroxide electrode. *Chem. Mater.* **16,** 3936–3948 (2004).
48. Cheng, J. *et al.* Preliminary study of single flow zinc-nickel battery. *Electrochem. Commun.* **9,** 2639–2642 (2007).
49. Shivkumar, R., Kalaignan, G. P. & Vasudevan, T. Performance characteristics of Zn/NiOOH cell. *Bull. Electrochem.* **15,** 347–349 (1999).
50. Adler, T. C. *et al.* Rechargeable zinc cell with alkaline electrolyte which inhibits shape change in zinc electrode. US Patent 5,453,336 (1995).
51. Luo, J. Y., Cui, W. J., He, P. & Xia, Y. Y. Raising the cycling stability of aqueous lithium-ion batteries by eliminating oxygen in the electrolyte. *Nat. Chem.* **2,** 760–765 (2010).
52. Wang, Y. G. *et al.* High performance hybrid supercapacitor based on graphene-supported $Ni(OH)_2$-nanowires and ordered mesoporous carbon CMK-5. *J. Electrochem. Soc.* **160,** A98–A104 (2013).

Acknowledgements

We acknowledge funding support from the Natural Science Foundation of China (21333002, 21373060), and Shanghai Science and Technology Committee (13JC1407900).

Author contributions

Y.W. conceived and designed the experiments. Y.W. and Y.X. directed the project. L.C., X. D. and Y. W. performed the experiments. L. C. and Y. W. co-wrote the paper. All authors discussed the results and commented on the manuscript.

Additional information

Competing financial interests: The authors declare no competing financial interests.

13

Network dynamics in nanofilled polymers

Guilhem P. Baeza[1], Claudia Dessi[1,2], Salvatore Costanzo[1,2], Dan Zhao[3], Shushan Gong[4], Angel Alegria[5], Ralph H. Colby[4], Michael Rubinstein[6], Dimitris Vlassopoulos[1,2] & Sanat K. Kumar[3]

It is well accepted that adding nanoparticles (NPs) to polymer melts can result in significant property improvements. Here we focus on the causes of mechanical reinforcement and present rheological measurements on favourably interacting mixtures of spherical silica NPs and poly(2-vinylpyridine), complemented by several dynamic and structural probes. While the system dynamics are polymer-like with increased friction for low silica loadings, they turn network-like when the mean face-to-face separation between NPs becomes smaller than the entanglement tube diameter. Gel-like dynamics with a Williams-Landel-Ferry temperature dependence then result. This dependence turns particle dominated, that is, Arrhenius-like, when the silica loading increases to ~31 vol%, namely, when the average nearest distance between NP faces becomes comparable to the polymer's Kuhn length. Our results demonstrate that the flow properties of nanocomposites are complex and can be tuned via changes in filler loading, that is, the character of polymer bridges which 'tie' NPs together into a network.

[1] Foundation for Research and Technology - Hellas (FORTH), Institute of Electronic Structure and Laser, Heraklion, Crete 70013, Greece. [2] Department of Materials Science and Technology, University of Crete, Heraklion, Crete 71003, Greece. [3] Department of Chemical Engineering, Columbia University, New York, New York 10027, USA. [4] Department of Materials Science and Engineering, Pennsylvania State University, University Park, Pennsylvania 16802, USA. [5] Departamento de Física de Materiales y Centro de Física de Materiales CSIC-UPV/EHU, Universidad del País Vasco UPV/EHU, San Sebastian E-20018, Spain. [6] Department of Chemistry, University of North Carolina, Chapel Hill, North Carolina 27599-3290, USA. Correspondence and requests for materials should be addressed to G.P.B. (email: guilhem.baeza@iesl.forth.gr).

The ability of nanoparticles (NPs) to substantially alter the macroscopic properties of polymers to which they are added is well established, but developing the ability to improve properties, for example, mechanical, remains an open challenge[1–17]. Restricting the discussion to spherical NPs, it is accepted that particle size and loading influence the local (segmental α-relaxation) dynamics of the matrix polymer[12,13,18,19], and there is consensus that the region in the neighbourhood of the NP, sometimes termed the 'bound' layer, has reduced segmental mobility, driven by attractive NP-polymer interactions. Long and co-workers go further and suggest that favourable NP-polymer interactions lead to the formation of a glassy (that is, dynamically frozen) bound polymer layer[14,20]. While the existence of a bound polymer layer has now been verified[12,21–23], it is also established that its segmental dynamics is not necessarily glassy[24,25]. There is therefore good understanding of segmental dynamics in these situations.

In contrast, an outstanding challenge is the consequence of these segmental dynamics on collective system relaxation, and hence mechanical reinforcement, especially at high NP loadings. There are two dominant trains of thought in the field: first, mechanical reinforcement is purely due to NP clusters[26] and second, there is a network formed with the NPs as the nodes and the adsorbed polymers as the network strands[27]. Long and co-workers further suggest that mechanical reinforement results when the glassy layers on adjacent NPs overlap, and the resulting network structure percolates[14,20,28,29]. Apart from the new feature of a glassy bound polymer layer, this last picture is analogous to the pure NP aggregation mechanism discussed above. Since it is not clear if bound glassy layers are always formed at attractive NP/polymer interfaces, their importance in mechanical reinforcement and how they can affect system dynamics of nanofilled polymers remains open.

We address this challenge here. In particular, we demonstrate that system dynamics at low NP loadings are polymer-like, but with an increased friction, presumably due to favourable NP-polymer interactions. Above a critical NP loading, specifically when the mean inter-NP spacing becomes smaller than the polymer's entanglement tube diameter, they become consistent with the formation of a polymer-mediated NP network. System dynamics are gel-like in this regime, but the temperature dependence of system relaxation still follows the Williams–Landel–Ferry (WLF) relation reminiscent of the polymer chains. This behaviour turns into a particle-dominated Arrhenius-like dependence at even higher loadings, that is, where the NP face-to-face distances are comparable to the Kuhn length of the chains (≈1 nm, ref. 30). Thus, when the NPs with their tightly bound polymer layers begin to overlap, system dynamics lose any obvious connections to those of the polymer chains that bridge the NPs. These novel findings unequivocally establish network formation as the cause of mechanical reinforcement in this class of polymer nanocomposites, and suggest that it could be due to either of the two mechanisms proposed in the literature in different regions of parameter space.

Results

Systems studied. We investigate canonical, favourably interacting mixtures of silica NPs (Nissan MEK-ST, of a diameter $d = 14 \pm 4$ nm) with poly(2-vinylpyridine) (P2VP, $T_g \approx 375$ K), of molecular weights $M_w = 554$, 219, 105 and 36 kg mol^{-1}, and respective radii of gyration R_g of 17.8, 11.2, 7.7 and 4.5 nm, using a suite of static and dynamic probes. These nanocomposites have an essentially uniform distribution of NPs in the polymer matrix at all loadings, a result that has been established using transmission electron microscopy (TEM; Supplementary Fig. 1)[12,13,31–33]. Using TEM alone to characterize NP dispersion is unsatisfactory due to the small sample sizes (typically several μm on a side) coupled to the thickness of the microtomed slices (typically 60 nm in our work). Thus, NPs which are at different 'depths' in a sample can appear to be overlapping in a TEM image. These issues are alleviated by the use of a scattering probe, such as ultrasmall angle X-ray scattering, which we previously used to show that the NPs do not cluster[32]; in addition, the mean distance between the NPs deduced from these measurements are consistent with uniform NP spatial dispersion. Consequently, these NPs are well-dispersed at all scales that we can characterize[32]. Moreover, while two-dimensional NMR (2D-NMR) shows that no chemical reactions occur during sample preparation or during the subsequent thermal annealing, Fourier transform infrared spectroscopy (FTIR) measurements indicate strongly favourable interactions between the NPs and the polymer[22], that is, strong hydrogen-bonds (H-bonds) between the silanol groups on the particle surface and the nitrogen atoms of the pyridine ring in the polymer backbone (see the FTIR and NMR sections, Supplementary Figs 2–5).

Segmental dynamics. Figure 1 shows differential scanning calorimetry (DSC) results of nanocomposite samples with the 554 kg mol^{-1} polymer matrix as a function of NP loading. Note that in the following text the volume fraction ϕ is the ratio of the volume of silica NPs to the total sample volume. Here we plot the specific heat capacity of the polymer[13] as a function of temperature. While the calorimetric glass transition temperature T_g of the P2VP matrix (375 K) is barely affected by NP loading, the transition becomes broader, especially for NP loadings of 16 vol% and larger. In fact, the breadth of this transition goes from 11 K for pure P2VP to 11.5 K for 16 vol% and 16 K for 31 vol% NP loading, that is, 1.5 times broader than the neat matrix, regardless of M_w (see inset of Fig. 1 and Supplementary Fig. 6). Apparently, segmental dynamics become more heterogeneous with increased NP loading, with significant changes occurring somewhere between 16–31 vol% NPs[30].

The local P2VP dynamics are also characterized using dielectric spectroscopy[12,13]. Figure 2 shows that NP loading has

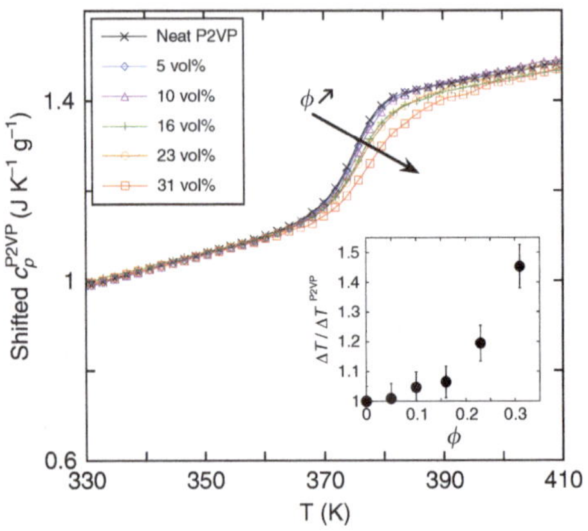

Figure 1 | Glass transition broadening in nanocomposites. Specific heat capacity of the polymer fraction as a function of the temperature for neat P2VP (554 kg mol^{-1}) and nanocomposites loaded with 5, 10, 16, 23 and 31 vol% silica. c_p^{P2VP} is calculated by subtracting the silica contribution from the nanocomposite's heat capacity as calculated in ref. 13. Note that the thermograms are arbitrarily shifted to illustrate the broadening of the glass transition. The inset is the breadth ΔT of the glass transition as a function of the filler fraction normalized by that of the neat P2VP. The error bars come from the temperature range considered around the glass transition to establish linear dependency of the heat capacity.

a strong influence on the strength of the segmental α-relaxation process, which decreases at high silica content; the NPs also broaden this relaxation towards lower frequencies (Fig. 2b), in good agreement with the DSC data. However, the mean relaxation time of the α-process, τ_α, remains essentially unaltered (see inset of Fig. 2b) as does the mean time of the more local, β-relaxation (see Supplementary Figs 7–9)[18]. The broadening of the α-process with increased NP loading is consistent with the DSC results and is attributed to the bound polymer layer formed due to strong NP-polymer interactions[12,14,18].

Collective dynamics. Viscoelastic master curves in the linear regime (storage G' and loss G'' moduli plotted against reduced angular frequency $a_T\omega$) obtained using time–temperature superposition (Supplementary Figs 10 and 11) at each NP loading are compiled in Fig. 3a for the 554 kg mol^{-1} matrix. (Supplementary Fig. 12 verifies that these data are collected in the regime of linear response.) Limited, single temperature (T = 453 K) data for nanocomposites based on shorter P2VP chains (219 and 36 kg mol^{-1}) are also shown for comparison (Fig. 3b,c, respectively). Up to 10 vol% NP (Fig. 3a), a response typical of viscoelastic liquids is found with the following additional observations: the entanglement plateau modulus (of pure P2VP) is barely affected by the addition of 5 vol% silica, whereas the clear increase at 10 vol% reflects the reinforcing action of NPs (for example, the G' at a frequency of 100 rad s^{-1}). Moreover, the local dynamics (the high-frequency moduli crossover, which is associated with the Rouse time of an entanglement strand for pure entangled polymers) is unaffected. The NPs thus simply serve to increase the friction, as might be expected by the venerable Einstein-Sutherland-Batchelor ideas.

With increasing NP loading, the ratio of high- to low-frequency moduli crossover increases (that is, the width of the intermediate plateau-like region extends to a wider range of frequencies, Fig. 3a). We find ratios of 1.83×10^4 for pure P2VP, 5.08×10^4 for 5 vol%, 1.56×10^5 for 10 vol% and 3.52×10^6 for 16 vol% NP. Above 16 vol%, it appears that the nanocomposites behave like gels over a wide frequency range down to the lowest frequencies probed (Fig. 3). Since these observations are independent of the molecular weight of the matrix chains (to first order), we cannot evoke the chain size (and its relationship to the inter-NP spacing) as being the cause of this effect. We remark that, at 16 vol%, the mean NP-NP face-to-face separation (about 12 nm, calculated following $d[\phi^{-1/3}-1]$) becomes comparable to the tube diameter of this polymer (which is estimated to be 13–15 nm from the molecular weight between entanglements, M_e, extracted from Supplementary Fig. 13). At 23 vol% the network becomes denser (we estimate the distance from the face of a NP to the face of its nearest particle is ~2.6 nm, while the mean distance between NP faces is ~9 nm, see Monte-Carlo simulations section in Supplementary Information) and the high-frequency crossover in the master curve (Fig. 3a) disappears, suggesting that the system rheology cannot be depicted through chain dynamics in the framework of the tube model. At these high NP loadings, it appears that adsorption interactions between the NPs and the polymers hinder chain motion, rather than the standard chain chain entanglements which occurs in the molten state[19,34].

More importantly, with increasing NP loading (from 16% to 31 vol%), the storage modulus in the intermediate frequency region exhibits a power law response with an approximate scaling $G' \sim \omega^{0.2}$, practically independent of the P2VP molar mass. The 31 vol% sample exhibits the same power law dependence over a wide frequency range (about 10 decades down to 10^{-2} rad s^{-1}). At lower frequencies (10^{-3} to 5×10^{-5} rad s^{-1}, see data obtained from creep measurements, Supplementary Fig. 14), however, there is a transition in behaviour with a self-similar response approximately characterized by a power law exponent of 0.5 (ref. 35). Finally, at the lowest frequency, G' tends towards a plateau while G'' keeps decreasing. A few facts that arise from this data are noted: the $G' \sim \omega^{0.2}$ behaviour which is observed over a large frequency range reflects the slow motion of short P2VP chain segments, which are adsorbed at the surface of the particles as 'trains' and require an exponentially distributed spectrum of desorption modes. Eventually there will be a change of slope due to predominance of desorption, but no flow (see Supplementary Information for further details). Additionally, the observed reinforcement (inset of Fig. 3a) is well fitted by the empirical Doolittle equation, that is, $\eta/\eta_0 = 1 + A\exp(B(\phi/(1-\phi)))$ (where η_0 and η are the viscosity of the pure polymer and the nanocomposite, respectively, ϕ is the NP volume fraction,

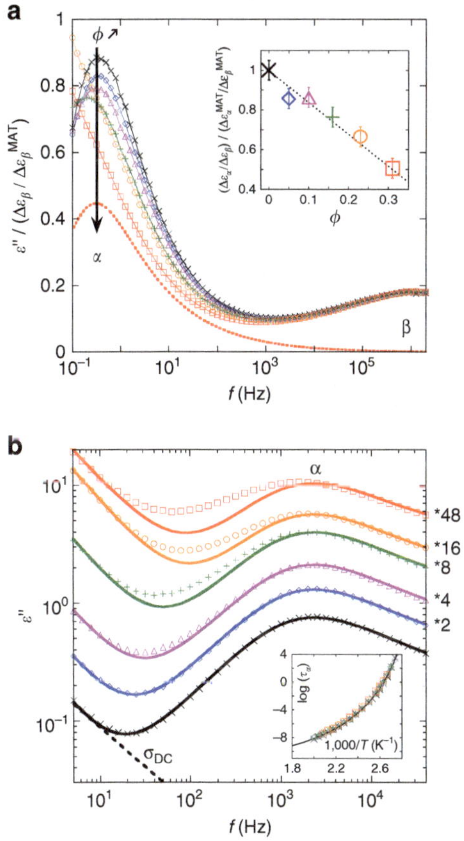

Figure 2 | Dielectric properties of nanocomposites. (**a**) Dielectric loss (ε'') reduced by the ratio of the β-process intensity over the corresponding matrix one, $(\Delta\varepsilon_\beta/\Delta\varepsilon_\beta^{MAT})$, as a function of frequency at T = 380 K for neat P2VP (cross) and nanocomposites loaded with 5 (diamond), 10 (triangle), 16 (plus), 23 (circle) and 31 (square) vol%. The red dashed line is the contribution of the α- process to the global ε'' in the 31 vol% sample. The inset is the ratio of the α- and β-relaxation strength normalized by the corresponding quantity for the pure matrix $(\Delta\varepsilon_\alpha/\Delta\varepsilon_\beta)/(\Delta\varepsilon_\alpha^{MAT}/\Delta\varepsilon_\beta^{MAT})$ as a function of the NP loading. The decreasing trend highlights the relative attenuation of the α-relaxation with the NP loading. (**b**) Similar measurements performed at T = 410 K have been vertically shifted for clarity (see numbers on the right side). Solid lines are fits to the data with a σ_{DC} contribution at low frequency and a Havriliak-Negami function with a fixed α-relaxation shape (symmetric and asymmetric), as found in the neat matrix in order to illustrate its broadening when NP content is increased. σ_{DC} (black dashed line) stands for the contribution of the ionic conductivity. Inset: segmental relaxation time $\log(\tau_\alpha)$ as a function of 1,000/T. The solid line is fit to the neat P2VP data with the WLF equation.

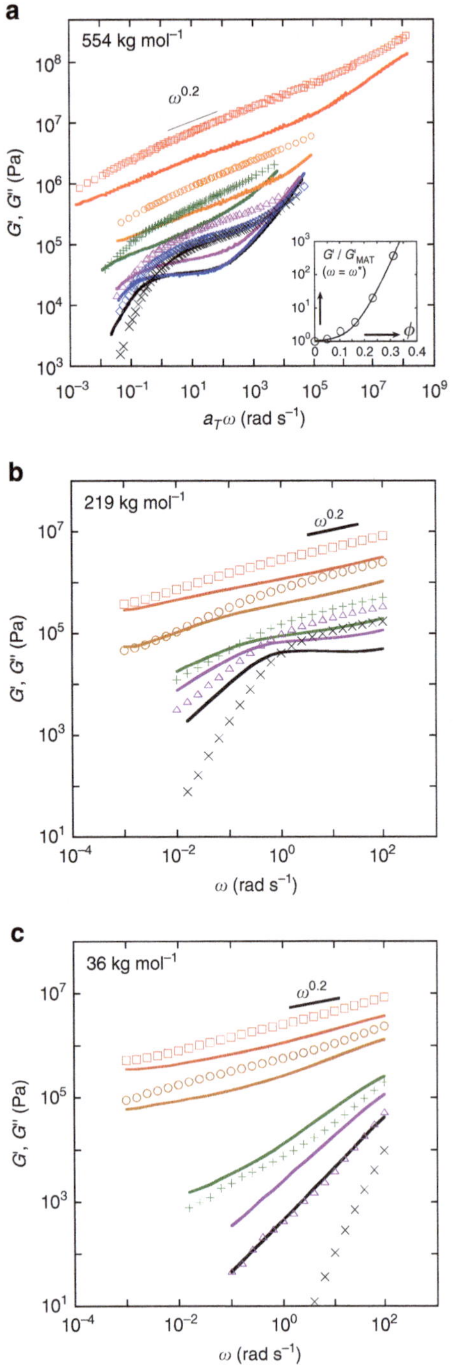

Figure 3 | Rheological behaviour of the nanocomposites. (**a**) Master curves of storage (G', symbols) and loss (G'', solid lines) moduli as a function of frequency at $T_{ref} = 473$ K for neat P2VP 554 kg mol^{-1} (cross) and nanocomposites loaded with 5 (diamond), 10 (triangle), 16 (plus), 23 (circle) and 31 (square) vol% in silica. The master curves were obtained by frequency-scale shifts only. Inset: relative increase of the storage modulus as a function of the silica volume fraction for ω where $\tan(\delta) = G''/G'$ reaches its minimum (Supplementary Fig. 10). The solid line is a fit to the Doolittle equation[41] with $A = 0.052$ and $B = 19.46$ (see text). (**b**,**c**) Respective measurements performed in different matrices at a single temperature $T = 453$ K for (**b**) $M_w = 219$ kg mol^{-1} and (**c**) $M_w = 36$ kg mol^{-1}. The legends are the same as that in **a**, where the symbols and solid lines respectively stand for G' and G''.

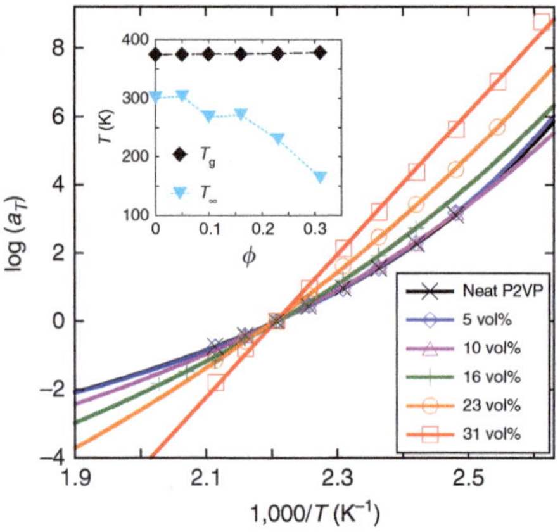

Figure 4 | Variations of shift factors with temperature and NP loading. Temperature dependence of the frequency-scale shift factor a_T used to build the viscoelastic master curves in Fig. 3a for the neat P2VP 554 kg mol^{-1} matrix and its associated nanocomposites. Solid lines are WLF fits to the data, except for the 31 vol% data, which is fit to an Arrhenius equation; fit parameters are provided in Supplementary Table 1. Inset: NPs volume fraction dependence of the glass transition temperature T_g and Vogel temperature T_∞ for the same samples. T_∞ has been extracted from fitting the data of the main figure with the Vogel-Fulcher-Tammann equation (see Supplementary Table 1).

and A, B are fit parameters), which is often used to describe the viscosity η of glass forming materials based on free volume considerations. Finally, a time sweep measurement (Supplementary Fig. 15) reveals that the nanocomposite loaded with 31 vol% silica shows extremely delayed relaxation (beyond the experimental measurement window) consistent with the extended frequency spectrum presented in Supplementary Fig. 14b. This suggests that the structure obtained is temporally evolving as in transient gels or glasses (this is the well known physical aging mechanism in kinetically arrested systems). These independent observations support the same idea: the nanocomposites show a dynamic transition when the NP loading is increased to 31 vol%. We shall speculate below on the structural origins of this transition.

The essential information on the temperature dependence of the measured chain-level dynamics is summarized in Fig. 4, which depicts the frequency-scale shift factors as a function of inverse temperature for the different NP loadings studied, using $T_{ref} = 453$ K (see also Supplementary Table 1 and Supplementary Fig. 16). Note that the temperature ranges we can explore are limited from below by the vitrification of the polymer ($\sim$375 K, independent of loading), and from above by thermal degradation, which occurs beyond 473 K. First, we note that the shift factors obtained from the different NP loadings do not superimpose. More interestingly, at 31 vol% a clear change from WLF to Arrhenius behaviour is observed, with the latter characterized by an unusually large activation energy (403 kJ mol^{-1}). Concominantly, the extrapolated Vogel temperature, T_∞, decreases strongly with NP loading and its distance from T_g is much larger at 31 vol% (inset of Fig. 4). In fact, the growing difference between T_g and T_∞ with increasing ϕ implies that polymer–polymer interactions become less important relative to global dynamics. While the Arrhenius temperature dependence may thus be rationalized by the experiments being performed at temperatures far above the Vogel temperature, this result appears at odds with the fact that these materials show very slow relaxation and age (gel/glassy behaviour). These facts can be reconciled by postulating that the material behaves akin to a percolating NP-polymer network, but with a slow local rearrangement of adsorbed chains (Supplementary Fig. 15); the Arrhenius dependence reflects the (apparently non-cooperative) activated exchange of adsorbed segments with bulk molecules. These experiments have been repeated on shorter P2VP chains ($M_w = 105\,kg\,mol^{-1}$; see Supplementary Figs 17 and 18) where similar results have been found.

The outstanding question then is the structural basis of the transition seen in the temperature dependence of system dynamics at a loading of 31 vol%. Computer simulations of hard sphere fluids have been used to calculate the mean face-to-face distances to the first, second and third nearest neighbouring NPs as a function of loading (Supplementary Tables 2 and 3). Between 23 and 31 vol% it is apparent that the distance to the closest neighbour drops to ~1.65 nm (i.e., about 1 to 2 Kuhn lengths). Under these circumstances, many of the nearest NP faces must accommodate only one Kuhn segment; in other words, the NPs come so close that they are bridged by one bound Kuhn segment, which by definition is adsorbed simultaneously on both NPs. It is the accompanying doubling of the (free) energy of desorption that we believe exponentially increases the lifetime of these bridges. Similarly, the second nearest NP is only 3 Kuhn segments away. This corresponds to one Kuhn segment adsorbed on each NP, with a third, strongly constrained internal segment. Since each of these segments adsorb/desorb effectively independent of each other, there should be no cooperative effects associated with their slow relaxations. It is these two facts, in combination, that we believe are responsible for the Arrhenius dependence that emerges under these larger NP loadings[36].

We now rationalize the large activation energy. It is true that, even though most of polymers exhibit an Arrhenius-like dependence of viscosity at temperatures far above T_g (typically at $T_g + 150\,K$ or $T_\infty + 200\,K$), the corresponding activation energy is usually much lower than that found in our case. For example, metallocene-based polyethylenes have an activation energy of $\sim 35\,kJ\,mol^{-1}$ (ref. 37). We conjecture that the large activation energy seen is a consequence of favourable silica-P2VP interactions (for the silica-P2VP bond see sections of FTIR and 2D-NMR in Supplementary Information). At high NP loadings (>10 vol%), this results in the percolation of NPs and hence a NP-polymer network forms[30]. For the 31 vol% loading, where face-to-face contacts are mediated by a Kuhn segment, we conjecture that the relaxation dynamics are dominated by the release of this doubly adsorbed Kuhn segment. With an average of 4–5 such 'doubly adsorbed' P2VP Kuhn segments per NP (a reasonable hypothesis), and a N–H bond activation energy of $\sim 45\,kJ\,mol^{-1}$ (ref. 38) (or $90\,kJ\,mol^{-1}$ for detachment of a Kuhn segment), this results in an energy of $\approx 400\,kJ\,mol^{-1}$, that is, close to what is experimentally found. The FTIR data provided in Supplementary Figs 3 and 4 supports this strong silica-P2VP bonding through the perturbation of the Si–OH bonds present at the NPs surface. Further evidence comes from viscoelastic measurements at higher strains (Supplementary Fig. 19) which indicate that the activation energy is reduced as strain increases, apparently due to some strain-induced debonding of the NP-polymer bound layer, in agreement with literature reports[39]. Finally, simple dissolution experiments corroborate this scenario. We find that while P2VP dissolves in chloroform, the high NP loading nanocomposites do not. On the other hand, both matrix and nanocomposite dissolve in THF which is a hydrogen bond acceptor, apparently due to substitution of silanol-P2VP by THF-P2VP hydrogen bonding.

In summary, in our system the average local segmental level dynamics (α-process, which presumably reflects the motions of the unadsorbed chain segments) show no change with the added NPs although it becomes broader at higher loadings, supporting the fundamental decorrelation of dynamics at different length scales as already observed by Ding and Sokolov[40]. Mechanical spectroscopy shows a clear transition in system dynamics, from that corresponding to the slower relaxation of chains (due to the presence of adsorbing NPs) at low NP loadings, to that of a network when the NP loading exceeds a critical value of ~10–16 vol% (with essentially no molecular weight dependence). Following past work[27], we deduce that a particle-dominated network comprised of silica NPs 'connected' through adsorbed P2VP chains forms at these higher NP loadings. Through geometrical arguments we deduce that this transition occurs when the mean face-to-face distance between the NPs becomes comparable to the entanglement tube diameter of the neat polymer ($a \approx 13$–$15\,nm$). More interestingly, the temperature dependence of system dynamics changes to Arrhenius-like for the 31 vol% sample, that is, when the closest face-to-face distance between two NPs becomes comparable to the Kuhn length of the polymer chains in question ($b \approx 1\,nm$). In these cases, the shortest mean bridge comprises only one adsorbed Kuhn segment, while the second shortest bridge is formed by three Kuhn segments; we propose that their sluggish desorption dynamics is reflected in the Arrhenius temperature dependence of the experimental relaxation data. These results demonstrate both the onset of network-like behaviour (gel-like in a wide range of frequency but not permanently percolated) in these systems, but, more pertinently, they show that its character can be readily tuned through variations of the NP loading. Our novel findings offer a unified explanation that network formation, that is NPs bridged by bound polymer segments, dominates the mechanical behavior of this class of polymer nanocomposites.

Methods

Nanocomposites preparation and processing. All materials were used as received. Methyl ethyl ketone (MEK, HPLC-grade, >99.7%) and pyridine (ACS agent, >99.0%) were purchased from Sigma Aldrich (USA). P2VP ($M_w = 554\,kg\,mol^{-1}$, $M_w/M_n = 1.12$; $M_w = 105\,kg\,mol^{-1}$, $M_w/M_n = 1.08$) was obtained from Polymer Source. P2VP ($M_w = 36\,kg\,mol^{-1}$, $M_w/M_n = 1.07$ and $M_w = 219\,kg\,mol^{-1}$, $M_w/M_n = 1.11$) was ordered from Scientific Polymer Products. The colloidal silica NPs (MEK-ST) and the antioxidant Irganox 1010 were donated by Nissan Chemical Industries and Ciba Specialty Chemicals, now BASF, respectively.

Nanocomposite films were prepared by co-casting composite dispersions of colloidal silica NPs and P2VP in a mixture of MEK and pyridine. First, the as-received Nissan silica suspensions were diluted with pyridine in a volume ratio of 4:1. In the meantime, as-received P2VP and antioxidant Irganox (0.2 wt% relative to P2VP) solutions in MEK were prepared. Following that, appropriate amounts of diluted silica dispersion were added to the polymer/Irganox solution, resulting in silica/P2VP formulations with a known weight ratio (see TGA; Supplementary Fig. 20). For high molar-mass polymer-based samples, especially at high silica loadings, additional small amount of pyridine was gradually added into the composite suspensions to dissolve the bridged silica/P2VP precipitants back into the solution. The resulting mixtures were vortex-shaken for 2 h and then probe ultrasonicated for 3 min using an Ultrasonic Processor (model GEX-750) operated at 24% of maximum amplitude with a pulse mode of 2 s sonication following by 1 s rest time. The solutions were then poured into a PTFE Petri dish and air dried in a fume hood for at least 5 days. Finally, in order to provide the identical sample history and completely remove all the residual solvents in the films, the as-cast samples were annealed following a well-defined procedure: 7 days at 353 K and then 10 days at 423 K, both under vacuum. The films were milled down and remoulded at 453 K and 1 bar in a hot press equipped with a vacuum mould, for 15 h (463 K and 2 bars for the 31 vol% sample). This additional step allowed specimens of 8 mm in diameter and 0.5–0.6 mm in thickness to be produced. DSC and dielectric spectroscopy tests have been subsequently performed to verify that this additional processing step did not change local segmental properties. The experimental methods are described in the Supplementary information.

References

1. Kumar, S. K., Jouault, N., Benicewicz, B. & Neely, T. Nanocomposites with polymer grafted nanoparticles. *Macromolecules* **46,** 3199–3214 (2013).
2. Kumar, S. K. & Krishnamoorti, R. Nanocomposites: structure, phase behavior, and properties. *Annu. Rev. Chem. Biomol. Eng.* **1,** 37–58 (2010).

3. Mackay, M. E. *et al.* General strategies for nanoparticle dispersion. *Science* **311,** 1740–1743 (2006).
4. Metzner, A. Rheology of suspensions in polymeric liquids. *J. Rheol.* **29,** 739–775 (1985).
5. Winey, K. I. & Vaia, R. A. Polymer nanocomposites. *MRS Bull.* **32,** 314–322 (2007).
6. Choi, J., Hore, M. J., Clarke, N., Winey, K. I. & Composto, R. J. Nanoparticle brush architecture controls polymer diffusion in nanocomposites. *Macromolecules* **47,** 2404–2410 (2014).
7. Choi, J. *et al.* Universal scaling of polymer diffusion in nanocomposites. *ACS Macro Lett.* **2,** 485–490 (2013).
8. Montes, H., Papon, A., Lequeux, F. & Guy, L. Particles in model filled rubber: dispersion and mechanical properties. *Eur. Phys. J. E* **31,** 263–268 (2010).
9. Koerner, H., Drummy, L. F., Benicewicz, B., Li, Y. & Vaia, R. A. Nonisotropic self-organization of single-component hairy nanoparticle assemblies. *ACS Macro Lett.* **2,** 670–676 (2013).
10. Baeza, G. P. *et al.* Multiscale filler structure in simplified industrial nanocomposite silica/SBR systems studied by SAXS and TEM. *Macromolecules* **46,** 317–329 (2012).
11. Baeza, G. P. *et al.* Mechanism of aggregate formation in simplified industrial silica styrene-butadiene nanocomposites: effect of chain mass and grafting on rheology and structure. *Soft Matter* **10,** 6686–6695 (2014).
12. Gong, S., Chen, Q., Moll, J. F., Kumar, S. K. & Colby, R. H. Segmental dynamics of polymer melts with spherical nanoparticles. *ACS Macro Lett.* **3,** 773–777 (2014).
13. Holt, A. P. *et al.* Dynamics at the polymer/nanoparticle interface in poly(2-vinylpyridine)/silica nanocomposites. *Macromolecules* **47,** 1837–1843 (2014).
14. Berriot, J., Montes, H., Lequeux, F., Long, D. & Sotta, P. Evidence for the shift of the glass transition near the particles in silica-filled elastomers. *Macromolecules* **35,** 9756–9762 (2002).
15. Harms, S. *et al.* Free volume of interphases in model nanocomposites studied by positron annihilation lifetime spectroscopy. *Macromolecules* **43,** 10505–10511 (2010).
16. Papon, A. *et al.* Solid particles in an elastomer matrix: impact of colloid dispersion and polymer mobility modification on the mechanical properties. *Soft Matter* **8,** 4090–4096 (2012).
17. Zhao, D. *et al.* Role of filler shape and connectivity on the viscoelastic behavior in polymer nanocomposites. *Macromolecules* **48,** 5433–5438 (2015).
18. Holt, A. P., Sangoro, J. R., Wang, Y., Agapov, A. L. & Sokolov, A. P. Chain and segmental dynamics of poly(2-vinylpyridine) nanocomposites. *Macromolecules* **46,** 4168–4173 (2013).
19. Schneider, G., Nusser, K., Willner, L., Falus, P. & Richter, D. Dynamics of entangled chains in polymer nanocomposites. *Macromolecules* **44,** 5857–5860 (2011).
20. Long, D. & Lequeux, F. Heterogeneous dynamics at the glass transition in van der Waals liquids, in the bulk and in thin films. *Eur. Phys. J. E* **4,** 371–387 (2001).
21. Jouault, N., Lee, D., Zhao, D. & Kumar, S. K. Block-copolymer-mediated nanoparticle dispersion and assembly in polymer nanocomposites. *Adv. Mater.* **26,** 4031–4036 (2014).
22. Jouault, N. *et al.* Bound polymer layer in nanocomposites. *ACS Macro Lett.* **2,** 371–374 (2013).
23. Papon, A. *et al.* Low-field NMR investigations of nanocomposites: polymer dynamics and network effects. *Macromolecules* **44,** 913–922 (2011).
24. Glomann, T. *et al.* Microscopic dynamics of polyethylene glycol chains interacting with silica nanoparticles. *Phys. Rev. Lett.* **110,** 178001 (2013).
25. Krutyeva, M. *et al.* Effect of nanoconfinement on polymer dynamics: surface layers and interphases. *Phys. Rev. Lett.* **110,** 108303 (2013).
26. Zhu, Z. Y., Thompson, T., Wang, S. Q., von Meerwall, E. D. & Halasa, A. Investigating linear and nonlinear viscoelastic behavior using model silica-particle-filled polybutadiene. *Macromolecules* **38,** 8816–8824 (2005).
27. Maier, P. G. & Goritz, D. Molecular interpretation of the Payne effect. *Kautsch. Gummi Kunstst.* **49,** 18–21 (1996).
28. Merabia, S., Sotta, P. & Long, D. R. A microscopic model for the reinforcement and the nonlinear behavior of filled elastomers and thermoplastic elastomers (Payne and Mullins effects). *Macromolecules* **41,** 8252–8266 (2008).
29. Berriot, J. *et al.* Filler–elastomer interaction in model filled rubbers, a ^{1}H NMR study. *J. Non-Cryst. Solids* **307,** 719–724 (2002).
30. Chen, Q. *et al.* Mechanical reinforcement of polymer nanocomposites from percolation of a nanoparticle network. *ACS Macro Lett.* **4,** 398–402 (2015).
31. Jouault, N., Zhao, D. & Kumar, S. K. Role of casting solvent on nanoparticle dispersion in polymer nanocomposites. *Macromolecules* **47,** 5246–5255 (2014).
32. Harton, S. E. *et al.* Immobilized polymer layers on spherical nanoparticles. *Macromolecules* **43,** 3415–3421 (2010).
33. Zhao, D., Schneider, D., Fytas, G. & Kumar, S. K. Controlling the thermomechanical behavior of nanoparticle/polymer films. *ACS Nano* **8,** 8163–8173 (2014).
34. Li, Y., Kröger, M. & Liu, W. K. Nanoparticle effect on the dynamics of polymer chains and their entanglement network. *Phys. Rev. Lett.* **109,** 118001 (2012).
35. Chen, Q., Tudryn, G. J. & Colby, R. H. Ionomer dynamics and the sticky Rouse model. *J. Rheol.* **57,** 1441–1462 (2013).
36. Klonos, P. *et al.* Morphology and molecular dynamics investigation of PDMS adsorbed on titania nanoparticles: effects of polymer molecular weight. *Eur. Polym. J.* **74,** 64–80 (2016).
37. Devendra, R., Hatzikiriakos, S. G. & Vogel, R. Rheology of metallocene polyethylene-based nanocomposites: influence of graft modification. *J. Rheol.* **50,** 415–434 (2006).
38. Hertl, W. & Hair, M. L. Hydrogen bonding between adsorbed gases and surface hydroxyl groups on silica. *J. Phys. Chem.* **72,** 4676–4682 (1968).
39. Blyler, L. & Haas, T. The influence of intermolecular hydrogen bonding on the flow behavior of polymer melts. *J. Appl. Polym. Sci.* **13,** 2721–2733 (1969).
40. Ding, Y. & Sokolov, A. P. Breakdown of time-temperature superposition principle and universality of chain dynamics in polymers. *Macromolecules* **39,** 3322–3326 (2006).
41. Doolittle, A. K. Studies in Newtonian Flow. II. The dependence of the viscosity of liquids on free-space. *J. Appl. Phys.* **22,** 1471–1475 (1951).

Acknowledgements

We thank Leon Serc (ETH Zurich) for help with FTIR. Enlightening discussions with Ulrich Jonas are gratefully acknowledged. Partial support has been provided by the EU FP7 (ETN Supolen GA-607937, Infrastructure ESMI GA-262348) and the Greek General Secretariat for Research and Technology (Thalis-380238 COVISCO). M.R. acknowledges financial support from the National Science Foundation under grants DMR-1309892, DMR-1436201 and DMR-1121107, the National Institutes of Health under grants P01-HL108808 and 1UH2HL123645 and the Cystic Fibrosis Foundation. D.Z., S.G., R.H.C. and S.K.K. gratefully acknowledge the National Science Foundation grant DMR-1408323 for financial support.

Authors contributions

D.Z. prepared and characterized the samples. G.P.B., C.D., S.C. and S.G. performed rheological measurements. G.P.B. and C.D. did IR and 2D-NMR measurements. G.P.B. did dielectric spectroscopy and DSC. S.K.K. performed the simulations. G.P.B., D.V. and S.K.K. wrote the paper. R.H.C., M.R. and A.A. contributed to interpretation and analysis. Project conceived by S.K.K. and D.V.

Additional information

Competing financial interests: The authors declare no competing financial interests.

14

Discovery of abnormal lithium-storage sites in molybdenum dioxide electrodes

Jeong Kuk Shon[1,2], Hyo Sug Lee[1], Gwi Ok Park[2,3], Jeongbae Yoon[3], Eunjun Park[4], Gyeong Su Park[1], Soo Sung Kong[2], Mingshi Jin[5], Jae-Man Choi[1], Hyuk Chang[1], Seokgwang Doo[1], Ji Man Kim[2,3], Won-Sub Yoon[3], Chanho Pak[6,*], Hansu Kim[4] & Galen D. Stucky[7,*]

Developing electrode materials with high-energy densities is important for the development of lithium-ion batteries. Here, we demonstrate a mesoporous molybdenum dioxide material with abnormal lithium-storage sites, which exhibits a discharge capacity of 1,814 mAh g^{-1} for the first cycle, more than twice its theoretical value, and maintains its initial capacity after 50 cycles. Contrary to previous reports, we find that a mechanism for the high and reversible lithium-storage capacity of the mesoporous molybdenum dioxide electrode is not based on a conversion reaction. Insight into the electrochemical results, obtained by *in situ* X-ray absorption, scanning transmission electron microscopy analysis combined with electron energy loss spectroscopy and computational modelling indicates that the nanoscale pore engineering of this transition metal oxide enables an unexpected electrochemical mass storage reaction mechanism, and may provide a strategy for the design of cation storage materials for battery systems.

[1] Samsung Advanced Institute of Technology, Samsung Electronics Co., Ltd., Suwon 443-803, Republic of Korea. [2] Department of Chemistry, Sungkyunkwan University, Suwon 440-746, Republic of Korea. [3] Department of Energy Science, Sungkyunkwan University, Suwon 440-746, Republic of Korea. [4] Department of Energy Engineering, Hanyang University, Seoul 133-791, Republic of Korea. [5] Key Laboratory of Natural Resource of the Changbai Mountain and Functional Molecular (Yanbian University), Ministry of Education, Department of Chemistry, Park Road 977, Yanji City, Jilin Province 133002, China. [6] Fuel Cell Group, Corporate R&D Center, Samsung SDI Co., Ltd., Yongin 446-577, Republic of Korea. [7] Department of Chemistry and Biochemistry, University of California, Santa Barbara, California 93106, USA. * These authors jointly supervised this work. Correspondence and requests for materials should be addressed to J.M.K. (email: jimankim@skku.edu) or to W.-S.Y. (email: wsyoon@skku.edu) or to H.K. (email: khansu@hanyang.ac.kr).

Lithium (Li)-ion batteries (LIBs) are a key-enabling technology for addressing the power and energy demands of electric vehicles and stationary electrical storage for renewable energy as well as mobile electronics[1]. However, the energy density of currently commercialized LIBs is already close to its technological limit[2]. In order to achieve the battery performance that all applications expect, much effort has been made to develop new electrode materials to improve both the energy density and cycle performance of LIBs[3]. The main goal of these research efforts are to enable energy densities that are higher than the theoretical limit predicted for current Li-ion intercalation batteries. These new batteries would have the desired physicochemical properties, especially high reversible capacity, structural flexibility and stability, high rate capability, low cost and environmental benignity[4,5]. In order to improve the performance of anode parts replacing graphite (theoretical capacity of 372 mAh g^{-1}), there has been extensive research on developing anode materials such as transition metal oxides, silicon- or tin-based metal alloys, and related composite configurations[3–11]. Since they follow Li-storage mechanisms such as conversion and alloying reactions with Li, which are different from the Li intercalation reaction in the graphite anode, these newly developed anode materials show higher capacity than the current anode systems. In particular, there had been several reports that some transition metal oxides showed the reversible capacity exceeding their theoretical capacity based on the conversion reaction. This abnormal capacity of transition metal oxides has been explained by various mechanisms based on the interfacial reaction: (i) reversible formation/dissolution of organic film at the interface between the electrolyte and the transition metal oxide[9,12]; (ii) interfacial charge storage between metal nanocrystals and Li salts[13,14]; and (iii) generation of LiOH on the surface of metal oxide and subsequent reversible reaction between LiOH and Li[15]. Hence, even though the origins of extra capacity of transition metal oxides are different from each other, one of the effective approaches is nano-engineering of transition metal oxides for LIBs[16–18]. This approach has the potential to significantly increase the reversible capacity and the rate capability of anode materials by enlarging the interfacial area between the electrode and electrolyte. It is also expected that nanostructured transition metal oxides should show much improved cycle performance by providing the mechanical/structural integrity against huge changes in volume and crystal structure.

Mesoporous materials are excellent nanoscale-engineered candidates for numerous applications because of their high surface areas, tunable pore sizes, adjustable framework thickness and compositions, and diverse surface properties[19–23]. Recently, various novel mesoporous metal oxides have been widely investigated as electrode materials for LIBs[24–28]. These studies have opened up a possibility for the development of anode materials with significantly improved Li-storage performance. The mesostructure can be readily designed to exhibit hierarchical mesoporosity to promote facile and fast Li-ion diffusion, and a uniform framework thickness of 10 nm or less so that there is a reduced diffusion length for solid-state Li transport. Interestingly, we have found that the mesoporous MoO_2 anode presented here gives a high Li-storage capacity (1,814 mAh g^{-1} at first cycle and 1,607 mAh g^{-1} after 50 cycles), which is much higher than its theoretical capacity based on the conversion reaction of MoO_2 with Li (838 mAh g^{-1}). As aforementioned, several research groups have newly proposed Li-storage mechanisms[12–15,29]; however, this extra Li-storage mechanism of transition metal oxide is still unclear, suggesting that the high capacity of the mesoporous MoO_2 in the present work probably results from a different Li-storage mechanism than those previously reported[12–15].

Here we report a mechanism for the high and reversible Li-storage capacity of the mesoporous MoO_2 anode based on various physicochemical analyses and computational modelling. To the best of our knowledge, this is the first demonstration of a Li-storage oxide anode material that utilizes two different Li-storage mechanisms composed of both a Li-ion intercalation and a metallic Li storage. This enables a structural and electronic configuration that is, in a sense, a hybrid between the Li intercalation compound anode and the metallic Li anode, with the inherent safety features of the former and energy density approaching that of the latter. We believe that understanding the origin and mechanism of this excellent performance will enable important advances in the design and creation of high-energy storage devices.

Results

Li-storage characteristics of mesoporous MoO_2. The mesoporous MoO_2 material, which exhibits regular mesopores (18.2 nm in diameter), highly crystalline frameworks (~7 nm thickness) and high surface areas of 115 $m^2 g^{-1}$, was successfully obtained from a mesoporous silica template (Fig. 1a–f and Supplementary Fig. 1). Electrochemical performances of the mesoporous MoO_2 electrode and a bulk MoO_2 (Aldrich, $S_{BET} = 0.23\ m^2 g^{-1}$) electrode are shown in Fig. 1g,h. There is an obvious difference in the Li-storage behaviours of the mesoporous MoO_2 and bulk MoO_2 electrodes. For the bulk MoO_2 electrode, the discharge (Li insertion) and charge (Li removal) capacities are 385 and 183 mAh g^{-1} (for the first cycle), respectively, indicating that 1.84 mol of Li per mol of MoO_2 can be stored into bulk MoO_2, and 0.87 mol of Li are reversibly released from the lithiated bulk MoO_2. On the other hand, the mesoporous MoO_2 delivered a reversible charge capacity of 1,308 mAh g^{-1} (for the first cycle) with an initial coulombic efficiency of 72.1% (discharge capacity of 1,814 mAh g^{-1}), which is much higher than those of the bulk MoO_2 electrode and corresponds to the reversible removal of 6.24 mol of Li per mol of MoO_2 from the fully lithiated mesoporous MoO_2. More importantly, the reversible Li-storage capacity of mesoporous MoO_2 (6.24 mol of Li per mol of MoO_2) at the first cycle exceeds the theoretical limit of Li-storage capacity through conversion reaction of MoO_2 with Li (4 mol of Li per mol of MoO_2), suggesting that a new Li-storage mechanism should be introduced to explain this unexpected Li-storage performance.

As shown in Fig. 1i,j, Supplementary Figs 2 and 3 and Supplementary Table 1, the Li-storage capacities of the MoO_2 electrodes are linearly correlated with their surface areas, even though the shapes in the small-angle X-ray diffraction (XRD) patterns and N_2 sorption isotherms of mesoporous MoO_2 materials areas are very similar (Supplementary Figs 4 and 5). In the case of mesoporous MoO_2 with surface area of 39 $m^2 g^{-1}$, the first and tenth reversible charge capacities are only 422 mAh g^{-1} (discharge capacity of 1,022 mAh g^{-1}) and 814 mAh g^{-1}, respectively, which are much lower than those of the mesoporous MoO_2 with surface area of 115 $m^2 g^{-1}$ (1,308 and 1,594 mAh g^{-1}, respectively). This result clearly indicates that the quality of the nanostructure of the mesoporous MoO_2 material is very crucial to achieve a high Li-storage capacity that exceeds the theoretical value. Differences in the synthesis of mesoporous MoO_2 materials with low and high surface areas are described in the Methods section. For more insight, scanning electron microscopic (SEM) images were obtained for the mesoporous MoO_2 materials with different surface areas (Supplementary Fig. 6). Interestingly, there are co-existing large particles in the cases of the low surface area materials (39 and 76 $m^2 g^{-1}$; Supplementary Fig. 6c,e), whereas entire

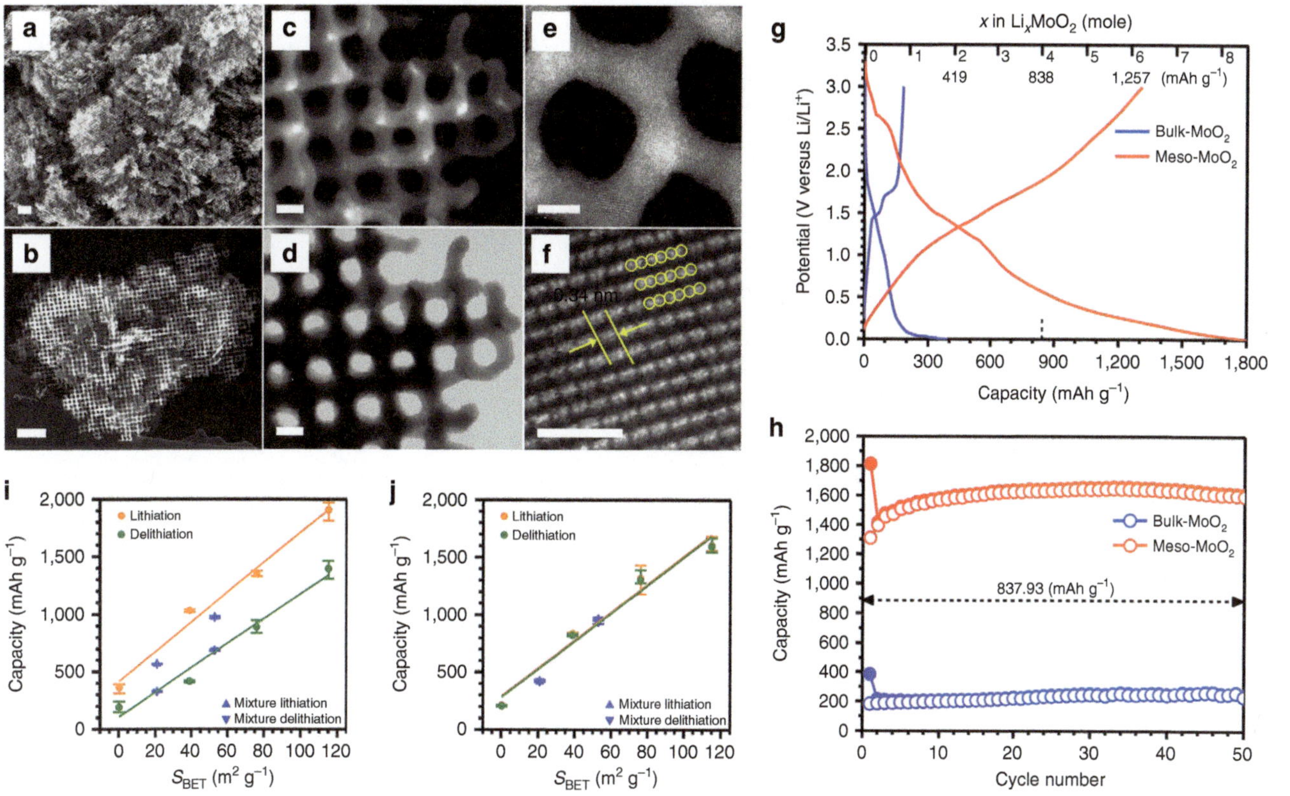

Figure 1 | Ordered mesoporous MoO_2 and Li-storage performance. (**a**,**b**) SEM, (**c**,**e**) STEM, (**d**) TEM and (**f**) HRTEM images of ordered mesoporous MoO_2 materials. Electrochemical performance of ordered mesoporous MoO_2 ($S_{BET}=115\,m^2\,g^{-1}$) and bulk MoO_2 (Aldrich, $S_{BET}=0.23\,m^2\,g^{-1}$): (**g**) voltage profiles and (**h**) cycle performances at current rate of 0.1 C in 1.3 M $LiPF_6$ (ethylene carbonate/diethyl carbonate (EC/DEC) = 3/7, by volume ratio). The value (837.93 mAh g^{-1}) represents the theoretical capacity for MoO_2, based on conversion reaction (4 mol of Li per 1 mol of MoO_2). Relationship of the Li-storage capacities of the MoO_2 electrodes with respective to their surface areas: (**i**) the first and (**j**) the tenth lithiation/delithiation. The galvanostatic lithiation/delithiation tests were replicated for three times. Error bars (**i**,**j**) represent the standard deviation of the mean about a series of the measured values. Straight line is a linear fit. Scale bars = 100 nm (**a**,**b**), 10 nm (**c**,**d**), 5 nm (**e**) and 1.5 nm (**f**).

particles of the mesoporous MoO_2 with the surface area of $115\,m^2\,g^{-1}$ exhibit highly ordered mesostructures (Fig. 1a and Supplementary Fig. 6f). For comparison, the Li-storage behaviours of two different mixtures of bulk MoO_2 and high-quality mesoporous MoO_2 materials (resulting in $S_{BET}=21\,m^2\,g^{-1}$ and $53\,m^2\,g^{-1}$, respectively; Supplementary Figs 4, 5 and 6b,d and Supplementary Table 1) were also investigated, indicating that the mixtures give similar trends of Li-storage capacities depending on their surface areas.

In order to gain insight on the physics and material science of Li storage in the present ordered mesoporous MoO_2, we performed *ex situ* XRD analyses during the lithiation and delithiation of the ordered mesoporous MoO_2. Figure 2a shows that Bragg peaks corresponding to MoO_2 phase shifted to lower angles with an increase in the amount of Li stored, and returned to their initial scattering angles during the delithiation step, indicating highly reversible Li intercalation of the MoO_2 host without collapse of its crystalline structure during cycling. High-resolution transmission electron microscopic (HRTEM) images (Fig. 2b) also show reversible changes of crystalline structure during the first cycle. We also observed reversible shifts of the Bragg peaks of MoO_2 without the breakdown of the crystal structure even during the second cycle (Supplementary Fig. 7). It should be noted that we could not detect the evolution of any other phase related to the conversion reaction of MoO_2, such as metallic Mo and Li_2O, even after lithiation was electrochemically completed. These *ex situ* XRD analyses suggest that the present ordered mesoporous MoO_2 does not follow the conventional conversion reaction mechanism, in which there is accommodation and release of Li ions by a Li intercalation mechanism. The XRD patterns obtained after the full lithiation (D – 0.0 V in Fig. 2a, >7 mol of Li per 1 mol of MoO_2) are very similar to the monoclinic Li_xMoO_2 phase, which means that the phase transition of crystal structures during the cycle is not the key for the unexpected high Li-storage capacity of the present mesoporous MoO_2 material. In the HRTEM images (Supplementary Fig. 8), we found that the highly crystalline framework of the pristine mesoporous MoO_2 material became a mixture of crystalline phase and amorphous phase upon the lithiation. This suggests that the amorphous phases formed between the crystalline domains may be critical for such high capacity of mesoporous MoO_2. *In situ* Mo *K*-edge X-ray absorption near edge structure (XANES; Fig. 2c and Supplementary Fig. 9) spectra for the ordered mesoporous MoO_2 electrode during lithiation show a clear shift of the absorption edge towards lower energy position, showing a change in the electronic structure of molybdenum from Mo^{4+} to a formal oxidation state close to Mo^{1+}. This result indicates formation of another domain (probably the amorphous phase shown in the HRTEM image; Supplementary Fig. 8), in addition to the Li-intercalated monoclinic Li_xMoO_2 phase during the lithiation. Furthermore, *in situ* Mo *K*-edge extended X-ray absorption fine structure (EXAFS; Fig. 2d) spectra for the ordered mesoporous MoO_2 show that the Mo–O and Mo–Mo peaks in the EXAFS spectra correspond to the coordination shells in MoO_2 structure, excluding any possibility of metallic Mo phase. Besides the extraordinarily high capacity, these contradictory XANES and EXAFS results against the conventional conversion

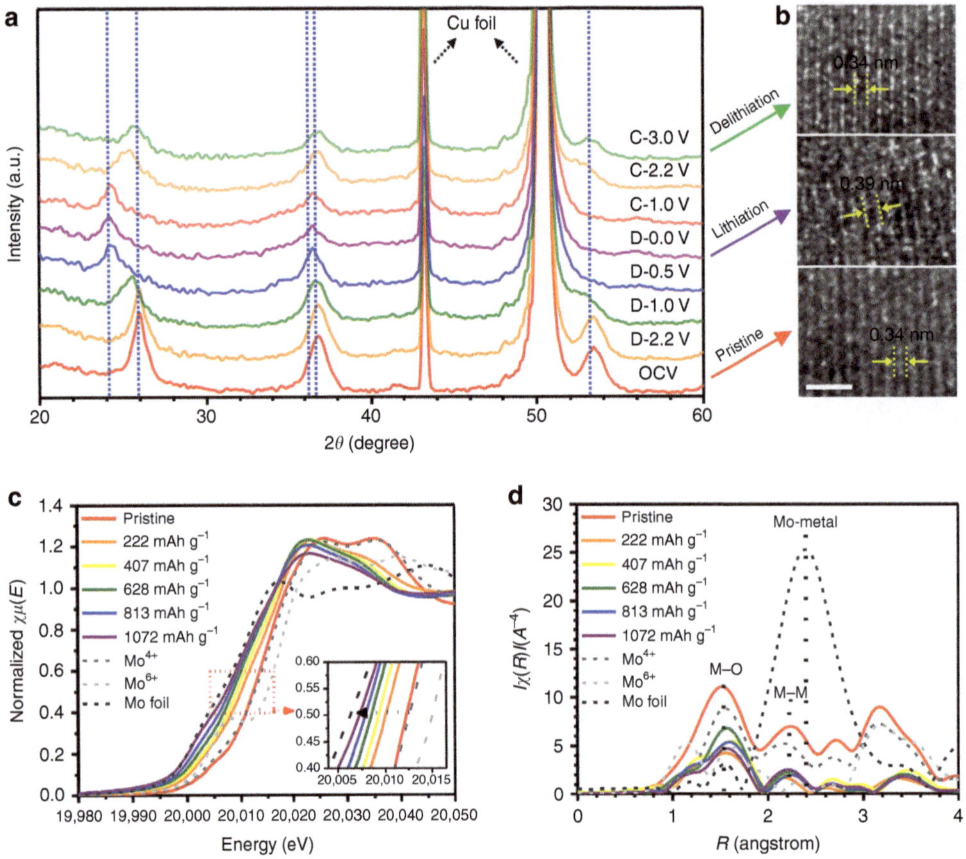

Figure 2 | Structure evolution of Li storage in ordered mesoporous MoO_2. (**a**) *Ex situ* XRD patterns of ordered mesoporous MoO_2 during the first cycle as a function of depth of discharge and charge. (**b**) HRTEM images of mesoporous MoO_2 before lithiation, after full lithiation and after full delithiation. Scale bar = 1.5 nm. (**c**) *In situ* XANES patterns and (**d**) *in situ* EXAFS patterns of ordered mesoporous MoO_2 electrode for the first lithiation.

reaction of transition metal oxide anode materials strongly suggest that another reaction model for Li storage into the ordered mesoporous MoO_2 should be considered in order to understand the origin of the extraordinary high Li-storage properties and the electrochemical reaction mechanism of the ordered mesoporous MoO_2 with Li.

Mechanism study on the Li-storage sites. In order to obtain a theoretical perspective into the atomic movement and redox behaviour that occur upon lithiation of MoO_2, we performed density functional theory (DFT) calculations on the Li-storage mechanism of the ordered mesoporous MoO_2. Figure 3a shows that the lithiation process of MoO_2 is almost the same for Li intercalation up to $Li_{1.5}MoO_2$ as evident from the small increase in unit cell volume (~20%) and the retention of initial crystal structure, that is, the monoclinic MoO_2 phase. However, further Li storage leads to phase separation of the lithiated MoO_2 into a Li-rich phase (the amorphous domain in Supplementary Fig. 8) and Li-intercalated MoO_2 crystalline phase (the crystalline domain in Supplementary Fig. 8). During the initial stage of this phase separation of lithiated MoO_2, we found that additional Li atoms are deposited as nearest neighbours to a position occupied by a former Li atom. This near-neighbour Li atom siting is more energetically favourable as a new position for a Li atom than is a random position (Supplementary Fig. 10). As shown in Fig. 3a, the further lithiation results in volume increase during cycling without expansion in the lithiated crystalline $Li_{1.5}MoO_2$ domain. This total volume expansion by formation of Li-rich phase is probably an important characteristic that enables efficient Li storage in the ordered mesoporous MoO_2. With an increase in the depth of the lithiation layer, only the Li-rich phase (formed between crystalline Li_xMoO_2 domains) reacts with Li ion, and the interatomic distance between Li atoms tends to decrease, suggesting that the electronic structure of the Li atoms turns into a metallic state. As seen in Supplementary Fig. 11, the calculated partial density of state of Li *s*-band clearly shows the change of the electronic states of Li into metallic states when $x > 1.5$, which suggests that the metallic Li insertion between the crystalline Li_xMoO_2 domains causes the change in electronic structure.

It should be noted that this change in the electronic state of Li stored into a metal oxide differs from the interfacial Li-storage mechanism that has been used to explain the extra capacity of the transition metal oxide anode based on the conversion reaction mechanism[14]. According to the interfacial storage mechanism, extra Li ions are stored on the side of Li_2O at the interface between Li_2O and transition metal, which are formed through the fully conversion reaction of the transition metal oxide[14]. However, the Li-storage mechanism proposed in the present work for ordered mesoporous MoO_2 does not follow the conversion reaction of MoO_2 into Mo and Li_2O, but rather retains the crystalline phases (MoO_2 and Li_xMoO_2 during the cycle), with metallic Li accommodated between the Li-ion-intercalated MoO_2 crystalline domains. This unexpected Li-storage mechanism for the extraordinarily high capacity of ordered mesoporous MoO_2 can be attributed to the nanoscale pore-engineered structure, which offers a much higher surface area with a high metal oxide density. The thin MoO_2 frameworks

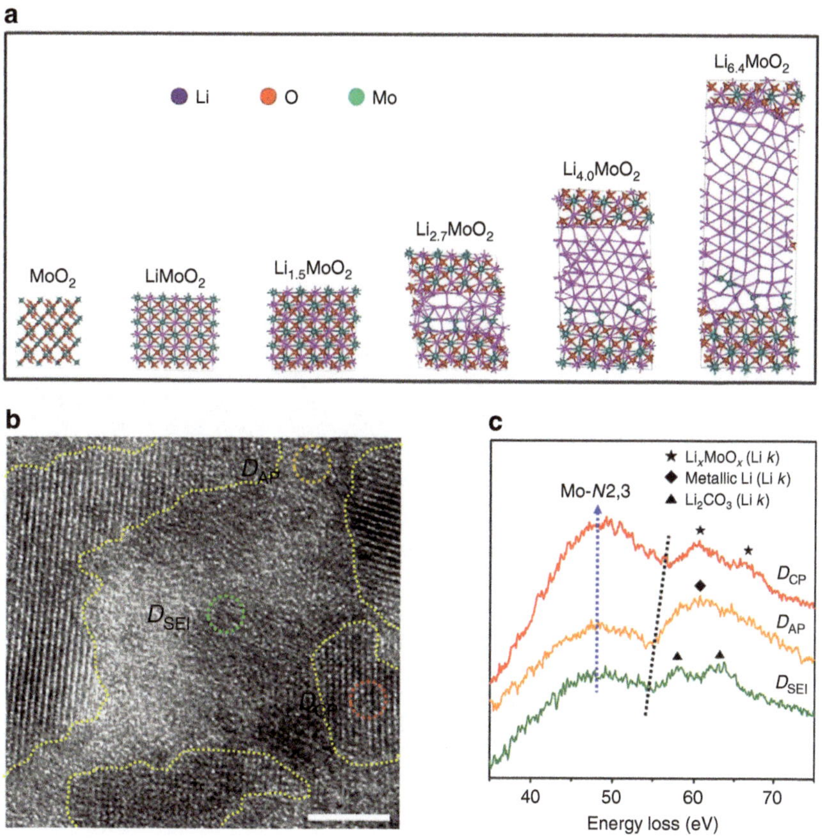

Figure 3 | DFT calculations and STEM-EELS spectra. (**a**) Snapshots of the ordered mesoporous MoO_2 electrode with the increase of Li inserted, calculated by DFT, and (**b**) top-view HRTEM image taken from a fully lithiated ordered mesoporous MoO_2, demonstrating the formation of three different phases: crystalline phase (D_{CP}), amorphous phase (D_{AP}) and SEI (D_{SEI}). Scale bar = 5 nm. (**c**) STEM-EELS spectra taken from areas D_{CP}, D_{AP} and D_{SEI} of the fully lithiated ordered mesoporous MoO_2. Dotted lines represent the peak position of Mo *N* edge (blue) and the onset of Li *K*-edge (black).

(~7 nm) greatly shorten the diffusion length for the Li ions, thus enabling the electrochemical reaction that cannot occur in the bulk analogue.

In order to obtain additional direct experimental evidences for the two-phase reaction model proposed here (formation of both Li-ion-intercalated MoO_2 crystalline phase and metallic Li-rich phase) along with the change of the electronic state of Li, we performed HRTEM and electron energy loss spectroscopy (EELS) studies on the fully lithiated and delithiated MoO_2 electrode. The HRTEM image in Fig. 3b clearly shows that there are three different phases formed in the fully lithiated MoO_2, that is, crystalline phase (D_{CP}), amorphous phase (D_{AP}) and solid electrolyte interface (SEI) phase (D_{SEI}). The EELS data in scanning TEM (STEM) mode (STEM-EELS) taken at the crystalline area (D_{CP}) show that the Li *K*-edge corresponds to Li_xMoO_2 (ref. 30), and this is closely relevant to the phase observed by *ex situ* XRD analyses, that is, Li-ion-intercalated MoO_2 in Fig. 2a. However, STEM-EELS spectrum (Fig. 3c) taken at the amorphous area (D_{AP}) shows that energy loss near edge structures of Li species are different from those taken at other areas (D_{CP} and D_{SEI}) and corresponds to metallic Li[31], implying that some of Li ions are accumulated as a form of metallic Li at the interface between nanosized crystalline Li_xMoO_2 domains. This Li storage as a metallic cluster resembles Li-storage mechanism of disordered carbon in the microspaces located at the edges of carbon clusters[32–35].

In addition, as shown in Supplementary Fig. 12, the *Z*-contrast image and EELS spectra of O *K*-edge taken at the crystalline phase (D_{CP}) and amorphous phase (D_{AP}) strongly support this finding because the area of amorphous phase (D_{AP}) shows darker contrast and lower oxygen intensity than the domain of the crystalline phase (D_{CP}). Another STEM-EELS result on the area of D_{SEI} clearly shows a peak for Li_2CO_3 (Supplementary Fig. 13), which is one of the main components of SEI formed on the anode in the LIBs. As expected, the delithiated mesoporous MoO_2 electrode does not contain any Li peaks, which suggests that it is reversibly returned to its initial state after the cycle (Supplementary Fig. 14). These EELS results strongly support the new reaction model of the ordered mesoporous MoO_2 described in the present work.

Cycle performance and rate capability. Recently, we have developed an *in operando* small-angle X-ray scattering (SAXS) technique to investigate the nanostructural changes of ordered mesoporous electrode materials[36]. As shown in Fig. 4a,b and Supplementary Fig. 15, there is no significant changes in the *in operando* SAXS data of mesoporous MoO_2 electrode during the lithiation and delithiation process (~20% reduction of peak intensity and ~25% increase of net volume change). These small changes in nanostructural properties indicate that the present mesoporous MoO_2 electrode seems to follow the intercalation mechanism with additional Li-storage sites rather than the conventional conversion mechanism[36]. Figure 4c–e compares TEM images of the ordered mesoporous MoO_2 structure during the first cycle. After complete lithiation, the diameter of the MoO_2 frameworks increases from ~7.8 to ~11.8 nm. In addition to the diameter increase, there also should be a longitudinal volume

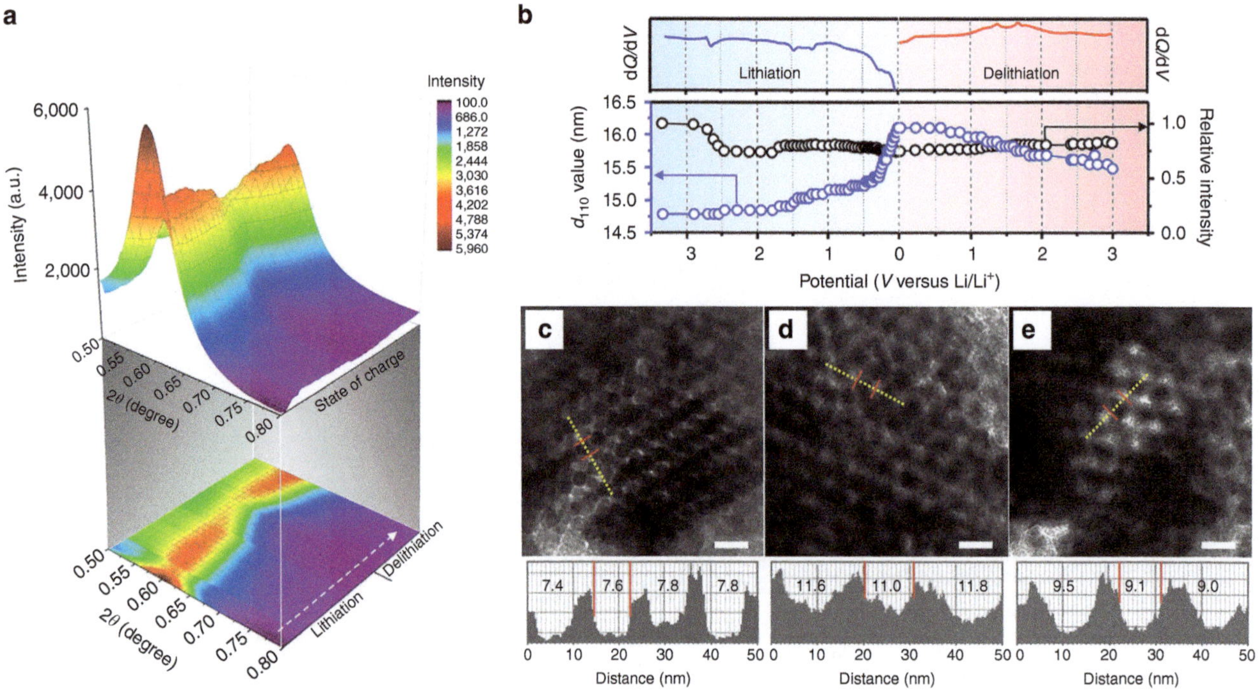

Figure 4 | Pore structure of mesoporous MoO_2 by *in operando* SAXS and *ex situ* TEM. (**a**) Color-coded 3D contour and projection map showing SAXS data collected from ordered mesoporous MoO_2 during *in operando* experiment. (**b**) The changes in lattice parameter and resolved peak relative intensity calculated from the (110) reflection with the corresponding dQ/dV plot. Representative TEM images and framework thickness of ordered mesoporous MoO_2 electrodes: (**c**) pristine, (**d**) lithiated and (**e**) delithiated. Scale bars = 20 nm.

expansion of MoO_2 frameworks upon the lithiation (not experimentally observable). As shown by the *ex situ* XRD patterns and the DFT calculation on the lithiated MoO_2 mesostructure (Figs 2a and 3a), the volume expansion can be attributed to the formation of both the Li-ion-intercalated crystalline phase and the amorphous metallic Li-rich phase during the lithiation. However, it is reasonable that the expansion associated with Li-ion intercalation is relatively smaller than that caused by the formation of the metallic Li-rich phase. This means that the volume expansion can be mainly attributed to the formation of the Li-rich phase, because the d_{110} spacing of crystalline MoO_2 in the XRD patterns and the HRTEM images of Fig. 2a,b expanded by 14.7% from the pristine state during the transformation from MoO_2 to Li_xMoO_2. It should be noted that, despite these volume changes during the first cycle, the ordered mesoporous MoO_2 retained its mesostructure even after complete lithiation and returned to the initial state after the complete delithiation (Fig. 4c–e). It is also important to note that this high capacity of ordered mesoporous MoO_2 electrode shows a significant improvement in capacity retention compared with the bulk material. As shown in Fig. 1h, the present mesoporous MoO_2 electrode retains 121.9% of the initial capacity after 50 cycles (from 1,308 to 1,594 mAh g^{-1}).

Cyclic voltammetry (CV) of the ordered mesoporous MoO_2 electrode reveals the same tendency of highly stable electrochemical storage and removal of Li up to 30 cycles (Supplementary Fig. 16). Li-storage capacity of the ordered mesoporous MoO_2 electrode shows the continuous increase with an increase of cycle numbers up to 20 cycles, and then reached the steady-state value (Supplementary Fig. 16b). Considering that this newly evolved couple of CV peaks could not be reported by other reports ever, the reduction ($\sim$0.2 V) and oxidation peak ($\sim$2.5 V), which can be also found at the differential capacity plots of the ordered mesoporous MoO_2 electrode (Supplementary Fig. 17), would be associated with Li-storage reaction mechanism to account for high capacity of the ordered mesoporous MoO_2 electrode presented here. It should be also noted that the rectangular CV profiles took shape with an increase of CV cycle numbers, indicating that Li-storage reaction in the ordered mesoporous MoO_2 might resemble the electrochemical reaction observed in the supercapacitors rather than in the typical rechargeable batteries. Considering that under-potential deposition of metal ions is one of the typical super-capacitive electrochemical reactions, the rectangular shape of CV profiles observed in the mesoporous MoO_2 electrode might be a circumstantial evidence to prove the storage of Li as a metallic phase in the ordered mesoporous MoO_2 electrode.

Mo *K*-edge EXAFS spectra for MoO_2 during electrochemical cycles are shown in Supplementary Fig. 18. We found that Mo–O peak from oxide phase is still major contribution in EXAFS spectra during cycles, even though the peak intensities are gradually reduced due to the cycle-induced structural disorder. It is notable that Mo–Mo peak from Mo metallic phase does not appear even after 10 cycles, clearly confirming no possibility to have the conversion reaction in MoO_2 electrode.

Rate capability is also very important for electrode materials in addition to Li-storage capacity. We modified the mesostructure of the ordered mesoporous MoO_2 by reducing the wall thickness from 7.9 to 7.0 nm and 5.5 to 4.0 nm (but for similar mesopore size of $\sim$18 nm, see Supplementary Figs 19–22 and Supplementary Tables 2 and 3), in order to examine the effects of framework thicknesses on the rate capability of mesoporous MoO_2. As shown in Supplementary Fig. 23, the rate capability was slightly enhanced by reducing the framework thickness of the mesoporous MoO_2 materials from 7.0 to 5.8 nm, which is probably by shortening the diffusion length for Li ion in the MoO_2 framework. However, neither further reduction (5.1 nm) nor increase (7.8 nm) gives a positive effect on the rate capability. Research is underway to find the optimized mesopore size and

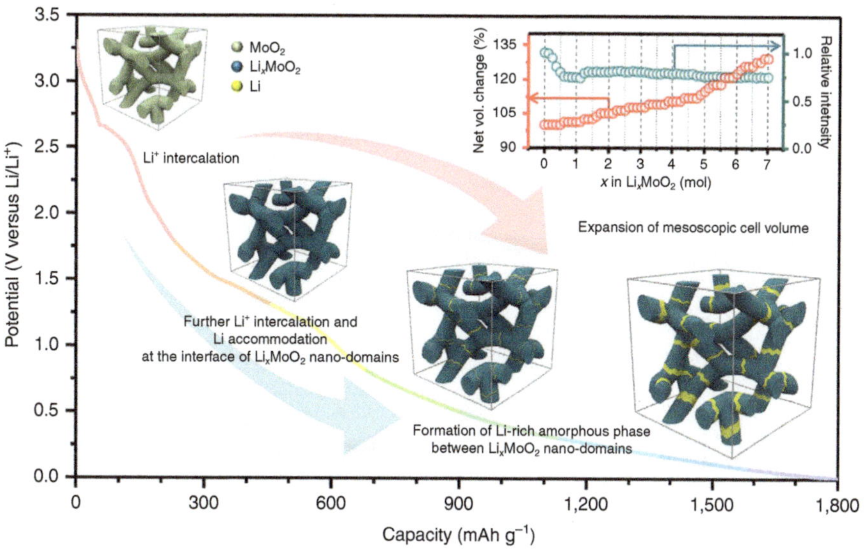

Figure 5 | Schematic diagram of the reaction pathways and mesoscale morphology. Schematic diagram of the reaction pathways and the resulting products of ordered mesoporous MoO_2 with respect to the amount of Li ion inserted.

framework thickness necessary to obtain the best electrochemical performance.

Discussion

We report a high-capacity Li-storage material, ordered mesoporous MoO_2, with a capacity that exceeds the storage capacity predicted by the conventional Li-storage mechanism such as Li intercalation and/or conversion reaction of transition metal oxide. We suggest a Li-storage mechanism consisting of a Li-ion intercalation reaction and the formation of a metallic Li-rich phase between the Li-ion-intercalated MoO_2 phase, based on first-principle calculations on the electronic structure of lithiated ordered mesoporous MoO_2, along with DFT calculations, *in situ* X-ray absorption spectroscopies and HRTEM analyses combined with EELS studies of the reacted ordered mesoporous MoO_2 (Fig. 5). In addition to the high capacity and excellent cycle performance of ordered mesoporous MoO_2, the proposed Li-storage mechanism clearly shows how the nanoscale engineering impacts the associated physicochemical properties of the material for electrochemical mass storage applications. The present results enable a structural and electronic configuration that is, in a sense, a hybrid between the Li-intercalated compound anode and the metallic Li anode, with the inherent safety features of the former and a high power density approaching to that of the latter. The results further provide a more complete understanding of possible Li-storage mechanisms for transition metal oxides, and thus make possible the further advancement of ultrahigh capacity anode materials for Li rechargeable batteries.

Methods

Mesoporous silica template. In the present work, mesoporous silica, KIT-6, with cubic *Ia3d* mesostructure was used as the silica template for preparing mesoporous MoO_2 materials. Pluronic triblock copolymer P123 ($EO_{20}PO_{70}EO_{20}$, MW = 5,800) was utilized as the structure-directing agent for synthesis of KIT-6. Typically, 90.0 g of P123 was dissolved in a mixture of 3,255 g of distilled water, 90 g of 1-BuOH (99.7 wt%, Aldrich) and 177 g of *c*-HCl (35 wt%, Aldrich). After stirring at 35 °C for 10 min, 193.5 g of tetraethylorthosilicate (98 wt%, Aldrich) was added to this solution under vigorous stirring. The resulting mixture was stirred for 24 h at 35 °C and subsequently kept in static condition at 100 °C for 24 h in an oven. The solid product was filtered, washed with double distilled water and dried at 100 °C overnight. The white powder, thus obtained, was washed with EtOH, dried at 80 °C for 12 h, and finally calcined under static air conditions at 550 °C for 3 h in order to remove the structure-directing agent. Mesopore size of the KIT-6 template calculated by Barrett–Joyner–Halenda method was 7.1 nm. KIT-6 materials with different pore sizes (5.1, 5.8 and 7.9 nm; see Supplementary Table 2 and Supplementary Fig. 19) were also synthesized using the above method, except for the hydrothermal treatment at 60, 80 and 140 °C, respectively.

Highly ordered mesoporous MoO_2 materials. The mesoporous MoO_2 materials were synthesized by the nano-replication method using the KIT-6 as the mesoporous silica templates. An amount of 6.49 g of ammonium molybdate tetrahydrate ($(NH_4)_6Mo_7O_{24}\bullet 4H_2O$, Aldrich) was dissolved in 8.4 g of distilled water. This precursor solution was infiltrated into 10.0 g of the calcined KIT-6 template by an incipient wetness method. After drying the composite at 80 °C for 24 h, the resulting material was heated to 500 °C under nitrogen atmosphere for 5 h for crystallization. After the heat treatment, the silica template was removed from the composite by a wet-etching process using a 20 wt% hydrofluoric acid (HF) solution. The resulting solid product was washed with distilled water and acetone several times, and then dried at 80 °C overnight in an oven. The mesoporous MoO_2 materials, synthesized from the KIT-6 with different pore sizes, are denoted as meso-$MoO_2 - x$ (where x means the synthesis temperature of KIT-6 templates). The mesoporous MoO_2 materials with surface area of 39 and 76 $m^2 g^{-1}$ were prepared by following the same synthesis method with the above procedure, except for the precursor infiltration conditions. The precursor solutions were prepared by dissolving 20.7 g of $(NH_4)_6Mo_7O_{24}\bullet 4H_2O$ in 15 g of distilled water and 8.11 g of $(NH_4)_6Mo_7O_{24}\bullet 4H_2O$ in 10.5 g of distilled water for the mesoporous MoO_2 with surface areas of 39 and 76 $m^2 g^{-1}$, respectively. Two mixed MoO_2 electrodes was simply prepared by physically mixing the bulk MoO_2 (Aldrich, $S_{BET} = 0.23\ m^2 g^{-1}$) and the highly ordered mesoporous MoO_2 (this work, $S_{BET} = 115\ m^2 g^{-1}$). The weight ratios between the bulk and mesoporous MoO_2 materials were 1:3 and 3:2 for the mixtures with surface areas of 21 $m^2 g^{-1}$ and 53 $m^2 g^{-1}$, respectively. The mixed MoO_2 materials were thoroughly hand-mixed with mortar and pestle.

Characterization. XRD patterns for the mesoporous silica and mesoporous MoO_2 powder were obtained using a Rigaku D/MAX-2200 Ultima equipped with Cu Kα radiation at 30 kV and 40 mA. *Ex situ* XRD patterns of electrodes during the cycles were recorded in reflection mode using an X'Pert PRO equipped with Cu Kα radiation at 40 kV and 40 mA in the 20–60° (2θ) angular range. *In situ* XANES and EXAFS spectra were collected on beamline 10 C at Pohang Accelerator Laboratory. Energy calibration was carried out using the first inflection point of the spectrum of Mo metal foil as a reference (that is, Mo *K*-edge = 20,000 eV). Reference spectra were simultaneously collected for each *in situ* spectrum using Mo metal foil. *In operando* SAXS experiments were carried out using BL 9 A U-SAXS beamline (Pohang Accelerator Laboratory) with a two-dimensional (2D) CCD detector (Rayonix SX165, USA) that was positioned 2 m away from the sample, which measured scattering in the 2θ range of 0.3–2.5 ($\lambda = 1.54$ Å). The size of the focused beam was 300 μm in diameter and the energy of the beam was 11 keV. The 2D patterns of the mesoporous samples during electrochemical cycling were recorded with a 1-s exposure time and 8 s detector readout time. The 2D patterns were scanned with the FIT2D software package to obtain the 1D patterns in the form of the intensity versus 2θ. The storage ring was operated at 3.0 GeV with a ring current of 300 mA. SEM images were taken using a LEO Supra 55 field-emission SEM operating at an accelerating voltage of 15 kV. Ultra-high-resolution SEM images were obtained using a Hitachi UHR S 5500 FE-SEM operating at 30 kV.

The main instruments we used in this study for microscopy were an aberration-corrected FEI TITAN field-emission TEM and a JEOL JEM 3010 TEM operated at an accelerating voltage of 300 kV. EELS analysis was performed in STEM mode using the field-emission TEM (TITAN) equipped with a high-resolution Gatan Tridieum 865ER imaging spectrometer. Because the MoO_2 electrodes during lithiation and delithiation were sensitive to air and moisture, all the sampling process was carried out in an argon-filled glove box. For TEM observations, all the specimens were moved to the instruments using argon-filled plastic tubes and carefully transferred to the instruments within 10 s in order to minimize the air-contacting time. N_2 adsorption–desorption isotherms were collected on a Micromeritics Tristar system at liquid N_2 temperature. All of the samples were completely dried under vacuum at 100 °C for 24 h before the measurement. The specific BET (Brunauer–Emmett–Teller) surface areas were calculated from the adsorption branches in the range of relative pressure (p/p_0) = 0.05–0.20. The pore size distribution curves were obtained by the Barrett–Joyner–Halenda method on the basis of the adsorption branches. Total pore volumes were measured at $p/p_0 = 0.99$.

Electrochemistry. Electrodes were prepared by coating the slurries that contain the mesoporous MoO_2 material as the active material powder, carbon black (Super-P, MMM) as the conducting agent and polyamideimide (Torlon 4,000 T, Solvay) as the binder in the weight ratio of 70:15:15 in *N*-methyl-2-pyrrolidone (Aldrich) solvent to produce an electrode slurry. The slurry was coated onto a Cu foil current collector using a Doctor-Blade technique. After the coating procedure, the electrodes were pressed and dried for 2 h at 200 °C under vacuum conditions. The electrodes thus obtained were cut into disks (12 mm in diameter). Coin-type cells (CR2016) were assembled in a dry room using Celgard 3,501 as the separator and Li foil as the counter and reference electrodes. On the basis of the current density of 0.1 C ($= 80\ mAh\ g^{-1}$), all cells were tested at within a fixed voltage window (0.001–3.0 V) using a battery cycle tester (TOSCAT 4,000 series, Tokyo, Japan).

Computational method. The first-principle calculations based on the DFT was performed using the VASP code. Initial structure of MoO_2 is the $3 \times 3 \times 1$ supercells including 36 Mo and 72 O atoms. The plane-wave cutoff energy was chosen to be 400 eV and appropriate *k*-points were chosen to ensure that the total energies are converged within a few meV. The exchange–correlation interactions between electrons were described by the generalized gradient approximation, and projector-augmented wave potentials were used for the description of ion–electron interactions. We used the conjugate gradient method for geometry optimizations, and the optimization procedure was truncated when the residual forces for the relaxed atoms were < 0.03 eV per Å. In the calculation of Li_xMoO_2, Li atoms were inserted to the position of the energetically most stable site. During the lithiation, the atomic position and cell structures are fully relaxed simultaneously.

References

1. Nishi, Y. The development of lithium ion secondary batteries. *Chem. Rec.* **1,** 406–413 (2001).
2. Tarascon, J.-M. & Armand, M. Issues and challenges facing rechargeable lithium battery. *Nature* **414,** 359–367 (2001).
3. Ellis, B. L., Lee, K. T. & Nazar, L. F. Positive electrode materials for Li-ion and Li-batteries. *Chem. Mater.* **22,** 691–714 (2010).
4. Jeong, G., Kim, Y.-U., Kim, H., Kim, Y.-J. & Sohn, H.-J. Prospective materials and applications for Li secondary batteries. *Energ. Environ. Sci.* **4,** 1986–2002 (2011).
5. Lee, K. T. & Cho, J. Roles of nanosize in lithium reactive nanomaterials for lithium ion batteries. *Nano Today* **6,** 28–41 (2011).
6. Huang, J. Y. *et al. In situ* observation of the electrochemical lithiation of a single SnO_2 nanowire electrode. *Science* **330,** 1515–1520 (2010).
7. Chan, C. K. *et al.* High-performance lithium battery anodes using silicon nanowires. *Nat. Nanotechnol.* **3,** 31–35 (2008).
8. Park, C.-M., Kim, J.-H., Kim, H. & Sohn, H.-J. Li-alloy based anode materials for Li secondary batteries. *Chem. Soc. Rev.* **39,** 3115–3141 (2010).
9. Poizot, P., Laruelle, S., Grugeon, S., Dupont, L. & Tarascon, J.-M. Nano-sized transition-metal oxides as negative-electrode materials for lithium-ion batteries. *Nature* **407,** 496–499 (2000).
10. Lee, Y. J. *et al.* Fabricating genetically engineered high-power lithium-ion batteries using multiple virus genes. *Science* **324,** 1051–1055 (2009).
11. Armand, M. & Tarascon, J.-M. Building better batteries. *Nature* **451,** 652–657 (2008).
12. Grugeon, S. *et al.* Particle size effects on the electrochemical performance of copoer oxides toward lithium. *J. Electochem. Soc.* **148,** A285–A292 (2001).
13. Li, H., Richter, G. & Maier, J. Reversible formation and decomposition of LiF clusters using transition metal fluorides an precursors and their application in rechargeable Li batteries. *Adv. Mater.* **15,** 736–739 (2003).
14. Jamnik, J. & Maier, J. Nanocrystallinity effects in lithium battery materials. *Phys. Chem. Chem. Phys.* **5,** 5215–5220 (2003).
15. Hu, Y.-Y. *et al.* Origin of additional capacities in metal oxide lithium-ion battery electrodes. *Nat. Mater.* **12,** 1130 (2013).
16. Cheng, F., Liang, J., Tao, Z. & Chen, J. Functional materials for rechargeable batteries. *Adv. Mater.* **23,** 1695–1715 (2011).
17. Bruce, P. G., Scrosati, B. & Tarascon, J.-M. Nanomaterials for rechargeable lithium batteries. *Angew. Chem. Int. Ed.* **47,** 2930–2946 (2008).
18. Liu, C., Li, F., Ma, L.-P. & Cheng, H.-M. Advanced materials for energy storage. *Adv. Mater.* **22,** E28–E62 (2010).
19. Kresge, C. T., Leonowicz, M. E., Roth, W. J., Vartuli, J. C. & Beck, J. S. Ordered mesoporous molecular sieves synthesized by a liquid-crystal template mechanism. *Nature* **359,** 710–712 (1992).
20. Zhao, D. *et al.* Triblock copolymer syntheses of mesoporous silica with periodic 50 to 300 angstrom pores. *Science* **279,** 548–552 (1998).
21. Joo, S. H. *et al.* Ordered nanoporous arrays of carbon supporting high dispersions of platinum nanoparticles. *Nature* **412,** 169–172 (2001).
22. Lee, H. I. *et al.* Spontaneous phase separation mediated synthesis of 3D mesoporous carbon with controllable cage and window size. *Adv. Mater.* **23,** 2357–2361 (2011).
23. Chang, H., Joo, S. H. & Pak, C. Synthesis and characterization of mesoporous carbon for fuel cell applications. *J. Mater. Chem.* **17,** 3078–3088 (2007).
24. Yue, W. *et al.* Syntheses, Li insertion and photoactivity of mesoporous crystalline TiO_2. *Adv. Func. Mater.* **19,** 2826–2833 (2009).
25. Ren, Y., Jardwick, L. J. & Bruce, P. G. Lithium intercalation into mesoporous anatase with an ordered 3D pore structure. *Adv. Mater.* **49,** 2570–2574 (2010).
26. Ren, Y., Armstrong, A. R., Jiao, F. & Bruce, P. G. Influence of size on the rate of mesoporous electrodes for lithium batteries. *J. Am. Chem. Soc.* **132,** 996–1004 (2010).
27. Shi, Y. *et al.* Ordered mesoporous metallic MoO_2 materials with highly reversible lithium storage capacity. *Nano Lett.* **9,** 4215–4220 (2009).
28. Sohn, J. K. *et al.* Nano-propping effect of residual silicas on reversible lithium storage over highly ordered mesoporous SnO_2 materials. *J. Mater. Chem.* **19,** 6727–6732 (2009).
29. Chen, C., Ding, N., Wang, L., Yu, Y. & Lieberwirth, I. Some new facts on electrochemical reaction mechanism for transition metal oxide electrodes. *J. Power Sources* **189,** 552–556 (2009).
30. Cosandey, F. in *Microscopy: Science, Technology, Applications and Education* (eds Méndez-Vilas, A. & Díaz, J.) Vol. 3, 1662–1666 (Formatex, 2010).
31. Hightower, A., Ahn, C. C., Fultz, B. & Rez, P. Electron energy-loss spectrometry on lithiated graphite. *Appl. Phys. Lett.* **77,** 238–240 (2000).
32. Mochida, I., Ku, C.-H. & Korai, Y. Anodic performance and insertion mechanism of hard carbons prepared from synthetic isotropic pitches. *Carbon N. Y.* **39,** 399–410 (2001).
33. Dahn, J. R., Zheng, T., Liu, Y. & Xue, J. S. Mechanisms for lithium insertion in carbonaceous materials. *Science* **270,** 590–593 (1995).
34. Mabuchi, A., Tokumitsu, K., Fujimoto, H. & Kasuh, T. Charge-discharge characteristics of the mesocarbon miocrobeads heat-treated at different temperatures. *J. Electrochem. Soc.* **142,** 1041–1046 (1995).
35. Mochida, I., Ku, C.-H., Yoon, S. H. & Korai, Y. Anodic performance and mechanism of mesophase-pitch-derived carbons in lithium ion batteries. *J. Power Sources* **75,** 214–222 (1998).
36. Park, G. O. *et al. In operando* monitoring of the pore dynamics in ordered mesoporous electrode materials by small angle X-ray scattering. *ACS Nano* **9,** 5470–5477 (2015).

Acknowledgements

This work was supported by Samsung Research Funding Center for Future Technology (SRFC-MA1401-03). We also thank partial supports obtained from the Energy Efficiency and Resources Core Technology Program of the Korea Institute of Energy Technology Evaluation and Planning (KETEP), which granted financial resource from the Ministry of Trade, Industry and Energy, Republic of Korea (no. 20132020000260), the research fund of Hanyang University (HY-2012-T), the US National Science Foundation (DMR 08-05148) and the National Research Foundation of Korea (NRF-2010-C1AAA001-2010-0029065 and the Mid-Career Researcher Program No. 2012R1A2A2A01010011).

Author contributions

J.M.K., W.-S.Y., C.P., H.K. and G.D.S. conceived, designed and coordinated the study. H.S.L. carried out theoretical calculations. J.K.S., G.O.P., J.Y., E.P., G.S.P., S.S.K., M.J., J.-M.C., H.C. and S.D. performed the experiment and acquired the data. J.M.K., H.S.L., J.K.S., W.-S.Y, C.P., H.K. and G.D.S. wrote the paper. All the authors participated in analysis of the experimental data and discussions of the results, as well as preparing the paper.

Additional information

Competing financial interests: The authors declare no competing financial interests.

15

Van der Waals interactions and the limits of isolated atom models at interfaces

Shigeki Kawai[1,2,3], Adam S. Foster[4,5], Torbjörn Björkman[4,6], Sylwia Nowakowska[2], Jonas Björk[7], Filippo Federici Canova[4,8], Lutz H. Gade[9], Thomas A. Jung[2,10] & Ernst Meyer[2]

Van der Waals forces are among the weakest, yet most decisive interactions governing condensation and aggregation processes and the phase behaviour of atomic and molecular matter. Understanding the resulting structural motifs and patterns has become increasingly important in studies of the nanoscale regime. Here we measure the paradigmatic van der Waals interactions represented by the noble gas atom pairs Ar–Xe, Kr–Xe and Xe–Xe with a Xe-functionalized tip of an atomic force microscope at low temperature. Individual rare gas atoms were fixed at node sites of a surface-confined two-dimensional metal-organic framework. We found that the magnitude of the measured force increased with the atomic radius, yet detailed simulation by density functional theory revealed that the adsorption induced charge redistribution strengthened the van der Waals forces by a factor of up to two, thus demonstrating the limits of a purely atomic description of the interaction in these representative systems.

[1] International Center for Materials Nanoarchitectonics, National Institute for Materials Science, 1-1, Namiki, Tsukuba, 305-0044 Ibaraki, Japan. [2] Department of Physics, University of Basel, Klingelbergstrasse 82, CH-4056 Basel, Switzerland. [3] Precursory Research for Embryonic Science and Technology (PRESTO), Japan Science and Technology Agency, 4-1-8, Honcho, Kawaguchi, 332-0012 Saitama, Japan. [4] COMP, Department of Applied Physics, Aalto University, PO Box 11100, FI-00076 Aalto, Finland. [5] Division of Electrical Engineering and Computer Science, Kanazawa University, Kanazawa 920-1192, Japan. [6] Physics/Department of Natural Sciences, Åbo Akademi University, 20500 Turku, Finland. [7] Department of Physics, Chemistry and Biology (IFM), Linköping University, Linköping 58183, Sweden. [8] Aalto Science Institute, Aalto University, PO Box 15500, FI-00076 Aalto, Finland. [9] Anorganisch-Chemisches Institut, Universität Heidelberg, Im Neuenheimer Feld 270, 69120 Heidelberg, Germany. [10] Laboratory for Micro- and Nanotechnology, Paul Scherrer Institute, CH-5232 Villigen, Switzerland. Correspondence and requests for materials should be addressed to S.K. (email: KAWAI.Shigeki@nims.go.jp) or to A.S.F. (email: adam.foster@aalto.fi).

The van der Waals (dispersive) interaction[1], arising from the temporal fluctuation of electronic charge inducing transient dipoles of variable strengths, is ubiquitous in nature and is of central importance for many physical and chemical phenomena. This interaction plays a key role in the condensation of atoms and molecules, friction and adhesion[2], and is crucial for the adsorption of molecules on inert surfaces such as graphite[3,4] and of π-conjugated molecules on metal surfaces[5]. While dispersion forces are inherently a quantum mechanical phenomenon, simple quasi-classical models have been applied very successfully[6], and remain popular for many critical systems[7,8]. However, the advances of first-principles quantum methods, being applied to increasingly complex systems, have made it essential to include accurate van der Waals contributions and thus gain deeper insight into their physical consequences. Nonlocal correlations, responsible for dispersion interactions, are not included in the most commonly used functionals[9], but more efficient approaches are constantly evolving[10–14]. Any new methods are usually rigorously tested against a set of known atomic structures or materials properties calculated using higher-order methods[15], but as the systems studied increase in size and include large substrates and adsorbates, their capability is still limited by the current computational power. Therefore, experimental approaches to measure interactions in complex systems remain complementary to advances in quantum physical simulations. For example, this combination has proved successful in the direct probing of the van der Waals interaction between two Rydberg atoms[16].

Atomic force microscopy (AFM) can directly measure the interaction between tip and sample down to the atomic scale via the frequency shift of an oscillating cantilever[17], and by integrating the measured frequency shift in the z-direction, the force and potential can be extracted quantitatively[18]. This technique has been successfully used to discriminate embedded Si, Sn and Pb atoms in a Si(111) surface[19] as well as In, Sn and Si atoms in a heterogeneous III–IV chain on a Si(100) surface[20], and also to study the complex interplay between deformations and electronic states in Pt–Pt and Cu–Cu metal contacts[21]. Alongside such reactive systems, studies of weak interactions involving inert graphite[22] and carbon nanotubes have been carried out[23]. Moreover, the role of van der Waals interactions between Au structures and aromatic hydrocarbons was explored by AFM tip-induced lifting of individual molecules[24,25]. However, direct force measurements of the atomic van der Waals potentials remain an unexplored topic to date. Rare gas atoms have long been providing a benchmark in van der Waals interactions due to their inert, monoatomic nature and the dominance of dispersion forces in their binding[26].

In this work, we directly measure the van der Waals interaction between two adsorbed rare gas atoms. Individual Ar, Kr and Xe atoms were adsorbed and stabilized at a nodal site of a molecular network on Cu(111), and their interaction with a Xe-functionalized tip of AFM was measured at low temperature. We found that the magnitude of the van der Waals interaction scales with the size of the rare gas atom on the surface, that is, in order of Xe–Xe > Kr–Xe > Ar–Xe. A detailed theoretical simulation generally corroborated the experimental findings, and provided insight into the role of tip-induced atomic relaxations and charge transfer processes that cause deviations from a pure van der Waals interaction between two individual atoms.

Results

Rare gas atoms stabilized by 2D MOF. In Fig. 1a, a schematic representation of the AFM experiment is depicted, with the interaction forces categorized into long-range and site-dependent short-range contributions. The system has four contributions; F_1 the tip-sample surface interaction, F_2 the Xe tip-sample surface interaction, F_3 the tip–rare gas atom on surface interaction and F_4 the Xe tip–rare gas atom on surface interaction. The F_4 contribution corresponds to a direct component between the two atoms, which at long range is given by the van der Waals interaction. To extract the latter components, reference measurements were carried out at an equivalent atomic site in absence of a surface adsorbed rare gas atom (denoted 'empty site' in Fig. 1a). By subtracting the two z-distance-dependent curves, the interaction force of $F_3 + F_4$ could be determined directly. F_1–F_4 have different power laws and origins of the Z distance. Among them, F_4 has the largest power law and the smallest Z distance for a given tip-sample distance. We first attempted to probe the force of Xe adsorbed on a flat Cu(111) surface (Supplementary Fig. 1)[27]. However, the high mobility of the rare gas atoms on the metal surface hampered a reliable measurement. In fact, the diffusion barrier was so low that the rare gas atom was often accidentally moved in the lateral direction by the tip-sample interaction (Supplementary Fig. 2).

To overcome these problems, we employed a surface-confined two-dimensional metal-organic framework (2D MOF) as an anchor net for the inert rare gas atoms. We formed the 2D MOF[28] by depositing the perylene derivative 4,9-diaminoperylene-quinone-3,10-diimine (DPDI) on a clean Cu(111) surface, and then annealing the sample at ~300 °C. Recent work has shown that a series of dehydrogenation reactions and subsequent coordination of the molecule's nitrogen atoms to Cu adatoms results in the honeycomb structure depicted in Fig. 1b (ref. 29). As a result of the adsorption geometry of adatoms and molecules on the surface, the 2D MOF has two kinds of coordination motifs, indicated as i and ii (see also Supplementary Fig. 3), each comprising three Cu adatoms and which are referred to as nodes in the honeycomb network. Figure 1c shows that Xe atoms, dosed onto the cold sample (<8 K), adsorb predominantly on node sites. By setting a clean

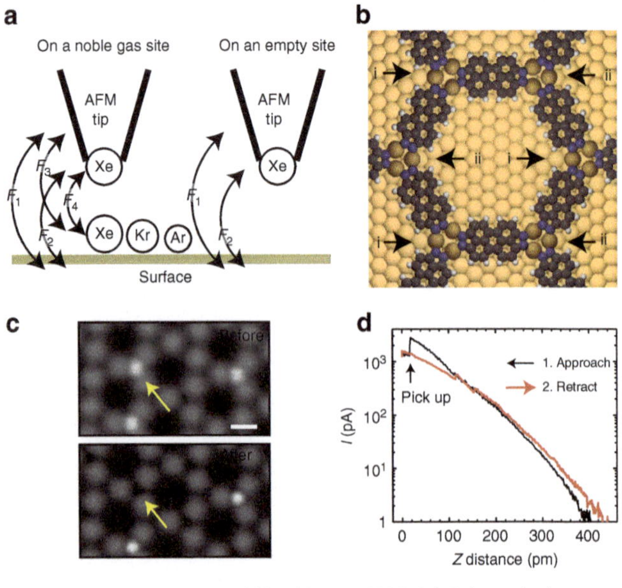

Figure 1 | Rare gas atom stabilized by 2D MOF. (**a**) Schematic drawing of the AFM measurement set-up. (**b**) Chemical structure of Cu-coordinated molecular network on Cu(111); the two symmetry-inequivalent nodal sites are indicated as i and ii. (**c**) STM topographies before and after the vertical manipulation of a Xe atom. (**d**) Distance-dependent curve of the tunnelling current in the process. Measurement parameters: I = 2 pA and V = 200 mV in **c**, and V = 2 mV in **d**. Scale bar, 1 nm.

Cu tip 400 pm closer to the Xe site than the imaging distance[30], the tip was reliably functionalized by the Xe atom (Fig. 1d).

Atomic scale imaging of the 2D MOF. Figure 2a displays the scanning tunnelling microscopy (STM) topography, taken with a Xe-functionalized tip. Each molecule appears as a featureless ball-like shape, reflecting the molecular electronic structure, as established in previous measurements[28]. To obtain more details of the molecular structure, we took advantage of the capabilities of AFM with functionalized tips[31–36]. In Fig. 2b, the frequency shift map taken at a constant height is shown, in which the chemical structure at the centre of the perylene core becomes apparent but the molecular units remain blurred at the nodes. Since the measurement was performed in the repulsive interaction region, the darker area (more negative frequency shift) corresponds to a larger tip–sample separation, that is, the nodes are closer to the Cu substrate and the nitrogen atoms are thus not resolved.

To enhance the resolution, we switched to a CO-functionalized tip, which is known to improve resolution compared with a Xe tip due to its smaller atomic radius and higher flexibility[37]. Figure 2c shows the frequency shift image, in which the chemical bonds in the perylene derivative now appear clearly. The carbon rings in perylene are apparently elongated in the direction of the short molecular axis due to the tilting effect of the CO molecule on the tip apex[31]. Even the C_3N_2 pentagonal rings of the triply dehydrogenated (3deh)-DPDI exo-ligands at the nodes are resolved as indicated by a white arrow. Since the nitrogen atoms are coordinated with Cu adatoms on the surface, the end groups of the molecule are closer to the Cu substrate than the perylene core, in agreement with X-ray standing wave measurements and theoretical calculations of the system[29]. The Cu adatoms coordinated to the nitrogen atoms are not visible in the images, although they have been identified indirectly by an STM fingerprint of the metal–organic coordination at particular bias voltages[29]. Closer inspection reveals that the two distinct nodal positions of the honeycomb network can be resolved by the adsorption of Xe atoms, which appear in STM images with different apparent sizes (Supplementary Fig. 4). This experimental result unambiguously supports the previous theoretical calculations and experiments[29].

Van der Waals force detection. To prepare the test bench for interatomic van der Waals force measurements, Ar, Kr and Xe gases were deposited on the substrate with the Cu-coordinated 3deh-DPDI network. Figure 3a shows a STM topography, in which the node sites are partially filled by Ar, Kr and Xe atoms. The observed apparent sizes of the rare gas atoms relate to the van der Waals radii ($R_{Ar} = 188$ pm, $R_{Kr} = 202$ pm and $R_{Xe} = 216$ pm), though they do not scale quantitatively since, for instance, the Ar atom is observed as about half the size of the Xe atom. As indicated above, two kinds of node sites can be differentiated via the apparent sizes of the same rare gas (Supplementary Fig. 4). We also observed that some atoms adsorbed to the pore sites and they could be easily manipulated by a scanning tip[38], whereas ones at the node sites were found to be highly stable.

In Fig. 3b, distance-dependent curves of the frequency shift are displayed, which were measured on the Xe-filled and -empty nodal sites by the Xe-functionalized tip. The reference site (empty node) was selected at the nearest symmetry-equivalent node, as indicated by the arrows in Fig. 3a. Before the distance-dependent measurement, the z-feedback was deactivated so that the relative z-distances of two curves were consistent. However, in the experiment, the absolute z-distance from the surface was undefined, and the origin of the z-distance therefore set *a posteriori* by equating the equilibrium force distance to the sum of the van der Waals radii (Ar–Xe = 404 pm, Kr–Xe = 418 pm and Xe–Xe = 432 pm).

On the Xe-filled node site, a strong site-dependent interaction is detected while no significant site-dependent force was observed on the corresponding empty node site. Above the empty site, the tip–sample separation is so large that only the macroscopic attractive interaction is present and the tip does not approach close enough to experience short-range interactions in the plotted tip–sample separation range (equivalent to $F_1 + F_2$, see Fig. 1a). By subtracting two curves measured at the Xe-filled and -empty node sites, we obtain the $F_3 + F_4$ distance-dependent curve. Since the origin of the z-axis in the F_3 distance-dependent curve is shifted by the size of Xe atom, its contribution is smaller than that of F_4. Further, the tip was ultra-sharpened by focused ion beam milling, which also contributed to the reduction of F_3 (Supplementary Fig. 5), giving a relation of $F_2 > F_3$. Thus, the maximum contribution of F_3 can be estimated via F_2 (Supplementary Fig. 6). We estimated the contribution of F_2 via fitting the different power laws in the z-distance-dependent frequency shifts on the empty node site and consequently found that the contribution of F_3 is <5% of F_4. Therefore, by subtracting two curves measured at the Xe-filled and -empty node sites, we could estimate the interatomic van der Waals interaction (F_4).

To compare a magnitude of F_4 for Ar–Xe, Kr–Xe and Xe–Xe atom pairs, the z-distance-dependent frequency shifts were measured with the same Xe-functionalized tip as shown in Fig. 3c. To obtain a more quantitative physical value, the force was extracted via the measured frequency shift with the aid of a numerical algorithm (Fig. 3d)[18]. The magnitude of the maximum attractive force between two noble gas atoms was found to be <35 pN, which is much smaller than what has been found in

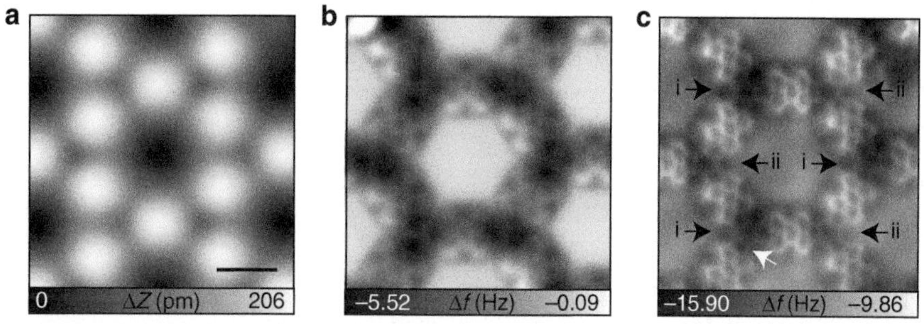

Figure 2 | Atomic scale observation. (**a**) STM topography and (**b**) AFM images taken with a Xe-functionalized tip and (**c**) a CO-functionalized tip. The two symmetry-inequivalent nodal sites are indicated as i and ii; the five-membered C_3N_2 of the 3deh-DPDI exo-ligands are indicated by a white arrow. Measurement parameters: $I = 2$ pA and $V = -200$ mV in **a**, and $A = 60$ pm and $V = 0$ mV in **b**,**c**. Scale bar, 1 nm.

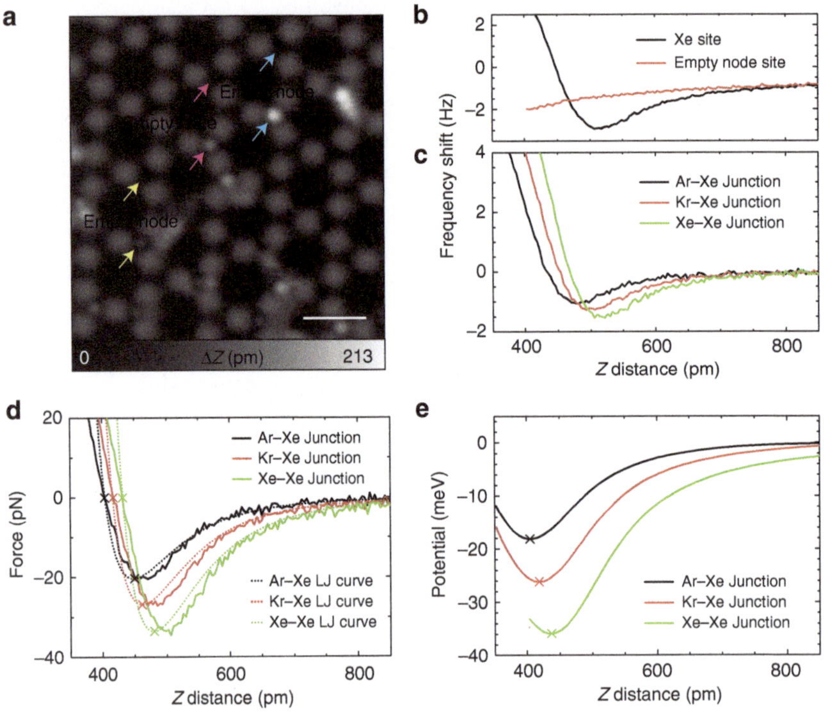

Figure 3 | van der Waals force detection. (**a**) STM topography of the Cu-coordinated 3deh-DPDI network after measuring interaction curves above Ar, Kr and Xe atoms. (**b**) Frequency shift curves taken at the equivalent node sites with and without a Xe atom, measured with a Xe-functionalized tip. (**c**) Subtracted distance-dependence curves of the frequency shift for Ar-Xe, Kr-Xe and Xe-Xe junctions. (**d**) F_4 interaction forces and Lennard-Jones fits for each system, extracted from the frequency shift signal and (**e**) the equivalent potential energy curves. Measurement parameters: $V_{tip} = 500$ mV and $I = 4$ pA in **a**, and $A = 38$ pm, $f = 23{,}063$ Hz and $Q = 52{,}044$, and $V = 1$ mV in **b**,**c**. Scale bar, 2 nm.

previous measurements on semiconductors and metals[19–21]. By integrating the force in the *z*-direction, the potential between Ar-Xe, Kr-Xe and Xe-Xe were obtained as shown in Fig. 3e.

In a first attempt to relate these curves to a simple physical model, the corresponding force versus distance relations calculated from Lennard-Jones models[6], where the equilibrium force distance and the depth of the force were adjusted to the experimental data, were determined and are shown with broken lines in Fig. 3d. From these, the depths of the potential wells were derived as $\varepsilon_{Ar\text{-}Xe} = 19$ meV, $\varepsilon_{Kr\text{-}Xe} = 26$ meV and $\varepsilon_{Xe\text{-}Xe} = 34$ meV, which are in reasonable agreement with the depths of the potential curves directly extracted via the measured frequency shift, that is, 18.1 meV for a Ar-Xe dimer, 26.1 meV for a Kr-Xe dimer and 35.9 meV for a Xe-Xe dimer. However, the shapes of the fitted force curves deviate from the measured ones, indicating that the simple and rigid Lennard-Jones fitting was not suitable for a precise representation of the experimental results.

Theoretical simulations. To demonstrate the dimer-specific relaxation of the adatom positions and the conclusions to be drawn on the interatomic potentials in detail, we used first-principles simulations to calculate potential energy curves using a Xe on copper tip (the calculated system is shown in the inset of Fig. 4) above the three different rare gas atoms adsorbed on the node and empty sites. As in the experiments, the curve above the empty site was subtracted from the other curves to remove background interactions (Supplementary Fig. 7). Figure 4 shows a comparison of the experimental short-range potential energy curves with the simulated tip-surface and bare dimer curves. For Xe-Ar, all three curves are reasonably close to each other. In the Xe-Kr curves, the simulated and measured tip-surface curves are also in very good agreement, but both significantly overestimate

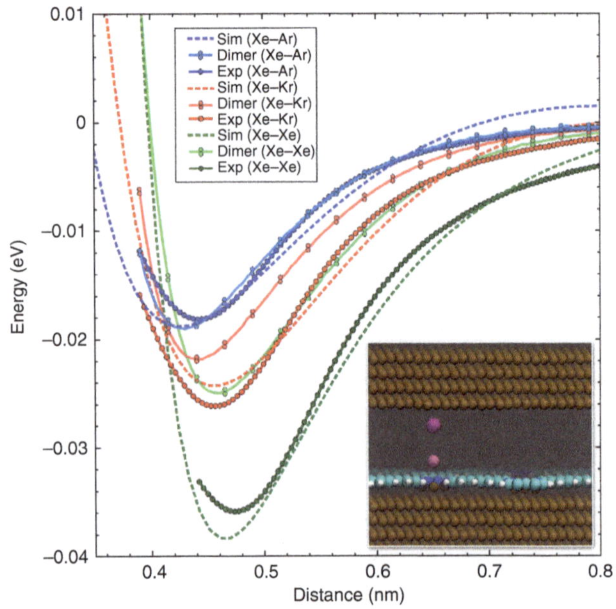

Figure 4 | Theoretical calculation. Comparison of simulated tip-surface and dimer potential energy curves with experimentally derived curves (the same experimental data as in Fig. 3e) for each rare gas atom pair. The *z*-position of the experimental curves has been normalized with respect to the simulated Xe-Kr tip-surface curve. The curves were fitted with polynomial functions. Inset: atomic snapshot of the simulated tip-surface system.

the calculated dimer potential energy. Finally, the Xe-Xe curves show the largest deviation from the dimer potential, with the simulated and measured tip-surface curves in reasonable agreement, but with a minimum potential energy which is almost double that of the dimer.

To establish possible explanations for the differences from the pure dimer potential energies, we analysed the atomic positions and electron density during the simulated tip–surface approaches. We found that major tip-induced displacements of the noble gas atoms occur at a separation of <0.5 nm, in the region of strongest interaction—in particular as the repulsive part of the potential is entered. In this region, the Xe-tip and atom–surface network bond lengths gradually reduce (up to $\sim$0.03–0.04 nm at a separation of 0.3 nm) as the tip approaches. There is no systematic difference between Ar, Kr and Xe in this regard, although the displacements will have different relative energetic costs in each system. Since the energy cost of these relaxations is very specific to the approach site, they cannot be cancelled by subtraction of the empty site potential curve.

There is also a clear systematic difference when comparing electron transfer in the three systems. As in the binding energy, and clearly linked to it, the magnitude of electron transfer between the investigated atoms and the surface network/metal tip correspondingly scales as Xe > Kr > Ar. Specifically, at a separation of 0.5 nm for each system, the Bader charge (integrating the electron density) on the surface noble atom is 0.01/0.03/0.09 e less than in the equivalent dimer configuration for Ar/Kr/Xe, respectively. A similar difference (0.03 e) for Xe on the tip is also observed for each system. Analysis of the nature of the electron density transfer (Supplementary Fig. 8) shows that this is dominated by transfer from the rare gas atom into bonding states between the same atom and the tip or surface, and to copper and/or nitrogen if resting on the nodes of the network. Although this is mitigated by subtraction of the empty site reference potential (which has charge transfer contributions) to some degree, the electron transfer is significantly smaller in the region of the strongest interaction for the empty site system.

The observed difference between pure van der Waals binding and our results on Xe to Ar/Kr/Xe dimers, along with the induced relaxation, certainly plays a significant role in the differences between the experimental and simulated tip–surface energy potentials and the pure atom–dimer curves. As the binding strength and electron transfer increases, so does the deviation, reducing the validity of subtraction of the empty site reference curve, such that ultimately the measured Xe–Xe interaction is almost double the bare dimer interaction, while this discrepancy is not observed for the Xe–Ar potential. Using an Ar-functionalized tip would mitigate the effect, but attempts to this end gave rise to severe issues with tip stability that precluded reliable measurements.

Despite these difficulties to correctly describe the depth and position of the potential minimum based on a simple dimer picture, the asymptotic behaviour of the interaction at large distances is very similar to the asymptotics of the dimer curves. For dimers, or more generally any two point-like polarizable objects, the interaction potential asymptotically at long range has a leading term from the van der Waals interaction, which takes the form C_6/r^6 as a function of the distance, r. The interaction strength is determined on the dynamical polarizabilities of the individual atoms[39]. The curves of Xe–Ar and Xe–Kr in this region fit very well to such a limiting $-C_6/r^{-6}$ behaviour expected for the interaction between two rare gas atoms (Supplementary Fig. 9), although the C_6 coefficients obtained in this way are very large, 2.5 and 3.4 times higher than the free-ion cases (Table 1). Since the tail of the interaction extends beyond a distance at which direct charge transfer between tip and substrate is to be expected, the most natural explanation of this enhancement is the notion that the change in polarizability mostly stems from the rare atom adsorbed on the surface and its interaction with the substrate. For the Xe–Xe interaction, the optimal asymptotic form is r^{-5}. For consistency of comparison, Table 1 gives the value of the C_6 coefficient from a fit to the non-optimal r^{-6} form, which also in this case is much larger than the free atom value, reflecting the much stronger binding. The deviation from the pure r^{-6} form indicates that the interaction has contributions also from sources other than the van der Waals interaction that stems from the polarizability of the objects. Our calculations show the formation of weak bond dipoles both at the tip and the substrate through charge transfer. Two constant dipoles should result in an interaction decaying as r^{-3}, which is not seen for Ar and Kr, so it seems as an unlikely contribution also for Xe. Furthermore, the interaction between dipoles oriented as in our calculations should repel, giving a positive energy contribution and thus decrease the interaction, which is the opposite of what we see. Performing a fit to a sum $-C_6r^{-6}-C_Nr^{-N}$ functions yields near-perfect fits for any $N=2$, 3, 4 and 5, and so is not helpful for disentangling these interactions. However, Xe is well known to be more prone to covalent interaction than both Kr and Ar, which may result in either electrostatic interaction due to the larger charge transfer to the substrate or a stronger covalent bond for the Xe–Xe dimer. Over the measured interval, we are not able to clearly distinguish these two scenarios for the Xe–Xe interaction.

Table 1 | Comparison of C_6 coefficients fitted to AFM measurements and theoretical reference[39].

Dimer	C_6 (atomic units)		Ratio
	AFM measured	Reference	
Xe-Ar	341	135.0	2.5
Xe-Kr	662	192.9	3.4
Xe-Xe	1,267*	287.5	4.4

AMF, atomic force microscopy.
Note that the value for Xe-Xe, marked with an asterisk, does not correspond to the optimal fit.

Discussion

We present a quantitative force measurement of van der Waals forces in the atomic-scale contacts of Argon–Xenon, Krypton–Xenon and Xenon–Xenon atoms. These rare gas atoms were stabilized in the nodal sites of a metal-coordinated two-dimensional molecular network, and the probing AFM tip was prepared by picking up a Xe atom from the nodal site. Systematic force measurements established the depths of the potential curves for Xe–Ar, Xe–Kr and Xe–Xe at 18.1, 26.1 and 35.9 meV, respectively, which is in agreement with the expected trend. The tails of the interactions of Xe–Ar and Xe–Kr were found to display van der Waals r^{-6} limiting behaviour; however, the effective strength of the interaction was found to depend on the influence of the tip and the substrate on the atomic polarizability, leading to a deviation from the behaviour of isolated noble gas atom. For Xe–Xe interaction, we find a deviation from the van der Waals power law, which we interpret as an effect of stronger covalent interaction between the Xe atoms. This approach offers a systematic way to characterize the interaction between any atomic and molecular species that can be fixed at a solid surface of a solid or be templated by a surface framework, and has the sensitivity to measure even the weakest forces. The combination with theory has allowed for a detailed analysis of the interaction components, clearly highlighting the role of 'classical' van der Waals interactions while identifying the limits of an isolated atom model.

Methods

Experimental measurement. All experiments were performed with Omicron STM/AFM with a qPlus configuration[40], operating at 4.8 K in ultrahigh vacuum

(UHV). A clean Cu(111) surface was *in situ* prepared by repeated cycles of standard sputtering and annealing. The W tip of a tuning fork sensor was *ex situ* sharpened by focused ion beam milling technique and was then *in situ* covered with Cu atoms by contacting to the sample surface. Perylene derivative, 4,9-diaminoperylene-quinone-3,10-diimine (DPDI) molecules[41] were deposited on the surface from a Knudsen cell, resistively heated at 200 °C. By annealing at ~300 °C, the molecular network was formed[28]. The resonance frequency of the self-oscillating qPlus sensor was detected by a digital lock-in amplifier (Nanonis: OC4 and Zurich Instruments: HF2LI and PLL). Xe, Kr and Ar gases were sequentially dosed via an UHV leak valve while the sample stayed in a cold microscope. In STM mode, the tip was biased, while the sample was electronically grounded. The topographic images were taken in a constant current mode. In AFM mode, the tip apex was terminated by a CO or Xe molecule and all images were taken at a constant height mode.

Theoretical simulation. All first-principles calculations in this work were performed using the periodic plane-wave basis Vienna ab initio simulation package (VASP) code[42,43] implementing the spin-polarized density functional theory. To accurately include van der Waals interactions in this system, we used the optB86B + vdW-DF functional[44–46], selected based on benchmarks (Supplementary Fig. 10) and previous work showing that it provides a sufficiently accurate description for all subsystems involved in the measurement. Projected augmented wave potentials were used to describe the core electrons[47], with a kinetic energy cutoff of 550 eV (with PREC = accurate). Systematic *k*-point convergence was checked for all systems, with sampling chosen according to the system size and the gamma point being used for the largest systems. This approach converged the total energy of all the systems to the order of meV. The properties of the bulk and surface of Cu, each noble atom dimer pair and their adsorption on copper, were carefully checked within this methodology, and excellent agreement was achieved with experiments. Bader charge analysis was used to estimate charge transfer in the simulations[48].

The geometry of the Cu-coordinated 3deh-DPDI network was taken from previous studies[29]. In general, the interaction strength with the network scaled as Xe > Kr > Ar (see Supplementary Table 1), which is consistent with the experiments and expected from previous studies[49]. To represent the tip, we used an equivalent (111) Cu slab to that under the DPDI network (see inset in Fig. 4), with the Xe tip atom adsorbed in its equilibrium position—since the tip is directly contacted to the copper surface in experiments, this seems a plausible model (good agreement with experimental studies of Xe adsorption on Cu (111) was achieved[50]—see Supplementary Table 1). Other, smaller copper tip models were considered, but were either very unstable or showed poorer agreement with the experimental curves. The interaction parameters were obtained by fitting the single parameter expression C_N/r^N to the asymptotic tail of the experimental data, defined as starting just above the inflection point of the curve. The appropriate power, *N*, was selected based on the quality of the fits as measured by the correlation coefficients.

References

1. Margenau, H. Van der Waals forces. *Rev. Mod. Phys.* **11,** 1–35 (1939).
2. Israelachvili, J. N. *Intermolecular and Surface Forces* (Academic Press, 2011).
3. Ortmann, F., Schmidt, W. G. & Bechstedt, F. Attracted by long-range electron correlation: adenine on graphite. *Phys. Rev. Lett.* **95,** 186101 (2005).
4. Björk, J. *et al.* Adsorption of aromatic and anti-aromatic systems on graphene through pi stacking. *J. Phys. Chem. Lett.* **1,** 3407–3412 (2010).
5. Mercurio, G. *et al.* Structure and energetics of azobenzene on Ag(111): Benchmarking semiempirical dispersion correction approaches. *Phys. Rev. Lett.* **104,** 036102 (2010).
6. Lennard-Jones, J. E. Processes of adsorption and diffusion on solid surfaces. *Trans. Faraday Soc.* **28,** 333–359 (1932).
7. Vega, C., Abascal, J. L. F., Sanz, E., MacDowell, L. G. & McBride, C. Can simple models describe the phase diagram of water? *J. Phys. Condens. Matter* **17,** S3283–S3288 (2005).
8. Vega, C., Abascal, J. L. F., Conde, M. M. & Aragones, J. L. What ice can teach us about water interactions: a critical comparison of the performance of different water models. *Faraday Discuss* **141,** 251–276 (2009).
9. Kuemmel, S. & Kronik, L. Orbital-dependent density functionals: theory and applications. *Rev. Mod. Phys.* **80,** 3–60 (2008).
10. Dion, M., Rydberg, H., Schroder, E., Langreth, D. & Lundqvist, B. Van der Waals density functional for general geometries. *Phys. Rev. Lett.* **92,** 246401 (2004).
11. Schwabe, T. & Grimme, S. Double-hybrid density functionals with long-range dispersion corrections: higher accuracy and extended applicability. *Phys. Chem. Chem. Phys.* **9,** 3397–3406 (2007).
12. Silvestrelli, P. L. Van der Waals interactions in DFT made easy by Wannier functions. *Phys. Rev. Lett.* **100,** 053002 (2008).
13. Tkatchenko, A., DiStasio, R., Car, R. & Scheffler, M. Accurate and efficient method for many-body van der Waals interactions. *Phys. Rev. Lett.* **108,** 236402 (2012).
14. Klimeš, J. & Michaelides, A. Perspective: advances and challenges in treating van der Waals dispersion forces in density functional theory. *J. Chem. Phys.* **137,** 120901 (2012).
15. Pyykkö, P. & Stanton, J. F. Introduction to the quantum chemistry 2012 issue. *Chem. Rev.* **112,** 1–3 (2012).
16. Béguin, L., Vernier, A., Chicireanu, R., Lahaye, T. & Browaeys, A. Direct measurement of the van der Waals interaction between two Rydberg atoms. *Phys. Rev. Lett.* **110,** 263201 (2013).
17. Lantz, M. A. *et al.* Quantitative measurement of short-range chemical bonding forces. *Science* **291,** 2580–2583 (2001).
18. Sader, J. E. & Jarvis, S. P. Accurate formulas for interaction force and energy in frequency modulation force spectroscopy. *Appl. Phys. Lett.* **84,** 1801–1803 (2004).
19. Sugimoto, Y. *et al.* Chemical identification of individual surface atoms by atomic force microscopy. *Nature* **446,** 64–67 (2007).
20. Setvín, M. *et al.* Chemical identification of single atoms in heterogeneous III-IV chains on Si(100) surface by means of nc-AFM and DFT calculations. *ACS Nano* **6,** 6969–6976 (2012).
21. Ternes, M. *et al.* Interplay of conductance, force, and structural change in metallic point contacts. *Phys. Rev. Lett.* **106,** 016802 (2011).
22. Albers, B. J. *et al.* Three-dimensional imaging of short-range chemical forces with picometre resolution. *Nat. Nanotechnol.* **4,** 307–310 (2009).
23. Ashino, M., Schwarz, A., Behnke, T. & Wiesendanger, R. Atomic-resolution dynamic force microscopy and spectroscopy of a single-walled carbon nanotube: characterization of interatomic van der Waals forces. *Phys. Rev. Lett.* **93,** 136101 (2004).
24. Aradhya, S. V., Frei, M., Hybertsen, M. S. & Venkataraman, L. Van der Waals interactions at metal/organic interfaces at the single-molecule level. *Nat. Mater.* **11,** 872 (2012).
25. Wagner, C. *et al.* Non-additivity of molecule-surface van der Waals potentials from force measurements. *Nat. Commun.* **5,** 5568 (2014).
26. Grochala, W. Atypical compounds of gases, which have been called 'noble'. *Chem. Soc. Rev.* **36,** 1632–1655 (2007).
27. Allers, W., Schwarz, A., Schwarz, U. D. & Wiesendanger, R. Dynamic scanning force microscopy at low temperatures on a noble-gas crystal: atomic resolution on the xenon(111) surface. *Europhys. Lett.* **48,** 276–279 (1999).
28. Stöhr, M. *et al.* Controlling molecular assembly in two dimensions: the concentration dependence of thermally induced 2D aggregation of molecules on a metal surface. *Angew. Chem. Int. Ed.* **44,** 7394–7398 (2005).
29. Matena, M. *et al.* On-surface synthesis of a two-dimensional porous coordination network: Unraveling adsorbate interactions. *Phys. Rev. B* **90,** 125408 (2014).
30. Eigler, D. M., Lutz, C. P. & Rudge, W. E. An atomic switch realized with the scanning tunnelling microscope. *Nature* **352,** 600–603 (1991).
31. Gross, L., Mohn, F., Moll, N., Liljeroth, P. & Meyer, G. The chemical structure of a molecule resolved by atomic force microscopy. *Science* **325,** 1110–1114 (2009).
32. Gross, L. *et al.* Organic structure determination using atomic-resolution scanning probe microscopy. *Nat. Chem.* **2,** 821–825 (2010).
33. Pavliček, N. *et al.* Atomic force microscopy reveals bistable configurations of dibenzo[a,h]thianthrene and their interconversion pathway. *Phys. Rev. Lett.* **108,** 086101 (2012).
34. Gross, L. *et al.* Bond-order discrimination by atomic force microscopy. *Science* **337,** 1326–1329 (2012).
35. Kawai, S. *et al.* Obtaining detailed structural information about supramolecular systems on surfaces by combining high-resolution force microscopy with ab initio calculations. *ACS Nano* **7,** 9098–9105 (2013).
36. Kawai, S. *et al.* Extended halogen bonding between fully fluorinated aromatic molecules. *ACS Nano* **9,** 2574–2583 (2015).
37. Mohn, F., Schuler, B., Gross, L. & Meyer, G. Different tips for high-resolution atomic force microscopy and scanning tunneling microscopy of single molecules. *Appl. Phys. Lett.* **102,** 073109 (2013).
38. Nowakowska, S. *et al.* Interplay of weak interactions in the atom-by-atom condensation of xenon within quantum boxes. *Nat. Commun.* **6,** 6071 (2015).
39. Kumar, A. & Thakkar, A. J. Dipole oscillator strength distributions with improved high-energy behavior: Dipole sum rules and dispersion coefficients for Ne, Ar, Kr, and Xe revisited. *J. Chem. Phys.* **132,** 074301 (2010).
40. Giessibl, F. J. High-speed force sensor for force microscopy and profilometry utilizing a quartz tuning fork. *Appl. Phys. Lett.* **73,** 3956–3958 (1998).
41. Gade, L. H. *et al.* Tetraaminoperylenes: their efficient synthesis and physical properties. *Chem. Eur. J.* **8,** 3732–3746 (2002).
42. Kresse, G. & Furthmuller, J. Efficiency of *ab-initio* total energy calculations for metals and semiconductors using a plane-wave basis set. *Comput. Mater. Sci.* **6,** 15–50 (1996).
43. Kresse, G. Efficient iterative schemes for ab initio total-energy calculations using a plane-wave basis set. *Phys. Rev. B* **54,** 11169–11186 (1996).

44. Klimeš, J., Bowler, D. R. & Michaelides, A. Chemical accuracy for the van der Waals density functional. *J. Phys. Condens. Matter* **22,** 022201 (2010).
45. Klimeš, J., Bowler, D. R. & Michaelides, A. Van der Waals density functionals applied to solids. *Phys. Rev. B* **83,** 195131 (2011).
46. Björkman, T. Testing several recent van der Waals density functionals for layered structures. *J. Chem. Phys.* **141,** 074708 (2014).
47. Blöchl, P. E. Projector augmented-wave method. *Phys. Rev. B* **50,** 17953–17979 (1994).
48. Tang, W., Sanville, E. & Henkelman, G. A grid-based Bader analysis algorithm without lattice bias. *J. Phys. Condens. Matter* **21,** 084204 (2009).
49. Vidali, G., Ihm, G., Kim, H.-Y. & Cole, M. W. Potentials of physical adsorption. *Surf. Sci. Rep.* **12,** 135–181 (1991).
50. Ruiz, V. G., Liu, W. & Tkatchenko, A. Density-functional theory with screened van der Waals interactions applied to atomic and molecular adsorbates on close-packed and non-close-packed surfaces. *Phys. Rev. B* **93,** 035118 (2016).

Acknowledgements

This work was supported in part by the Japan Science and Technology Agency (JST) 'Precursory Research for Embryonic Science and Technology (PRESTO)' for a project of 'Molecular technology and creation of new functions', by the Swiss National Science Foundation, by the Swiss Nanoscience Institute, and by COST Action MP1303 'Understanding and Controlling Nano and Mesoscale Friction'. A.S.F. has been supported by the Academy of Finland through its Centres of Excellence Program project no. 915804 and EU project PAMS (contract no. 610446), and acknowledge use of the CSC, Helsinki for computational resources.

Author contributions

S.K. and S.N. conducted experiments. S.K. carried out data analysis. L.H.G. synthesized the precursor molecule. A.S.F., T.B. and J.B. conducted theoretical calculations. S.K., A.S.F., T.B., S.N., J.B., F.F.C., L.H.G., T.A.J. and E.M. contributed to writing the manuscript. S.K. planned the project.

Additional information

Competing financial interests: The authors declare no competing financial interests.

16

Balancing surface adsorption and diffusion of lithium-polysulfides on nonconductive oxides for lithium-sulfur battery design

Xinyong Tao[1,2,*], Jianguo Wang[3,*], Chong Liu[2], Haotian Wang[2], Hongbin Yao[2], Guangyuan Zheng[2], Zhi Wei Seh[2], Qiuxia Cai[3], Weiyang Li[2], Guangmin Zhou[2], Chenxi Zu[2] & Yi Cui[2,4]

Lithium-sulfur batteries have attracted attention due to their six-fold specific energy compared with conventional lithium-ion batteries. Dissolution of lithium polysulfides, volume expansion of sulfur and uncontrollable deposition of lithium sulfide are three of the main challenges for this technology. State-of-the-art sulfur cathodes based on metal-oxide nanostructures can suppress the shuttle-effect and enable controlled lithium sulfide deposition. However, a clear mechanistic understanding and corresponding selection criteria for the oxides are still lacking. Herein, various nonconductive metal-oxide nanoparticle-decorated carbon flakes are synthesized via a facile biotemplating method. The cathodes based on magnesium oxide, cerium oxide and lanthanum oxide show enhanced cycling performance. Adsorption experiments and theoretical calculations reveal that polysulfide capture by the oxides is via monolayered chemisorption. Moreover, we show that better surface diffusion leads to higher deposition efficiency of sulfide species on electrodes. Hence, oxide selection is proposed to balance optimization between sulfide-adsorption and diffusion on the oxides.

[1]College of Materials Science and Engineering, Zhejiang University of Technology, Hangzhou 310014, China. [2]Department of Materials Science and Engineering, Stanford University, Stanford, California 94305, USA. [3]College of Chemical Engineering, Zhejiang University of Technology, Hangzhou 310014, China. [4]Stanford Institute for Materials and Energy Science, SLAC National Accelerator Laboratory, Menlo Park, California 94025, USA. *These authors contributed equally to this work. Correspondence and requests for materials should be addressed to Y.C. (email: yicui@stanford.edu).

Rechargeable lithium–sulfur (Li–S) batteries have recently become one of the more exciting energy storage systems due to the low-cost and high-specific energy of sulfur cathodes[1–27]. Although there have been significant developments for designing state-of-the-art Li–S batteries in the past two decades, the practical application is still hindered by many material challenges, including dissolution of intermediate lithium polysulfides (Li_2S_x, $x>3$) in the electrolyte[28], large volumetric expansion (80%) of sulfur upon lithiation[6], and poor electronic/ionic conductivity of sulfur and lithium sulfide (Li_2S) (ref. 6). To date, tremendous efforts have been made to solve the above problems by constructing advanced composite cathode materials. One effective strategy is the encapsulation of sulfur to prevent the leakage of active materials and suppress the shuttle effect of high-order Li_2S_x (refs 3,6). Oxides[6], carbon[3,29], polymers[30] and metals[31] are proved to be good matrix materials for the encapsulation of sulfur[19]. The second approach is the controllable deposition of the discharge product Li_2S, which is an ionic and electronic insulator[4]. The detaching and irreversible phase transformation of Li_2S is considered as the main reason for capacity fading[4,32]. The third strategy is using Li_2S as a starting cathode material, which undergoes volumetric contraction instead of the expansion in the case of sulfur[20]. In addition, Li_2S-based cathodes can be paired with lithium metal-free anodes such as graphite, silicon and alloys[33], thus suppressing the dendrite growth and the corresponding safety concerns of lithium-metal anodes.

All the previous research reveals the importance of understanding the sulfide species interaction with the matrix materials. Our earlier work pointed out that the usual carbon substrates interact with Li_2S_x weakly but the polar group enabled strong interaction with Li_2S_x, which can facilitate the Li_2S_x trapping and promote the attachment of solid Li_2S_2 and Li_2S and improve the cycling stability of Li–S batteries[34]. Many similar examples followed-up using polymers[35,36], oxides[4,19], sulfides[20], functionalized graphene[18,26,27,37,38], metal organic framework[39] and nitrogen doped carbon[36,40], which all have polar surfaces to adsorb Li_2S_x species. Results from this study seem to suggest that the stronger the binding, the better the Li–S batteries. In addition, it was recognized that using conducting materials such as indium tin oxide[4] and Ti_4O_7 (ref. 19) is preferable due to the electron transfer needed to induce electrochemical reaction. Our recent work showed that indium tin oxide decorated carbon nanofibres can enhance the redox kinetics of Li_2S_x, realize the controllable deposition of Li_2S and improve the electrochemical performance of Li–S batteries[4].

Besides the conductive matrix material, there are indeed abundant insulating materials to trap Li_2S_x. But there is a fundamental problem here: insulated materials cannot transport electrons. Trapping Li_2S_x on insulating materials would cause them to be accumulated in electronically inactive areas and reduce the capacity retention. However, some studies have shown improvements of the electrochemical properties after the decoration of the electrode with poor conductive oxides such as MnO_2 (ref. 16), $Mg_{0.6}Ni_{0.4}O$ (ref. 12), Al_2O_3 (ref. 41) and La_2O_3 (ref. 42).

The above background research has motivated us to hypothesize that surface diffusion of Li_2S_x species on solid substrates can play an important role in Li–S battery electrochemical performance. This is particularly important for insulating solid materials with strong adsorption of Li_2S_x. The competition between the adsorption and diffusion of the Li_2S_x adsorbates on solid substrates can be very important, yet has been overlooked for Li–S batteries.

Usually, the nonconductive metal oxides work together with the carbon matrix to improve the conductivity of sulfur cathodes. For the modified carbon matrix with some nonconductive metal oxides nanostructures (Fig. 1), there is no direct electron transfer between these oxides and Li_2S_x species. Because of the poor conductivity of the metal oxide, the absorbed Li_2S_x should be transferred from the surface of the oxide to the conductive carbon substrate to undergo the electrochemical reaction. Therefore the competitive surface diffusion and adsorption of sulfur species must play key roles in the Li–S batteries. If the metal oxide has weak Li_2S_x capture capability (Fig. 1a), a large amount of Li_2S_x can diffuse away from the carbon matrix, resulting in serious shuttle effect and uncontrollable deposition of Li_2S. When the diffusion of sulfur species from the surface of oxide to carbon is difficult (Fig. 1c), the electrochemical reaction of Li_2S_x and the corresponding growth of Li_2S on the oxide/carbon is impeded. Therefore, the balance optimization between Li_2S_x adsorption and diffusion on the metal oxides surface is necessary (Fig. 1b).

In this work, various nonconductive metal oxide (MgO, Al_2O_3, CeO_2, La_2O_3 and CaO) nanoparticles-decorated carbon flakes are synthesized via a facile and generic biotemplating method using Kapok trees fibres as both the template and the carbon source. The sulfur cathodes based on MgO, CeO_2 and La_2O_3 show higher capacity and better cycling stability. Adsorption test, microstructure analysis and electrochemical performance evaluation combined with density functional theory (DFT) calculations reveal that better surface diffusion leads to higher deposition efficiency of sulfide species. A comprehensive-oxide-selection criteria referring to the strong binding, high surface area and good surface diffusion properties is proposed.

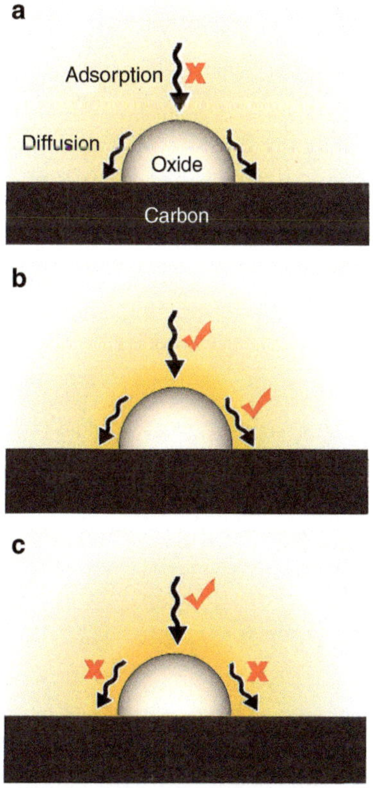

Figure 1 | Schematic of the Li_2S_x adsorption and diffusion on the surface of various nonconductive metal oxides. (**a**) The metal oxide with weak Li_2S_x adsorption capability; only few Li_2S_x can be captured by the oxide; (**b**) the metal oxide with both strong adsorption and good diffusion, which is favourable for the electrochemical reaction and the controllable deposition of sulfur species; (**c**) the metal oxide with strong bonding but without good diffusion; the growth of Li_2S and the electrochemical reaction on the oxide/C surface is impeded.

Results

Sulfide capture by metal oxides. To reveal the role of metal oxides in Li-S batteries, five kinds of pure metal oxide-nanoparticles were prepared by a generic Pechini sol-gel method[13]. 1,3-dioxolane (DOL) and dimethoxyethane (DME) are commonly used solvent in the Li-S battery electrolyte[4]. Therefore, 0.005 M Li_2S_8 in DOL/DME (1:1, v-v) was prepared for the adsorption test of sulfides. Figure 2a shows the camera image of the adsorption test using different mass of oxide samples with the same total surface area. The colour of the solution containing Al_2O_3 and CeO_2 is lighter than the others, indicating better adsorption of these two metal oxide nanoparticles. Inductively coupled plasma-optical emission spectroscopy (ICP-OES) results reveal that Al_2O_3 and CaO show the biggest and the smallest absorption capability, respectively (Fig. 2b). In addition, it was found that the adsorption capacity increases slightly with the rise of temperature (Fig. 2b and Supplementary Table 1), which is one essential characteristic of chemisorption. The measured Li_2S_8 adsorption quantity of metal oxides is in the range of 2.78–4.94 μmol m^{-2}, close to the simulated monolayer adsorption capacity ranging from 2.76 to 4.88 μmol m^{-2} (Fig. 2b). Therefore, the Li_2S_8 capture involves monolayer adsorption, which is another well-known characteristic of chemisorption.

In order to better understand the absorption mechanism, DFT calculation was performed to reveal the corresponding adsorption energy and sites (Fig. 2c,d and Supplementary Figs 2–6). Considering both low-order and high-order, Li_2S_x are important discharge products of Li-S batteries, we choose both Li_2S and Li_2S_8 as the prototype for modelling. Figure 2c shows the optimized geometries of the most stable Li_2S on CeO_2(111), Al_2O_3(110), La_2O_3(001), MgO(100) and CaO(100) surfaces. On the entire surface, the most favourable binding site of Li_2S is two Li atoms bonding with the oxygen atom of metal oxide (Fig. 2c). On Al_2O_3(110), the Li of Li_2S is the bridge site of two oxygen atoms, on other four metal oxides, the Li is on the atop site of oxygen. On MgO(100) CaO(100) and La_2O_3(001) surfaces, the sulfur of Li_2S is away from the metal oxides surface. Sulfur is bonding with oxygen on CeO_2(111), in which the sulfur-oxygen distance is 1.70 Å while with Al on Al_2O_3(110), in which the Al-S distance is 2.21 Å. The adsorption energy of Li_2S on CeO_2(111), Al_2O_3(110), La_2O_3(001), MgO(100) and CaO(100) surfaces is −6.33, −7.12, −0.5.85, −5.71 and −5.49 eV, respectively. The optimized stable configurations of Li_2S_8 on five different surfaces are also shown in Fig. 2d and Supplementary Figs 2–6. Although the Li of Li_2S_8 has similar bonding sites on the metal oxide surface as that of Li_2S, the optimized most stable configuration of Li_2S_8 on each metal oxide is quite different. A lot of initial geometries of Li_2S_8 on each metal oxide have been considered in our calculations. It is found that after optimization, the structures have big change due to the interaction between Li_2S_8 and metal oxides. For both Li_2S and Li_2S_8 on metal oxides, the bonding between Li and oxygen plays a major role. In addition, only two or three sulfur atoms of Li_2S_8 are bonding with the oxide surface. The adsorption energies of Li_2S_8 have similar trend on these metal surfaces with Li_2S, while are much weaker than Li_2S. The calculated adsorption energies of both Li_2S and Li_2S_8 are in agreement with the experimental adsorption test results of Li_2S_8 on the oxide nanoparticles (Fig. 2b and Supplementary Table 1).

Biotemplated fabrication of oxides/carbon nanostructures. Although most of these oxides have remarkable adsorption behaviour for sulfide species, they are poor electronic conductors. Conductivity is one of the most important factors affecting the performance of Li-S batteries. Therefore, we fabricated metal-oxides (MgO, Al_2O_3, CeO_2, La_2O_3 and CaO) nanoparticles anchored on porous carbon nanoflakes to form an electronic conductive oxide/carbon nanocomposite. To fabricate the nano-composite, Kapok fibres (KFs)[43] were used as both the template and the carbon source (Fig. 3a). KFs are low-cost and high-yield agriculture products derived from the fruits of Kapok tree, which is chemically composed of 64% cellulose, 13% lignin and 23% pentosan[43]. In addition, the KFs have unique hollow

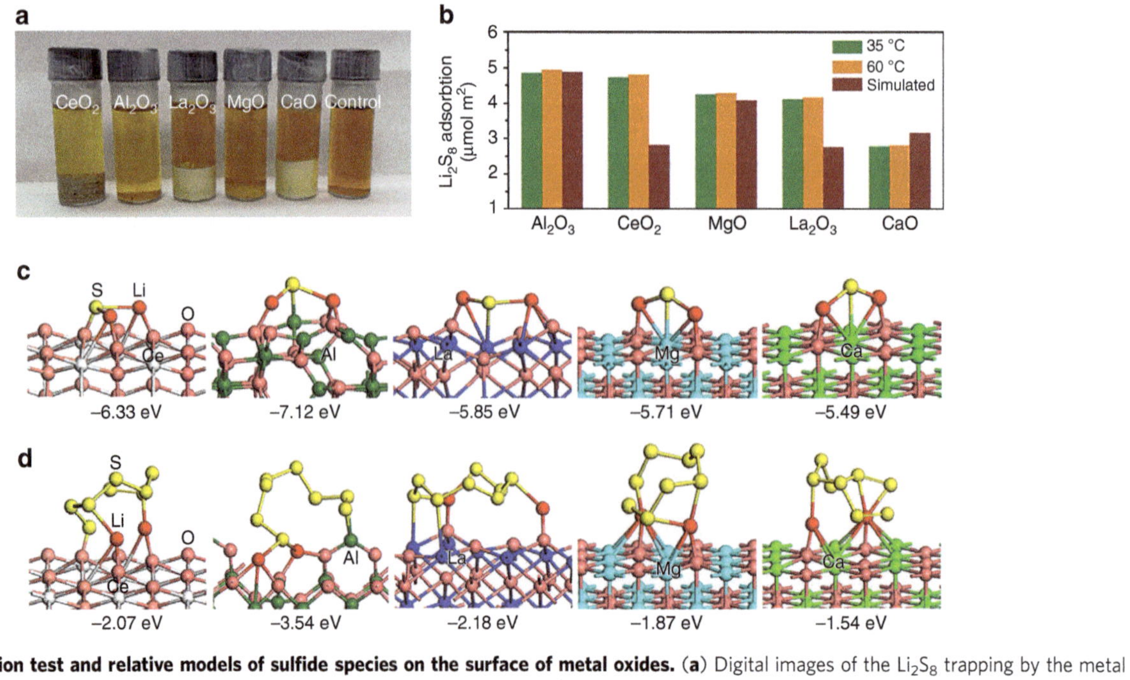

Figure 2 | Adsorption test and relative models of sulfide species on the surface of metal oxides. (**a**) Digital images of the Li_2S_8 trapping by the metal oxide nanoparticles in DOL/DME (1:1, v-v) solution. (**b**) Experimental and simulated adsorption amount of Li_2S_8 on different metal oxides. The simulated adsorption was based on the monolayer adsorption model. (**c**) Optimized geometries of the most stable Li_2S on CeO_2(111), Al_2O_3(110), La_2O_3(001), MgO(100) and CaO(100) surfaces. (**d**) Optimized geometries of most stable Li_2S_8 on the metal oxide surface.

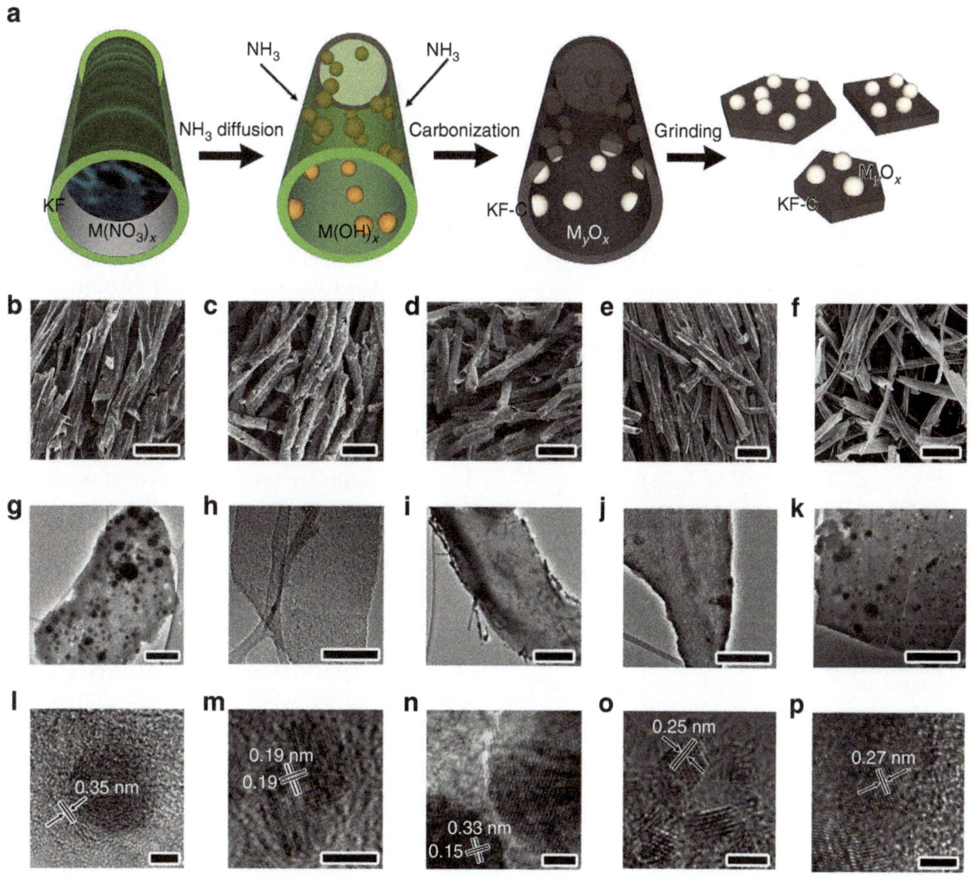

Figure 3 | Fabrication and microstructures of oxides/carbon nanostructures. (**a**) Schematic illustration of synthesis of oxides/carbon using the kapok tree fibres (KF) as both template and carbon sources. (**b-f**) SEM images of Al_2O_3/C, CeO_2/C, La_2O_3/C, MgO/C and CaO/C composites, respectively (scale bar = 50 μm). (**g-k**) The corresponding TEM images of Al_2O_3/C, CeO_2/C, La_2O_3/C, MgO/C and CaO/C, respectively (scale bar = 300 nm). (**l-p**) The corresponding HRTEM images of Al_2O_3/C, CeO_2/C, La_2O_3/C, MgO/C and CaO/C, respectively (scale bar = 3 nm).

lumens with a thin wall thickness ≤1 μm, which enables good sorption capacity through capillary force. Therefore, metal nitrate solution can be easily absorbed into the KFs (Fig. 3a). When the NH_3 gas diffuses into the KFs through the porous cell wall and the open end, metal hydroxides nanoparticles will be formed on the surface of the cell wall due to the confinement effect of KF template. After drying at 90 °C for 12 h, carbonization at 850 °C for 1.5 h and the following facile grinding, carbon nanoflakes decorated with metal oxide nanoparticles can be obtained (Fig. 3a). Figure 3b–f shows the scanning electron microscopy (SEM) images of Al_2O_3/C, CeO_2/C, La_2O_3/C, MgO/C and CaO/C nanocomposites, respectively. The composite remains the macromorphology of the original Kapok tree fibres, which have a unique hollow structure with a large lumen and a thin fibre wall. After simple grinding, the delicate carbon microtubes are converted to carbon nanoflakes (Fig. 3g–k and Supplementary Figs 7–9).

Moreover, transmission electron microscopy (TEM) images show that abundant oxide nanostructures are located on the carbon matrix (Fig. 3g–k). Figure 3l is the representative high resolution TEM (HRTEM) image of the Al_2O_3/C sample, showing the amorphous structure of Al_2O_3 and partially graphitized structure of C. The interlayer spacing is about 0.35 nm, corresponding to (002) planes of graphite. Figure 3m shows the [001] zone axis HRTEM image of a typical CeO_2 nanoparticle in CeO_2/C composite. The lattice spacing of 0.19 nm in Fig. 3m can be attributed to {220} crystal planes of the face centred cubic (fcc) phase CeO_2. The TEM image of La_2O_3/C is showed in Fig. 3i, suggesting that abundant rod-shaped nanoparticles are distributed in the carbon matrix. The corresponding HRTEM image in Fig. 3n shows that the La_2O_3 particle is single crystalline. The lattice spacings (0.15 and 0.33 nm) and the interplanar angle match the finger print of (004) and (100) planes of hexagonal phase La_2O_3. HRTEM image of MgO/C sample in Fig. 3o indicates lattice fringes with regular spacing of 0.25 nm, which can be indexed to (111) planes of fcc MgO. Figure 3p shows the HRTEM image of CaO/C sample. The fringes with spacing of 0.27 nm is corresponding to (111) planes of fcc CaO. These TEM results indicate that the Kapok tree fibre can act as ideal and general template for the synthesis of oxide nanostructures due to the confinement effect of fibre substrate.

Electrochemical performance of composite cathodes. Thermal diffusion method was used to fabricate the sulfur/M_xO_y/C composite[19]. The mass loading of the electrode ranges from 0.7 to 1.2 mg cm^{-2}. The electrolyte was 1 M Lithium bis (trifluoromethanesulphonyl)imide in DOL and DME, with $LiNO_3$ as additive to passivate the lithium anode. Figure 4a shows the representative charge–discharge curves of the composite electrode based on different oxide/carbon nanostructures at a current rate of 0.1 C (1 C = 1,672 mA g^{-1}). All the discharge curves show two typical discharge plateaus at 2.35 and 2.10 V, which can be assigned to the formation of high-order and low-order Li_2S_x, respectively[9]. No obvious difference can be found for the potential of the discharge plateaus. However, the CaO/C and

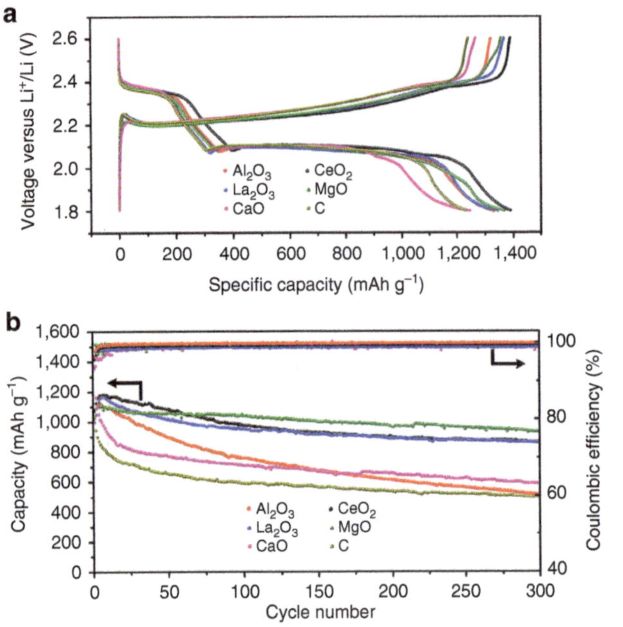

Figure 4 | Charge-discharge curves and cycling performance of the sulfur composite electrodes. (**a**) Representative charge-discharge profiles of the composite electrodes based on different oxide/carbon nanostructures at 0.1 C. (**b**) Specific capacity and the corresponding Coulombic efficiency of the composite electrodes upon prolonged 300 charge-discharge cycles at 0.5 C.

C composite cathodes show higher charge over-potentials than those of CeO_2/C, La_2O_3/C, Al_2O_3/C and MgO/C composite electrodes (Fig. 4a). The specific discharge capacities of Al_2O_3/C, CeO_2/C, La_2O_3/C, MgO/C, CaO/C and C at 0.1 C rate are measured to be 1,330, 1,388, 1,345, 1,368, 1,246 and 1,230 mAh g^{-1}, respectively (Fig. 4a). It can be found that the CaO/C and C composite cathodes show relatively lower discharge capacity. The high over-potential and low discharge capacity may result from the serious dissolution of Li_2S_x in electrolyte (shuttle effect), which causes the active material loss and the increase of electrolyte viscosity. ICP-OES test (Supplementary Table 2) based on the same mass of Al_2O_3/C, CeO_2/C, La_2O_3/C, MgO/C, CaO/C and C samples reveal that both CaO/C and C have poorer Li_2S_x capture capability compared with other samples.

Besides the specific capacity, cycling performance is one of the most important characteristics for Li-S batteries. Therefore, all the cathodes were subject to prolonged cycling. Figure 4b shows the discharge capacity and the corresponding Coulombic efficiency of the cathodes upon prolonged 300 cycles at 0.5 C. The representative charge/discharge curves (Fig. 4a) and the cycling performance (Fig. 4b) indicate that Al_2O_3/C, CeO_2/C, La_2O_3/C and MgO/C composite electrode show high specific capacity in the first several cycles. However, the composite electrodes show distinct capacity retention capability. The Al_2O_3/C, CaO/C and C cathodes exhibit obvious capacity fading especially in the first 100 cycles. Compared with Al_2O_3/C, CaO/C and C cathodes, the electrode based on CeO_2/C, La_2O_3/C and MgO/C show better cycling performance and the MgO/C cathode is the best among all the samples. The capacity decay per cycle is 0.171, 0.066, 0.047, 0.034, 0.136 and 0.170% for Al_2O_3/C, CeO_2/C, La_2O_3/C, MgO/C, CaO/C and C, respectively. Considering the serious capacity, decay mainly happens in the first 50 cycles, the average Columbic efficiency in the first 100 cycles was calculated (Fig. 4b). Al_2O_3/C, CeO_2/C, La_2O_3/C, MgO/C, CaO/C and C cathodes show 99.6, 99.1, 98.7, 99.4, 98.3 and 98.8% columbic efficiency. Lower Columbic efficiency of Li-S batteries resulted from the significant Li_2S_xdissolution, which cause the loss of S material and shuttle effect[19]. This can be also supported by the ICP-OES results, revealing that Al_2O_3/C has the best capture capability and CaO/C and C have poorer capture capability for Li_2S_8(Supplementary Table 2). Although Al_2O_3/C cathodes possess high initial discharge capacity and good Columbic efficiency, the rate of capacity decay is higher than those of CeO_2/C, La_2O_3/C and MgO/C cathodes. The distinct Columbic efficiency for Al_2O_3 (99.6%) cathode from both CaO/C (98.3%) and C (98.8%) cathodes may imply different capacity decay mechanism.

Analysis of capacity failure mechanism of composite cathodes. In order to further reveal the detailed capacity decay mechanism, some batteries were disassembled after 100 cycles at 0.5 C to observe the morphology evolution of the cathode materials by SEM. Figure 5a–f shows low magnification SEM images of the cycled electrodes based on Al_2O_3/C, CeO_2/C, La_2O_3/C, MgO/C, CaO/C and C nanostructures, respectively. In contrast to CeO_2/C (Fig. 5b), La_2O_3/C (Fig. 5c) and MgO/C (Fig. 5d) electrodes with uniform and flat surface, the Al_2O_3/C (Fig. 5a), CaO/C (Fig. 5e) and C (Fig. 5f) cathodes show high surface roughness. Some carbon nanoflakes with well-defined profile can be observed on the surface of the Al_2O_3/C (Fig. 5a), CaO/C (Fig. 5e) and C (Fig. 5f) cathodes after 100 cycles. In addition, some cracks and pinholes can be found in the CaO/C and carbon cathodes, which may result from the significant dissolution and loss of sulfur. Figure 5g–l and m–r shows the top-view and cross-sectional SEM images of cycled Al_2O_3/C, CeO_2/C, La_2O_3/C, MgO/C, CaO/C and C cathodes, respectively. Only a thin, uniform and dense Li_2S film can be found on the surface of Al_2O_3/C (Fig. 5g,m). Abundant Li_2S particles with irregular shape were formed between the Al_2O_3/C nanoflakes (Supplementary Fig. 1). Similar phenomenon can also be found in the CaO/C (Fig. 5k,q) and C (Fig. 5l,r) cathodes. Some characteristic stripes (Fig. 5k,l) and distinct fracture surface (Fig. 5q,r) indicate that the Li_2S particles are detached from the oxides/C matrix and may become electrochemically inactive. This is due to the uncontrollable precipitation of Li_2S on the non-polar or weakly polar surface[6], which leads the further decay of capacity. In contrast, it is not easy to identify the oxide/carbon nanoflakes from the CeO_2/C (Fig. 5h), La_2O_3/C (Fig. 5i) and MgO/C (Fig. 5j) cathodes. To obtain the cross-sectional morphology of the cathodes, the electrode materials were scraped from the Al foil current collector and mounted on a copper foil with rough surface for SEM observation. CeO_2/C (Fig. 5n), La_2O_3/C (Fig. 5o) and MgO/C (Fig. 5p) nanoflakes were wrapped by thick Li_2S layer. Supposing that the average thickness of carbon nanoflakers is 450 nm, the thickness of Li_2S layer deposited on the surface of CeO_2/C, La_2O_3/C and MgO/C will be 1,400, 1,677 and 1,573 nm, respectively. After mechanical scrapping and mounting processes, no detachment can be observed, indicating that there is good cohesion between the Li_2S layer and the oxide/C nanoflakes. These results are consistent with our DFT calculation results in Fig. 2. Because the Li_2S is a poor ionic and electronic conductor, the good combination of Li_2S with the conductive matrix must be favourable for the reversible electrochemical reaction in the following charging process[4]. Therefore, the better cycling performance of CeO_2/C, La_2O_3/C and MgO/C cathodes can be attributed to the controllable precipitation of Li_2S on the polar surface of carbon matrix.

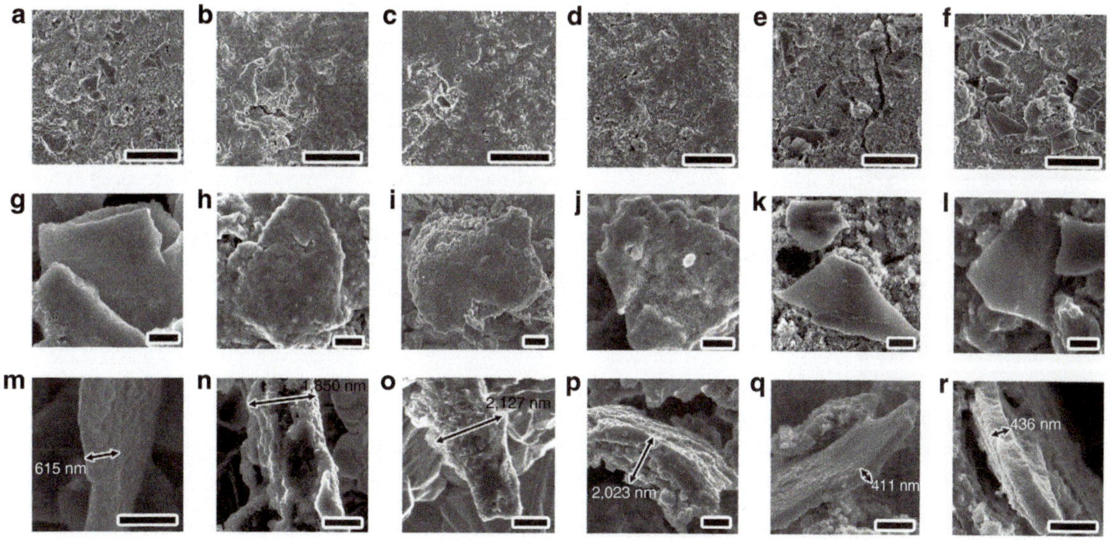

Figure 5 | Morphology of the discharged composite electrodes after cycling. (**a-f**) SEM images of the cycled composite electrodes based on Al_2O_3/C, CeO_2/C, La_2O_3/C, MgO/C, CaO/C and carbon nanostructures, respectively. (**g-l**) Top view and (**m-r**) cross-sectional SEM images show typical Al_2O_3/C, CeO_2/C, La_2O_3/C, MgO/C, CaO/C and C nanostructures after cycling, respectively. Scale bars = 10μm (**a-f**) and 1μm (**g-r**).

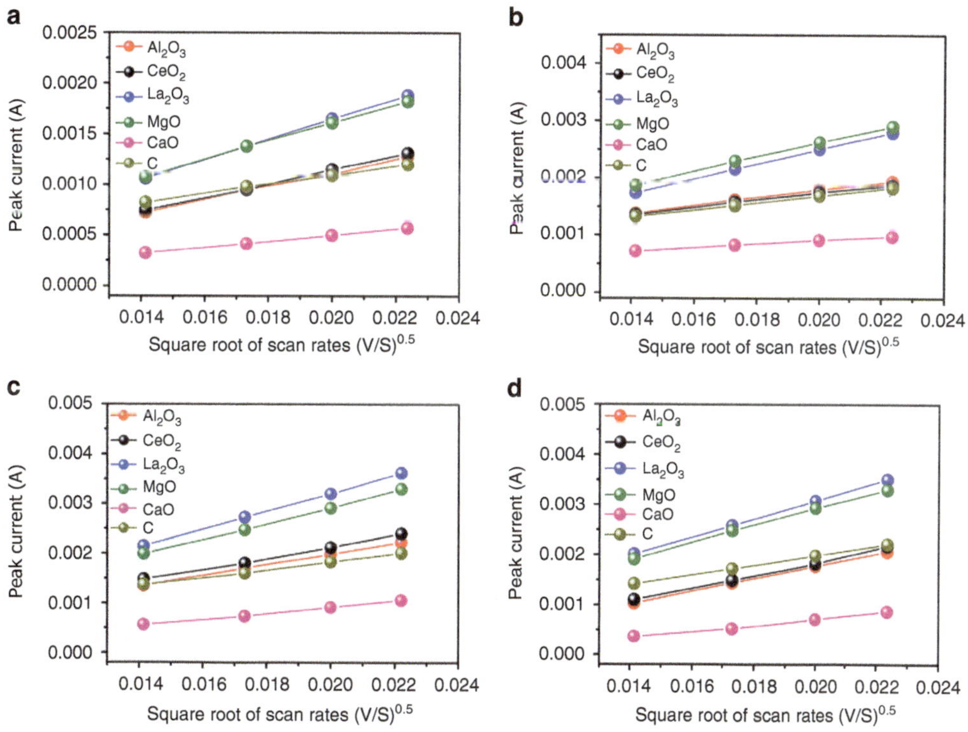

Figure 6 | Lithium ion diffusion properties of the electrode at various voltage scan rates. Plot of CV peak current of (**a**) the cathodic reaction 1 ($S_8 \rightarrow Li_2S_4$), (**b**) the cathodic reaction 2 ($Li_2S_4 \rightarrow Li_2S$), (**c**) the anodic reaction 1 ($Li_2S \rightarrow Li_2S_4$) and (**d**) the anodic reaction 2 ($Li_2S_4 \rightarrow S_8$) versus the square root of scan rates.

By now, some important questions arise: why Li_2S shows different deposition behaviour on various oxide/C surfaces? Which kind of oxide/C surface is favourable for the controllable precipitation of Li_2S? Because the nucleation and initial growth sites of Li_2S are located on the surface of oxide/C, the growth behaviour of Li_2S must be related to surface chemical properties of oxide/C matrix. First, the oxide/C surface should absorb high-order Li_2S_x owing to its strong adsorption ability (Fig. 2d), which acts as the sulfur source for the growth of Li_2S. The absorbed Li_2S_x must be transferred from the oxide surface to conductive carbon surface to enable the electrochemical reactions due to the insulating properties of these metal oxides. Therefore, the distribution and structure of Li_2S will be affected by the surface diffusion properties of Li_2S_x species on the oxide/C substrate.

Lithium ion diffusion properties and mechanism of the cathodes. Although it is very difficult to get the diffusivity of Li_2S_x on the oxide surface from the electrochemical measurement, the lithium diffusivity in the whole Li-S batteries can offer the important information about Li_2S_x surface diffusion because of that the most favourable binding site of sulfides is two Li atoms bonding with metal oxide (Fig. 2c,d). In order to explore the lithium diffusion properties, we performed cyclic voltammetry (CV) measurements under different scanning rates ranging from 0.2 to 0.5 mV s^{-1}. As shown in Fig. 6, all cathodic and anodic peak currents are linear with the square root of scan rates, from which the lithium diffusion performance can be estimated using the classical Randles Sevcik equation[44]:

$$I_p = (2.69\times10^5)n^{1.5}aD^{0.5}Cv^{0.5}\Delta C_o, \quad (1)$$

where, I_p is the peak current, n is the number of electrons per reaction species, a is the active electrode area, D is the lithium ion diffusion coefficient, ΔC_o is the Li concentration change corresponding to the electrochemical reaction. The n, a and ΔC_o are constant in our battery system. The slopes of curves in Fig. 6a–d are positively correlated to the corresponding lithium ion diffusion, which indicates that the sulfur composite cathode based on both CaO/C and C have lower diffusivity. Abundant high viscosity Li_2S_x dissolved in the electrolyte and the poor Li_2S_x capture capability of both CaO/C and C is believed to be the main reason for their low diffusivity. Compared with CaO/C, CeO_2/C, Al_2O_3/C and C electrodes, La_2O_3/C and MgO/C samples show better diffusion properties and the measured diffusivity of Al_2O_3/C cathode is comparable with the CeO_2/C.

The diffusion of lithium on the surface of various metal oxides has been investigated by the DFT calculation (Fig. 7). Because the most favourable binding site of sulfides is two Li atoms bonding with metal oxide (Fig. 2c,d), the calculated Li ion diffusion can also indicate the diffusivity of sulfides species on the surface of the oxide. On MgO(100), CaO(100) and La_2O_3(001) surfaces, the diffusions of Li in different dimensions can be realized on three equivalent adsorption sites (Fig. 7). Among the three kinds of surfaces, the diffusion barrier of Li on CaO(100) is largest. The space group of MgO is same with CaO, however, the diffusion barrier of Li on MgO(100) is about 0.45 eV lower than that on CaO(100). The suitable adsorption energies of sulfide species and small diffusion barriers of Li on MgO will lead to the formation of abundant Li_2S particles on MgO/C surfaces, which are responsible for the best cycling performance of MgO/C cathodes. On CeO_2(111) surfaces, the large diffusion barrier is 0.66 eV, which is similar with that on La_2O_3(001) surfaces. This may explain why CeO_2/C and La_2O_3/C cathodes show similar cycling performance. Among the five kinds of metal oxides surfaces, the largest diffusion barrier (1.22 eV) of Li is found to be on Al_2O_3(110), which is about three times of that on MgO(100). It is seen that sulfide species can strongly adsorb, however, difficult to diffuse on Al_2O_3. Although Al_2O_3 has the strongest Li_2S_8 adsorption (Fig. 2), the slow diffusion of Li_2S_x indicated that Al_2O_3 may not be a good additive for sulfur cathode.

Discussion

Based on our experimental results, we clarify three functions of these oxides. The first basic function of these metal oxides is the Li_2S_x adsorption. Although many literatures have reported the Li_2S_x capture capacities, the detailed absorption mechanism is still unclear. Our DFT calculation and temperature swing adsorption experiments (Fig. 2) confirm that the monolayer chemisorption is dominant during the Li_2S_x capture. The second function of these metal oxides, especially some nonconductive oxides, is the Li_2S_x transfer station, which transports the Li_2S_x from the poorly conductive oxide surface to high conductive carbon matrix to ensure the full electrochemical conversion. The third function is to induce the controlled growth of Li_2S species on the surface of the composite instead of random deposition. Many reports proved that the uncontrolled deposition will result in electrochemically inactive large agglomerations of Li_2S. The subsequent detachment of Li_2S from the oxide/carbon matrix is the main capacity decay mechanism, which can be supported by the SEM observation in Fig. 5. The SEM observations, diffusion test and DFT calculations revealed that the deposition of Li_2S on the surface of oxide/carbon matrix may be influenced by the lithium ion diffusion properties on the surface of metal oxides, which has not yet been identified. Surface diffusion properties will affect the distribution and growth of Li_2S.

Based on these functions of nonconductive metals oxides, we can propose an oxide selection criterion for the Li-S batteries.

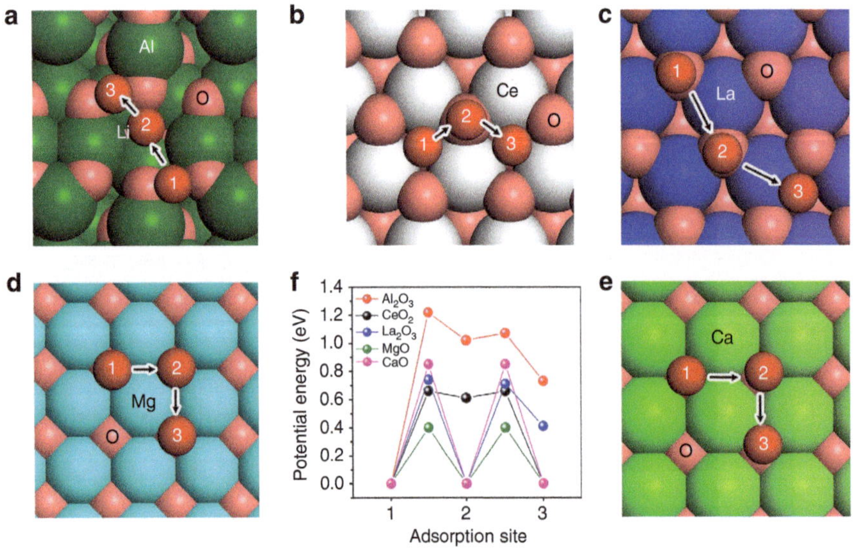

Figure 7 | Lithium diffusion mechanism on the surface of various metal oxides. (**a–e**) Minimum energy path for lithium ion diffusion on Al_2O_3(110), CeO_2(111), La_2O_3(001), MgO(100) and CaO(100) surfaces, respectively. (**f**) Potential energy profiles for Li^+ diffusion along different adsorption sites on the oxide surface.

Because the first role of oxides is adsorption, the binding between the sulfides species and the matrix should be strong, which can both suppress the shuttle effect and enable the full utilization of active materials. Considering that the Li_2S_x capture is the monolayer chemisorption and the adsorption amount will depend on the surface area of oxides, uniformly distributed oxides nanostructures with high surface area are essential. Although strong binding and high surface area are preconditions, the surface diffusion properties of oxides are also very important, which affect the distribution and structure of Li_2S. An optimized balance between Li_2S_x adsorption and surface diffusion is favourable for the sulfide species to deposit on the surface of oxide/carbon matrix, keep active during the cycling and ensure the final good cycling performance of batteries. In addition, some other factors such as electric conductivities, chemical stability and lithiation/delithiation of the oxides also need to be considered.

In conclusion, a series of nonconductive metal oxides (MgO, Al_2O_3, CeO_2, La_2O_3 and CaO) nanoparticles anchored on porous carbon nanoflakes have been synthesized successfully via a facile and generic biotemplating method using Kapok trees fibres as both the template and the carbon source. The composite cathode materials based on the MgO/C, La_2O_3/C and CeO_2/C nanoflakes show higher capacity and better cycling performance. Moreover, the working mechanisms of these oxides were revealed by adsorption test, microstructure analysis, electrochemical performance evaluation and DFT calculations. In addition, the comprehensive oxide selection criteria referring to the strong binding, high surface area and good surface diffusion properties were proposed for the first time. We believe that our proposed selection criteria can be generalized to other matrix materials for high performance Li–S batteries such as metal sulfides, metal nitrides, metal chlorides, and so on.

Methods

Preparation of metal oxide nanoparticles. A generic Pechini sol–gel method[13] was used to synthesize the pure metal oxides. 0.01 mol metal nitrates of the desired metals were dissolved in 5 ml of deionized water under stirring. 0.015 mol citric acid was then added in to the prepared solution to chelate the metal ions and heated to 200 °C to form a dry and porous gel. After the calcination of the gel at 850 °C for 1 h, pure metal oxide (Al_2O_3, CeO_2, La_2O_3, MgO and CaO) nanoparticles were obtained.

Biotemplated fabrication of metal oxide/carbon nanoflakes. Owing to the poor electric conductivity of Al_2O_3, CeO_2, La_2O_3, MgO and CaO, a series or metal oxide decorated porous carbon flakes were fabricated using the Kapok tree fibres as both the template and the carbon sources. To start with, 0.01 mol metal nitrates of the desired metals were dissolved in 40 ml of deionized water to form transparent solution. Commercial Kapok tree fibre was washed with acetone and ethanol to remove the surface wax and dipped in to the nitrate solution. After 2 h soaking, the nitrate loaded fibres were separated from the solution and introduced into ammonia atmosphere to ensure the formation of the metal hydroxides, followed by the drying at 100 °C for 10 h to remove the water. Then the dried fibres were inserted into the tube furnace and calcined at 850 °C for 1 h with 100 sccm continuous flow of argon. After cooling, the obtained carbon microtubes remaining the macromorphology of original Kapok tree fibres were easily grinded into carbon nanoflakes decorated with metal oxide nanoparticles.

Adsorption test. To prepare Li_2S_8 solution, stoichiometric Li_2S and sulfur were dissolved in 1,3-dioxolane/1,2-dimethoxyethane solution (1:1 in volume) and stirred at 80 °C for 24 h. Then the Li_2S_8 solution was diluted to 0.005 M for the capture test. Before the adsorption test, all the oxide and oxide/carbon samples were dried under vacuum at 80 °C for 24 h. After the immersion of the samples in 0.005 M Li_2S_8 at different temperatures, 20 μl of the solution was transferred for the ICP-OES test[45].

Characterization. The morphology and the microstructure were studied using SEM (FEI, XL30 Sirion) and TEM (FEI, Tecnai G2 F20 X-TWIN). The specific surface area was characterized from nitrogen adsorption–desorption measurement (Micromeritics, ASAP 2020). ICP-OES was conducted using a Thermo Scientific ICAP 6300 Duo View Spectrometer.

Electrochemical measurements. The sulfur cathode materials were prepared via a facile thermal diffusion method[19]. First, sulfur and grinded oxide/carbon nanoflakes with a weight ratio of 1:3 were mixed with appropriate amount of CS_2 solution. After the evaporation of CS_2, the mixture were pressed and heated at 155 °C for 12 h under vacuum to obtain the composite cathode materials. To fabricate the 2,023 type coin cells for the electrochemical measurements, the synthesized composites were grinded again and mixed with conductive carbon black and polyvinylidene fluoride binder in N-methyl-2-pyrrolidinone (70:20:10 by weight) to form a slurry. After a 12 h magnetic stirring, the slurry was coated onto the aluminium foil and dried overnight at 60 °C under vacuum. The sulfur content was in the range of 63–70 wt% and the mass loading of the electrodes ranges from 0.7 to 1.2 mg cm^{-2}. 1 M lithium bis(trifluoromethanesulphonyl)imide dissolved in a mixture of 1,3-dioxolaneand dimethoxymethane (1:1 by volume) with 0.1 M $LiNO_3$ additive was used as the electrolyte. Galvanostatic cycling was performed on Arbin or MTI testers with the potential range of 1.8–2.6 V versus Li/Li^+ at ambient temperature.

DFT calculations. All calculations were performed using the Vienna *ab initio* Simulation Package code[46] based on density functional theory. The projector augmented wave potentials were used to describe the interaction between ions and electrons[47,48]. Nonlocal exchange correlation energy was evaluated using the Perdew-Burke-Ernzehof functional. The electron wave function was expanded using plane waves with an energy cutoff of 400 eV. All structures were optimized with force convergence criterion of 10 meV Å^{-1}. All oxide surfaces were created based on the corresponding optimized bulk unit cell, which were in good agreement with the experimental values. A 4 × 4 × 1 Monkhorst–Pack k-point mesh and a vacuum slab of about 15 Å was inserted between the surface slabs for all the metal oxide models. The cell parameter is 8.40 × 8.07 × 25.00 Å for Al_2O_3(110), 11.67 × 11.67 × 22.94 Å for CeO_2(111), 11.82 × 11.82 × 25.00 Å for La_2O_3(001), 8.99 × 8.99 × 20.00 Å for MgO(100) and 10.25 × 10.25 × 20.00 Å for CaO(100). DFT + U approach, where U is an empirical parameter for on site electronic correlations, was used in the calculation of CeO_2(111) with a U value of 5.0 eV (ref. 49). The adsorption energies (E_a) for S and Li_2S on the metal oxide surfaces are defined as $E_a = E_{total} - E_{ads} - E_{suf}$, where E_{total} is the total energy of the adsorbed system, E_{ads} is the energy of the adsorbate in vacuum and E_{suf} is the energy of the optimized clean surface slab.

References

1. Bruce, P. G., Freunberger, S. A., Hardwick, L. J. & Tarascon, J. M. Li-O_2 and Li-S batteries with high energy storage. *Nat. Mater.* **11**, 19–29 (2012).
2. Yamin, H., Gorenshtein, A., Penciner, J., Sternberg, Y. & Peled, E. Lithium sulfur battery-oxidation reduction-mechanisms of polysulfides in the solutions. *J. Electrochem. Soc.* **135**, 1045–1048 (1988).
3. Ji, X. L., Lee, K. T. & Nazar, L. F. A highly ordered nanostructured carbon-sulphur cathode for lithium-sulphur batteries. *Nat. Mater.* **8**, 500–506 (2009).
4. Yao, H. B. *et al.* Improving lithium–sulphur batteries through spatial control of sulphur species deposition on a hybrid electrode surface. *Nat. Commun.* **5**, 3943 (2014).
5. Su, Y. S. & Manthiram, A. Lithium-sulphur batteries with a microporous carbon paper as a bifunctional interlayer. *Nat. Commun.* **3**, 1166 (2012).
6. Seh, Z. W. *et al.* Sulphur-TiO_2 yolk-shell nanoarchitecture with internal void space for long-cycle lithium-sulphur batteries. *Nat. Commun.* **4**, 1331 (2013).
7. Guo, J. C., Xu, Y. H. & Wang, C. S. Sulfur-impregnated disordered carbon nanotubes cathode for lithium-sulfur batteries. *Nano Lett.* **11**, 4288–4294 (2011).
8. Demir-Cakan, R. *et al.* Li-S batteries: simple approaches for superior performance. *Energy Environ. Sci.* **6**, 176–182 (2013).
9. Yang, Y., Zheng, G. Y. & Cui, Y. Nanostructured sulfur cathodes. *Chem. Soc. Rev.* **42**, 3018–3032 (2013).
10. Zhou, G. M. *et al.* A graphene-pure-sulfur sandwich structure for ultrafast, long-life lithium-sulfur batteries. *Adv. Mater.* **26**, 625–631 (2014).
11. Kim, J. S., Hwang, T. H., Kim, B. G., Min, J. & Choi, J. W. A lithium-sulfur battery with a high areal energy density. *Adv. Funct. Mater.* **24**, 5359 (2014).
12. Zhang, Y. G., Zhao, Y., Yermukhambetova, A., Bakenov, Z. & Chen, P. Ternary sulfur/polyacrylonitrile/$Mg_{0.6}Ni_{0.4}O$ composite cathodes for high performance lithium/sulfur batteries. *J. Mater. Chem. A* **1**, 295–301 (2013).
13. Tao, X. Y. *et al.* Highly mesoporous carbon foams synthesized by a facile, cost-effective and template-free Pechini method for advanced lithium-sulfur batteries. *J. Mater. Chem. A* **1**, 3295–3301 (2013).
14. Zhou, G. M. *et al.* A flexible sulfur-graphene-polypropylene separator integrated electrode for advanced Li-S batteries. *Adv. Mater.* **27**, 641–647 (2015).
15. Wang, J., He, Y.-S. & Yang, J. Sulfur-based composite cathode materials for high-energy rechargeable lithium batteries. *Adv. Mater.* **27**, 569–575 (2015).
16. Liang, X. *et al.* A highly efficient polysulfide mediator for lithium-sulfur batteries. *Nat. Commun.* **6**, 5682 (2015).
17. Zhao, M.-Q. *et al.* Unstacked double-layer templated graphene for high-rate lithium-sulphur batteries. *Nat. Commun.* **5**, 3410 (2014).

18. Wang, Z. *et al.* Enhancing lithium-sulphur battery performance by strongly binding the discharge products on amino-functionalized reduced graphene oxide. *Nat. Commun.* **5,** 5002 (2014).
19. Tao, X. Y. *et al.* Strong sulfur binding with conducting Magneli-phase Ti_nO_{2n-1} nanomaterials for improving lithium-sulfur batteries. *Nano Lett.* **14,** 5288–5294 (2014).
20. Seh, Z. W. *et al.* Two-dimensional layered transition metal disulphides for effective encapsulation of high-capacity lithium sulphide cathodes. *Nat. Commun.* **5,** 5017 (2014).
21. Qiu, Y. C. *et al.* High-rate, ultra long cycle-life lithium/sulfur batteries enabled by nitrogen-doped graphene. *Nano Lett.* **14,** 4821–4827 (2014).
22. Huang, C. *et al.* Manipulating surface reactions in lithium-sulphur batteries using hybrid anode structures. *Nat. Commun.* **5,** 3015 (2014).
23. Suo, L., Hu, Y.-S., Li, H., Armand, M. & Chen, L. A new class of Solvent-in-Salt electrolyte for high-energy rechargeable metallic lithium batteries. *Nat. Commun.* **4,** 1481 (2013).
24. Kim, H., Lee, J., Ahn, H., Kim, O. & Park, M. J. Synthesis of three-dimensionally interconnected sulfur-rich polymers for cathode materials of high-rate lithium-sulfur batteries. *Nat. Commun.* **6,** 7278 (2015).
25. Chen, H. W. *et al.* Monodispersed sulfur nanoparticles for lithium sulfur batteries with theoretical performance. *Nano Lett.* **15,** 798–802 (2015).
26. Chen, H. W. *et al.* Rational design of cathode structure for high rate performance lithium-sulfur batteries. *Nano Lett.* **15,** 5443–5448 (2015).
27. Wang, B., Alhassan, S. M. & Pantelides, S. T. Formation of large polysulfide complexes during the lithium-sulfur battery discharge. *Phys. Rev. Appl.* **2,** 034004 (2014).
28. Mikhaylik, Y. V. & Akridge, J. R. Polysulfide shuttle study in the Li/S battery system. *J. Electrochem. Soc.* **151,** A1969–A1976 (2004).
29. Wang, H. L. *et al.* Graphene-wrapped sulfur particles as a rechargeable lithium-sulfur battery cathode material with high capacity and cycling stability. *Nano Lett.* **11,** 2644–2647 (2011).
30. Yang, Y. *et al.* Improving the performance of lithium-sulfur batteries by conductive polymer coating. *ACS Nano.* **5,** 9187–9193 (2011).
31. Tao, X. Y. *et al.* Decoration of sulfur with porous metal nanostructures: an alternative strategy for improving the cyclability of sulfur cathode materials for advanced lithium-sulfur batteries. *Chem. Commun.* **49,** 4513–4515 (2013).
32. Cheon, S. E. *et al.* Rechargeable lithium sulfur battery—II. Rate capability and cycle characteristics. *J. Electrochem. Soc.* **150,** A800–A805 (2003).
33. Yang, Y. *et al.* New nanostructured Li_2S/silicon rechargeable battery with high specific energy. *Nano Lett.* **10,** 1486–1491 (2010).
34. Zheng, G. Y. *et al.* Amphiphilic surface modification of hollow carbon nanofibers for improved cycle life of lithium sulfur batteries. *Nano Lett.* **13,** 1265–1270 (2013).
35. Li, W. Y. *et al.* Understanding the role of different conductive polymers in improving the nanostructured sulfur cathode performance. *Nano Lett.* **13,** 5534–5540 (2013).
36. Zhou, W. D., Xiao, X. C., Cai, M. & Yang, L. Polydopamine-coated, nitrogen-doped, hollow carbon sulfur double-layered core-shell structure for improving lithium sulfur batteries. *Nano Lett.* **14,** 5250–5256 (2014).
37. Wang, C. *et al.* Macroporous free-standing nano-sulfur/reduced graphene oxide paper as stable cathode for lithium-sulfur battery. *Nano Energy* **11,** 678–686 (2015).
38. Zhou, G. M., Paek, E., Hwang, G. S. & Manthiram, A. Long-life Li/polysulphide batteries with high sulphur loading enabled by lightweight three-dimensional nitrogen/sulphur-codoped graphene sponge. *Nat. Commun.* **6,** 7760 (2015).
39. Zheng, J. M. *et al.* Lewis acid-base interactions between polysulfides and metal organic framework in lithium sulfur batteries. *Nano Lett.* **14,** 2345–2352 (2014).
40. Zhou, G. M., Zhao, Y. B. & Manthiram, A. Dual-confined flexible sulfur cathodes encapsulated in nitrogen-doped double-shelled hollow carbon spheres and wrapped with graphene for Li-S batteries. *Adv. Energy Mater.* **5,** 1402263 (2015).
41. Dong, K., Wang, S. P., Zhang, H. Y. & Wu, J. P. Preparation and electrochemical performance of sulfur-alumina cathode material for lithium-sulfur batteries. *Mater. Res. Bull.* **48,** 2079–2083 (2013).
42. Sun, F. G. *et al.* A high-rate lithium-sulfur battery assisted by nitrogen-enriched mesoporous carbons decorated with ultrafine La_2O_3 nanoparticles. *J. Mater. Chem. A* **1,** 13283–13289 (2013).
43. Chaiarrekij, S., Apirakchaiskul, A., Suvarnakich, K. & Kiatkamjornwong, S. Kapok I: characteristcs of Kapok fiber as a potential pulp source for papermaking. *Bioresources* **7,** 475–488 (2012).
44. Li, P. *et al.* Lithium sodium vanadium phosphate and its phase transition as cathode material for lithium ion batteries. *Electrochim. Acta* **180,** 120–128 (2015).
45. Diao, Y., Xie, K., Xiong, S. & Hong, X. Analysis of polysulfide dissolved in electrolyte in discharge-charge process of Li-S battery. *J. Electrochem. Soc.* **159,** A421–A425 (2012).
46. Kresse, G. & Furthmuller, J. Efficiency of ab-initio total energy calculations for metals and semiconductors using a plane-wave basis set. *Comput. Mater. Sci.* **6,** 15–50 (1996).
47. Blochl, P. E. Projector augmented-wave method. *Phys. Rev. B* **50,** 17953–17979 (1994).
48. Kresse, G. & Joubert, D. From ultrasoft pseudopotentials to the projector augmented-wave method. *Phys. Rev. B* **59,** 1758–1775 (1999).
49. Nolan, M., Parker, S. C. & Watson, G. W. The electronic structure of oxygen vacancy defects at the low index surfaces of ceria. *Surf. Sci.* **595,** 223–232 (2005).

Acknowledgements

Y.C. acknowledges the support from the Assistant Secretary for Energy Efficiency and Renewable Energy, Office of Vehicle Technologies of the US Department of Energy under the Battery Materials Research (BMR) Program. X.Y.T. and J.G.W. acknowledge support from the Natural Science Foundation of Zhejiang Province (grants LR13E020002) and the National Natural Science Foundation of China (51002138 and 21136001).

Author contributions

Y.C. and X.Y.T. conceived the idea. X.Y.T. designed the experiments, synthesized the materials and performed electrochemical test. J.G.W. and Q.X.C. carried out the DFT calculation and analyzed the result. C.L., H.T.W., G.Y.Z. and other authors conducted materials characterization. All authors discussed the electrochemical results and the whole paper.

Additional information

Competing financial interests: The authors declare no competing financial interests.

17

Nickel-vanadium monolayer double hydroxide for efficient electrochemical water oxidation

Ke Fan[1,2], Hong Chen[1], Yongfei Ji[3], Hui Huang[4], Per Martin Claesson[4], Quentin Daniel[1], Bertrand Philippe[5], Håkan Rensmo[5], Fusheng Li[1], Yi Luo[3] & Licheng Sun[1,6]

Highly active and low-cost electrocatalysts for water oxidation are required due to the demands on sustainable solar fuels; however, developing highly efficient catalysts to meet industrial requirements remains a challenge. Herein, we report a monolayer of nickel-vanadium-layered double hydroxide that shows a current density of 27 mA cm^{-2} (57 mA cm^{-2} after ohmic-drop correction) at an overpotential of 350 mV for water oxidation. Such performance is comparable to those of the best-performing nickel-iron-layered double hydroxides for water oxidation in alkaline media. Mechanistic studies indicate that the nickel-vanadium-layered double hydroxides can provide high intrinsic catalytic activity, mainly due to enhanced conductivity, facile electron transfer and abundant active sites. This work may expand the scope of cost-effective electrocatalysts for water splitting.

[1] Department of Chemistry, Organic Chemistry, KTH Royal Institute of Technology, 10044 Stockholm, Sweden. [2] State Key Laboratory of Advanced Technology for Materials Synthesis and Processing, Wuhan University of Technology, Wuhan 430070, China. [3] Division of Theoretical Chemistry and Biology, School of Biotechnology, KTH Royal Institute of Technology, SE-106 91 Stockholm, Sweden. [4] Department of Chemistry, Surface and Corrosion Science, KTH Royal Institute of Technology, SE-10044 Stockholm, Sweden. [5] Department of Physics and Astronomy, Uppsala University, Box 516, 751 20 Uppsala, Sweden. [6] State Key Laboratory of Fine Chemicals, DUT-KTH Joint Education and Research Center on Molecular Devices, Dalian University of Technology (DUT), Dalian 116024, China. Correspondence and requests for materials should be addressed to L.S. (email: lichengs@kth.se).

Water splitting is considered one of the most promising strategies to produce chemical fuels such as hydrogen. The half reaction of the water splitting process, water oxidation, remains the bottleneck of the whole process at present. Therefore, developing highly efficient water oxidation catalysts is crucial. Some precious metal-based electrocatalysts, such as IrO_2 and RuO_2, have shown excellent performance for water oxidation; however, they suffer from high-cost and relative scarcity of precious metals, which limits their applications. Although some first-row transition metal oxides (for example, NiO_x, $NiFeO_x$, CoO_x and MnO_x) had been developed as low-cost electrocatalysts for water oxidation, most of them still cannot compete with IrO_2 and RuO_2[1,2]. Recently, the earth-abundant Ni–Fe double-layered hydroxide (NiFe-LDH) catalysts have attracted attention[3–7]. From being first reported as an advanced electrocatalyst coupled with carbon nanotubes for water oxidation[8], it is nowadays known as one of the most active catalysts with a low overpotential and high electrolysis current. Since, then tremendous efforts have been devoted to further improve the activity of NiFe-LDH, such as exfoliation[9] and hybridization[6,8,10,11], to the extent that LDH catalysts can now outperform IrO_2 in alkaline media[6,8,9]; however, the aforementioned methods are still too complicated for large-scale applications.

It is already known that Fe(III) incorporated in Ni(II)-based LDH is the key aspect for the high catalytic performance, although the role of Fe in LDH is still ambiguous[5,6,12]. Besides Fe(III), cobalt is also commonly incorporated in nickel hydroxides to construct NiCo-LDHs for water oxidation[13–16], and the resulting NiCo-LDHs show promising catalytic activities; however, the performance is relatively low compared with the reported NiFe-LDHs under identical conditions[9]. Besides, more earth-abundant metal elements have been incorporated into $Ni(OH)_2$ to explore novel LDHs for water oxidation, for example, recently Koper and co-workers[17] investigated a series of Ni-based double hydroxides with Cr, Mn, Fe, Co, Cu and Zn for water oxidation, and among these candidates NiFe-LDH still appears as the most promising and shows the best activity. Up until now, there has been no reported earth-abundant metal element that can outperform Fe incorporated Ni-based LDHs. Searching for an earth-abundant metal to form efficient Ni-based LDH comparable to NiFe-LDH is still the state-of-the-art in this area of energy research.

In this work, we incorporate another earth-abundant element into $Ni(OH)_2$: vanadium, and succeed in forming NiV-LDH as an efficient catalyst for the water oxidation reaction. A simple one-step hydrothermal method is employed to synthesize NiV-LDH. Without need for exfoliation or hybridization with other materials, the resulting monolayer NiV-LDH catalyst exhibits comparable activity to the best-performing NiFe-LDH for water oxidation in alkaline electrolyte.

Results

Monolayer of $Ni_{0.75}V_{0.25}$-LDH. Figure 1a shows the typical X-ray diffraction patterns of pure $Ni(OH)_2$ and $Ni_{0.75}V_{0.25}$-LDH. As can be seen, pure $Ni(OH)_2$ was successfully synthesized in the simple hydrothermal system without addition of V sources, and the X-ray diffraction pattern supports the formation of pure hexagonal α-$Ni(OH)_2$ (JCPDS 380715), which exhibits a layered structure constructed from $[NiO_6]$ coordinated octahedra connected by sharing their edges. After incorporation of V into the structure of α-$Ni(OH)_2$, no obvious change can be observed in the X-ray diffraction spectrum, indicating $Ni_{0.75}V_{0.25}$-LDH is the isomorphous compound as α-$Ni(OH)_2$ with the layered structure. The X-ray diffraction pattern of prepared bare V-based hydroxide without Ni source (Supplementary Fig. 1) shows very low crystallinity that is significantly different from those of the above Ni-based materials. Figure 1b–d shows the Ni 2*p*, V 2*p* and O 1*s* X-ray photoelectron spectroscopy (XPS) core-level spectra of the $Ni_{0.75}V_{0.25}$-LDH powder. The Ni 2*p* spectrum presents two main structures, resulting from the spin–orbit splitting of the *p* orbital that are assigned as Ni $2p_{3/2}$ (850 – 870 eV region) and Ni $2p_{1/2}$ (870 – 890 eV region). The binding energy difference between the 3/2 and 1/2 components is ~17.4 eV. Ni $2p_{3/2}$ presents a main peak at ~855.4 eV with an intense satellite structure at ~861.2 eV, (~872.8 and ~879.6 eV, respectively, for Ni2$p_{1/2}$), this signature is characteristic of Ni^{2+} (refs 18–21). The spectra of O 1*s* and V 2*p* are presented in Fig. 1c, the O 1*s* signal originates from $Ni_{0.75}V_{0.25}$-LDH, but its main contributions is from the fluorine doped tin oxide (FTO) glass substrate used with an O1*s*-binding energy ~530 eV. The V 2*p* core-level spectrum is also decomposed into V $2p_{3/2}$ and V $2p_{1/2}$ due to the spin–orbit splitting that are separated by ~7.5 eV. A closer view on the V $2p_{3/2}$ region is shown in Fig. 1d. Three components can be distinguished at ~515.4 eV (in blue), ~516.3 eV (in light grey) and ~517.5 eV (in grey) and are in good agreement with respectively V^{3+}, V^{4+} and V^{5+} (refs 22–24). This result indicates that V^{3+} was partially oxidized by oxygen to V^{4+} and V^{5+} during the synthesis.

Typical transmission electron microscopy (TEM) images of α-$Ni(OH)_2$ and $Ni_{0.75}V_{0.25}$-LDH are shown in Fig. 2. Interestingly, as shown in Fig. 2a,b, the pure α-$Ni(OH)_2$ shows a narrow nanosheet morphology with the size of few tens of nanometre. Aggregated by these nanosheet crystals, a porous sphere structure with the diameter of few micrometres was obtained, as observed by scanning electron microscopy (SEM; Supplementary Fig. 2a). After adding the vanadium source to the initial solution for hydrothermal reaction, $Ni_{0.75}V_{0.25}$-LDH presents a three-dimensional morphology assembled by ultrathin nanosheets as shown in Fig. 2c,d. Selected area electron diffraction pattern in the inset of Fig. 2d confirms the hexagonal phase of $Ni_{0.75}V_{0.25}$-LDH. The atomic force microscopy (AFM) and height profile of $Ni_{0.75}V_{0.25}$-LDH in Fig. 2e,f show that the nanosheet is ultrathin with thickness of ~0.9 nm, indicating the obtained nanosheet is monolayered. To make a comparison, $Ni_{0.75}Fe_{0.25}$-LDH was also prepared by a similar protocol in the literature[25] and the SEM image of the products obtained is shown in Supplementary Fig. 2b and c. Some large aggregation plates in $Ni_{0.75}Fe_{0.25}$-LDH can be observed, showing that the size of aggregations by nanosheets in $Ni_{0.75}Fe_{0.25}$-LDH is bigger than $Ni_{0.75}V_{0.25}$-LDH. However, the TEM, AFM images and height profile of $Ni_{0.75}Fe_{0.25}$-LDH in Supplementary Fig. 2d–f exhibit that the nanosheet of $Ni_{0.75}Fe_{0.25}$-LDH is also ultrathin with thickness of ~1.2 nm. Meanwhile, differing from the above nanosheet-based structure, the bare V-based and Fe-based hydroxides (β-FeOOH) without Ni content show mainly 'nanostick' morphologies after the hydrothermal processes (Supplementary Fig. 3).

Oxygen evolution catalysis. It is known that the composition can affect the water oxidation performance of the catalysts significantly. First, we investigated the electrocatalytic activities of NiV-LDHs and NiFe-LDHs on glassy carbon (GC) electrodes with different Ni content to optimize the composition (the measurements were carried out without ohmic-drop correction unless noted otherwise). All the amount of catalyst loadings on GC electrodes were 0.143 mg cm^{-2}. As shown in Supplementary Fig. 4, the Ni content plays an essential role for the catalytic activity in this study (the electrocatalytic activities of pure $Ni(OH)_2$, bare V-based and Fe-based hydroxides are very low, so they are not shown here for clarity). When the molar ratio of Ni/X (X = V or Fe) is 3:1, that is, $Ni_{0.75}V_{0.25}$-LDH and $Ni_{0.75}Fe_{0.25}$-LDH, the best water oxidation performances are

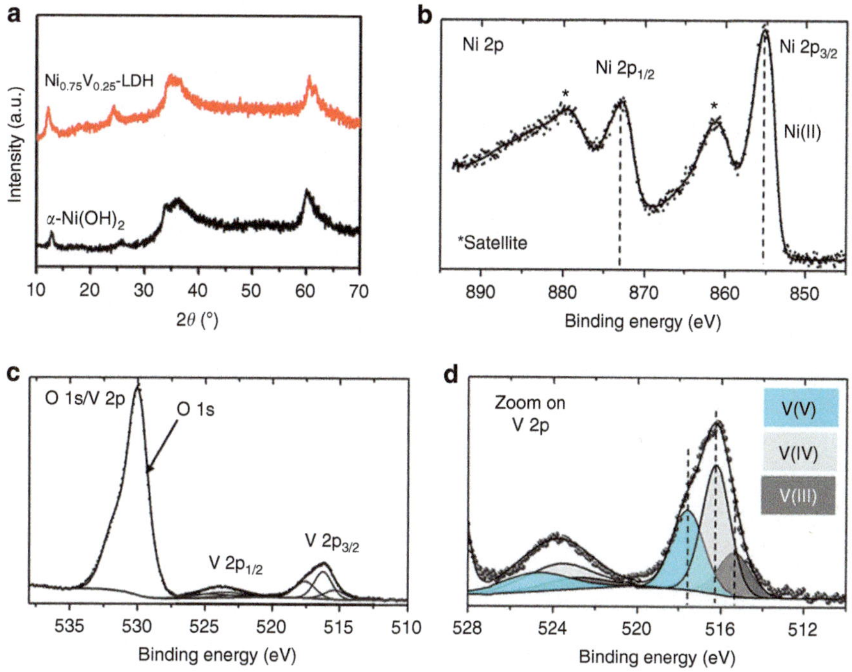

Figure 1 | X-ray diffraction and XPS. (**a**) X-ray diffraction patterns of α-$Ni(OH)_2$ and $Ni_{0.75}V_{0.25}$-LDH. XPS measurements ($h\nu$ = 1,486.6 eV) on the $Ni_{0.75}V_{0.25}$-LDH powder deposited on a FTO conductive glass: (**b**) Ni 2*p* (**c**) O 1s and V 2*p* core-level spectra, and (**d**) zoom on the V 2*p* core-level spectrum.

achieved with the highest catalytic current density (Supplementary Fig. 4a–c) and turnover frequency (TOF) (Supplementary Fig. 4d), and the lowest required overpotential (Supplementary Fig. 4e) in NiV-LDH and NiFe-LDH series, respectively (note that for NiFe-LDH, this optimized molar ratio of Ni/Fe is in a good agreement with the literature reported previously[6]). Either a lower or higher mole ratio of the two metal elements will decrease the catalytic performances of LDHs. The atomic percentage of Ni, V and Fe in $Ni_{0.75}V_{0.25}$-LDH and $Ni_{0.75}Fe_{0.25}$-LDH has been measured by energy-dispersive X-ray spectroscopy (EDS; Supplementary Fig. 5), showing the molar ratios of Ni/V (3.29) and Ni/Fe (2.93) are close to 3:1 that was used in the corresponding solution stoichiometries. The catalytic performances of $Ni_{0.75}V_{0.25}$-LDH and $Ni_{0.75}Fe_{0.25}$-LDH are very reproducible (Supplementary Fig. 6) and compared in Fig. 3. A sharp onset catalytic current density can be observed at low overpotential $\eta \sim 250$ and 300 mV for $Ni_{0.75}V_{0.25}$-LDH and $Ni_{0.75}Fe_{0.25}$-LDH, respectively. It is apparent that $Ni_{0.75}V_{0.25}$-LDH exhibits a better catalytic activity than $Ni_{0.75}Fe_{0.25}$-LDH, suggesting that substituting Fe entirely by V in Ni-based LDH can improve the water oxidation performance in our case. The improved catalytic activity of $Ni_{0.75}V_{0.25}$-LDH is not only shown via the higher catalytic current density of linear scan voltammogram (LSV) curves, but also reflected in its lower Tafel slope (Fig. 3b). $Ni_{0.75}V_{0.25}$-LDH catalyst exhibits a Tafel slope of $\sim 50\,mV\,dec^{-1}$, which is smaller than that of $Ni_{0.75}Fe_{0.25}$-LDH catalyst ($\sim$ 64 $mV\,dec^{-1}$). Tafel slopes can be influenced by mass transport and electron transport[26,27]. Different scan rate ranging from 1 to 5 $mV\,s^{-1}$ for LSV have been done, which resulted in negligible change of current densities for $Ni_{0.75}Fe_{0.25}$-LDH and $Ni_{0.75}V_{0.25}$-LDH catalysts, indicating sufficiently fast mass transport for both LDHs[26,27]. Therefore, it is reasonable to conclude that the lower Tafel slope of $Ni_{0.75}V_{0.25}$-LDH is likely ascribed to the facile electron transport through the layers of the catalyst.

At η of 350 mV, $Ni_{0.75}V_{0.25}$-LDH can achieve current density of $27.0 \pm 1.6\,mA\,cm^{-2}$, which is more than twice higher than the one of $Ni_{0.75}Fe_{0.25}$-LDH ($11.7 \pm 1.5\,mA\,cm^{-2}$). TOF was also calculated at an overpotential of 350 mV in 1 M KOH, assuming all the metal sites were electrochemically active in the water oxidation reaction. $Ni_{0.75}V_{0.25}$-LDH also shows the better TOF of $0.054 \pm 0.003\,s^{-1}$ compare with $0.021 \pm 0.003\,s^{-1}$ for $Ni_{0.75}Fe_{0.25}$-LDH. It is worth noting that the above TOFs in this study are obviously underestimated, since some of the metal sites are electrochemically non-accessible. Meanwhile, the overpotential η required to achieve 10 $mA\,cm^{-2}$ current density, which is approximately the current expected at the anode in a 10% efficient solar water splitting device under 1-sun illumination[28], was also evaluated. $Ni_{0.75}Fe_{0.25}$-LDH and $Ni_{0.75}V_{0.25}$-LDH show required overpotential 0.347 ± 0.002 and 0.318 ± 0.003 V, respectively, exhibiting the lower required overpotential of $Ni_{0.75}V_{0.25}$-LDH.

In addition, we compared the durability of $Ni_{0.75}Fe_{0.25}$-LDH and $Ni_{0.75}V_{0.25}$-LDH in 1 M KOH (Fig. 4). When biased galvanostatically at 10 $mA\,cm^{-2}$, both of $Ni_{0.75}Fe_{0.25}$-LDH and $Ni_{0.75}V_{0.25}$-LDH electrodes show considerable stability, the slightly decrease of the required potential for $Ni_{0.75}V_{0.25}$-LDH electrode may be due to the amorphazation during the anodic conditioning process[26,29]. Obviously, $Ni_{0.75}V_{0.25}$-LDH electrode requires less overpotential than $Ni_{0.75}Fe_{0.25}$-LDH, showing its comparable stability and catalytic performance to $Ni_{0.75}Fe_{0.25}$-LDH for water splitting.

Discussion

To understand the reason behind the high catalytic performance of $Ni_{0.75}V_{0.25}$-LDH for water splitting, electrochemical active surface areas (ECSA) of $Ni_{0.75}V_{0.25}$-LDH and $Ni_{0.75}Fe_{0.25}$-LDH were obtained from cyclic voltammetry (CV) curves in 1 M KOH and compared. Figure 5a shows typical CV curves of $Ni_{0.75}V_{0.25}$-LDH with different scan rates. By plotting the $\Delta J = (J_a - J_c)$ at 0.25 V versus Ag/AgCl against the scan rate, the linear slope that is twice the double layer capacitance (C_{dl}) can be obtained, and is normally used to represent the corresponding ECSA[1,9,11,26].

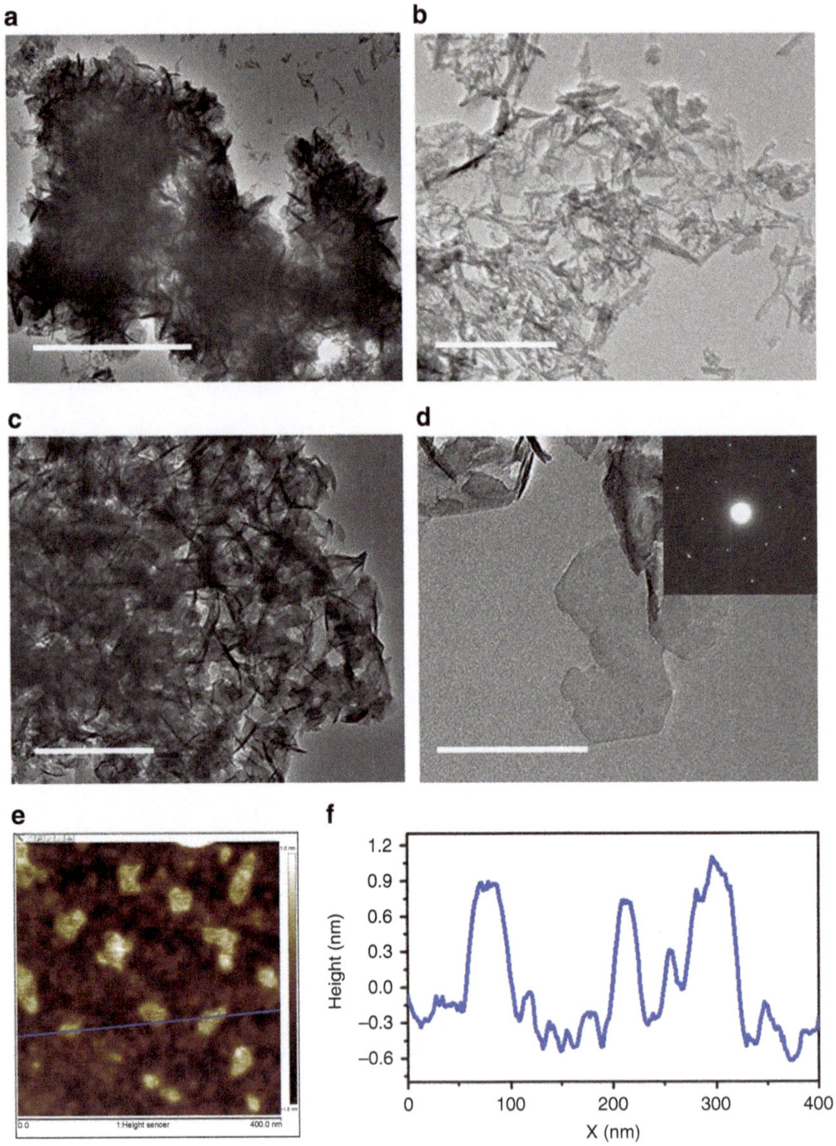

Figure 2 | TEM and AFM. (**a**,**b**) TEM images of α-$Ni(OH)_2$. Scale bar, 1 μm (**a**); 200 nm (**b**); (**c**,**d**) TEM images of $Ni_{0.75}V_{0.25}$-LDH (inset of **d**: selected area electron diffraction pattern). Scale bar, 200 nm (**c**); 100 nm (**d**); (**e**) AFM image and (**f**) height profile of $Ni_{0.75}V_{0.25}$-LDH nanosheets.

Active surface area is a very important factor for catalysts in water oxidation reaction, as it is well known that an increase of active surface area often leads to enhancement of the catalytic activity. The linear slope (ECSA) of $Ni_{0.75}Fe_{0.25}$-LDH is $0.199 \pm 0.018\,mF\,cm^{-2}$, while $Ni_{0.75}V_{0.25}$-LDH has a higher linear slope of $0.270 \pm 0.030\,mF\,cm^{-2}$, which means $Ni_{0.75}V_{0.25}$-LDH has a more electroactive surface than $Ni_{0.75}Fe_{0.25}$-LDH (Fig. 5b). The bigger ECSA should be attributed to the smaller size of aggregations by nanosheets in $Ni_{0.75}V_{0.25}$-LDH (as shown in Supplementary Fig. 2). This can contribute to partial enhancement of water oxidation reaction presented here. Nevertheless, it is important to note that comparing to $Ni_{0.75}Fe_{0.25}$-LDH, $Ni_{0.75}V_{0.25}$-LDH has just slightly ~36% higher ECSA due to the similar nanostructure, with such small increase of ECSA, more than twofold of current density at $\eta = 350\,mV$ can be achieved. This improvement of the catalytic activity cannot be merely ascribed to the slightly increased surface area of the catalysts. This result suggests that $Ni_{0.75}V_{0.25}$-LDH may have higher intrinsic catalytic activity for water oxidation reaction.

To verify this point, the ECSAs of NiV-LDH with different Ni content were further examined in details, as shown in Fig. 6a–c. The slopes (ECSA) of $Ni_{0.25}V_{0.75}$-, $Ni_{0.5}V_{0.5}$-, $Ni_{0.75}V_{0.25}$- and $Ni_{0.83}V_{0.17}$-LDH electrodes were 0.123 ± 0.009, 0.166 ± 0.012, 0.270 ± 0.030 and $0.144 \pm 0.015\,mF\,cm^{-2}$, respectively, in Fig. 6a. Obviously, the largest ECSA can be achieved when the Ni content was 0.75 in NiV-LDHs, as shown in Fig. 6b as well. The relationship of ECSA with the current density of different NiV-LDHs is exhibited in Fig. 6c. One can see that the current density of NiV-LDH increases with ECSA, further confirming the active surface area of catalyst is an important contributor for the enhancement of water oxidation reaction. However, investigating this result carefully, the ECSA has more than twofold increase from ~0.123 to $0.270\,mF\,cm^{-2}$, while the corresponding current density increases from ~4.0 to $27.0\,mA\,cm^{-2}$, showing significantly almost seven times improvement. This is a strong clue that incorporating V into Ni hydroxide can not only change the active surface area, but also enhance the intrinsic catalytic property. As a comparison, we also investigated current density-dependent ECSA of different NiFe-LDHs in Fig. 6d.

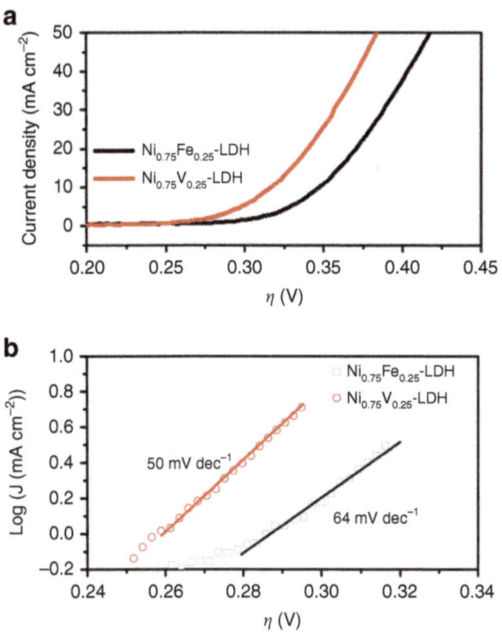

Figure 3 | Linear scan voltammogram (LSV) curves and Tafel plots. (**a**) LSV curves and (**b**) Tafel plots (with ohmic-drop correction) of $Ni_{0.75}Fe_{0.25}$-LDH and $Ni_{0.75}V_{0.25}$-LDH.

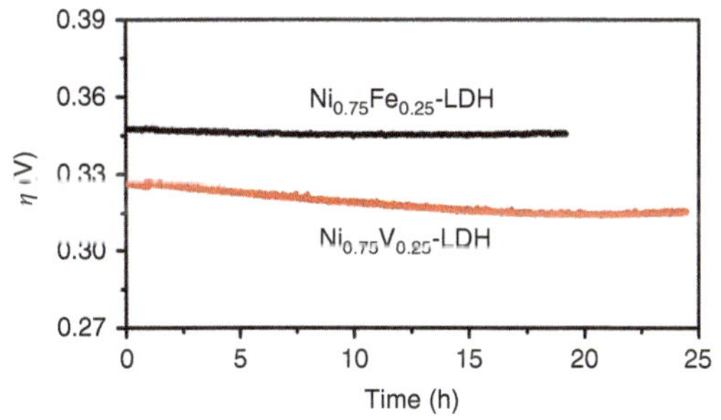

Figure 4 | Long-term stability. Chronopotentiometry curves at current density of 10 mA cm^{-2} of $Ni_{0.75}Fe_{0.25}$-LDH and $Ni_{0.75}V_{0.25}$-LDH.

For NiFe-LDHs, the current density increases with ECSA as well; however, comparing with the significantly enhanced current density of NiV-LDH, the current density of NiFe-LDH just almost linearly increases to 1.6-fold (from ~7.45 to 11.76 mA cm^{-2}) with 1.3-fold increase of ECSA (from ~0.157 to 0.199 mF cm^{-2}). This linear correlation indeed indicates that the improvement of catalytic activity of NiFe-LDHs is mainly due to the increase of actively accessible surface area. We attempted to make a comparison of the catalytic activities between NiFe-LDH and NiV-LDH with the same ECSA value. The linear correlation in Fig. 6d for NiFe-LDH is used to predict the current density of NiFe-LDH with 0.270 mF cm^{-2} of ECSA (the blue solid square in Fig. 6d), which is supposed to have the same ECSA with $Ni_{0.75}V_{0.25}$-LDH catalyst (the red solid circle in Fig. 6d). The expected current density of NiFe-LDH described above (~20.0 mA cm^{-2}) is still lower than $Ni_{0.75}V_{0.25}$-LDH (~27.0 mA cm^{-2}), although they have been estimated to have the same ECSA value. These results suggest in addition to the increased accessible surface area, probably the V in the Ni-based LDHs can also lead to the increased intrinsic catalytic activity than NiFe-LDHs in our case.

Although it is difficult to correctly normalize the catalytic current to the number of active sites in the LDH electrocatalysts to compare the intrinsic catalytic performances accurately, the kinetic parameters of the Tafel slope and onset potential can be selected to reflect the intrinsic activity of catalysts[30,31]. Thus, we investigated the Tafel slopes and onset potentials of $Ni_{0.75}Fe_{0.25}$-LDH and $Ni_{0.75}V_{0.25}$-LDH in Fig. 3 again. The Tafel slope is only affected by the kinetics of the reaction, which involves the type of active sites but not its quantity, surface area, morphology or electrical resistance[30]. Therefore, the lower Tafel slope of $Ni_{0.75}V_{0.25}$-LDH (~50 mV dec^{-1}) than that of $Ni_{0.75}Fe_{0.25}$-LDH (~64 mV dec^{-1}) indicates the higher catalytic activity and superior type of the active sites in $Ni_{0.75}V_{0.25}$-LDH. In addition, $Ni_{0.75}V_{0.25}$-LDH also shows lower onset potential (~250 mV overpotential) than that of $Ni_{0.75}Fe_{0.25}$-LDH (~300 mV overpotential), implying the facile kinetics of $Ni_{0.75}V_{0.25}$-LDH for the water oxidation reaction, which is consistent with the Tafel slope investigation. These results indicate that $Ni_{0.75}V_{0.25}$-LDH possesses superior intrinsic catalytic activity in our case.

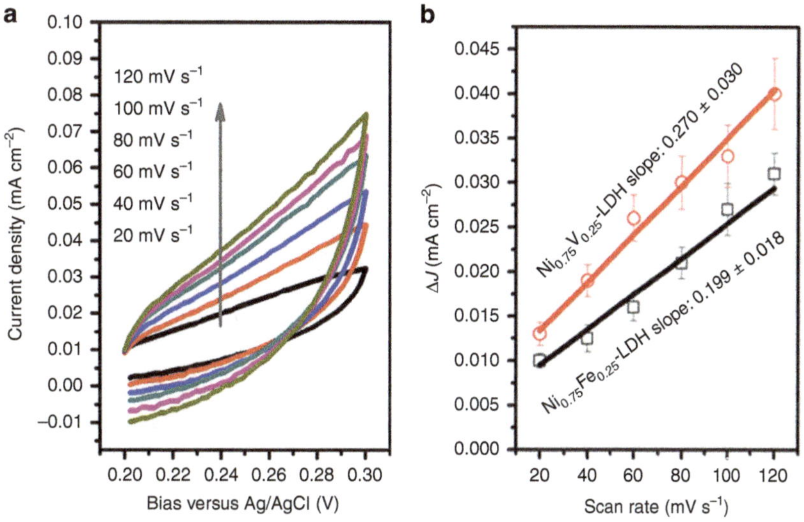

Figure 5 | ECSA of $Ni_{0.75}Fe_{0.25}$-LDH and $Ni_{0.75}V_{0.25}$-LDH. (**a**) Typical cyclic voltammetry curves of $Ni_{0.75}V_{0.25}$-LDH electrode in 1 M KOH with different scan rates; (**b**) ΔJ ($= J_a - J_c$) of $Ni_{0.75}Fe_{0.25}$-LDH and $Ni_{0.75}V_{0.25}$-LDH plotted against scan rates. All the error bars represent the s.d.'s of three replicate measurements. The slopes ($2C_{dl}$) were used to represent ECSA. The unit of slopes is mF cm^{-2}.

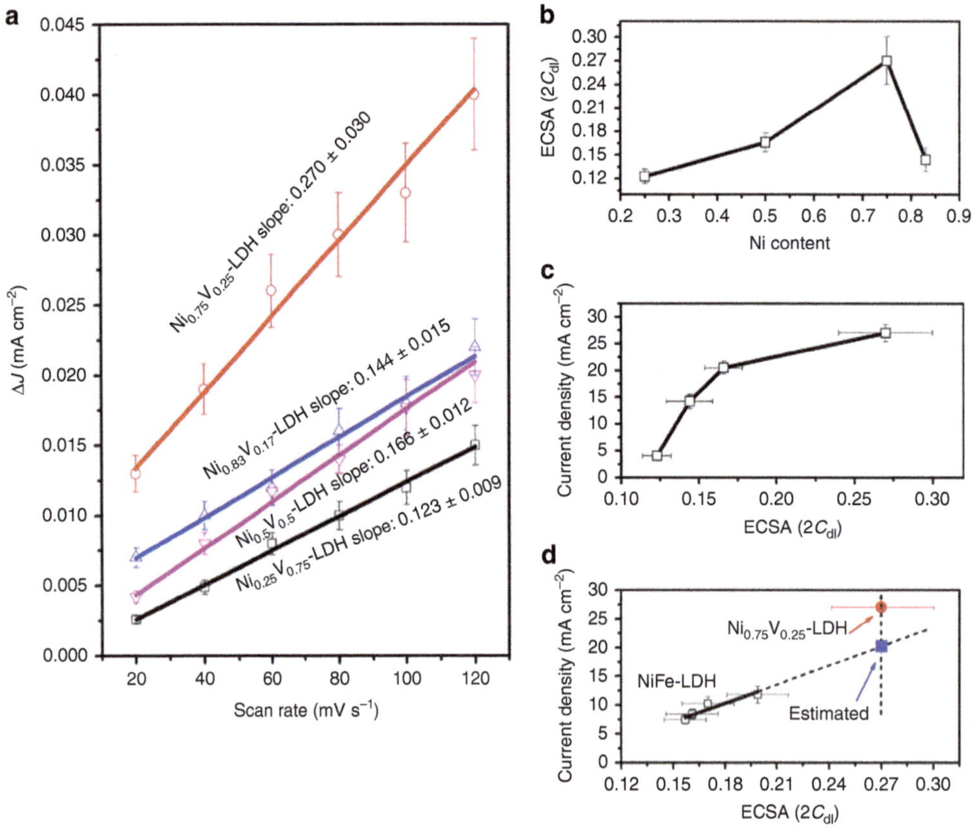

Figure 6 | ECSA of LDHs. (**a**) ΔJ ($=J_a - J_c$) of NiV-LDH with different Ni contents plotted against scan rates. The unit of slopes is mF cm^{-2}; (**b**) ECSA ($2C_{dl}$) of NiV-LDH with different Ni contents; (**c**) current density at 350 mV overpotential plotted against ECSA ($2C_{dl}$) of NiV-LDH with different Ni contents; and (**d**) current density at 350 mV overpotential plotted against ECSA ($2C_{dl}$) of NiFe-LDH with different Ni contents (black open squares), estimated the activity of NiFe-LDH (blue solid square) with the same ECSA of $Ni_{0.75}V_{0.25}$-LDH (red solid circle). All the error bars represent the s.d.'s of three replicate measurements.

The higher intrinsic catalytic activity of $Ni_{0.75}V_{0.25}$-LDH may be due to the higher conductivity than $Ni_{0.75}Fe_{0.25}$-LDH. To test the conductivity, electrochemical impedance spectroscopies (EIS) of $Ni_{0.75}Fe_{0.25}$-LDH and $Ni_{0.75}V_{0.25}$-LDH electrodes were carried out in the three-electrode configuration in 1 M KOH. The Nyquist diagrams of both electrodes show an apparent semicircle in the high frequency range (Fig. 7a), which should be mainly associated with charge transfer resistance (R_{ct}) in the LDH catalysts[32,33]. The diameter of the semicircle in Nyquist diagram of $Ni_{0.75}V_{0.25}$-LDH decreases comparing to that of $Ni_{0.75}Fe_{0.25}$-LDH, indicating lower charge transfer resistance, that is, improved conductivity in LDH catalysts. A reported electrical analogue was used to fit the EIS data[32], and the R_{ct} of $Ni_{0.75}V_{0.25}$-LDH is estimated to be $\sim 62\,\Omega\,cm^2$, lower than that of $Ni_{0.75}Fe_{0.25}$-LDH ($\sim 94\,\Omega\,cm^2$). This is also reflected in the Bode plots in Fig. 7b, where $Ni_{0.75}V_{0.25}$-LDH shows smaller resistance than $Ni_{0.75}Fe_{0.25}$-LDH. It is known that the LDHs suffer from their low conductivity, in the current case, the higher electron conductivity of $Ni_{0.75}V_{0.25}$-LDH benefits the charge transfer efficiently, which is also in a good agreement with the lower Tafel slope in Fig. 3b.

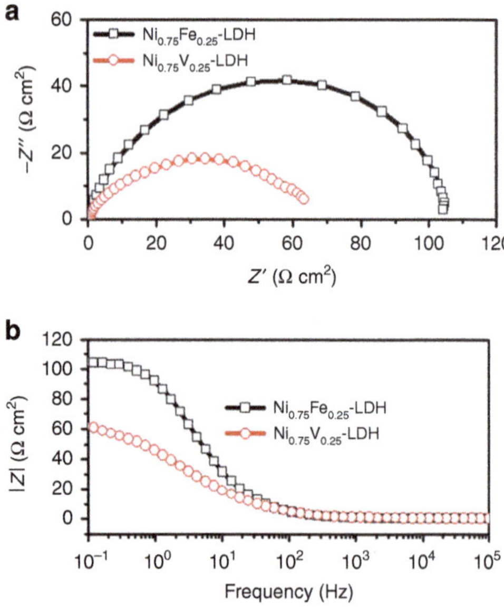

Figure 7 | EIS measurements. (**a**) Nyquist diagram and (**b**) Bode plots of $Ni_{0.75}Fe_{0.25}$-LDH and $Ni_{0.75}V_{0.25}$-LDH with bias of 350 mV overpotential.

On the other hand, comparing to the relatively planar structure in $Ni_{0.75}Fe_{0.25}$-LDH, the slightly more three-dimensional structured $Ni_{0.75}V_{0.25}$-LDH can expose more edges of the nanosheets to the electrolyte (see TEM images in Fig. 2c,d and Supplementary Fig. 2). The edges are expected to contain open coordination sites that can be the active sites for water oxidation. Therefore, from this point, $Ni_{0.75}V_{0.25}$-LDH exposes more active sites than $Ni_{0.75}Fe_{0.25}$-LDH in electrolyte, while maintaining a similar active surface area. This may be another important factor contributing to the highly catalytic activity of $Ni_{0.75}V_{0.25}$-LDH. Meanwhile, investigating Fig. 6c again, it is also noted that the increases in ECSA produce large changes in current density up to

ECSA = 0.17 ($2C_{dl}$), but after that the trend is much less pronounced. This change of trend of the activity dependent on ECSA for NiV-LDHs indicates that there should be other factors that can influence the activity of NiV-LDHs besides ECSA. It was reported that the catalytic performance of the layered catalyst for water splitting more strongly depends on the edge-state length than ECSA. The catalytic activity can be linear to the edge-state length[34]. Therefore, as shown in Fig. 6c, a plausible explanation is that it can be speculated that at small ECSA (< 0.17 in our case), the active sites on the edge-state length can be exposed to the electrolyte sufficiently, increasing with ECSA, and dominate the catalytic activity, so the modest increases in ECSA can produce large changes in current density by the edge-state length. However, when the ECSA increases to a large value, due to the aggregation, ESCA will cover numerous edges of the nanosheets from the electrolyte, which makes ECSA show more significant effect than the edge-state length, resulting in a slower increase rate. This result also indicates the important role that the edge state of LDHs plays in the catalytic activity. More investigations into the detailed mechanism are in progress.

To further understand the advantages of $Ni_{0.75}V_{0.25}$-LDH, density functional theory (DFT) calculations were performed based on the following mechanism for water oxidation:

$$H_2O + * \rightarrow *OH_2 \quad (1)$$

$$*OH_2 \rightarrow *OH + H^+ + e \quad (2)$$

$$*OH \rightarrow *O + H^+ + e \quad (3)$$

$$H_2O + *O \rightarrow *OOH + H^+ + e \quad (4)$$

$$*OOH \rightarrow O_2 + * + H^+ + e \quad (5)$$

Here '*' presents the adsorption site that is usually on the top of the doping element. The calculation of the reaction free energy with the zero-point energy and entropy corrections followed the same procedure in ref. 35. Reaction free energies of reaction 1–5 are denoted as ΔG_1–ΔG_5, ΔG_5 is defined as 4.92 eV–ΔG_1–ΔG_2–ΔG_3–ΔG_4 to avoid the calculation of the energy of O_2 molecule. The overpotential η is defined in equation (1):

$$\eta = \max\{\Delta G_2, \Delta G_3, \Delta G_4, \Delta G_5\}/e - 1.23\,V \quad (6)$$

The optimized structures of the intermediates in the free-energy landscape are shown in Fig. 8. The intermediate *OH and *OOH bind to the surface through oxygen with a single bond. The difference of their binding energies has been shown to be a constant (3.2 ± 0.2 eV) for various materials, such as metals and oxides[36]. In our case, the constant is calculated to be 3.22 eV. Because this binding energy difference equals to the sum of the reaction free energies of reaction 3 and 4 ($\Delta G_3 + \Delta G_4$)[35,37], the rate-limiting step is either the formation of *O from *OH (ΔG_3) or formation of *OOH from *O (ΔG_4). The calculated free-energy landscape shows that the rate-limiting step is the formation of *OOH when the sample is doped with V with $\eta = 0.62$ V, whereas the rate-limiting step becomes the formation of *O when the sample is doped with Fe with $\eta = 1.28$ V. Because $\Delta G_3 + \Delta G_4 = 3.22$ eV, the lower limit of the η can be reached when $\Delta G_3 = \Delta G_4 = 1/2 \times 3.22$ eV = 1.61 eV if a proper element was doped. Then $\eta = 1.61 - 1.23 = 0.38$ V, which suggests it has a large space to further optimize the activity of Ni-based LDH by doping with other elements.

Besides being better than the prepared NiFe-LDH in our case, $Ni_{0.75}V_{0.25}$-LDH shows superior catalytic performance than other reported LDHs as well, to the best of our knowledge. Supplementary Table 1 compares the activity of $Ni_{0.75}V_{0.25}$-LDH with recently reported state-of-the-art LDH catalysts without exfoliation or coupling with other materials, which were loaded on GC electrodes for water splitting, including NiFe-, NiCo-, ZnCo-, CoCo- and MnCo-LDHs. Although the catalytic performances of LDH electrodes are strongly dependent on the preparation methods in literatures, to make a general overview, the current density at 350 mV overpotential and the corresponding mass activity ($A\,g^{-1}$) are still used to compare the intrinsic activities between different LDH catalysts. To make the comparison more reasonable, ohmic-drop correction was also performed for $Ni_{0.75}V_{0.25}$-LDH, as most of the literature has done (Supplementary Fig. 7)[8,9,26,29,38]. In our case, without ohmic-drop correction, at 350 mV overpotential, $Ni_{0.75}V_{0.25}$-LDH can achieve ~27 $mA\,cm^{-2}$ current density with ~190 $A\,g^{-1}$ mass activity. After ohmic-drop correction, the current density and mass activity can be as high as ~57 $mA\,cm^{-2}$ and ~400 $A\,g^{-1}$, respectively. Those results are even better than some exfoliated NiFe-, NiCo- and CoCo-LDHs. It is worthy to note that $Ni_{0.75}V_{0.25}$-LDH shows even higher activity than MnCo-LDH (~43 $mA\,cm^{-2}$ and ~301 $A\,g^{-1}$) before long-term anodic conditioning that was reported as one of the most active LDH catalyst for water oxidation[26].

It is worth pointing out that although $Ni_{0.75}V_{0.25}$-LDH outperforms $Ni_{0.75}Fe_{0.25}$-LDH in our case, we cannot claim $Ni_{0.75}V_{0.25}$-LDH has higher catalytic activity than all the NiFe-LDHs in the literature. For instance, Louie and Bell reported NiFe-LDH that requires < 300 mV overpotential to deliver current densities of 10 $mA\,cm^{-2}$ with a Tafel slope of

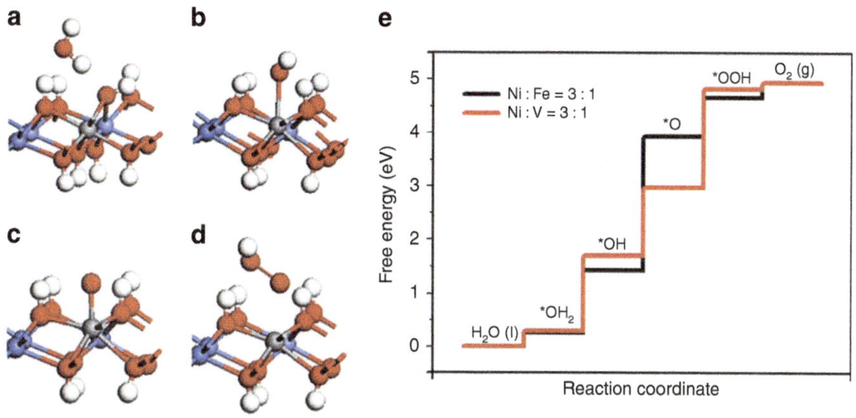

Figure 8 | DFT calculation. Adsorption geometries of the intermediates H_2O, *OH, *O and *OOH in **a**,**b**,**c** and **d**, respectively. The red, blue, white, grey atoms represent the O, Ni, H and V atoms, respectively. The adsorption structures are similar to these when one Ni is substituted by Fe instead of V; (**e**) the free-energy landscape.

~40 mV dec^{-1} (with ohmic-drop correction), which outperforms our $Ni_{0.75}V_{0.25}$-LDH probably due to the different preparation method[12]. Nevertheless, from Supplementary Table 1, our $Ni_{0.75}V_{0.25}$-LDH is still highly efficient among the listed catalysts, and comparable to the best-performing NiFe-LDHs[12], which indicates our $Ni_{0.75}V_{0.25}$-LDH is one of the best water oxidation catalysts.

To confirm the O_2 evolution and Faradaic efficiency of $Ni_{0.75}V_{0.25}$-LDH electrode, the experimental and theoretical O_2 evolution amount by $Ni_{0.75}V_{0.25}$-LDH at a constant oxidative current of 1 mA in 1 M KOH were performed as shown in Supplementary Fig. 8. The experimental O_2 evolution determined by gas chromatography exhibited 101 ± 7% Faradaic efficiency, when the electrolysis time was 60 min. Moreover, to further explore the highly catalytic activity of $Ni_{0.75}V_{0.25}$-LDH, we drop-casted $Ni_{0.75}V_{0.25}$-LDH suspension onto Ni foam as well, which can supply excellent conductivity and very high surface area for the electrode, and tested the catalytic activity of resulted electrode in 1 M KOH for water oxidation. The $Ni_{0.75}V_{0.25}$-LDH on Ni foam exhibits the promising activity for practical applications, with ~44 mA cm^{-2} at 350 mV overpotential, and just ~300 mV overpotential required to reach 10 mA cm^{-2} (Supplementary Fig. 9) without ohmic-drop correction. Such catalytic activities of $Ni_{0.75}V_{0.25}$-LDH on Ni foam indicate attractive prospects for large-scale and practical applications.

In summary, $Ni_{0.75}V_{0.25}$-LDH has been prepared by a simple hydrothermal method, which is monolayered and shows high catalytic activity for the water oxidation reaction comparable to the best-performing NiFe-LDHs. The high intrinsic catalytic activity of $Ni_{0.75}V_{0.25}$-LDH is mainly due to the good conductivity, facile electron transfer and abundant active sites in the nanolayers of $Ni_{0.75}V_{0.25}$-LDH, showing potential to be one of the most effective Ni-based LDH electrocatalysts. Our study reveals the promising catalytic properties of vanadium incorporated Ni-based LDHs, and expands the scope of non-precious metal catalysts with the highly intrinsic activity for the water oxidation reaction.

Methods

Preparation of NiV- and NiFe-LDHs. A one-step hydrothermal method was employed to synthesize the bulk LDHs. In brief, for preparation of NiV-LDH, different mole ratios of Ni/V solution (1:0, 5:1, 3:1, 1:1, 1:3 and 0:1 for the synthesis of pure $Ni(OH)_2$, $Ni_{0.83}V_{0.17}$-LDH, $Ni_{0.75}V_{0.25}$-LDH, $Ni_{0.5}V_{0.5}$-LDH, $Ni_{0.25}V_{0.75}$-LDH and bare V-based hydroxide, respectively) was obtained by mixing $NiCl_2$ and VCl_3 in 80 ml H_2O, while the total amount of metal ions (Ni^{2+} + V^{3+}) was kept to 3.2 mmol. Afterwards, 0.3 g of urea was added, and the above mixture solution was transferred to a stainless-steel Teflon-lined autoclave, and heated in an oven at 120 °C for 12 h. After cooling the autoclave to room temperature, the resulting powder was washed by deionized water and ethanol three times, collected and then dried at 70 °C overnight. As a reference, for preparation of NiFe-LDH, the same processes were followed except $FeCl_3$ was used as iron source instead of VCl_3, which was analogous to the literature[25].

Electrode preparation. A measure of 5 mg of the obtained LDH powders were dispersed in the mixture solution of 1 ml H_2O, 0.25 ml 2-propanol and 10 µl 5% Nafion (ethanol solution) by sonication for >1 h. A measure of 2.5 µl of the above suspension were drop-casted to a pre-polished GC electrode (diameter: 3 mm), and dried at 50 °C for 15 min to evaporate the solvent.

Structure and surface characterization. X-ray diffraction measurements were carried out on Bruker X-ray diffraction diffractometer D5000 with Cu Kα radiation. SEM images were taken on JEOL JSM-7000F instrument with energy-dispersive X-ray spectroscopy (EDS). TEM images were taken on JEOL JEM2100 TEM. Tapping mode AFM was performed to examine the surface morphology by a Dimension Icon AFM (Bruker, Santa Barbara, USA) on silicon wafer. Rectangular cantilevers with approximate dimensions of 125 µm in length and 40 µm in width (BudgetSensors Tap300Al-G) were used to perform the tapping mode experiments. The NanoScope Analysis software (version 1.50, Bruker) was used to analyse the recorded AFM data. A first-order polynomial-flattening algorithm was employed to remove surface tilt from height images. XPS measurements were conducted with an in-house spectrometer (PHI 5500) using monochromatized Al Kα radiation (1,486.6 eV). The pressure in the analysis chamber was ~5 × 10^{-9} mbar during the measurement. Core peaks were analysed using a nonlinear Shirley-type background. The peak positions and areas were optimized by a weighted least-squares fitting method using 70% Gaussian/30% Lorentzian lineshapes. The powder was grinded and deposited on a conductive FTO glass. An electron flood gun was used to compensate the charging effects. The XPS spectra were energy calibrated by setting the adventitious carbon peak to 285 eV.

Electrochemical measurements. Electrolysis experiments were carried out in a polytetrafluoroethylene cell with an Autolab potentiostat with GPES electrochemical interface (Eco Chemie) in a standard three-electrode configuration, which was composed of working electrode (LDHs deposited on GC electrodes), counter electrode (Pt net) and reference electrode (Ag/AgCl). The electrolyte was 1 M KOH, and the applied potentials were converted with respect to reversible hydrogen electrode (RHE), $E_{RHE} = E_{Ag/AgCl} + 0.059\ pH + 0.197\ V$, and overpotential $\eta = E_{RHE}-1.23\ V$.

First, before all the electrochemical measurements, a galvanostatic measurement at a fixed current density of 5 mA cm^{-2} was performed until a stable potential was obtained. Then LSV were measured from 0.2 to 0.7 V versus Ag/AgCl with a slow scan rate of 2 mV s^{-1}. By plotting overpotential η against log (J) from LSV curves, Tafel slopes can be obtained. To test the stability of LDHs, a galvanostatic measurement at a fixed current density (J) of 10 mA cm^{-2} was performed. ECSA were measured by CV at the potential window 0.2–0.3 V versus Ag/AgCl, with different scan rates of 20, 40, 60, 80, 100 and 120 mV s^{-1}. By plotting the $\Delta J = (J_a - J_c)$ at 0.25 V versus Ag/AgCl against the scan rate, the linear slope that is twice of the double layer capacitance (C_{dl}) is used to represent ECSA. All the above measurements were carried out without ohmic-drop correction unless noted otherwise.

Turnover frequency. The TOF of LDH catalysts were calculated according to the following equation (2)[9],

$$TOF = \frac{JA}{4Fm} \tag{7}$$

where J is the current density at a given overpotential, for example, in our cases $\eta = 350$ mV, A is the surface area of the electrode (0.07 cm^2), F is Faraday constant (96,485 s A mol^{-1}) and m is the number of moles of the metal on the electrodes. In our cases, we assumed all the metal sites were actively involved in the electrochemical reaction.

To compare the conductivities of LDHs, EISs were carried out in 1 M KOH with a three-electrode configuration, the frequency ranged from 0.1 to 10^5 Hz, with an a.c. amplitude of 10 mV and overpotential bias of 350 mV.

The measurements of O_2 were performed in an air-tight H shape cell, which was divided by a glass frit to two chambers. The working electrode, the Ag/AgCl reference electrode and a magnetic stirring bar were inserted in one chamber of the cell, the Pt counter electrode was inserted in the other chamber. The cell was filled with 1 M KOH and degassed with helium for >1 h. The headspace of the compartment containing the working electrode was 23.6 ml. The electrolysis was carried out with a constant oxidation current of 1 mA for 60 min. A measure of 500 µl of the gas sample in the compartment containing the working electrode was transferred by a specific syringe to the gas chromatography, (Shimadzu) where the amount of O_2 evolution was determined. The Faradaic efficiency was determined from the total amount of charge Q (C) passed through the cell and the total amount of the produced O_2 n_{O2} (mol): Faradaic efficiency = $4F \times n_{O2}/Q$, where F is the Faraday constant, assuming the four electrons are needed to produce one oxygen molecule.

DFT calculation. Calculations were carried out with DFT implanted in the Vienna Ab initio Simulation Package (VASP)[39–42] to give a better understanding for the superior activity of $Ni_{0.75}V_{0.25}$-LDH. Perdew-Burke-Ernzerhof (PBE)[43] exchange-correlation functional and projector augmented-wave (PAW)[44] pseudo-potential were adopted. An energy cutoff of 400 eV was applied for the plane-wave basis set. A 2 × 2 supercell was used with one of the Ni atom substituted by Fe or V. A 7 × 7 Monkhorst–Pack K-point grid was applied for the sampling of Brillouin zone. To describe the transition metal elements, DFT + U[45] method have been used with the U values from ref. 35, U − J = 3.8 eV for Ni^{2+} and U − J = 4.3 eV for Fe^{3+} and U − J = 3.4 eV for V^{3+} species. One proton of the material is removed to create the metal ion at 3+ oxidation state. The structures were optimized until the maxima force on the atoms was smaller than 0.02 eV/Å.

References

1. McCrory, C. C. L., Jung, S., Peters, J. C. & Jaramillo, T. F. Benchmarking heterogeneous electrocatalysts for the oxygen evolution reaction. *J. Am. Chem. Soc.* **135**, 16977–16987 (2013).

2. Galán-Mascarós, J. R. Water oxidation at electrodes modified with earth-abundant transition-metal catalysts. *ChemElectroChem.* **2,** 37–50 (2015).
3. Chen, G. *et al.* Interfacial effects in iron-nickel hydroxide–platinum nanoparticles enhance catalytic oxidation. *Science* **344,** 495–499 (2014).
4. Lu, Z. *et al.* Three-dimensional NiFe layered double hydroxide film for high-efficiency oxygen evolution reaction. *Chem. Commun.* **50,** 6479–6482 (2014).
5. Trotochaud, L., Young, S. L., Ranney, J. K. & Boettcher, S. W. Nickel–iron oxyhydroxide oxygen-evolution electrocatalysts: the role of intentional and incidental iron incorporation. *J. Am. Chem. Soc.* **136,** 6744–6753 (2014).
6. Friebel, D. *et al.* Identification of highly active Fe sites in (Ni,Fe)OOH for electrocatalytic water splitting. *J. Am. Chem. Soc.* **137,** 1305–1313 (2015).
7. Gong, M. & Dai, H. A mini review of NiFe-based materials as highly active oxygen evolution reaction electrocatalysts. *Nano Res.* **8,** 23–39 (2015).
8. Gong, M. *et al.* An advanced Ni–Fe layered double hydroxide electrocatalyst for water oxidation. *J. Am. Chem. Soc.* **135,** 8452–8455 (2013).
9. Song, F. & Hu, X. Exfoliation of layered double hydroxides for enhanced oxygen evolution catalysis. *Nat. Commun.* **5,** 4477 (2014).
10. Long, X. *et al.* A strongly coupled graphene and FeNi double hydroxide hybrid as an excellent electrocatalyst for the oxygen evolution reaction. *Angew. Chem. Int. Ed.* **53,** 7584–7588 (2014).
11. Ma, W. *et al.* A superlattice of alternately stacked Ni–Fe hydroxide nanosheets and graphene for efficient splitting of water. *ACS Nano* **9,** 1977–1984 (2015).
12. Louie, M. W. & Bell, A. T. An investigation of thin-film Ni–Fe oxide catalysts for the electrochemical evolution of oxygen. *J. Am. Chem. Soc.* **135,** 12329–12337 (2013).
13. Valdez, R., Grotjahn, D. B., Smith, D. K., Quintana, J. M. & Olivas, A. Nanosheets of Co-(Ni and Fe) layered double hydroxides for electrocatalytic water oxidation reaction. *Int. J. Electrochem. Sci.* **10,** 909–918 (2015).
14. Liang, H. *et al.* Hydrothermal continuous flow synthesis and exfoliation of NiCo layered double hydroxide nanosheets for enhanced oxygen evolution catalysis. *Nano Lett.* **15,** 1421–1427 (2015).
15. Burke, M. S., Kast, M. G., Trotochaud, L., Smith, A. M. & Boettcher, S. W. Cobalt-iron (oxy)hydroxide oxygen evolution electrocatalysts: the role of structure and composition on activity, stability, and mechanism. *J. Am. Chem. Soc.* **137,** 3638–3648 (2015).
16. Liang, J. *et al.* Topochemical synthesis, anion exchange, and exfoliation of Co-Ni layered double hydroxides: a route to positively charged Co-Ni hydroxide nanosheets with tunable composition. *Chem. Mater.* **22,** 371 378 (2010).
17. Diaz-Morales, O., Ledezma-Yanez, I., Koper, M. T. M. & Calle-Vallejo, F. Guidelines for the rational design of Ni-based double hydroxide electrocatalysts for the oxygen evolution reaction. *ACS Catal.* **5,** 5380–5387 (2015).
18. Tan, Z. A. *et al.* Solution-processed nickel acetate as hole collection layer for polymer solar cells. *Phys. Chem. Chem. Phys.* **14,** 14217–14223 (2012).
19. De Jesus, J. C., González, I., Quevedo, A. & Puerta, T. Thermal decomposition of nickel acetate tetrahydrate: an integrated study by TGA, QMS and XPS techniques. *J. Mol. Catal. A: Chem.* **228,** 283–291 (2005).
20. Mansour, A. N. Characterization of NiO by XPS. *Surf. Sci. Spectra* **3,** 231–238 (1994).
21. Biesinger, M. C. *et al.* Resolving surface chemical states in XPS analysis of first row transition metals, oxides and hydroxides: Cr, Mn, Fe, Co and Ni. *Appl. Surf. Sci.* **257,** 2717–2730 (2011).
22. Chen, H. *et al.* 3D open-framework vanadoborate as a highly effective heterogeneous pre-catalyst for the oxidation of alkylbenzenes. *Chem. Mater.* **25,** 5031–5036 (2013).
23. Biesinger, M. C., Lau, L. W. M., Gerson, A. R. & Smart, R. S. C. Resolving surface chemical states in XPS analysis of first row transition metals, oxides and hydroxides: Sc, Ti, V, Cu and Zn. *Appl. Surf. Sci.* **257,** 887–898 (2010).
24. Silversmit, G., Depla, D., Poelman, H., Marin, G. B. & De Gryse, R. Determination of the V2p XPS binding energies for different vanadium oxidation states (V^{5+} to V^{0+}). *J. Electron Spectrosc.* **135,** 167–175 (2004).
25. Luo, J. *et al.* Water photolysis at 12.3% efficiency via perovskite photovoltaics and Earth-abundant catalysts. *Science* **345,** 1593–1596 (2014).
26. Song, F. & Hu, X. Ultrathin cobalt–manganese layered double hydroxide is an efficient oxygen evolution catalyst. *J. Am. Chem. Soc.* **136,** 16481–16484 (2014).
27. Ma, T. Y., Dai, S., Jaroniec, M. & Qiao, S. Z. Graphitic carbon nitride nanosheet–carbon nanotube three-dimensional porous composites as high-performance oxygen evolution electrocatalysts. *Angew. Chem. Int. Ed.* **53,** 7281–7285 (2014).
28. Walter, M. G. *et al.* Solar water splitting cells. *Chem. Rev.* **110,** 6446–6473 (2010).
29. Nai, J. *et al.* Efficient electrocatalytic water oxidation by using amorphous Ni-Co double hydroxides nanocages. *Adv. Energy. Mater.* **5,** 1401880 (2015).
30. Huynh, M., Shi, C., Billinge, S. J. L. & Nocera, D. G. Nature of activated manganese oxide for oxygen evolution. *J. Am. Chem. Soc.* **137,** 14887–14904 (2015).
31. Smith, R. D. L., Prévot, M. S., Fagan, R. D., Trudel, S. & Berlinguette, C. P. Water oxidation catalysis: electrocatalytic response to metal stoichiometry in amorphous metal oxide films containing iron, cobalt, and nickel. *J. Am. Chem. Soc.* **135,** 11580–11586 (2013).
32. Lopes, T., Andrade, L., Ribeiro, H. A. & Mendes, A. Characterization of photoelectrochemical cells for water splitting by electrochemical impedance spectroscopy. *Int. J. Hydrogen Energ.* **35,** 11601–11608 (2010).
33. Long, X., Xiao, S., Wang, Z., Zheng, X. & Yang, S. Co intake mediated formation of ultrathin nanosheets of transition metal LDH-an advanced electrocatalyst for oxygen evolution reaction. *Chem. Commun.* **51,** 1120–1123 (2015).
34. Jaramillo, T. F. *et al.* Identification of active edge sites for electrochemical H2 evolution from MoS_2 nanocatalysts. *Science* **317,** 100–102 (2007).
35. Liao, P., Keith, J. A. & Carter, E. A. Water oxidation on pure and doped hematite (0001) surfaces: prediction of Co and Ni as effective dopants for electrocatalysis. *J. Am. Chem. Soc.* **134,** 13296–13309 (2012).
36. Man, I. C. *et al.* Universality in oxygen evolution electrocatalysis on oxide surfaces. *ChemCatChem.* **3,** 1159–1165 (2011).
37. Calle-Vallejo, F. & Koper, M. T. M. First-principles computational electrochemistry: achievements and challenges. *Electrochim. Acta* **84,** 3–11 (2012).
38. Li, Y., Zhang, L., Xiang, X., Yan, D. & Li, F. Engineering of ZnCo-layered double hydroxide nanowalls toward high-efficiency electrochemical water oxidation. *J. Mater. Chem. A* **2,** 13250–13258 (2014).
39. Kresse, G. & Hafner, J. Abinitio molecular-dynamics for liquid-metals. *Phys. Rev. B* **47,** 558–561 (1993).
40. Kresse, G. & Hafner, J. Ab-initio molecular-dynamics simulation of the liquid-metal amorphous-semiconductor transition in germanium. *Phys. Rev. B* **49,** 14251–14269 (1994).
41. Kresse, G. & Furthmuller, J. Efficient iterative schemes for ab initio total-energy calculations using a plane-wave basis set. *Phys. Rev. B* **54,** 11169–11186 (1996).
42. Kresse, G. & Furthmuller, J. Efficiency of ab-initio total energy calculations for metals and semiconductors using a plane-wave basis set. *Comp. Mat. Sci.* **6,** 15–50 (1996).
43. Perdew, J. P., Burke, K. & Ernzerhof, M. Generalized Gradient Approximation Made Simple. *Phys. Rev. Lett.* **77,** 3865–3868 (1996).
44. Blöchl, P. E. Projector augmented-wave method. *Phys. Rev. B* **50,** 17953–17979 (1994).
45. Dudarev, S. L., Botton, G. A., Savrasov, S. Y., Humphreys, C. J. & Sutton, A. P. Electron-energy-loss spectra and the structural stability of nickel oxide: an LSDA + U study. *Phys. Rev. B* **57,** 1505–1509 (1998).

Acknowledgements

We thank Beatrice Johansson from KTH, Dr Valentina Leandri, Chao Xu and Sareh Ahmed from Uppsala University for their help in characterizations of materials. We acknowledge the financial support of this work by Swedish Energy Agency, the Knut and Alice Wallenberg Foundation, the Swedish Research Council, the National Natural Science Foundation of China (21120102036, 91233201) and the National Basic Research Program of China (973 program, 2014CB239402).

Author contributions

K.F. and H.C. contributed equally to this paper. K.F. designed the experiments, fabricated and measured the devices, and wrote the paper. H.C. carried out the X-ray diffraction, SEM and TEM characterization, analysed the data and helped to revise the paper. Y.J. and Y.L. did the DFT calculation; H.H. and P.C. carried out the AFM measurement; and B.P. and H.R. carried out and analysed the XPS measurements. Q.D. and F.L. contributed to the discussion of the paper. L.S. supervised the project and wrote the paper.

Additional information

Competing financial interests: The authors declare no competing financial interests.

18

A zwitterionic gel electrolyte for efficient solid-state supercapacitors

Xu Peng[1], Huili Liu[1], Qin Yin[1], Junchi Wu[1], Pengzuo Chen[1], Guangzhao Zhang[1], Guangming Liu[1], Changzheng Wu[1] & Yi Xie[1]

Gel electrolytes have attracted increasing attention for solid-state supercapacitors. An ideal gel electrolyte usually requires a combination of advantages of high ion migration rate, reasonable mechanical strength and robust water retention ability at the solid state for ensuring excellent work durability. Here we report a zwitterionic gel electrolyte that successfully brings the synergic advantages of robust water retention ability and ion migration channels, manifesting in superior electrochemical performance. When applying the zwitterionic gel electrolyte, our graphene-based solid-state supercapacitor reaches a volume capacitance of 300.8 F cm^{-3} at 0.8 A cm^{-3} with a rate capacity of only 14.9% capacitance loss as the current density increases from 0.8 to 20 A cm^{-3}, representing the best value among the previously reported graphene-based solid-state supercapacitors, to the best of our knowledge. We anticipate that zwitterionic gel electrolyte may be developed as a gel electrolyte in solid-state supercapacitors.

[1] Hefei National Laboratory for Physical Sciences at the Microscale, iChEM (Collaborative Innovation Centre of Chemistry for Energy Materials), Hefei Science Center (CAS), and CAS Key Laboratory of Mechanical Behavior and Design of Materials, University of Science and Technology of China, Hefei, Anhui 230026, China. Correspondence and requests for materials should be addressed to C.W. (email: czwu@ustc.edu.cn) or to G.L. (email: gml@ustc.edu.cn).

The development of highly efficient energy storage devices has greatly satisfied growing energy demands for our daily life, of which supercapacitors have emerged as high-performance energy storage devices for long operating lifetimes and high power densities[1–3]. Recently, gel electrolytes have attracted increasing attention in solid-state supercapacitors, due to their capability to fulfil multiple roles of electrolyte, separator and binder in solid-state supercapacitors[4–6]. Developing gel electrolytes accelerates the evolution of solid-state supercapacitors from traditional sandwich-type supercapacitors, to flexible, transparent and planar supercapacitors (micro-supercapacitors)[7,8], and thus offering power support to flexible and even transparent electronics[9,10]. Generally, gel electrolytes are made of a polymeric material as matrix and an electrolyte salt to provide mobile ions[4]. Nowadays, non-aqueous gel electrolytes, such as block-copolymer-based gel electrolyte and silica-based gel electrolyte, have been developed dissolved in ionic liquid to solve the evaporation problem, achieving the enhancement in the ion migration rate and good mechanical strength, and making a great advance in electrochemical performance[11–14]. Aqueous gel electrolytes are dominantly based on polyvinyl alcohol (PVA) matrix, such as PVA/H_2SO_4 (refs 15,16), PVA/KOH (refs 17,18) and PVA/LiCl (refs 19,20). The good abilities of PVA gel electrolyte, with a wide range of pH values like the aqueous electrolyte solution and serving as an elastic coating with a certain mechanical strength to avoid structure degradation of electrode materials, render excellent performance of PVA gel electrolytes for solid-state supercapacitors[21]. Despite the fact that PVA gel electrolytes offer convenience to fabricate solid-state supercapacitors, the development of aqueous polymeric gel electrolytes is still at a primary stage and the inner electrochemical mechanism remains to be explored. This leaves plenty of room to improve electrochemical performance of solid-state supercapacitors via chemical design of gel electrolytes.

Polyzwitterions are a type of charged polymer with robust water retention ability coming from the presence of a zwitterionic group in a repeat unit[22], serving as a potential gel electrolyte catering for highly efficient solid-state supercapacitors. Polyzwitterions can be highly hydrated due to the strong interactions between the charged groups and water molecules, which make them suitable as a kind of superabsorbent polymeric material[23]. Also, their zwitterionic nature makes the cationic and anionic counterions of polyzwitterions easily separated during ion migration, ensuring a high ionic conductivity[24,25]. Moreover, polyzwitterions can form physical gels through the dipole–dipole interactions between the zwitterionic groups, thereby rendering the polyzwitterionic gel a certain mechanical strength[26]. In contrast to conventional polyelectrolytes, polyzwitterions usually exhibit the so-called 'anti-polyelectrolyte' effect, thus it has a good solubility in aqueous solutions with a high salt concentration[27,28]. Besides, the charged and polar groups associated with polyzwitterions can strengthen the adhesion between the gel electrolyte and the electrodes, so that the polyzwitterionic gel electrolyte can be employed as a polymer binder to hold the electrodes together. More importantly, zwitterions have been used as electrolytes for high ionic conductivity and high lithium ion transfer number[29–32]. For example, zwitterionic gel electrolyte with 66.1 mS cm^{-1} with the help of high ion mobility of H^+ was reported recently[33]. But there still presents great demand for introducing them to energy storage applications, especially for enhancing electrochemical performance via the electrolyte. In short, the combined advantages of water retention ability, high ion conductivity, reasonable mechanical strength and anti-polyelectrolyte effect renders polyzwitterions as a promising gel electrolyte for energy storage devices.

In this work, we develop a class of zwitterionic gel electrolyte, realising superior electrochemical performance in solid-state supercapacitors. In our case, the zwitterionic nature of poly (propylsulfonate dimethylammonium propylmethacrylamide) (PPDP) not only offers this gel electrolyte robust water retention ability at the solid-state through a combination of about eight water molecules around the charged groups but also brings ion migration channels to the electrolyte ions, leading to better electrochemical performance. When applying PPDP as a gel electrolyte, our graphene-based solid-state supercapacitor reaches a volume capacitance of 300.8 F cm^{-3} at 0.8 A cm^{-3}, with a rate capacity of only 14.9% capacitance loss as current density increases from 0.8 to 20 A cm^{-3}, representing the best value among the previously reported graphene-based solid-state supercapacitors, to the best of our knowledge.

Results

Chemical structures of PPDP/LiCl gel electrolyte. PPDP is a kind of polyzwitterion bearing both positively charged quaternary ammonium group and negatively charged sulfonate group on the same monomeric unit. In this regard, PPDP is virtually neutral in a whole due to its zwitterionic character, as illustrated in Fig. 1a. Owing to the strong electrostatic interactions between the charged groups and water molecules, PPDP is highly hydrated by the surrounding water molecules with a robust water retention ability, showing a potential for superior electrochemical performance when supercapacitors operate at solid state. Meanwhile, the ion migration channel can be developed within

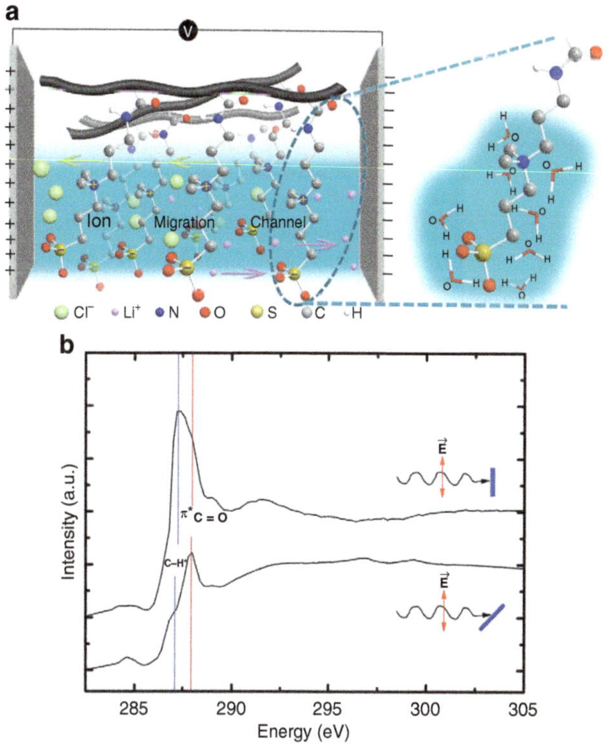

Figure 1 | Zwitterionic PPDP under external electric field. (**a**) Schematic illustration of the PPDP gel electrolyte applied on electrodes. The ion migration channel is formed by applying external electric field. PPDP is strongly hydrated by water molecules with robust water retention ability due to the electrostatic interactions between the zwitterionic groups and water molecules. (**b**) Angular-dependent C K-edge X-ray absorption near-edge spectroscopy (XANES) of zwitterionic PPDP sample after applying external field, with linearly polarized soft X-ray beam.

the hydration layer along the polyzwitterion chains between two electrodes by applying external electric field due to the robust water retention ability of PPDP gel electrolyte. Such a kind of ion migration channel strongly boosts the efficiency of ion transport in the gel electrolyte. Therefore, the zwitterionic nature of PPDP makes the electrolyte ions of cationic Li^+ and anionic Cl^- in gel electrolyte easily separated without overcoming the strong electrostatic attractions between the charged groups and the counterions during the ion migration and readily transferred onto the surface of graphene electrodes to realize the equivalent electric double-layer capacitance (EDLC), thereby enhancing the capacitance of solid-state supercapacitors.

Although the PPDP main chains are randomly dispersed in the gel electrolyte, the ion migration channel could be formed along the aligned zwitterionic side groups within the PPDP gel electrolyte induced by the external electric field. To verify the aligned zwitterionic groups in gel electrolyte under the external electric field, the angular-dependent X-ray absorption near-edge spectroscopy (XANES) is employed to probe the orientation of the molecular bonds, taking advantage of the polarized nature of the synchrotron radiation sources[34]. Figure 1b shows the total electron yield XANES spectra measured at the normal director ($\theta = 90^o$) and $\theta = 45^o$ while the PPDP sample was applied a potential. The spectra were normalized to the incident photon flux and further processed by levelling the pre- and post-edges[35]. The sample contains a lot of the carbon bonds, such as C–C, C–N, C=O, C–H and so on. So, the C K-edge XANES spectra present rich absorption features between 282 and 295 eV. Therein, the two main features of the XANES spectra at ~287.3 and 287.9 eV present the significant angular dependence. For E-vector of the incident soft X-ray parallel to the surface of the substrate ($\theta = 90^o$), the XANES presents a strong feature at 287.3 eV and a weak one at 287.9 eV. While in the case of oblique incidence of the X-ray beam ($\theta = 45^o$), the former feature becomes weaker and the later becomes stronger. According to the reported literatures[36], these two main features can be assigned to the excitations of the C–H antibond orbital (C–H*) and the C=O π antibond orbital ($\pi^*_{C=O}$). It is known that the C–H* molecular orbital is along the direction of the C–H axis and $\pi^*_{C=O}$ orbital lies two sides of the C=O axis. Therefore, the angular dependences of both XANES features indicate that the C–H and C=O bonds should be parallel to the substrate plane when a potential was applied. Since the main chains of the sample are fixedly attached on the surface of the substrate, these significant orientations of the C–H and C=O bonds point out that the zwitterionic side groups of the sample become ordered (vertical to the external field) in the electrochemical process. This fact indicates that the zwitterionic group can be aligned in some extent by the external electric field, accompanied by the formation of an ion migration channel in the gel electrolyte at solid state.

Electrochemical performances of solid-state supercapacitor. The PPDP gel electrolyte with the ion migration channel and robust water retention ability significantly enhances the electrochemical performances of the as-fabricated graphene-based supercapacitor, as shown in Figs 2 and 3. To verify the superior electrochemical performances of the PPDP gel electrolyte, we conducted graphene-based symmetric supercapacitors and tested in a two-electrode configuration. Cyclic voltammetry (CV) was first studied with a potential range of 0–1.0 V at the scan rate of 10, 50, 100 and 400 mV s^{-1}. As shown in Fig. 2a–c and Supplementary Fig. 1, the nearly rectangular CV curves indicated nearly ideal capacitive behaviours of graphene electrodes. Note that the resulting rectangle areas from the PPDP gel electrolyte are substantially larger than those for the PVA gel electrolyte, revealing that specific capacitance values of graphene-based supercapacitors applying PPDP gel electrolyte are much higher than those of PVA gel electrolyte. Considering that there is no pseudocapacitive material in electrode, the capacitance enhancement is mainly due to the better EDLC behaviour. Moreover, galvanostatic charge–discharge (CD) measurements were performed at the current density of 0.8, 1, 2, 4, 8 and 20 A cm^{-3} (Fig. 2e,f; Supplementary Fig. 2). The discharging times of PPDP sample were 384.0 s for liquid state and 376.0 s for solid state at the current density of 0.8 A cm^{-3}, longer than those of PVA sample 250.4 s at liquid state and 242.0 s at solid state. The CD time of PPDP gel electrolyte sample continues to provide higher capacitance than that of PVA gel electrolyte sample. The as-fabricated graphene-based solid-state supercapacitor applying PPDP gel electrolyte reaches a volume capacitance of 300.8 F cm^{-3} at 0.8 A cm^{-3}, recording the best value among the previously reported graphene-based solid-state supercapacitors to the best of our knowledge.

Accordingly, the as-fabricated supercapacitors applying PPDP gel electrolyte yielded specific capacitances of 300.8, 298.2, 292.4, 279.6, 270.4 and 256.0 F cm^{-3} at current densities of 0.8, 1, 2, 4, 8 and 20 A cm^{-3} at solid state, respectively, obviously larger than those of the as-fabricated supercapacitors applying PVA gel electrolyte, as summarized in Fig. 3a. Meanwhile, the as-fabricated supercapacitor applying PPDP gel electrolyte exhibited better rate capacity, with only 14.9% capacitance loss when the current density increases by a factor of 25 (from 0.8 to 20 A cm^{-3}) compared with 23.6% loss for the as-fabricated supercapacitor applying PVA gel electrolyte. The PPDP gel electrolyte shows more advantages than the PVA gel electrolyte: (i) The capacitance loss of the PPDP gel electrolyte (1.4%) is much less than that of the PVA gel electrolyte (7.6%) after the electrolyte is transformed from liquid to solid state. (ii) The capacitance of as-fabricated supercapacitor applying PPDP gel electrolyte is 55.4% higher than the PVA gel electrolyte, as shown in Fig. 3b. The enhanced capacitance and rate capacity can be attributed to the unique property of PPDP gel electrolyte, which synergizes the effects of robust water retention ability and the ion migration channel with higher ion conductivity, and thus enhancing the capacitance, rate capacity and durability of the as-fabricated supercapacitor. In a word, applying the zwitterionic gel electrolyte, the graphene-based solid-state supercapacitor reaches a volume capacitance of 300.8 F cm^{-3} at 0.8 A cm^{-3} with a rate capacity of only 14.9% capacitance loss as the current density increases from 0.8 to 20 A cm^{-3}. Cycling performance is a critically important characteristic for evaluating the stability of solid-state supercapacitors. Owing to the same electrode material in our as-fabricated symmetric supercapacitors, cycling performance is largely attributed to the long-term stability of gel electrolytes (Supplementary Figs 3 and 4; Supplementary Note 1). We tested the 10,000 times of charging and discharging cycles at 4 A cm^{-3}. The PPDP gel electrolyte-based supercapacitor delivers a remarkable cycling stability of 103% retention up to 10,000 cycles, better than 91.8% retention of PVA gel electrolyte-based supercapacitor, as shown in Fig. 3c. These splendid electrochemical performances make such a kind of zwitterionic gel electrolyte attractive for fabricating the next generation of solid-state supercapacitors.

Mechanism of PPDP gel electrolyte. To explore the physical mechanism of PPDP gel electrolyte, static contact angle and viscoelastic properties were tested to verify its favourable wetting behaviour and reasonable mechanical strength that is vital for construction of graphene-based solid-state supercapacitors.

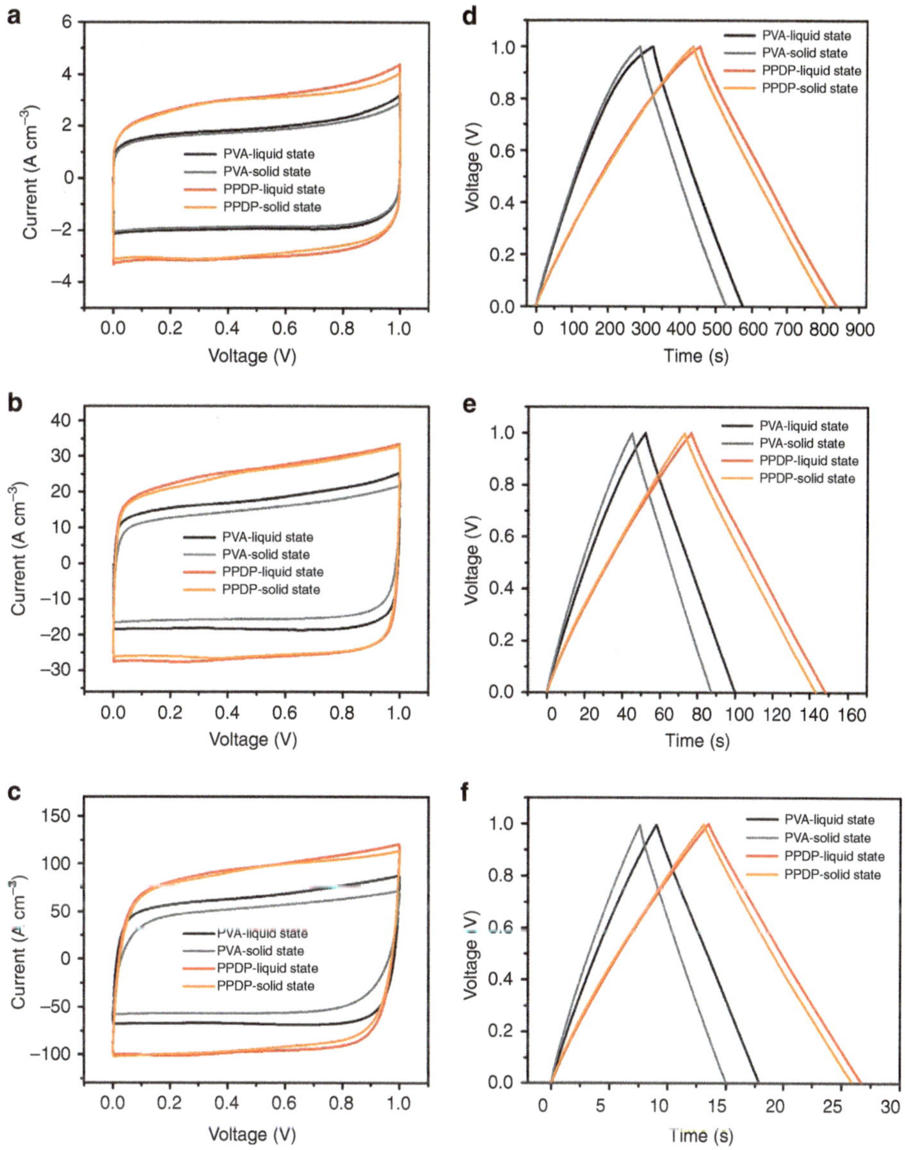

Figure 2 | Electrochemical performance of graphene-based supercapacitors. (**a–c**) CV curves of graphene-based solid-state supercapacitors applying PPDP and PVA gel electrolytes at liquid state and solid state, the scan rates were 10, 100 and 400 mV s^{-1}. (**d–f**) Galvanostatic charge-discharge curves of graphene-based solid-state supercapacitors applying PPDP and PVA electrolytes at liquid state and solid state at the current density of 0.8, 4 and 20 A cm^{-3}.

The affinity between the electrode and the gel electrolyte is crucial for supercapacitors because EDLC is strongly dependent on the surface contact areas between the electrode and the gel electrolyte. The wetting behaviour of the gel electrolytes on the graphene electrode is shown in Fig. 4a. For the initially prepared liquid-state electrolytes, the static contact angles are 72° and 61° for the PVA and PPDP electrolytes, respectively. The contact angles, respectively, decrease to 37° and 22° for the PVA and PPDP gel electrolytes after 24 h when the liquid state is transformed into the solid state due to evaporation. These facts suggest that the PPDP gel electrolyte can more effectively penetrate into the multilayer graphene electrodes and enhance the contact areas between the gel electrolyte and the working electrode compared with the PVA gel electrolyte as shown in the inset of Fig. 4a, due to the relatively low viscosity and high fluidity of the PPDP gel electrolyte than that of the PVA gel electrolyte (Supplementary Table 1). This characteristic of PPDP gel electrolyte would bring superior capacitance for solid-state supercapacitors. An ideal gel electrolyte not only can provide a superior electrochemical performance but also can serve as a binder with a certain mechanical strength to hold two electrodes together. Figure 4b shows the viscoelastic properties of the PPDP gel electrolyte. At the liquid state of PPDP electrolyte, the storage modulus (G') is smaller than the loss modulus (G''). In contrast, the G' dominates over G'' at the solid state, indicating that PPDP actually forms physically crosslinked gel due to the dipole–dipole interaction between the zwitterionic groups (Supplementary Fig. 5). The inter-chain hydrogen bonding between the amide parts of PPDP may also contribute to the formation of physical gel. Therefore, such a kind of gel with a storage modulus of 140 Pa provides enough mechanical strength to hold the two electrodes together, which can be supported from the formation of freestanding solid thin film by the PPDP gel electrolyte at a low water content as shown in the inset of Fig. 4b. Likewise, PVA gel electrolyte also exhibits similar viscoelastic properties at liquid and solid states (Supplementary Fig. 6; Supplementary Note 2).

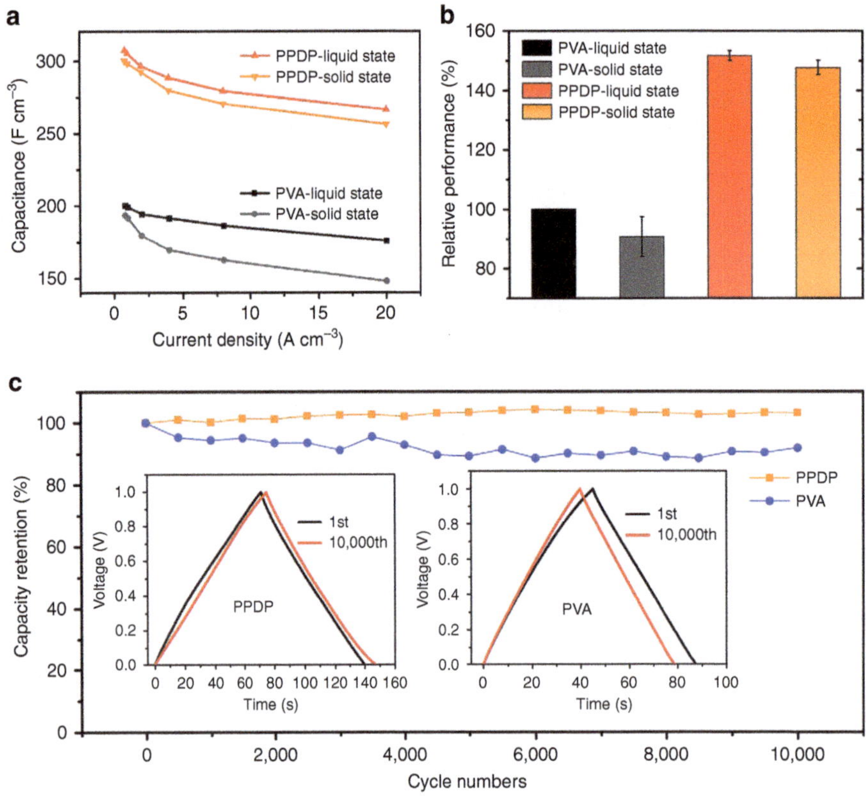

Figure 3 | Capacitance and cycling performance of graphene-based supercapacitors. (**a**) Comparison of specific capacitance values for graphene-based supercapacitors between PPDP and PVA gel electrolytes at different current densities. (**b**) Relative performance of PVA and PPDP electrolytes in liquid and solid states, where the error bars are obtained based on the capacitances at different current densities. (**c**) Cycling performance of graphene electrodes applying PPDP or PVA gel electrolyte at solid state. Inset: comparison of CD curves between the 1st CD cycle and the 10,000th cycle for graphene-based solid-state supercapacitors applying PPDP (left of panel) and PVA (right of panel) gel electrolyte at 4 A cm^{-3}.

To explore the chemical mechanism of PPDP gel electrolyte, differential scanning calorimetry (DSC) and electrochemical impedance spectroscopy were performed to verify the unique characteristics of robust water retention ability and efficient ion migration channel. The hydration capacity obtained from the DSC measurements demonstrates that PPDP has high water retention ability (Fig. 4c). Specifically, no endothermic peak can be observed in the thermogram during the heating of PPDP without water from −35 to 60 °C, suggesting that the polyzwitterion itself does not contribute to the thermal transition behaviour. Similarly, no thermal transition is observed during the heating process when the mole ratio of H_2O to PDP equals 6:1 and 7:1, indicating that all the water molecules around one zwitterionic group (about eight water molecules) in the hydration shell are tightly bound to the polyzwitterion through the electrostatic interactions with the positively charged $N^+(CH_3)_3$ and negatively charged SO_3^- and no freezable water exists in the system. However, an obvious endothermic peak is observed as the mole ratio of H_2O to PDP increases to 8:1, which means that the freezable water can be detected in the system when all binding sites of the polyzwitterion are saturated by water molecules[37]. Therefore, one PDP unit is strongly hydrated by seven to eight water molecules, as highlighted in Fig. 1. By contrast, Supplementary Fig. 7 and Supplementary Note 3 show that only one to two water molecules are tightly bound to one PVA monomeric unit via hydrogen bonding between the hydroxide group and the water molecules. Similar results are also observed in the low-field nuclear magnetic resonance (NMR) measurements (Supplementary Fig. 8; Supplementary Note 4). Consequently, the PPDP gel electrolyte gives rise to a higher water retention ability than PVA gel electrolyte at solid state (Supplementary Fig. 9; Supplementary Table 2; Supplementary Note 5). To confirm the efficient ion migration channel of the PPDP gel electrolyte, we performed electrochemical impedance spectroscopy tests using a two-electrode configuration from 100 mHz to 100 kHz. Usually, the intersection of the curve at the real part reflects the equivalent series resistance at the high frequency (100 kHz), which is contributed from the resistance of both electrolyte and electrode, while the linear region corresponds to the Warburg diffusion process (W), reflecting the ion diffusion into the electrode materials. As the Nyquist plots shown in Fig. 4d, the equivalent series resistance of PPDP-solid-state sample was 40.54 Ω, much smaller than that of PVA-solid-state sample of 122.6 Ω. Meanwhile, the slopes of PPDP gel electrolyte at liquid state and solid state are apparently larger than those of PVA samples at the low frequency, revealing that the transport and diffusion of ions into electrode materials of PPDP gel electrolyte are much better than those of PVA gel electrolyte, which answers for the superior electrochemical performance of the zwitterionic gel electrolyte.

The ion migration channel along with the robust water retention ability of PPDP gel electrolyte greatly facilitates the ion transport during charging and discharging process. To further evaluate the energy efficiency of the as-fabricated graphene-based solid-state supercapacitors applying PPDP gel electrolyte, energy density and power density were calculated from CD curves. The Ragone plot of the as-fabricated graphene-based solid-state supercapacitors was shown in Fig. 5a. The as-fabricated solid-state supercapacitor applying PPDP gel electrolyte delivers a high energy density of 41.78 Wh l^{-1} at a power density of

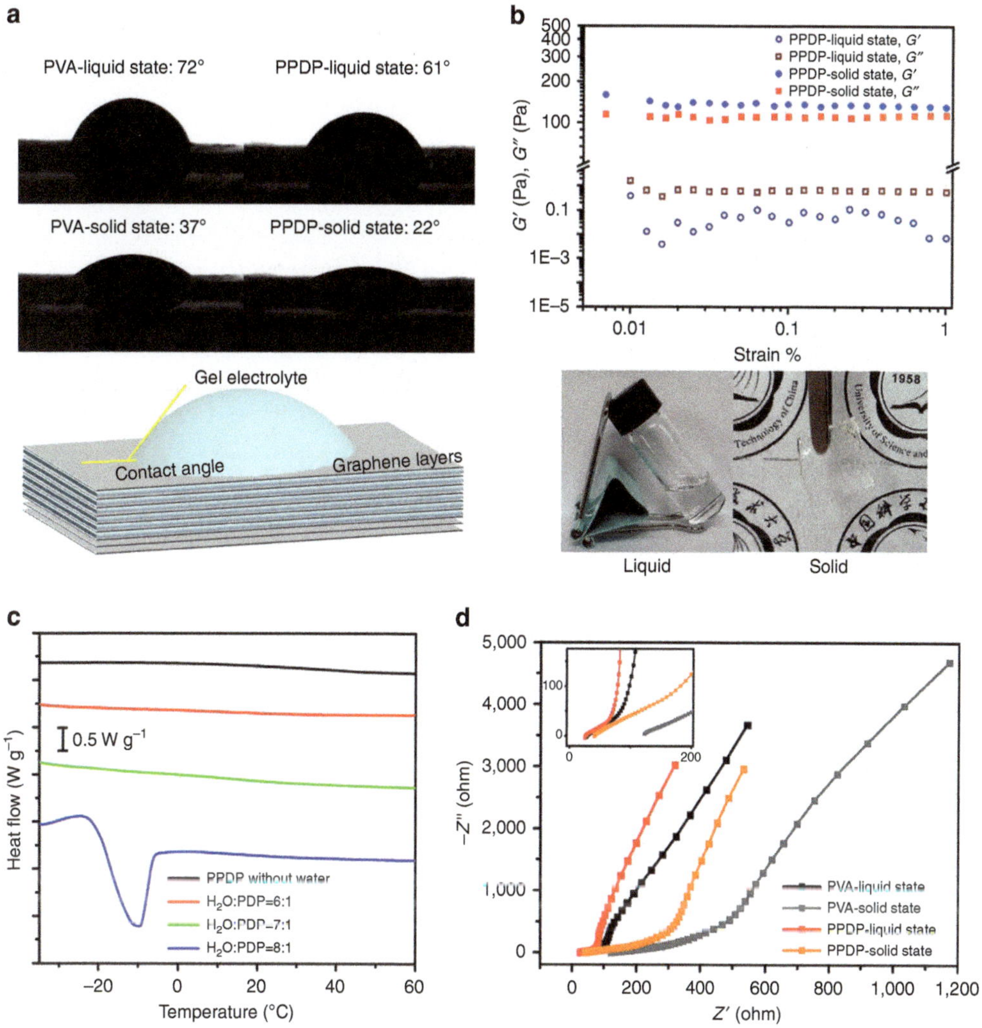

Figure 4 | Physical and chemical mechanisms of PPDP gel electrolyte. (**a**) Static contact angles of PVA and PPDP gel electrolytes on the graphene electrode. Inset is the schematic illustration of the penetration of gel electrolyte into the multilayer graphene electrode. (**b**) Viscoelastic properties of the PPDP gel electrolyte at liquid and solid states. PPDP gel electrolyte is demonstrated at liquid state in a vial and at solid state with a freestanding solid thin film. (**c**) DSC thermograms of PPDP at different water contents. (**d**) Electrochemical impedance spectroscopy (EIS) of graphene-based solid-state supercapacitors applying PPDP and PVA gel electrolytes at liquid state and solid state.

400 $W l^{-1}$, superior than 26.89 $Wh l^{-1}$ of the as-fabricated solid-state supercapacitor applying PVA gel electrolyte, which is also excellent among other previously reported state-of-the-art graphene-based solid-state supercapacitors[8,17,38], revealing that polyzwitterions especially PPDP would be a promising candidate to serve as gel electrolyte in solid-state supercapacitors. Besides, the feasibility of the as-fabricated solid-state supercapacitors is demonstrated in Fig. 5b. Generally, an effective way to enhance the potential or energy density can be achieved using serial and parallel assemblies. Compared with single supercapacitor, two as-fabricated supercapacitors in series show twice potential of 2.0 V, and two as-fabricated supercapacitors in parallel show almost twice CD time, indicating that the as-fabricated solid-state supercapacitor has good practical value. Meanwhile, there is an only slight difference in CV curves of the as-fabricated solid-state supercapacitor between normal and under bent status (90°), indicating that PPDP gel electrolyte has a certain mechanical strength with good flexibility and stability (Fig. 5c), which is suitable for establishing ultraflexible and even transparent energy storage devices and other electronics. The similar test about ionic conductivity of the PPDP gel electrolyte under different bending states is shown in Supplementary Fig. 10 and Supplementary Note 6. In addition, our proposed zwitterionic PPDP gel electrolyte also has general applicability for solid-state supercapacitors under acidic and basic conditions (Supplementary Fig. 11; Supplementary Note 7). Furthermore, the PPDP gel electrolyte has better self-discharging performance than that of the PVA gel electrolyte (Supplementary Fig. 12; Supplementary Note 8).

Discussion

Zwitterionic gel electrolyte has been proven to be a promising gel electrolyte catering for solid-state supercapacitors. The zwitterionic nature of PPDP as gel electrolyte not only offers robust water retention ability but also brings ion migration channels to the electrolyte ions, leading to superior electrochemical performances. When applying PPDP as a gel electrolyte, the as-fabricated graphene-based solid-state supercapacitor reaches a volume capacitance of 300.8 $F cm^{-3}$ at 0.8 $A cm^{-3}$, with a rate capacity of only 14.9% capacitance loss when the current density increases from 0.8 to 20 $A cm^{-3}$, recording the best value among the previously reported graphene-based solid-state supercapacitors, to the best of our knowledge. Moreover, the high water retention ability of the zwitterionic

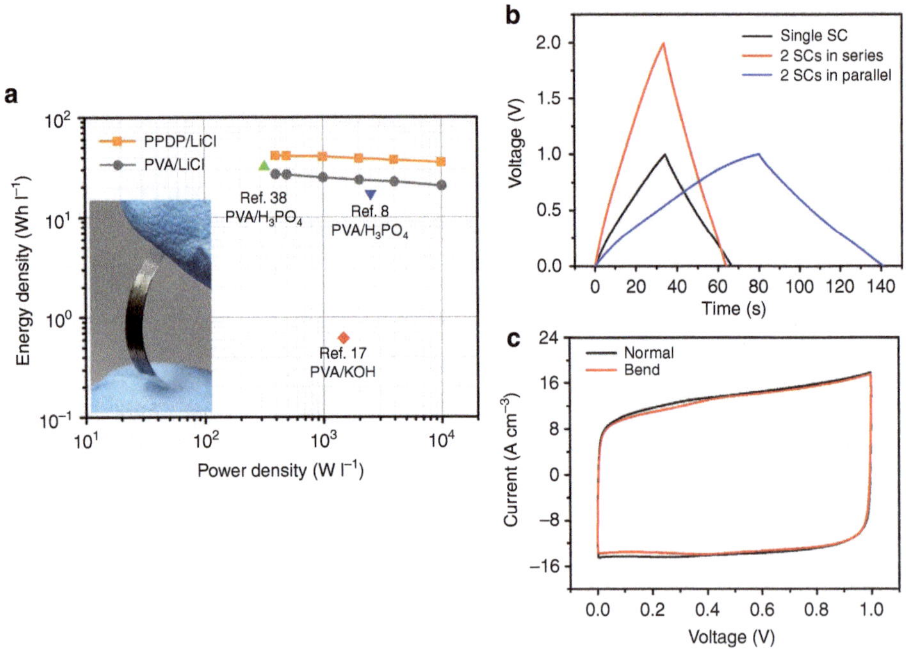

Figure 5 | Applications of PPDP gel electrolyte. (**a**) A typical Ragone plot of the as-fabricated graphene-based solid-state supercapacitors. Inset is a prototype of graphene-based solid-state supercapacitor applying PPDP gel electrolyte with a planar configuration. (**b**) Galvanostatic charge-discharge curves of single supercapacitor (SC) and two SCs connected in series and in parallel at 8 A cm^{-3}. (**c**) Comparison of CV curves between normal and bent status (90°), the scan rate is 50 mV s^{-1}.

groups brings a robust cyclability, the capacitance remaining 103% retention after undergoing 10,000 CD cycles. We anticipate that zwitterionic gel electrolyte will be a promising gel electrolyte for the next generation of solid-state supercapacitors.

Methods

Preparation of gel electrolytes. The PVA/LiCl gel electrolyte was synthesized according to the procedure reported in previous literature[20]. An amount of 4.0 g PVA powder was put into 40 ml distilled water with stirring at 95 °C. After the PVA powder was completely dissolved, 4.0 g LiCl•H_2O was added into the solution under vigorous stirring. When PVA/LiCl turned to be transparent and clear gel, it was cooled down to room temperature and the gel electrolyte was successfully prepared. The PPDP/LiCl gel electrolyte was prepared as follows. An amount of 1.0 g propylsulfonate dimethylammonium propylmethacrylamide (PDP), 1.0 mg 4,4′-azobis(4-cyanovaleric acid) (ACVA) and 4.0 ml distilled water were added into a round-bottom reactor. After three times of frozen–degassing–thawing cycles, the free-radical polymerization was conducted at 70 °C for 10 h. The polymerization is completed and cooled down to room temperature. Afterwards, the as-prepared PPDP was mixed homogeneously with 6.0 ml aqueous solution including 1.0 g LiCl•H_2O under vigorous stirring to prepare the zwitterionic gel electrolyte, PPDP physical gel was successfully prepared.

To compare the electrochemical performances between PVA and PPDP gel electrolytes, the optimal initial inventory mass ratio of H_2O:polymer:LiCl•H_2O is fixed at 10:1:1 for the both two gel electrolytes as the concentration of LiCl•H_2O has important influences on the physical state, ionic conductivity and specific capacitance of the gel electrolytes (Supplementary Figs 13 and 14; Supplementary Notes 9 and 10). And the link between ionic conductivity of gel electrolyte and the capacitance of solid-state supercapacitors is discussed in Supplementary Fig. 15 and Supplementary Note 11. The role of LiCl in enhancing the water retention of gel electrolytes is clarified in Supplementary Fig. 16 and Supplementary Note 12.

Fabrication of graphene electrodes and solid-state supercapacitors. Graphene was obtained from the chemical reduction of graphene oxides. Graphene oxide was synthesized in a modified Hummer's method[39]. Then, graphene was reduced using hydrazine according to a reported method[40]. The graphene thin film on the surface of cellulose acetate membrane was obtained by a vacuum filtration process, followed by tightly pressing thin film onto polyethylene terephthalate (PET) substrate and slowly peeling off the cellulose acetate membrane, leading to freestanding graphene thin film on PET. Then, the graphene thin film was scraped along the channel to fabricate working electrodes, while two columns of Au current collectors were thermally evaporated on each side of the working electrodes. Finally, the solid-state supercapacitor was finally established after coating PPDP/LiCl gel electrolyte on the parallel region filling the channel between two working electrodes. The working electrodes were prepared with the same thickness and the surface areas of graphene as well as the gel electrolyte were applied onto the graphene with the same amounts for PVA and PPDP gel electrolytes, so that the electrochemical behaviour between PVA and PPDP gel electrolytes can be compared fairly. After the gel electrolyte became solid state, the graphene-based solid-state supercapacitor was fabricated. The volume of the graphene electrodes was calculated through multiplying the thickness by areas. The exact thickness of the graphene electrodes was ~250 nm determined by the field-emission scanning electron microscopy on a FEI Sirion-200 SEM instrument (Supplementary Fig. 17). The distance between the two electrodes is ~150 μm (Supplementary Fig. 18) and the exact surface area of each graphene electrode is 10 × 1 mm (Supplementary Fig. 19; Supplementary Note 13).

Electrochemical tests. Electrochemical performances of the graphene-based supercapacitors were measured in a two-electrode system by CV and galvanostatic CD at an electrochemical station (CHI 760E). Potential range set for CV and CD tests was from 0 to 1.0 V. The as-fabricated graphene electrodes were directly used as the working electrodes for the electrochemical tests. Electrochemical impedance measurements of the supercapacitors were performed in the same configuration from 100 mHz to 100 kHz with a Zahner IM6 electrochemical workstation. The cycling stability of the as-fabricated supercapacitors was performed at room temperature with a sweep charge and discharge rate at the current density of 4 A cm^{-3} for 10,000 cycles.

Other characterizations. Proton NMR (^1H NMR) spectra were recorded on a Bruker AV400 spectrometer, using deuterium water as the solvent and tetramethylsilane as an internal standard. The ^1H NMR spectra of PPDP and PVA were shown in Supplementary Fig. 20 and Supplementary Note 14. The number-average molar mass ($M_n = 1.87 \times 10^5$ g mol^{-1}, $M_w/M_n \sim 1.6$) of PPDP was determined by size-exclusion chromatography (Waters 1,515), using monodisperse poly(ethylene glycol) as the standard and a 1.0 M NH_4NO_3 solution as the eluent with a flow rate of 1.0 ml min^{-1}. The viscoelastic property of PPDP gel electrolyte was determined by a rheological measurement on a TA AR-G2 rheometer. The storage modulus (G') and loss modulus (G'') were measured as a function of strain with 1 Hz frequency at 25 °C. The DSC scans were conducted on a TA Instruments Q2000 scanning calorimeter at a heating rate of 5 °C min^{-1}. The angular-dependent C K-edge XANES spectra, using the linearly polarized X-ray beam, were measured at the beamline BL12B of National Synchrotron Radiation Laboratory (NSRL, Hefei) in the total electron yield mode by collecting the sample drain current under a vacuum better than 10^{-7} Pa. The beam from the bending magnet was monochromatized utilizing a varied line-spacing plane grating and refocused by a toroidal mirror. The energy range is 100–1,000 eV with an energy resolution of ca. 0.2 eV. The PPDP sample was applied CV from 0 to 1.0 V for 1 h at the scan rate of 10 mV s^{-1} and underwent freeze-drying process before the measurement of XANES spectra. Low-field NMR measurements were performed by Bruker Minispec MQ-20 NMR analyser operating at a resonance frequency of 20 MHz (0.47 T). Sample was inserted in the NMR probe. Spin–Spin relaxation time, T2, was measured using the Carr-Purcell-Meiboom-Gill (CPMG) sequence.

References

1. Simon, P., Gogotsi, Y. & Dunn, B. Where do batteries end and supercapacitors begin? *Science* **343,** 1210–1211 (2014).
2. Simon, P. & Gogotsi, Y. Materials for electrochemical capacitors. *Nat. Mater.* **7,** 845–854 (2008).
3. Peng, X., Peng, L., Wu, C. & Xie, Y. Two dimensional nanomaterials for flexible supercapacitors. *Chem. Soc. Rev.* **43,** 3303–3323 (2014).
4. Lu, X., Yu, M., Wang, G., Tong, Y. & Li, Y. Flexible solid-state supercapacitors: design, fabrication and applications. *Energy Environ. Sci.* **7,** 2160–2181 (2014).
5. Sun, G. *et al.* Hybrid fibers made of molybdenum disulfide, reduced graphene oxide, and multi-walled carbon nanotubes for solid-state, flexible, asymmetric supercapacitors. *Angew. Chem. Int. Ed.* **127,** 4734–4739 (2015).
6. Li, L. *et al.* A flexible quasi-solid-state asymmetric electrochemical capacitor based on hierarchical porous V_2O_5 nanosheets on carbon nanofibers. *Adv. Energy Mater.* **5,** 1500753 (2015).
7. Feng, J. *et al.* Metallic few-layered VS_2 ultrathin nanosheets: high two-dimensional conductivity for in-plane supercapacitors. *J. Am. Chem. Soc.* **133,** 17832–17838 (2011).
8. Peng, L. *et al.* Ultrathin two-dimensional MnO_2/graphene hybrid nanostructures for high-performance, flexible planar supercapacitors. *Nano Lett.* **13,** 2151–2157 (2013).
9. Liu, Z., Xu, J., Chen, D. & Shen, G. Flexible electronics based on inorganic nanowires. *Chem. Soc. Rev.* **44,** 161–192 (2015).
10. Wang, X. *et al.* Flexible energy-storage devices: design consideration and recent progress. *Adv. Mater.* **26,** 4763–4782 (2014).
11. Kang, Y. *et al.* All-solid-state flexible supercapacitors fabricated with bacterial nanocellslose papers, carbon nanotubes, and triblock-copolymer ion gels. *ACS Nano* **6,** 6400–6406 (2012).
12. Kang, Y., Chung, H., Han, C. & Kim, W. All-solid-state flexible supercapacitors based on papers coated with carbon nanotubes and ionic-based gel electrolytes. *Nanotechnology* **23,** 065401 (2012).
13. Lee, J., Kim, W. & Kim, W. Stretchable carbon nanotube/ion-gel supercapacitors with high durability realized through interfacial microroughness. *ACS Appl. Mater. Interface* **6,** 13578–13586 (2014).
14. Kim, D., Shin, G., Kang, Y., Kim, W. & Ha, J. Fabrication of a stretchable solid-state micro-supercapacitor array. *ACS Nano* **7,** 7975–7982 (2013).
15. Cao, X. *et al.* Reduced graphene oxide-wrapped MoO_3 composites prepared by using metal–organic frameworks as precursor for all-solid-state flexible supercapacitors. *Adv. Mater.* **27,** 4695–4701 (2015).
16. Wang, G. *et al.* Solid-state supercapacitor based on activated carbon cloths exhibits excellent rate capability. *Adv. Mater.* **26,** 2676–2682 (2014).
17. Wang, X. *et al.* Fiber-based flexible all-solid-state asymmetric supercapacitors for integrated photodetecting system. *Angew. Chem. Int. Ed.* **53,** 1849–1853 (2014).
18. Xie, J. *et al.* Layer-by-layer β-$Ni(OH)_2$/graphene nanohybrids for ultraflexible all-solid-state thin-film supercapacitors with high electrochemical performance. *Nano Energy* **2,** 65–74 (2013).
19. Zhu, C. *et al.* All metal nitrides solid-state asymmetric supercapacitors. *Adv. Mater.* **27,** 4566–4571 (2015).
20. Wu, C. *et al.* Two-dimensional vanadyl phosphate ultrathin nanosheets for high energy density and flexible pseudocapacitors. *Nat. Commun.* **4,** 2431 (2013).
21. Wang, G. *et al.* LiCl/PVA gel electrolyte stabilizes vanadium oxide nanowire electrodes for pseudocapacitors. *ACS Nano* **6,** 10296–10302 (2012).
22. Kudaibergenov, S., Jaeger, W. & Laschewsky, A. in *Supramolecular Polymers Polymeric Betains Oligomers* **201,** 157–224 (Springer (2006).
23. Lalani, R. & Liu, L. Electrospun zwitterionic poly(sulfobetaine methacrylate) for nonadherent, superabsorbent, and antimicrobial wound dressing applications. *Biomacromolecules* **13,** 1853–1863 (2012).
24. Tiyapiboonchaiya, C. *et al.* The zwitterion effect in high-conductivity polyelectrolyte materials. *Nat. Mater.* **3,** 29–32 (2004).
25. Cardoso, J., Huanosta, A. & Manero, O. Ionic conductivity studies on salt-polyzwitterion systems. *Macromolecules* **24,** 2890–2895 (1991).
26. Vasantha, V. A., Jana, S., Parthiban, A. & Vancso, J. G. Water swelling, brine soluble imidazole based zwitterionic polymers - synthesis and study of reversible UCST behaviour and gel-sol transitions. *Chem. Commun.* **50,** 46–48 (2014).
27. Lowe, A. B. & McCormick, C. L. Synthesis and solution properties of zwitterionic polymers. *Chem. Rev.* **102,** 4177–4190 (2002).
28. Wang, T., Wang, X., Long, Y., Liu, G. & Zhang, G. Ion-specific conformational behavior of polyzwitterionic brushes: exploiting it for protein adsorption/desorption control. *Langmuir* **29,** 6588–6596 (2013).
29. Narita, A. *et al.* Lithium ion conduction in an organoborate zwitterion-LiTFSI mixture. *Chem. Commun.* 1926–1928 (2006).
30. Matsumi, N., Sugai, K., Miyake, M. & Ohno, H. Polymerized ionic liquid via hydroboration polymerization as single ion conductive polymer electrolytes. *Macromolecules* **39,** 6924–6927 (2006).
31. Brown, R. H. *et al.* Effect of ionic liquid on mechanical properties and morphology of zwitterionic copolymer menbranes. *Macromolecules* **43,** 790–796 (2010).
32. Yoshizawa, M., Hirao, M., Ito-Akita, K. & Ohno, H. Ion conduction in zwitterionic-type molten salts and their polymers. *J. Mater. Chem.* **11,** 1057–1062 (2001).
33. Zhou, T., Gao, X., Dong, B., Sun, N. & Zheng, L. Poly(ionic liquid) hydrogels exhibiting superior mechanical and electrochemical properties as flexible electrolytes. *J. Mater. Chem. A* **4,** 1112–1118 (2016).
34. Stöhr, J. NEXAFS spectroscopy. *Springer Ser. Surf. Sci.* **392,** 276–290 (1992).
35. Zhong, J. *et al.* Bio-nano interaction of proteins adsorbed on single-walled carbon nanotubes. *Carbon* **47,** 967–973 (2009).
36. Li, Z., Zhang, L., Resasco, D., Mun, B. & Requejo, F. Angle-resolved X-ray absorption near edge structure study of vertically aligned single-walled carbon nanotubes. *Appl. Phys. Lett.* **90,** 103115 (2007).
37. Wu, J., Lin, W., Wang, Z., Chen, S. & Chang, Y. Investigation of the hydration of nonfouling material poly(sulfobetaine methacrylate) by low-field nuclear magnetic resonance. *Langmuir* **28,** 7436–7441 (2012).
38. Niu, Z. *et al.* All-solid-state flexible ultrathin micro-supercapacitors based on graphene. *Adv. Mater.* **25,** 4035–4042 (2013).
39. Marcano, D. *et al.* Improved synthesis of graphene oxide. *ACS Nano* **4,** 4806–4814 (2010).
40. Park, S. *et al.* Hydrazine-reduction of graphite- and graphene oxide. *Carbon* **49,** 3019–3023 (2011).

Acknowledgements

This work was financially supported by the National Basic Research Program of China (2015CB932302 and 2012CB933802), the National Natural Science Foundation of China (21222101, U1432133, U1532265, 21331005, 11321503, J1030412, 21374110, 21574121, 21234003 and 91127042), National Program for Support of Top-notch Young Professionals the Chinese Academy of Sciences (XDB01020300), the Fok Ying-Tong Education Foundation, China (grant no.141042), Users with Excellence of Hefei Science Center (2015HSC-UE008) and the Fundamental Research Funds for the Central Universities (WK2060190027 and WK2340000066). We thank Professor Wangsheng Chu and Professor Wensheng Yan in National Synchrotron Radiation Laboratory, University of Science and Technology of China for assistance and discussion with the sample measurement by angular-dependent XANES test.

Author contributions

C.W. conceived the idea and experimentally realized the study, co-wrote the paper and supervised the project and is responsible for the infrastructure and project direction. X.P. and H.L. experimentally realized the idea and wrote the paper; J.W. and P.C. carried out the examples synthesis, characterization and data analysis. G.L. and G.Z. are responsible for the design of zwitterionic gel electrolyte and co-wrote the paper. Y.X. supervised the whole experimental procedure and co-wrote the paper. All the authors discussed the results, commented on and revised the manuscript. X.P. and H.L. contributed equally to this research.

Additional information

Competing financial interests: The authors declare no competing financial interests.

19

Atomically ordered solute segregation behaviour in an oxide grain boundary

Bin Feng[1], Tatsuya Yokoi[2], Akihito Kumamoto[1], Masato Yoshiya[2,3], Yuichi Ikuhara[1,3,4] & Naoya Shibata[1]

Grain boundary segregation is a critical issue in materials science because it determines the properties of individual grain boundaries and thus governs the macroscopic properties of materials. Recent progress in electron microscopy has greatly improved our understanding of grain boundary segregation phenomena down to atomistic dimensions, but solute segregation is still extremely challenging to experimentally identify at the atomic scale. Here, we report direct observations of atomic-scale yttrium solute segregation behaviours in an yttria-stabilized-zirconia grain boundary using atomic-resolution energy-dispersive X-ray spectroscopy analysis. We found that yttrium solute atoms preferentially segregate to specific atomic sites at the core of the grain boundary, forming a unique chemically-ordered structure across the grain boundary.

[1] Institute of Engineering Innovation, The University of Tokyo, Tokyo 113-8656, Japan. [2] Department of Adaptive Machine System, Osaka University, Osaka 565-0871, Japan. [3] Nanostructures Research Laboratory, Japan Fine Ceramics Center, Nagoya 456-8587, Japan. [4] WPI advanced Institute for Materials Research, Tohoku University, Sendai 980-8577, Japan. Correspondence and requests for materials should be addressed to N.S. (email: shibata@sigma.t.u-tokyo.ac.jp).

Grain boundary (GB) segregation has been widely studied due to its importance in technological applications[1–11]. The enrichment of dopant atoms into GBs markedly affects the local structure, chemistry and thus governing overall properties of the materials[1–5,7–10]. Controlling GB segregation thus offers a promising strategy for materials engineering. To this end, an understanding towards the fundamental mechanisms of GB segregation is crucially important. Based on the Gibbs adsorption isotherm in thermodynamics, phenomenological understanding of GB segregation has been well established[12]. By further applying statistical mechanics with the consideration of site change energy between dopant and host atoms in the vicinity of the GB sites, the McLean model provided a general insight into GB segregation phenomena[12]. Although these theories give a general concept of GB segregation, atomistic mechanisms of how dopant atoms actually segregate to complex GB structures is still not fully clarified. This problem severely limits our ability to control GB segregation and resultant properties of many technologically important materials.

Recently, it has become possible to directly visualize the 'non-soluble' impurity atoms segregated at GBs by using atomic-resolution scanning transmission electron microscopy (STEM) and spectroscopy[1–3,6,10,13–16]. The atomistic site-specific manner of impurity atom segregation has been proved for several ceramics, establishing a clear relationship between local GB non-solute segregation and resultant physical properties[1,3,10]. However, on the GB solute segregation, in which the dopant atoms can reside both at grain boundaries and within grain interiors, it has been extremely difficult to directly observe its structure and chemistry at atomic dimensions. At the same time, STEM–energy-dispersive X-ray spectroscopy (EDS) capabilities have steadily progressed. Ultrasensitive silicon drift detectors (SDDs) with much higher count rates allow the use of fine STEM probe, so that high quality, quantitative atomic-scale EDS mapping is now becoming possible[17,18]. This method may give us a chance to directly probe GB solute segregation phenomena by revealing preferred GB segregation sites of solute atoms.

Here, we report an atomic-scale STEM–EDS mapping of a GB in yttria-stabilized zirconia (YSZ), which is one of the most important materials for electrolytes in solid oxide fuel cells[19]. In this study, a symmetric tilt GB with $\Sigma3[110]/\{111\}$ orientation is selected as an initial attempt. It has been confirmed that Y atoms do segregate to this GB, but its structure is very coherent and simple[20], which is ideal for TEM observations and theoretical calculations. Contrary to the previous models of continuous dopant solute segregation in GB, we show that Y segregation behaviour is atomic-site specific at the core of the GB. The detailed segregation process will be discussed in combination with results from large-scale Monte Carlo (MC) simulations.

Results

Characterization of site-dependent GB solute segregation. Figure 1a shows a high angle annular dark-field (HAADF) STEM image of the $\Sigma3[110]/\{111\}$ GB. The atomic column intensity in HAADF STEM image is approximately proportional to Z^2 (Z represents the atomic number)[21], so the bright contrast in the image represents the cation columns intermixed with Zr ($Z=40$) and Y ($Z=39$) atoms along this viewing direction. The GB atomic arrangement is in good agreement with the previous high-resolution TEM study[20]. However, it is impossible to distinguish Zr and Y atoms only from the HAADF STEM image.

To selectively observe Y atoms at the GB, a series of STEM–EDS mapping were performed. The low magnification STEM–EDS mapping revealed that Y atoms are homogeneously distributed inside the bulk regions, while they are enriched (Zr atoms decrease) at the GB vicinity (see Supplementary Fig. 1). Atomic-resolution STEM–EDS mapping on the bulk region is shown in Fig. 1b (Zr–K line) and Fig. 1c (Y–K line). The peak Y–K intensities are always found on the Zr atomic columns, and are almost uniform throughout the bulk region. This result indicates that the Y atoms homogeneously substitute the Zr atomic sites in the bulk regions.

Figure 2a,b shows atomic-resolution STEM–EDS mappings around the GB core. Contrary to the bulk regions, the Y and Zr maps clearly reveal the formation of characteristic chemically ordered segregation structure. To better describe the segregation structure, we denote the cation planes as follows: cation mirror plane to be M_c, the right-hand side of the M_c plane to be R_n planes and the left-hand side to be L_n planes. Here, n is the sequential number as shown in the bottom of Fig. 2c,d. For the Y map, an apparent intensity increase can be found on the L_1 and R_1 planes, while for the Zr map, intensities decrease at these two planes. The intensity variation is more highlighted in the corresponding intensity profile normalized by the total Zr and Y counts, shown in Fig. 2c,d. These figures also visualize the subtle intensity variation across the GB. For the Y intensity line profile in Fig. 2d, slight increases in the L_3, L_5 and R_3 planes were found, while clear decreases in the M_C, L_2 and R_2 planes and slight intensity decreases in the L_4 and R_4 plane intensities were noted. On the other hand, for the Zr intensity line profile in Fig. 2c, an increase in the M_C, L_2, R_2, L_4 and R_4 plane intensities can be recognized, while intensity decreases in the L_1, R_1, L_3, R_3 and L_5 plane were evident. For the L_6, R_5 and R_6 planes, both the Zr and Y intensities are almost constant, indicating the GB segregation structure extends as far as five or six layers adjacent to the M_C plane. Note that the Y segregation is slightly stronger in the left-hand side of the M_c plane.

The present atomic-resolution STEM–EDS mappings clearly reveal that Y segregation formed chemically ordered structure across the GB within the range of about 3 nm. Surprisingly, Y atoms are obviously depleted in some atomic column layers (that is, M_c, L_2 and R_2). This indicates that Y atoms do not simply

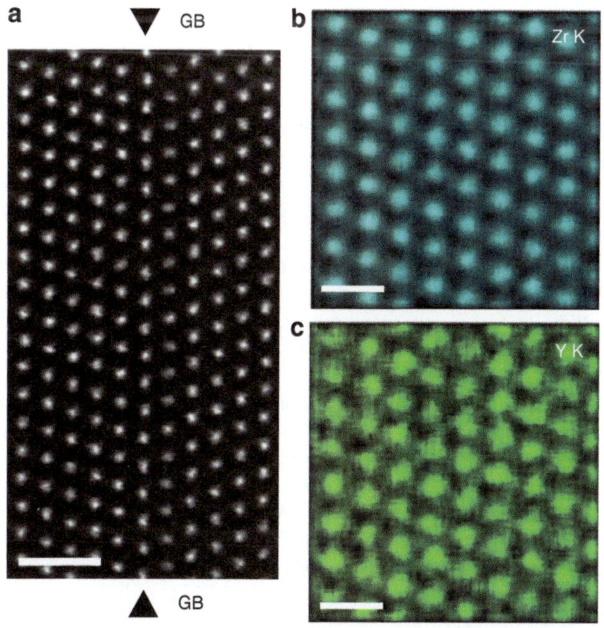

Figure 1 | Atomic structure of the grain boundary. (**a**) HAADF-STEM image of the GB. (**b,c**) Atomic resolution EDS elemental maps for (**b**) Zr K map, (**c**) Y K maps in the bulk region of the YSZ along [110] direction. Scale bar is 1 nm in (**a**) and 0.5 nm in (**b,c**).

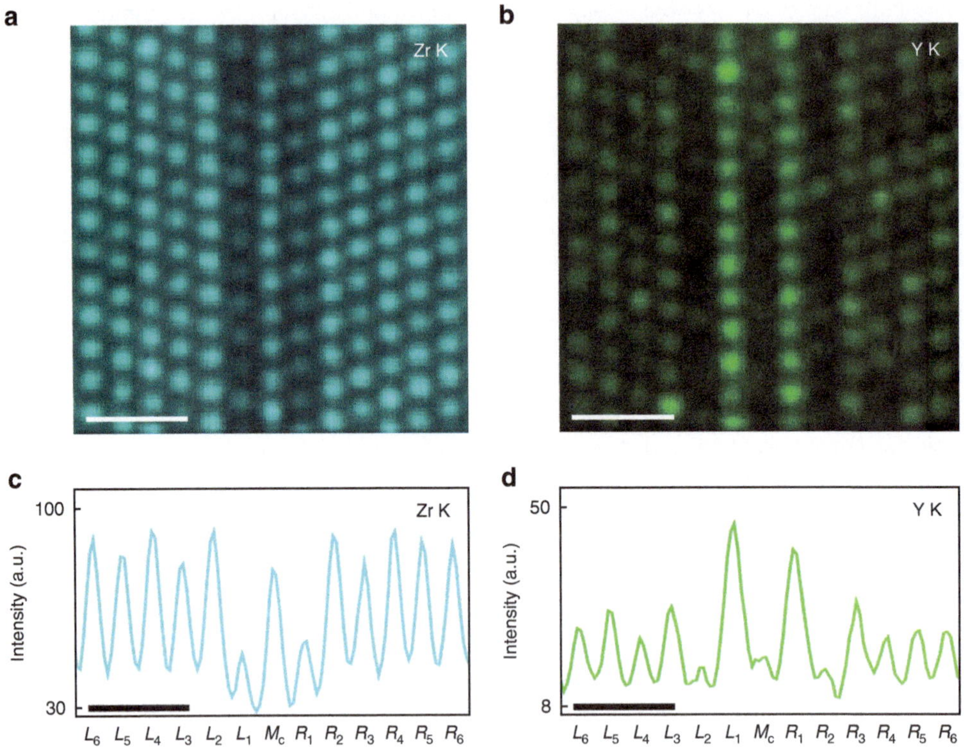

Figure 2 | Energy-dispersive X-ray spectroscopy maps and intensity profiles. (**a**,**b**) EDS elemental maps for (**a**) Zr K map and (**b**) Y K map. (**c**,**d**) Intensity profile (normalized by the total Zr and Y counts) by summing the X-ray counts in the maps in the direction parallel to the GB for (**c**) Zr K and (**d**) Y K. To show the subtle chemical variation more clearly, different scales are used for the two intensity profiles. Atomic layer number is listed in the bottom part of each figure. Scale bar is 1 nm.

substitute all the cation sites around the GB. The unique site-dependent segregation pattern can be summarized as follows. Y atoms strongly segregate to the {111} atomic planes of L_1 and R_1, and slightly to the L_n and R_n plane with odd number n. On the other hand, Y atoms are depleted in the M_C plane and the L_n and R_n planes with even number n.

Theoretical prediction of GB solute segregation behaviour. In order to understand the observed characteristic segregation structure, systematic MC simulations combined with static lattice calculations were performed. Figure 3 shows the predicted equilibrium GB structure with Y segregation. The site occupancies of Zr and Y atoms across the GB are also shown in Fig. 3b,c, respectively. The chemically ordered segregation structure is well reproduced in the simulation. Zr atoms are almost fully substituted by Y atoms in the R_1 and L_1 planes and partially in the L_3 and R_3 planes. The chemical ordering almost disappears in the L_6 and R_6 planes. It is worth mentioning that, similar to the experimental result, the segregation level on the left-hand side of the GB is predicted to be slightly higher than the right-hand side.

Based on the above simulation, we try to reveal the atomic-scale segregation processes. Here, GB atomic configurations are successively evaluated by the MC simulations varying Y concentrations from the lower to higher doping level (see Supplementary Fig. 2 and Supplementary Note 1). Figure 4 shows energetically favorable atomic configuration of the GB with different Y concentrations. Figure 4a shows the pristine model. It is seen that the plane distance between the M_C and L_1 planes is slightly wider than other plane distances. This is because Zr sites in the L_1 plane are forced to have sevenfold coordination in this GB, while all other Zr sites are in eightfold coordination, resulting in the slight asymmetric GB structure[20]. At the initial stage of segregation modeled in Fig. 4b, Y ions preferentially occupy the L_1 plane. This 'monolayer' segregation structure can be related to the bond distortions at the L_1 plane sites due to the structurally formed sevenfold coordination sites. The occupation of Y ions at these sites effectively relieves total bond distortion and thus internal energy of the whole system. This way of site-dependent segregation is also reported in the previous studies of non-solute segregations[1–3,10,13,22].

However, the present case involves more complicated factors for the segregation, namely the interactions between Y and O vacancies (V_O), as discussed below. We labeled anion planes as m_a, r_n and l_n (shown in the top of Fig. 4a) following the same manner as in the cation planes, where the anion mirror plane m_a is located in between M_c and L_1 planes. As preferential Y segregation to the L_1 plane proceeds, V_O segregate to the r_1 plane, which are located in the second-nearest neighbor sites relative to the Y ions in the L_1 plane. According to the previous studies, this geometric relationship between Y ions and V_O is energetically favorable in the bulk of YSZ[23,24]. Therefore, the point defects interactions are likely to have significant effects on determining the preferred atomic arrangement at the GB. The accumulation of Y and V_O causes larger oxygen bond distortion in the r_2 plane (Fig. 4b), which further accelerates segregation of both Y (on R_1) and V_O (on r_2) in an ordered manner. V_O preferentially segregates to the r_1 and r_2 planes, and consequently the two oxygen planes merged into form one oxygen plane (Fig. 4c). Then, the V_O also start to segregate towards the bulk part of l_2 and r_5 sites at the same time, which further induces the V_O segregation to the l_3 and r_6 sites and Y segregation to the L_3 and R_3 sites, forming the second-nearest neighbor pair Y–Vo configuration. In this way, Y atoms continue to segregate (Fig. 4d) and finally the GB reaches

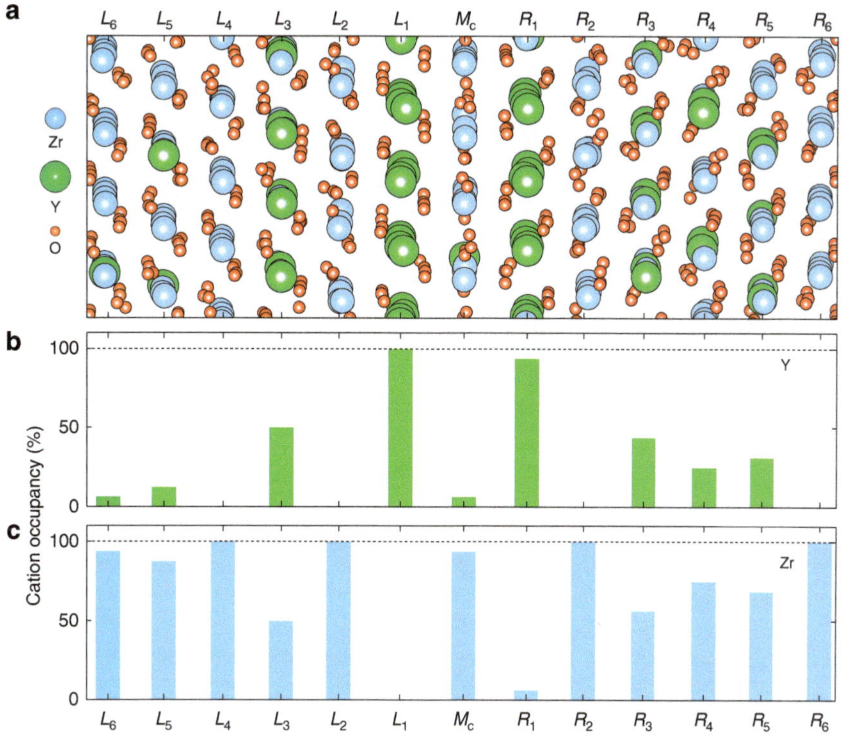

Figure 3 | Theoretical atomic structure at the grain boundary. (**a**) Atomic configuration of Y segregated GB predicted by theoretical calculations, along [110] direction. (**b**,**c**) atomic occupancy calculated from the predicted model structure for (**b**) Y atoms and (**c**) Zr atoms.

the unique segregation structures shown in Fig. 3. Note that we have also experimentally confirmed the presence of V_O in the GB by EDS (see Supplementary Fig. 3), which further support our conclusions.

Discussion

The present results provide atomic-scale insights into the GB solute segregation phenomena. Our study clearly reveals preferred segregation sites at the GB core even in the case of solute segregation. The interaction between Y and V_O governs the subsequent segregation behaviour. The equilibrium segregation distribution is thus determined by these defect interactions coupled with GB geometries. It is worth mentioning that chemically ordered segregation structure is similar to a phase of $Y_2Zr_2O_7$ oxides with pyrochlore structure[25,26]. This indicates that solute segregation might induce a phase transition in the vicinity of GB. Previous studies suggest that extremely severe synthesis conditions such as high-pressure are indispensable for forming ordered $Y_2Zr_2O_7$ with the pyrochlore structure, but this structure has not been experimentally fabricated so far[25]. Therefore, we have provided a new possibility here: through artificial design of GB geometries and tailoring of GB segregation, new phases with novel properties may be created at GBs.

Atomistic identification of subtle chemical difference at a GB, combined with theoretical calculations, shed light on the understanding towards solute segregation behaviours in GBs. It is demonstrated that GB solute segregation is driven by defect-GB interactions and the subsequent defect–defect interactions. These findings will be crucial for obtaining a complete picture of structure-property relationships at GBs in future. Furthermore, our results raise the possibility that segregation-induced GB phase transition is occurring, which might be a promising avenue to create novel phases inside materials that cannot be created by the current processing techniques. In addition, our approach may have broader utility in characterizing subtle chemical change at the atomic scale in materials science, solid-state physics and chemistry.

Methods

Experimental procedure. The YSZ bicrystal containing model GB was first fabricated by joining two YSZ (10 mol% Y_2O_3) single crystals at 1,600 °C for 15 h in air, respectively[20]. TEM specimens were prepared by mechanical polishing and Ar-ion beam milling. The sample thickness is about 30 nm. The GB were observed using STEM (JEM-ARM200CF, JEOL Co. Ltd) operated at 200 keV. STEM-EDS mappings were acquired by scanning the beam near the GB, using the NSS3 software developed by the Thermo Fisher Scientific Inc. The STEM EDS system is equipped with double SDD-EDS detectors and the solid angle for the whole collection system is about 1.7 sr. The probe size was 1.2 Å with the probe current of about 60 pA. Elemental maps were extracted with selected EDS energy for each element: Zr (using K_α of 15.776 keV and K_β of 17.668 keV) and Y (using K_α of 14.958 keV and K_β of 16.738 keV) were summed and the combined map is shown in the main text. The standard deviation of the integrated intensity profile are below ± 2% of the total intensity in each sites. A typical EDS spectrum obtained is shown in the Supplementary Fig. 1. In order to get high enough signals for the Zr and Y–K spectrum, the total acquisition time is about 48 min with the dwell time of 20 μs. The present results are reproducible and little affected by the experimental conditions used in this study (see Supplementary Figs 4 and 5 and Supplementary Note 2).

Computational procedure. A Σ3 [110]/{111} symmetric tilt GB model was constructed, which contains 512 Zr and 1024 O ions in the undoped model. The cell size is 96 × 14 × 13 Å and longest in the *x*-axis direction perpendicular to GB planes. The separation of 48 Å between the GB planes introduced by the periodic boundary condition is sufficient so that minimum-energy atomic configurations are unaffected by interactions between the GBs and point defects thereby. Y ions are substituted for Zr and charge compensating V_O are created for maintaining the charge neutrality. The dopant levels vary ranging from low to high dopant contents to pursue the process of point defect segregation. Total-energy calculations were carried out by static lattice methods implemented in the general utility lattice program (GULP) code[27] for all calculations. The Buckingham potential form with empirical parameters[26] was employed to evaluate short-range interactions. The parameters have been validated for various tilt GB models in previous studies[20,28,29]. The energy cutoff for short-range interactions was set to 15.0 Å for MC simulations and 20.0 Å for structural optimization. In order to obtain minimum-energy atomic configurations, we carried out site-interchange MC

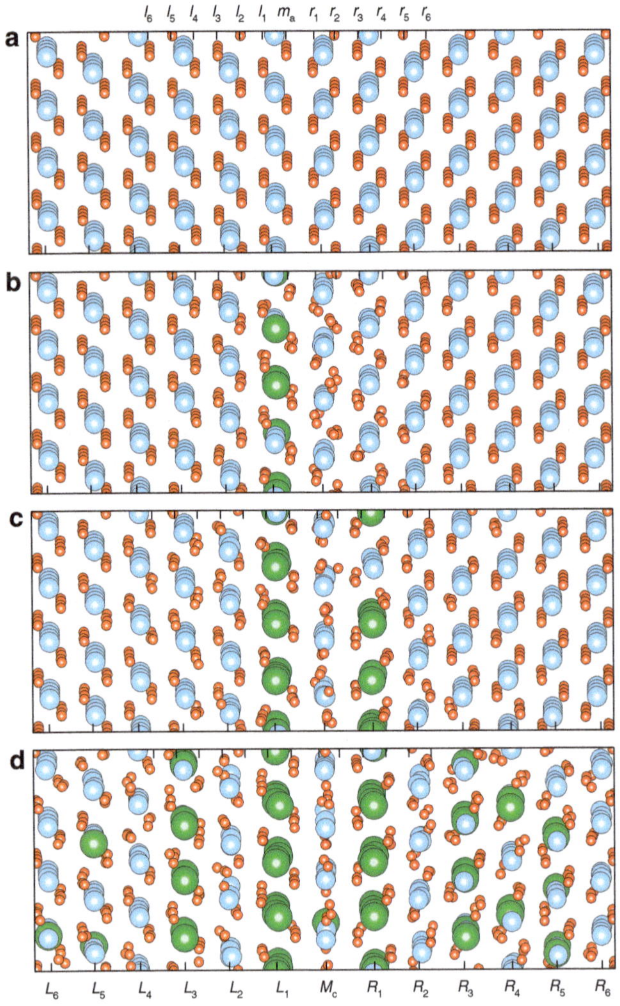

Figure 4 | Energetically favourable atomic arrangement. (**a**) The undoped model, (**b**) 6 Y, (**c**) 24 Y and (**d**) 60 Y atoms are introduced and O atoms are removed for maintaining charge neutrality.

simulations with Metropolis algorithm[28,30] for cations (Y and Zr) and oxygen (O and *Vo*), respectively at the same time, and subsequently all the ions were fully relaxed after each MC step. The favorable configurations were obtained varying dopant concentrations. The segregation energy and a comparison with the energetics obtained by the first-principles calculations can be found in the Supplementary Table 1 and Supplementary Note 3.

References

1. Buban, J. P. *et al.* Grain boundary strengthening in alumina by rare earth impurities. *Science* **311,** 212–215 (2006).
2. Duscher, G., Chisholm, M. F., Alber, U. & Ruhle, M. Bismuth-induced embrittlement of copper grain boundaries. *Nature Mater.* **3,** 621–626 (2004).
3. Sato, Y. *et al.* Role of Pr segregation in acceptor-state formation at ZnO grain boundaries. *Phys. Rev. Lett.* **97,** 106802 (2006).
4. Klie, R. F. *et al.* Enhanced current transport at grain boundaries in high-T_c superconductors. *Nature* **435,** 475–478 (2005).
5. Shibata, N. *et al.* Observation of rare-earth segregation in silicon nitride ceramics at subnanometer dimensions. *Nature* **428,** 730–733 (2004).
6. Ikuhara, Y. Grain boundary atomic structures and light-element visualization in ceramics: combination of Cs-corrected scanning transmission electron microscopy and first-principles calculations. *J. Electron Microsc.* **60,** S173–S188 (2011).
7. Guo, X. & Waser, R. Electrical properties of the grain boundaries of oxygen ion conductors: acceptor-doped zirconia and ceria. *Prog. Mater. Sci.* **51,** 151–210 (2006).
8. Lu, K., Lu, L. & Suresh, S. Strengthening materials by engineering coherent internal boundaries at the nanoscale. *Science* **324,** 349–352 (2009).
9. Lee, W. *et al.* Oxygen surface exchange at grain boundaries of oxide ion conductors. *Adv. Funct. Mater.* **22,** 965–971 (2012).
10. Nie, J., Zhu, Y. M., Liu, J. Z. & Fang, X. Y. Periodic segregation of solute atoms in fully coherent twin boundaries. *Science* **340,** 957–960 (2013).
11. Becher, P. F. *et al.* Debonding of interfaces between beta-silicon nitride whiskers and Si-Al-Y oxynitride glasses. *Acta Mater.* **44,** 3881–3893 (1996).
12. Sutton, A. P. & Balluffi, R. W. *Interface in Crystalline Materials* (Oxford University Press, 1995).
13. Wang, Z. *et al.* Atom-resolved imaging of ordered defect superstructures at individual grain boundaries. *Nature* **479,** 380–383 (2011).
14. Shibata, N. *et al.* Atomic-scale imaging of individual dopant atoms in a buried interface. *Nature Mater.* **8,** 654–658 (2009).
15. Huang, P. Y. *et al.* Grains and grain boundaries in single-layer graphene atomic patchwork quilts. *Nature* **469,** 389–392 (2011).
16. Lei, Y., Ito, Y., Browning, N. D. & Mazanec, T. J. Segregation effects at grain boundaries in fluorite-structured ceramics. *J. Am. Ceram. Soc.* **85,** 2359–2363 (2002).
17. Allen, L. J., D'Alfonso, A. J., Freitag, B. & Klenov, D. O. Chemical mapping at atomic resolution using energy-dispersive x-ray spectroscopy. *MRS Bull.* **37,** 47–52 (2012).
18. Chu, M. W., Liou, S. C., Chang, C. P., Choa, F. S. & Chen, C. H. Emergent Chemical Mapping at atomic-column resolution by energy-dispersive X-ray spectroscopy in an aberration-corrected electron microscope. *Phys. Rev. Lett.* **104,** 196101 (2010).
19. Steele, B. C. H. & Heinzel, A. Materials for fuel-cell technologies. *Nature* **414,** 345–352 (2001).
20. Shibata, N., Oba, F., Yamamoto, T. & Ikuhara, Y. Structure, energy and solute segregation behaviour of [110] symmetric tilt grain boundaries in yttria-stabilized cubic zirconia. *Philos. Mag.* **84,** 2381–2415 (2004).
21. Pennycook, S. & Boatner, L. A. Chemically sensitive structure-imaging with a scanning transimission electron microscope. *Nature* **336,** 565–567 (1988).
22. Yan, Y. *et al.* Impurity-induced structural transformation of a MgO grain boundary. *Phys. Rev. Lett.* **81,** 3675–3678 (1998).
23. Devanathan, R., Weber, W., Singhal, S. & Gale, J. Computer simulation of defects and oxygen transport in yttria-stabilized zirconia. *Solid State Ionics.* **177,** 1251–1258 (2006).
24. Zacate, M. O., Minervini, L., Bradfield, D. J., Grimes, R. W. & Sickafus, K. E. Defect cluster formation in M_2O_3-doped cubic ZrO_2. *Solid State Ionics* **128,** 243–254 (2000).
25. Pizara, M., Grimes, R. W., Minervini, L., Maguire, J. F. & Sickafus, K. E. Oxygen migration in $A_2B_2O_7$ pyrochlores. *Solid State Ionics* **140,** 201–208 (2001).
26. Minervini, L., Grimes, R. W. & Sickafus, K. E. Disorder in pyrochlore oxides. *J. Am. Ceram. Soc.* **83,** 1873–1878 (2000).
27. Gale, J. D. GULP: a computer program for the symmetry-adapted simulation of solids. *J. Chem. Soc., Faraday Trans.* **93,** 629–637 (1997).
28. Yokoi, T., Yoshiya, M. & Yasuda, H. Nonrandom point defect configurations and driving force transitions for grain boundary segregation in trivalent cation doped ZrO_2. *Langmuir* **30,** 14179–14188 (2014).
29. Yoshiya, M. & Oyama, T. Impurity and vacancy segregation at symmetric tilt grain boundaries in Y_2O_3-doped ZrO_2. *J. Mater. Sci.* **46,** 4176–4190 (2011).
30. Metropolis, N., Rosenbluth, A. W., Rosenbluth, M. N., Teller, A. H. & Teller, E. Equation of state calculations by fast computing machines. *J. Chem. Phys.* **21,** 1087–1092 (1953).

Acknowledgements

This work is supported by a Grant-in-Aid for Scientific Research on Innovative Areas 'Nano Informatics' (grant no. 25106003) from the Japan Society for the Promotion of Science (JSPS) and 'Nanotechnology Platform' (project no. 12024046) of the Ministry of Education, Culture, Sports, Science and Technology (MEXT), Japan. This work was partially supported by Grants-in-Aid for Scientific Research (A) (15H02290) and Grant-in-Aid for Scientific Research (B) (26289234) from the JSPS.

Author contributions

B.F. carried out experiments and wrote the paper. T.Y. performed theoretical calculations. A.K. supported the STEM experiments. M.Y. discussed and advised the theoretical calculations. Y.I. discussed and advised the experiments. N.S. discussed the results and directed the entire study. All authors read and commented on the manuscript.

Additional information

Competing financial interests: The authors declare no competing financial interests.

20

Direct evidence of two interatomic relaxation mechanisms in argon dimers ionized by electron impact

Xueguang Ren[1,2], Elias Jabbour Al Maalouf[3], Alexander Dorn[1] & Stephan Denifl[3]

In weakly bound systems like liquids and clusters electronically excited states can relax in inter-particle reactions via the interplay of electronic and nuclear dynamics. Here we report on the identification of two prominent examples, interatomic Coulombic decay (ICD) and radiative charge transfer (RCT), which are induced in argon dimers by electron collisions. After initial ionization of one dimer constituent ICD and RCT lead to the ionization of its neighbour either by energy transfer to or by electron transfer from the neighbour, respectively. By full quintuple-coincidence measurements, we unambiguously identify ICD and RCT, and trace the relaxation dynamics as function of the collisional excited state energies. Such interatomic processes multiply the number of electrons and shift their energies down to the critical 1–10 eV range, which can efficiently cause chemical degradation of biomolecules. Therefore, the observed relaxation channels might contribute to cause efficient radiation damage in biological systems.

[1] Max-Planck-Institut für Kernphysik, Saupfercheckweg 1, 69117 Heidelberg, Germany. [2] Physikalisch-Technische Bundesanstalt, Bundesallee 100, 38116 Braunschweig, Germany. [3] Institut für Ionenphysik und Angewandte Physik, Universität Innsbruck, Technikerstrasse 25, 6020 Innsbruck, Austria. Correspondence and requests for materials should be addressed to A.D. (email: Alexander.Dorn@mpi-hd.mpg.de).

One of the most fundamental questions in physics, chemistry and biology is the way of relaxation of atomic and molecular systems after initial (electronic) excitation. This relaxation results from the behaviour of nature to reach the lowest energetic state. For the transition from isolated atoms and molecules in the gas phase to the weakly bound systems (van der Waals clusters and liquids), a wealth of new energy transfer phenomena emerges like the recently discovered interatomic Coulombic decay (ICD), which is enabled only by the presence of the environment[1,2]. In ICD, first predicted by Cederbaum *et al.*[3] on theoretical grounds, an ionized or excited species relaxes by transferring its excess energy to the neighbour, where it leads to ejection of a low-energy electron and two repulsive ions at a distance of few Ångströms (Coulomb explosion).

Experimentally ICD was confirmed for the first time in 2003 (ref. 4). From that time onwards many experimental studies have been carried out using light sources such as synchrotron radiation[5] or free electron lasers[6]. In contrast, ICD studies involving ionization or excitation induced in clusters by particles like ions[7–9] or electrons are scarce[10–12] though, in principle, electron impact offers the advantage of populating optically forbidden states. However, experimentally one must face a major problem: in the outgoing scattering channel three electrons are present, two ejected ones and the scattered one, which lost an unknown fraction of its energy due to the excitation event. In addition two energetic ions are produced.

All previous experiments based on electron impact relied on the detection of two ions formed and on the determination of the kinetic energy release (KER; representing the sum of kinetic energies of the fragment ions formed in the outgoing channel). However, in this case the assignment to ICD is not ambiguous, since other competitive but fundamental relaxation mechanisms like radiative charge transfer (RCT)[13,14] may occur. In the latter case, one atom in the dimer becomes doubly ionized and subsequently an electron is transferred from the neutral atom, which is accompanied by release of a photon. Since in this case also two fragment ions are formed, which repel each other, the final observable (the KER) is very similar to the one for an ICD process.

In the present electron ionization study of argon dimers we have solved this problem by measuring all five charged particles in the outgoing reaction channel, that is, detecting both resulting ions and all three electrons. This full-coincidence experiment allows an unambiguous identification of the relaxation process accompanied by the electron ionization process. We show that both relaxation mechanisms are present and elucidate the ratio between both mechanisms. Moreover, we find a markedly enhanced production of low-energy electrons in the relaxation of argon dimer. The results suggest that the observed decay processes may have a significant influence on the radiation damage of biological tissues like DNA[15,16].

Results

KER of fragment ions. A multi-particle coincidence spectrometer (reaction microscope) designed for electron impact experiments[10,17] (Fig. 1) is employed to measure the momentum vectors and, consequently, the kinetic energies for all the final-state electrons and fragment ions. Figure 2 shows the KER spectra for $Ar^+ + Ar^+$ fragment ions measured at increasing projectile energies of $E_0 = 37$, 43 and 67 eV. For the highest impact energy (67 eV) the overall shape of the ion yield showing two peaks at 3.8 and 5.2 eV is in good agreement with theoretical predictions[18] and previous experimental studies determining the KER[10,19]. In a previous electron impact study on argon dimers Yan *et al.*[11] ascribed the peak at lower KERs to ICD and the other peak to RCT. This assignment was based on a simple comparison of the possible timescales of these processes since no information on the

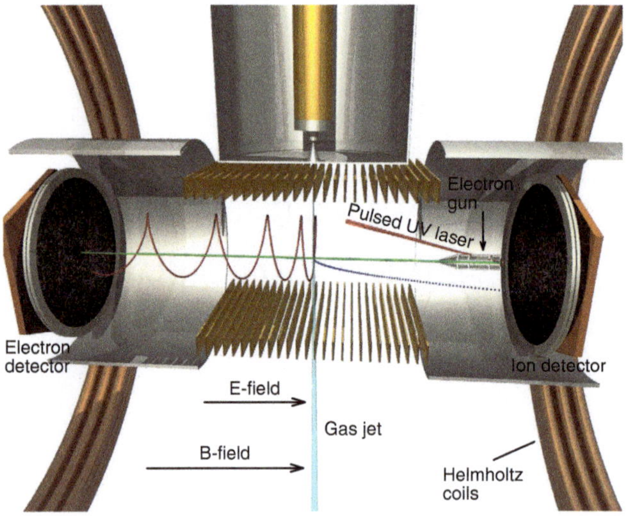

Figure 1 | Experimental set-up. Schematic view of the reaction microscope used to perform electron-scattering experiments on argon dimers. The pulsed projectile electron beam (horizontal green line) is crossing an argon supersonic gas jet (vertical light blue line). The projectiles are produced by photoemission from a tantalum photocathode illuminated by a pulsed ultraviolet (UV) laser beam (red conic line). The electrons and ions emerging from a collision in the reaction volume are extracted by means of a homogeneous electric field (E-field) created by a series of parallel electrodes and detected by two facing microchannel plate detectors. A pair of Helmholtz coils generates an uniform magnetic field (B-field), which forces the electrons to move along cyclotron trajectories (exemplary red spiral line), and thus confines their motion in transversal extension.

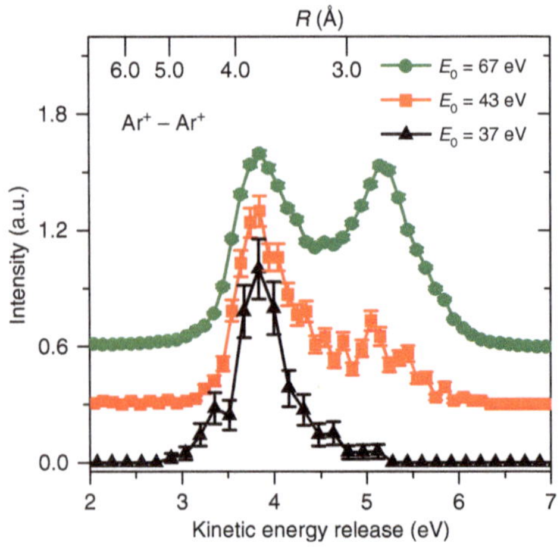

Figure 2 | Measured kinetic energy of Ar^+ ion pairs. KER spectra for the production of two repulsive Ar^+ ions after break-up of the dimer at projectile electron energies of $E_0 = 37$ (blue triangles), 43 (red squares) and 67 eV (green circles). The spectra are normalized at the 3.8-eV peak, and for better visibility the data for $E_0 = 43$ and 67 eV are shifted upwards by 0.3 and 0.6 a.u. (arbitrary units of signal intensity), respectively. The top scale shows the internuclear distance R at the instant when Coulomb explosion starts, that is, when the second Ar atom is ionized. R is determined from the associated Coulomb repulsion potential, which in good approximation corresponds to the observed KER. The error bars of the data points are defined as s.d. and are calculated as square root of the experimental count number.

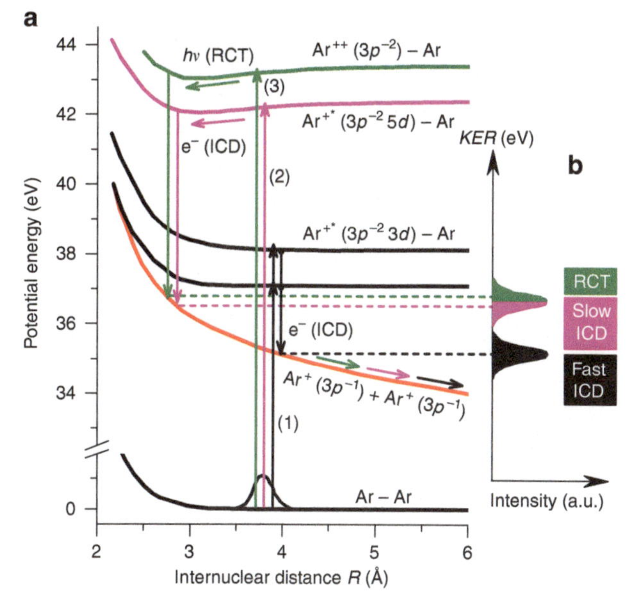

Figure 3 | Schematic view of potential energy curves and transitions. (a) Illustration of the three interatomic relaxation processes following ionizing collisions of electrons and argon dimers. In the fast ICD ((1) and blue arrows) low-lying satellite states are populated. The lifetime of these states with respect to ICD decay to the repulsive $Ar^+ - Ar^+$ final state is short. Thus, no vibrational motion to smaller internuclear distance is possible and the resulting KER is relatively low. For the slow ICD process ((2) and magenta arrows) higher-lying satellite states are populated, which have longer lifetime and allow vibrational motion. Here ICD occurs at shorter internuclear distance resulting in higher KER. For the process of RCT ((3) and green arrows), one dimer atom is doubly ionized ($Ar^{++} - Ar$). The dimer decays to the $Ar^+ - Ar^+$ state by transferring an electron from the neighbour atom to Ar^{++} accompanied by photon emission. This process preferentially takes place at shorter R and, thus, also contributes to the higher KER peak. The potential energy curves shown are adapted from refs 18,20. (**b**) The expected relative KER peak positions for the three different processes.

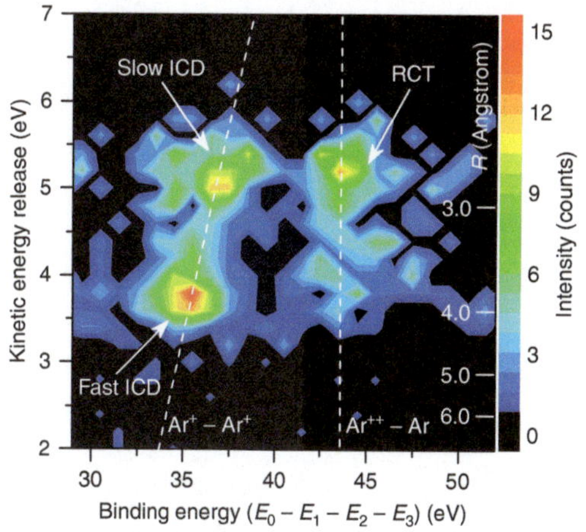

Figure 4 | Quintuple-coincidence data (2 ions + 3 electrons) for $E_0 = 67$ eV. Correlation map between binding energy (E_B) and KER. The horizontal axis shows E_B, the difference of incident electron energy and the sum energy of all the three outgoing electrons. The *y* axis shows the KER of the two Ar^+ ions. Regions ascribed to ICD and RCT are indicated. The right scale shows the internuclear distance *R*.

electrons was collected in the experiment. However, a recent theoretical study by Miteva *et al.*[18] has drawn a different picture suggesting that ICD should also contribute to the higher KER peak. Possible ionization and relaxation mechanisms are schematically shown in Fig. 3. They are the following: (1) fast ICD, for which vertical Franck–Condon transitions populate low-lying satellite states of the form ($Ar^{+*}(3p^{-2}\ nl) - Ar$) with shallow potential minima (between 36 and 130 meV binding energy) close to the equilibrium internuclear distance (R_{eq}) of the neutral argon dimer at 3.8 Å. They will quickly decay at R_{eq} via ICD to the repulsive $Ar^+ - Ar^+$ final state due to the short lifetime ($\sim$100 fs) compared with the vibrational periods ($\sim$1 ps) and form the low-KER peak at 3.8 eV; (2) slow ICD, where higher ionization satellites are populated, which have deeper potential wells (250–340 meV) being shifted to shorter R around 3 Å (ref. 18). These have one order of magnitude shorter vibrational periods ($\sim$200 fs) than their ICD lifetimes[18]. Hence vibrational motion can occur, resulting in ICD preferentially occurring at shorter R forming the higher KER peak; (3) RCT for population of the states ($Ar^{++} - Ar$) above the double ionization threshold of the argon atom (43.4 eV) (ref 20). The system can decay to the $Ar^+ - Ar^+$ state by transferring an electron from the neighbour atom to Ar^{++} accompanied by photon emission. Also this radiative process is rather slow (approximately nanosecond regime)[13] enabling vibrational motion, and it preferentially takes place at shorter R, and thus contributes to the higher KER peak.

Our observation of only one peak at 3.8 eV KER for $E_0 = 37$ eV in Fig. 2, where only the lowest states are reached, confirms the interpretation that these states decay fast and lead to low KER. For increasing impact energy 43 eV a second peak at higher KER emerges due to the population of the higher-lying states. We purely attribute the yield to the slow ICD process since the RCT channel is absent at this energy. The projectile energy of 67 eV is sufficient for double ionization of one argon atom and, therefore, to open the RCT channel in addition to the ICD channels. The KER spectra show increasing intensity at 5.2 eV, which might be due to both RCT and the increased cross-sections for the slow ICD process.

Two-dimensional energy correlation map. To clearly identify and separate the underlying relaxation processes above the threshold of both processes we carried out the fully quintuple-coincidence experiment of argon dimer at $E_0 = 67$ eV. From this we obtain a two-dimensional correlation map between the KER and the binding energy (E_B), as shown in Fig. 4. Here we define E_B as the initial projectile energy E_0 minus the sum energy of all three outgoing electrons ($E_B = E_0 - (E_1 + E_2 + E_3)$). E_B is the energy of the state reached after both initially bound electrons are ejected, that is, for ICD the $Ar^+ - Ar^+$ repulsive states and for RCT the intermediate state $Ar^{++} - Ar$ (Fig. 3). Accordingly, in the two-dimensional map we identify the peak at about $E_B = 35$ eV and 3.8 eV KER to the fast ICD process and the peak at about $E_B = 37$ eV and 5.2 eV KER to the slow ICD process. Also the linear relation $E_B = 2 \cdot IP_{Ar} + KER$ between E_B, that is, the point reached at the repulsive $Ar^+ - Ar^+$ state and KER indicated by the dashed line with slope 1 is seen in the data (IP_{Ar}: Ar ionization energy). In contrast, the region at higher E_B is characterized by transitions to doubly ionized states coupled with a neutral argon atom $Ar^{++} - Ar$, that is, these are precursor states that will lead via RCT to the $Ar^+ - Ar^+$ state. The additional E_B observed here is taken by the emitted photon. This result shows clearly that the RCT relaxation mechanism is present in the clusters once the energy of the electron is above the threshold since energetically only the $Ar^+ - Ar^+$ ion ground state channel is open.

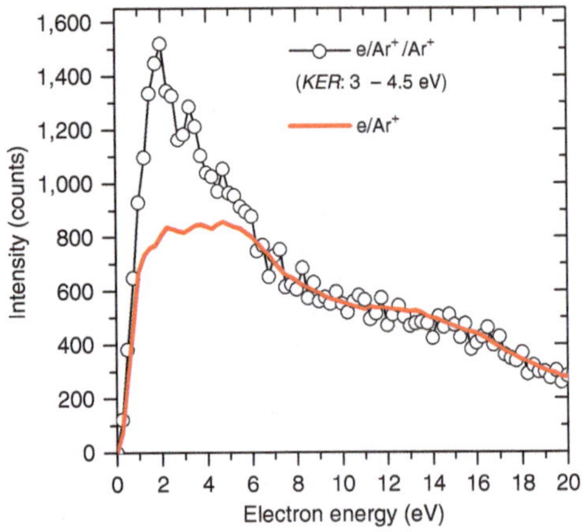

Figure 5 | Measured electron energy distributions for $E_0 = 67$ eV. Black open circles: the electron energy spectrum for ionization and fragmentation of argon dimers measured in $e^-/Ar^+/Ar^+$ triple coincidences. The data are summed over the KER range from 3 to 4.5 eV . Red solid line: corresponding spectrum for ionization of argon monomers measured in e^-/Ar^+ double coincidences. The spectra were normalized to each other in the energy range from 10 to 15 eV.

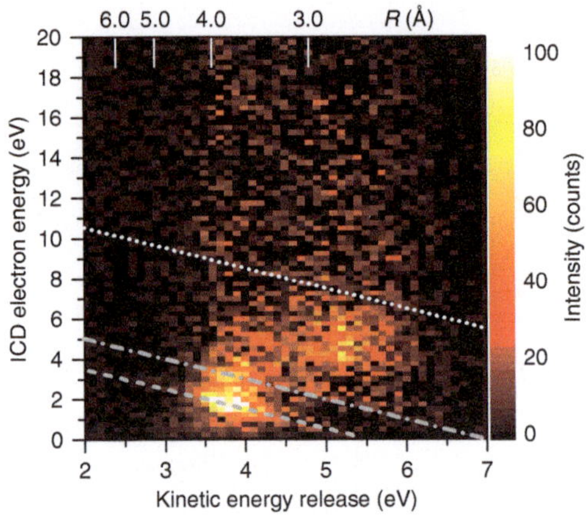

Figure 6 | Measured two-dimensional energy map of electron and fragment ions for $E_0 = 67$ eV. Correlation map between ICD electron energy (E_{ICD}) and KER. The upper scale shows the internuclear distance R in Ångströms. Dashed line: position of ICD initial states $Ar^{+*}(3p^{-2}(^1D)4p\ ^2P,^2F)$ and $Ar^{+*}(3p^{-2}\ (^1D)\ 3d\ ^2P,^2D)$. Dash-dotted line: position of states $Ar^{+*}(3p^{-2}\ (^1D)\ 3d\ ^2S)$ and $Ar^{+*}(3p^{-2}\ (^3P)\ 5s\ ^2P)$. Dotted line: maximum ICD electron energy for the $Ar^{+*}(3p^{-2}\ nl)$ series.

ICD electron energy spectra. To pin down the ICD signature for the low-E_B region in Fig. 4 we identify the ICD electrons and show that their energy is correlated to the KER. In Fig. 5 the measured electron energy spectrum for the low-KER peak is shown (open circles). It contains a strong contribution from slow electrons produced in the initial ionization process (red line). It is obtained by the simultaneous recording of ionization of the Ar monomers in the target where ICD is absent and normalizing the resulting spectrum to the dimer spectrum in the energy range above 10 eV where no ICD electrons contribute (see Methods section). Clearly, an enhanced production of low-energy electrons (<8 eV) is observed in the electron energy spectrum of dimers compared with the result for monomers, which is the signature of ICD electron. The same procedure can be done for the high-KER electron spectrum (not shown here). If the monomer spectrum is subtracted from the total electron energy spectrum the pure ICD contribution correlated to the KER is obtained (Fig. 6). Here a clear correlation between ICD electron energy and ion KER is seen since for a particular ICD initial state Ar^{+*} with energy E_{ICD}^{ini} the relation must hold: $E_{ICD}^{ini} = 2 \cdot IP_{Ar} + E_{ICD} + KER$, where E_{ICD} is the ICD electron energy. Moreover low ICD electron energies are observed for the fast ICD process at KER = 3.8 eV and higher ICD electron energies occur for slow process consistent with our discussion above. For the strongest line at $E_{ICD} \approx 1.8$ eV $\left(E_{ICD}^{ini} = 37.1\ \text{eV}\right)$ five terms of the two configurations $Ar^{+*}(3p^{-2}\ 4p)$ and $Ar^{+*}(3p^{-2}\ 3d)$ contribute. The second visible line at $E_{ICD} \approx 3.2$ eV $\left(E_{ICD}^{ini} = 38.5\ \text{eV}\right)$ belongs to two terms of the configurations $Ar^{+*}(3p^{-2}\ 3d)$ and $Ar^{+*}(3p^{-2}\ 5s)$. For slow ICD channels a band of contributions is found at KER = 5.2 eV and $E_{ICD} = 3.0$–6.0 eV, which can be assigned to a series of about 20 terms of the configurations $Ar^{+*}(3p^{-2}\ nl)$ with $nl = 4d$, $4f$, $5d$ and $6s$ (ref. 21).

Separation of ICD and RCT contributions to the KER spectra. Since we can separate ICD and RCT contributions, we may plot the KER spectrum for both relaxation mechanisms independently (Fig. 7). In ICD we observe two KER peaks at 3.8 and 5.2 eV, which agrees well with the calculations by Miteva *et al.*[18] and the dynamics discussed above. The RCT yield exhibits a pronounced

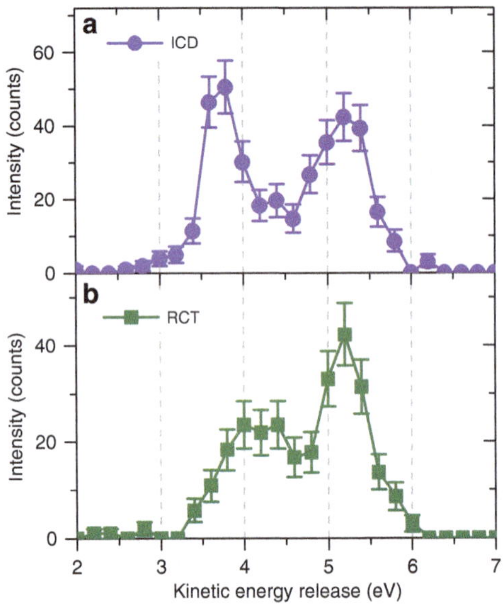

Figure 7 | Measured KER distributions. The KER (**a**) for the ICD process and (**b**) for the RCT process. The error bars of the data points are defined as s.d. and are calculated as square root of the experimental count number.

peak at 5.2 eV and a second peak (or an extended intensity) ranging down to 3.5 eV. This curve is consistent with simulated results for RCT obtained by Matsumoto *et al.*[22] and is result of the long radiative lifetime of the doubly ionized state much longer than the vibrational period. The KER curve for RCT is the result of the weighted contributions of excited vibrational states and the R-dependent transition probability. In addition, we determine the relative importance of ICD to RCT at 67 eV to be 14 to 10.

Role of sequential ionization. Finally, we consider direct Franck–Condon transitions to the repulsive $Ar^+ - Ar^+$ state from the neutral dimer ground state, that is, the situation when the scattered electron consecutively kicks out one electron from each atom of the dimer. This process can give a contribution to the lower KER peak at 3.8 eV since it occurs at R_{eq} and it has identical binding energy as the ICD process. We do not find any signature for this process as, for example, a preferential production of $Ar^+ - Ar^+$ pairs for an internuclear axis alignment along the projectile beam. From the cross-section for single ionization and the interatomic distance we estimate the probability for this process to be below 20% of the probability for ICD at $E_0 = 67$ eV. Any possible contribution is eliminated largely from the electron spectrum in Fig. 5 by subtracting the direct single ionization part as discussed above.

Discussion

Our work provides a detailed picture of the various relaxation mechanisms in small clusters utilizing full-coincidence measurements of all reaction products. We observe direct evidence of ICD and RCT in the decay of argon dimers ionized by electron impact. Recent years ICD processes were paid particular attention, since ICD is expected to be a general phenomenon in clusters and will be present also in hydrogen-bonding systems exposed to radiation, for example, liquid water[2]. It was proposed in refs 23,24 that ICD is an abundant source for low-energy electrons in radiation damage of biological tissue[19,25–27]. Such low-energy electrons with energies below 10 eV are known to be partly responsible for the chemical transformation of DNA[15,16] (in addition to the action of radicals), which leads to degradation of DNA and the starting point of cancer development. Since secondary electrons initially formed by ionizing radiation in cells have energies up to 100 eV (ref. 28), and drive many processes in aqueous systems[29], electron-induced ICD processes as identified here provide the possibility to shift the energy distribution towards the more critical lower values[30].

Methods

Experimental technique. The experiment was performed at the Max-Planck-Institut für Kernphysik in Heidelberg. A dedicated reaction microscope was used with two multichannel plate detectors with delay line position read-out for the coincident measurements of all reaction products (Fig. 1)[17,31]. In this apparatus a pulsed electron beam is crossed with an argon gas jet (2 mm diameter) containing a small fraction of about 2% dimers. The argon dimers were produced by supersonic expansion of pure argon gas through a 30-μm nozzle at room temperature and a typical stagnation pressure of 2 bar. The pulsed electron beam is emitted from an electron gun consisting of a tantalum photocathode, which is irradiated by ultraviolet-light pulses of 0.5 ns duration and electrostatic focusing lens elements. The energy width of the electron pulses is about 0.5 eV. The compact gun (8 mm × 50 mm = diameter × length) is placed in the centre of the ion drift region of the reaction microscope in a shielded housing such that it has negligible influence on the ion trajectories. An axial magnetic field (6.9 Gauss) guides the emitted collimated electron beam to the crossing zone with the gas jet and further to the beam dump, which is a central bore in the electron detector. Charged collisional products (electrons as well as ions) are extracted by means of a homogeneous electric field of 0.9 V cm^{-1} and projected onto two position- and time-sensitive multi-hit detectors. The guiding magnetic field also increases the transversal acceptance for electrons by restricting their transversal motion to cyclotron circles. The momentum vectors of the emitted species are determined by the impact positions on the detectors together with the corresponding times of flight. The acceptance angle for detection of electrons up to the energy of 15 eV is almost 4π, where small forward and backwards angles are excluded due to the presence of the beam dump in the electron detector. Ions originating from a Coulomb explosion have significantly higher momentum than the electrons observed, and require higher fields for efficient detection. Therefore, after 400 ns when the electrons have reached the detector the electric extraction field is ramped up to 20 V cm^{-1} for extraction of the ions. In this way full ion acceptance is reached up to a KER of 4 eV. For higher energies the decreasing ion acceptance is corrected for by scaling of the ion yield data with an energy dependent factor, which is obtained from simulating the spectrometer properties. The electron beam and spectrometer properties are calibrated using single ionization of atomic argon, which is the major constituent of the target gas jet. Here a binding energy resolution for the argon 3*p* orbital of about 2 eV is achieved. For the three electrons coincidence used to separate ICD and RCT a binding energy resolution of about 2.7 eV is estimated.

Data reduction. In our experiment, monomer ionization is simultaneously recorded with dimer ionization and fragmentation. Since the ICD processes are absent for monomers, the monomer data can be considered as a reference to determine the continuous electron energy distribution due to scattered projectiles and ejected electrons from the ionizing collision. Thus, this contribution can be subtracted from the electron spectra recorded for the dimers to reveal the ICD electrons solely present for dimers. As one example the ionized electron energy spectra from monomer and dimer are presented in Fig. 5 in which the dimer spectrum is integrated over the KER range from 3 to 4.5 eV. The electron energy distributions agree well in the higher-energy region (>8 eV) because here the ICD process does not contribute[18]. Therefore, we normalized both spectra at the energy range from 10 to 15 eV. Comparing with the spectrum of monomer (e^-/Ar^+) the dimer result ($e^-/Ar^+/Ar^+$) shows clear enhancement in the low-energy region (<8 eV) due to the presence of the ICD processes. Thus, the pure ICD electron spectrum is obtained and presented in Fig. 6 as a correlation diagram between electron energy and KER, where for each KER (±0.05 eV) the electron energy spectrum of monomer is normalized to the spectrum of dimer in the energy range from 10 to 15 eV to reduce the background. From the measured ICD electron energy (E_{ICD}) and KER we can determine the binding energy of the ICD initial state as given below:

$$E^{ini}_{ICD} = E^{R\to\infty}_{Ar^+-Ar^+} + E_{ICD} + KER \quad (1)$$

where E^{ini}_{ICD} is the binding energy of the ICD initial state and $E^{R\to\infty}_{Ar^+-Ar^+} = 31.5$ eV corresponds to the energy of the $Ar^+ - Ar^+$ state at R close to infinity.

References

1. Sisourat, N. *et al.* Ultralong-range energy transfer by interatomic Coulombic decay in an extreme quantum system. *Nat. Phys.* **6**, 508–511 (2010).
2. Thürmer, S. *et al.* On the nature and origin of dicationic, charge-separated species formed in liquid water on X-ray irradiation. *Nat. Chem.* **5**, 590–596 (2013).
3. Cederbaum, L. S., Zobeley, J. & Tarantelli, F. Giant intermolecular decay and fragmentation of clusters. *Phys. Rev. Lett.* **79**, 4778–4781 (1997).
4. Marburger, S., Kugeler, O., Hergenhahn, U. & Möller, T. Experimental evidence for interatomic Coulombic decay in Ne clusters. *Phys. Rev. Lett.* **90**, 203401 (2003).
5. Jahnke, T. Interatomic and intermolecular Coulombic decay: the coming of age story. *J. Phys. B: At. Mol. Opt. Phys.* **48**, 082001 (2015).
6. Schnorr, K. *et al.* Time-resolved measurement of interatomic Coulombic decay in Ne_2. *Phys. Rev. Lett.* **111**, 093402 (2013).
7. Iskandar, W. *et al.* Interatomic Coulombic decay as a new source of low energy electrons in slow ion-dimer collisions. *Phys. Rev. Lett.* **114**, 033201 (2015).
8. Kim, H. K. *et al.* Enhanced production of low energy electrons by alpha particle impact. *Proc. Natl Acad. Sci. USA* **108**, 11821 (2011).
9. Kim, H. K. *et al.* Orientation dependence in multiple ionization of He_2 and Ne_2 induced by fast, highly charged ions: probing the impact-parameter-dependent ionization probability in 11.37-MeV/u S14+ collisions with He and Ne. *Phys. Rev. A* **89**, 022704 (2014).
10. Pflüger, T., Ren, X. & Dorn, A. Electron-impact-induced dissociation of small argon clusters. *Phys. Rev. A* **91**, 052701 (2015).
11. Yan, S. *et al.* Observation of interatomic Coulombic decay and electron-transfer-mediated decay in high-energy electron-impact ionization of Ar_2. *Phys. Rev. A* **88**, 042712 (2013).
12. Grieves, G. A. & Orlando, T. M. Intermolecular Coulomb decay at weakly coupled heterogeneous interfaces. *Phys. Rev. Lett.* **107**, 016104 (2011).
13. Johnsen, R. & Biondi, M. A. Measurements of radiative charge-transfer reactions of doubly and singly charged rare-gas ions with rare-gas atoms at thermal energies. *Phys. Rev. A* **18**, 996 (1978).
14. Saito, N. *et al.* Evidence of radiative charge transfer in argon dimers. *Chem. Phys. Lett.* **441**, 16–19 (2007).
15. Boudaïffa, B., Cloutier, P., Hunting, D., Huels, M. A. & Sanche, L. Resonant formation of DNA strand breaks by low-energy (3 to 20 eV) electrons. *Science* **287**, 1658–1660 (2000).
16. Neustetter, M., Aysina, J., Ferreira da Silva, F. & Denifl, S. The effect of solvation on electron attachment to pure and hydrated pyrimidine clusters. *Angew. Chem. Int. Ed.* **54**, 9124–9126 (2015).
17. Ren, X. *et al.* An (e, 2e + ion) study of low-energy electron-impact ionization and fragmentation of tetrahydrofuran with high mass and energy resolutions. *J. Chem. Phys.* **141**, 134314 (2014).
18. Miteva, T. *et al.* Interatomic Coulombic decay following resonant core excitation of Ar in argon dimer. *J. Chem. Phys.* **141**, 064307 (2014).
19. Kimura, M. *et al.* Efficient site-specific low-energy electron production via interatomic Coulombic decay following resonant Auger decay in argon dimers. *Phys. Rev. A* **87**, 043414 (2013).

20. Stoychev, S. D., Kuleff, A. I., Tarantelli, F. & Cederbaum, L. S. On the doubly ionized states of Ar_2 and their intra- and interatomic decay to Ar_2^{3+}. *J. Chem. Phys.* **128,** 014307 (2008).
21. Moore, C. E. Atomic Energy Levels as Derived from the Analyses of Optical Spectra, Volume I. 1H to 23V, NSRDS-NBS. **35,** 358 (1971).
22. Matsumoto, J. *et al.* Asymmetry in multiple-electron capture revealed by radiative charge transfer in Ar dimers. *Phys. Rev. Lett.* **105,** 263202 (2010).
23. Mucke, M. *et al.* A hitherto unrecognized source of low-energy electrons in water. *Nat. Phys.* **6,** 143–146 (2010).
24. Jahnke, T. *et al.* Ultrafast energy transfer between water molecules. *Nat. Phys.* **6,** 139–142 (2010).
25. Trinter, F. *et al.* Resonant Auger decay driving intermolecular Coulombic decay in molecular dimers. *Nature* **505,** 664 (2014).
26. Gokhberg, K., Kolorenc, P., Kuleff, A. I. & Cederbaum, L. S. Site- and energy-selective slow-electron production through intermolecular Coulombic decay. *Nature* **505,** 661 (2014).
27. O'Keeffe, P. *et al.* The role of the partner atom and resonant excitation energy in interatomic Coulombic decay in rare gas dimers. *J. Phys. Chem. Lett.* **4,** 1797–1801 (2013).
28. Pimblott, S. M. & LaVerne, J. A. Production of low-energy electrons by ionizing radiation. *Radiat. Phys. Chem.* **76,** 1244–1247 (2007).
29. Bruce, C. *et al.* Role of water in electron-initiated processes and radical chemistry: issues and scientific advances. *Chem. Rev.* **105,** 355–390 (2005).
30. Alizadeh, E., Orlando, T. M. & Sanche, L. Biomolecular damage induced by ionizing radiation the direct and indirect effects of low-energy electrons on DNA. *Annu. Rev. Phys. Chem.* **66,** 379–398 (2015).
31. Ullrich, J. *et al.* Recoil-ion and electron momentum spectroscopy: reaction-microscopes. *Rep. Prog. Phys.* **66,** 1463 (2003).

Acknowledgements

This work was supported by DFG (FOR1789) and FWF (I1015). S.D. thanks Reinhard Dörner, Lorenz Cederbaum, Uwe Hergenhahn, Kirill Gokhberg Till Jahnke and Bernd Winter for discussions within the research group FOR1789.

Author contributions

X.R. and A.D.: experiment design and set-up; X.R., E.J.A.M. and A.D.: experiment data acquisition; X.R.: data analysis; X.R., A.D., S.D. and E.J.A.M.: interpretation of data and manuscript preparation.

Additional information

Competing financial interests: The authors declare no competing financial interests.

Super-resolution microscopy reveals structural diversity in molecular exchange among peptide amphiphile nanofibres

Ricardo M.P. da Silva[1,2,3], Daan van der Zwaag[2], Lorenzo Albertazzi[2,4], Sungsoo S. Lee[5], E.W. Meijer[2] & Samuel I. Stupp[1,5,6,7,8]

The dynamic behaviour of supramolecular systems is an important dimension of their potential functions. Here, we report on the use of stochastic optical reconstruction microscopy to study the molecular exchange of peptide amphiphile nanofibres, supramolecular systems known to have important biomedical functions. Solutions of nanofibres labelled with different dyes (Cy3 and Cy5) were mixed, and the distribution of dyes inserting into initially single-colour nanofibres was quantified using correlative image analysis. Our observations are consistent with an exchange mechanism involving monomers or small clusters of molecules inserting randomly into a fibre. Different exchange rates are observed within the same fibre, suggesting that local cohesive structures exist on the basis of β-sheet discontinuous domains. The results reported here show that peptide amphiphile supramolecular systems can be dynamic and that their intermolecular interactions affect exchange patterns. This information can be used to generate useful aggregate morphologies for improved biomedical function.

[1] Simpson Querrey Institute for BioNanotechnology (SQI), Northwestern University, Chicago, Illinois 60611, USA. [2] Laboratory of Macromolecular and Organic Chemistry and Institute for Complex Molecular Systems, Eindhoven University of Technology, Eindhoven MB 5600, The Netherlands. [3] Craniofacial Development & Stem Cell Biology, King's College London, London, SE1 9RT, UK. [4] Nanoscopy for Nanomedicine Group, Institute for Bioengineering of Catalonia (IBEC), Barcelona 08028, Spain. [5] Department of Materials Science and Engineering, Northwestern University, Evanston, Illinois 60208, USA. [6] Department of Chemistry, Northwestern University, Evanston, Illinois 60208, USA. [7] Department of Medicine, Northwestern University, Chicago, Illinois 60611, USA. [8] Department of Biomedical Engineering, Northwestern University, Evanston, Illinois 60208, USA. Correspondence and requests for materials should be addressed to E.W.M. (email: e.w.meijer@tue.nl) or to S.I.S. (email: s-stupp@northwestern.edu).

Reversible supramolecular interactions are ubiquitous in nature, controlling the self-assembly of ordered functional structures that need to be dynamic to perform their biological functions. One-dimensional cytoskeletal filaments such as actin and tubulin are typical examples of structures that use dynamics to mediate the adaptive behaviour of cells, resulting in cell motility, shape change, cell division, signalling and muscular contraction at larger length scales[1–6]. Artificial supramolecular materials could offer this bio-inspired dynamic behaviour, thus allowing enhanced interaction with natural systems and increased biomedical functionality. Since many natural processes are carefully regulated, optimization of an artificial supramolecular material requires a detailed understanding of its dynamic properties, for example its exchange kinetics.

Molecular mixing experiments typically assess exchange kinetics by utilizing Förster resonance energy transfer (FRET) between a pair of donor and acceptor fluorophores[7,8], and alternatively, radio-labelled molecules[9] or time-resolved small-angle neutron scattering[10,11]. Although these ensemble experiments can provide the timescale of the processes, they cannot distinguish different mechanisms. Moreover, they fail to detect the structural diversity among fibres, or within an individual fibre, for example, the occurrence of segregated domains. Local variations of molecular composition can have important biological implications, as they can greatly influence the signalling potency through multivalency effects[12–14], and thus understanding the exchange heterogeneity is important to design materials in which function is connected with dynamics for adaptive or responsive behaviour[15].

Super-resolution techniques are powerful tools to reveal the spatial distribution of molecules at the nanoscale, but these techniques have thus far been mainly applied to imaging fine details of cellular structures[16,17]. For instance, a resolution of ~20 nm can be achieved using stochastic optical reconstruction microscopy (STORM)[18], which is an order of magnitude below the diffraction limit and near the molecular scale. The enhanced resolution is achieved through the accurate localization of single fluorescent molecules; to identify individual fluorophores only a sparse subset of labels should be active at any given time. This sparse population of fluorophores is obtained using probes that can be photo-switched to a temporary non-fluorescent 'off' state by light. By repeatedly activating different subsets and overlaying the resulting localizations, an image can be reconstructed. We have previously reported how to apply STORM to probe the dynamics of supramolecular fibres[19]. Using two-colour STORM and quantitative image analysis, we were able to resolve the monomer distribution along the fibre backbone. By following the monomer distribution during the molecular exchange process, we were able to infer the exchange mechanism.

Peptide amphiphiles (PAs) that self-assemble into high aspect ratio objects offer exciting opportunities for regenerative medicine and other therapeutic applications[20–24]. As illustrated in Fig. 1a, this class of molecules is composed of an unbranched alkyl chain linked to a peptide segment, which can be further subdivided in several domains. A segment with propensity to form β-sheets is conjugated to the alkyl tail, followed by a charged segment for solubility. Additional domains can be conjugated to the canonical structure to introduce biofunctionality in the nanofibres. Whereas the hydrophobic collapse of the aliphatic tails induces self-assembly, experimental[25,26] and theoretical[27] evidence suggests that the formation of directional hydrogen bonds within the β-sheet domain is an additional important component of the driving force for assembly of the molecules into one-dimensional filamentous shapes. The facile incorporation of multiple bioactive signals at controlled concentrations[14,28], together with their structural resemblance with extracellular matrix fibres makes PA assemblies useful as bioactive artificial extracellular matrix components for cell signalling. Furthermore, they are also intrinsically biocompatible and biodegradable and can therefore disappear easily after fulfilling their biological functions. PAs have been extensively studied as a platform for applications that include bone, cartilage, enamel, neuronal regeneration, angiogenesis for ischaemic disease, targeted drug delivery and cancer therapeutics[20–24,29].

Ensemble measurements of PA dynamics revealed partial self-healing using rheological techniques[30] or the existence of a critical aggregation concentration[31], while kinetic spectroscopy showed pathway selection of PAs into different morphologies[32,33]. However, these approaches do not consider structural heterogeneity and its effect on dynamic molecular exchange at the level of individual filaments. In this paper, we use STORM to image individual PA nanofibres, addressing the distribution of molecules along the fibre during exchange and therefore analysing diversity among supramolecular nanofibres.

Results

Nanofibre design and preparation. The PA molecule studied in this work is shown in Fig. 1a. It consists of a palmitic acid tail, six alanines in the β-sheet-forming region, followed by three glutamic acids as charged solubilizing moieties. This PA molecule self-assembles in water, forming nanofibres around 7 nm in diameter and lengths in the range of micrometres as observed by cryogenic transmission electron microscopy (cryoTEM; Fig. 1e). Amino acids with different propensities to form β-sheets affect considerably the internal order of PA assemblies, as well as nanofibre stiffness[26]. Since alanine has a weaker tendency to form β-sheets than valine[34], placing this amino acid in the β-sheet-forming segment was expected to yield relatively dynamic PA fibres. Circular dichroism spectroscopy of this PA is consistent with the typical β-sheet conformation found for other PAs[26], at physiological pH and ionic strength (Fig. 1c). To perform imaging by fluorescence microscopy, PA molecules were labelled with cyanine dyes, namely Cy3 and Cy5 (Fig. 1b). Water soluble sulfonated dyes with a net charge of −1 were used to preserve the amphiphilic asymmetry of the PA molecule and, since sulfonate groups are fully ionized in the vicinity of physiological pH, their behaviour is insensitive to pH in the range of interest. The Cy3 and Cy5 dyes have been chosen for their suitable photophysical properties for STORM imaging; moreover, they constitute a good FRET pair with a Förster radius of 50 Å (ref. 35). Single-colour labelled nanofibres were created by mixing a stock solution of either Cy3-PA or Cy5-PA with a stock solution of unlabelled PA, lyophilization, brief co-dissolution in trifluoroacetic acid (TFA) and immediate TFA removal, lyophilization and final reconstitution in the aqueous working solution to form labelled supramolecular aggregates. The degree of labelling was accurately controlled by premixing the different PAs at the desired ratios. Fluorescently labelled nanofibres revealed a morphology that was indistinguishable from their non-labelled counterparts, as shown by cryoTEM (Fig. 1f).

Ensemble molecular exchange kinetics. The timescale of molecular exchange in PA-based nanofibres was measured using FRET kinetic experiments. Two sets of PA nanofibres were separately pre-assembled from a mixture of non-labelled PAs with either Cy3-PA (0.5%) or Cy5-PA (0.5%), as illustrated in Fig. 2a. Next, the two solutions were mixed and the FRET ratio, defined as the ratio between the fluorescence intensities of Cy5 acceptor and Cy3 donor, was monitored over time. As shown in Fig. 2b, the FRET ratio increases with time, reaching a plateau after several hours. This means that PA molecules are able to migrate between

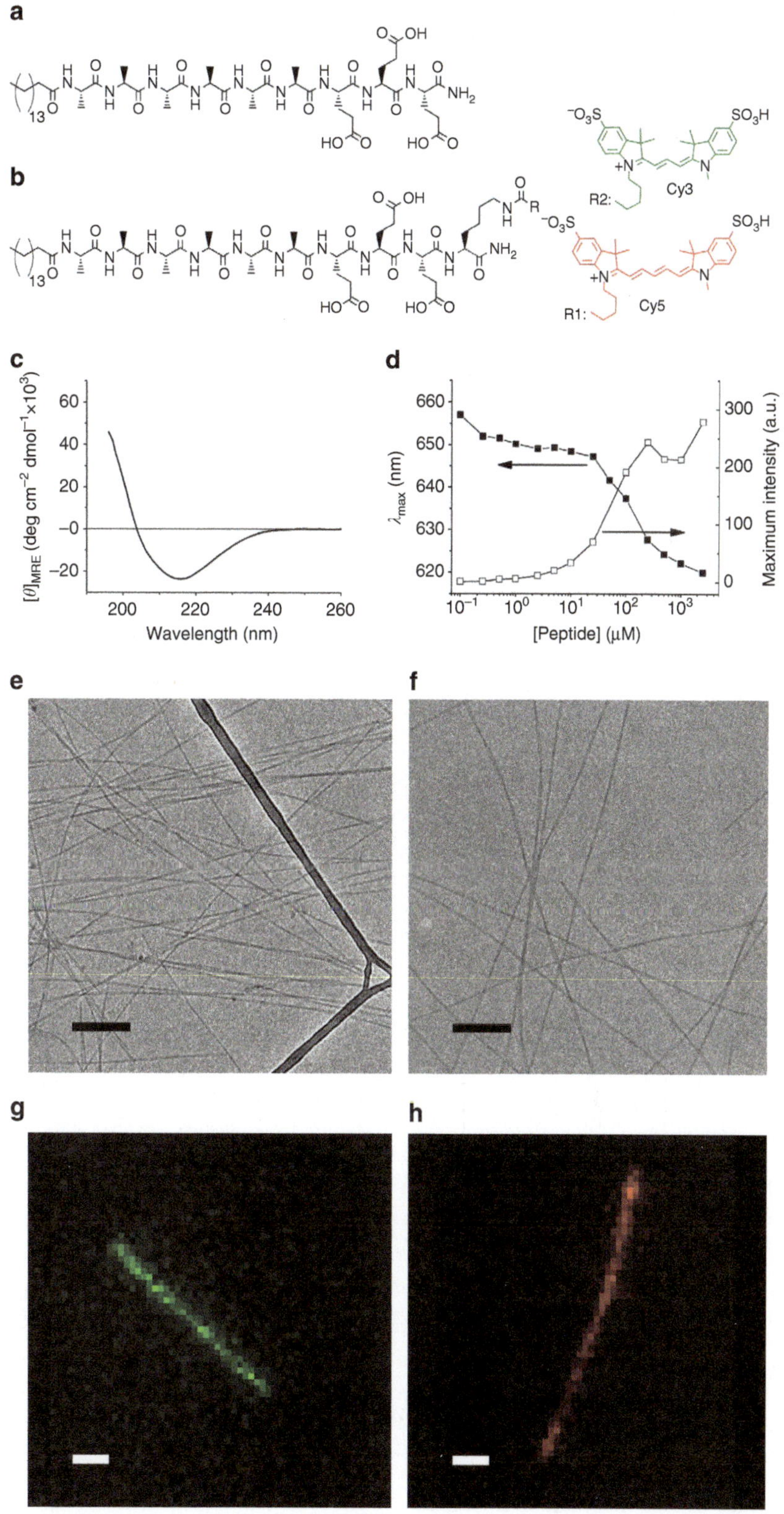

Figure 1 | PA self-assembly. Molecular structure of (**a**) non-labelled PA and (**b**) PA molecules labelled with photo-switchable sulfonated cyanine dyes, namely Cy3 (green) and Cy5 (red). (**c**) Circular dichroism spectrum and (**d**) Nile Red assay of non-labelled PA. CryoTEM images of nanofibres self-assembled at pH 7.5 and NaCl 150 mM (**e**) from non-labelled PA alone and (**f**) from a molecular mixture of non-labelled and Cy5-labelled PAs (scale bar, 200 nm). Diffraction-limited fluorescence microscopy images of Cy3- (**g**) and Cy5-labelled (**h**) PA nanofibres (scale bar, 1 μm).

nanofibres, resulting in mixed fluorophore fibres. Figure 2b depicts kinetic experiments at different temperatures, showing that the exchange rate is remarkably faster at 37 °C than at 20 °C. On the other hand, we observed a limited effect of concentration on the exchange rate (Fig. 2c), proving that the exchange process is not diffusion limited in this concentration range. These results resemble observations on self-assembled polymeric micelles, in which unimer expulsion and insertion is thought to be the

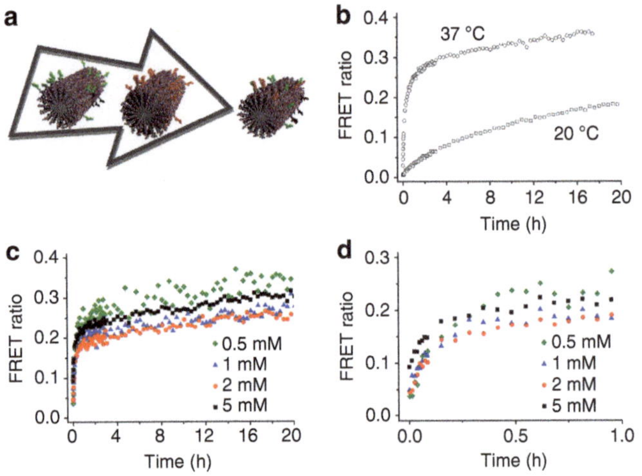

Figure 2 | Molecular exchange by FRET. (**a**) Schematic representation of a molecular exchange kinetic measurement. The molecular exchange progress over time is estimated by means of FRET ratio (dividing Cy5 by Cy3 fluorescence intensities), either at (**b**) a constant PA concentration (0.5 mM) or at (**c**,**d**) a constant temperature (37 °C), where **d** shows the FRET ratio for short timescale.

rate-determining step of system dynamics[10,11], as well as other synthetic supramolecular polymers[7]. So, while FRET measurements provide useful information about the timescale of the exchange, FRET is an ensemble technique, and it does not provide the spatially resolved information required to elucidate the mechanism of exchange.

Imaging PA nanofibres. Due to its high spatial resolution and multicolour imaging ability, STORM microscopy can be used to investigate the spatial details of the exchange process in PA nanofibres. Cy3- or Cy5-labelled PA nanofibres have been immobilized on a glass slide by physisorption and subsequently have been imaged using STORM (see the 'Methods' Section for details). The resulting images show well-reconstructed aggregates (Fig. 3a,b), indicating that both dyes display appropriate photoswitching behaviour when associated to PA structures. Since the actual width of the nanofibre is below the STORM resolution, the apparent thickness of the fibre can be used to provide an estimate for the experimental resolution of these measurements, approximately equal to 50 nm (Supplementary Fig. 1). The fibre characteristics observed by STORM, for example, rigidity and fibre length, match those reported by the cryoTEM images in Fig. 1e,f.

To analyse the monomer distribution inside PA nanofibres, which is crucial to understand the exchange mechanism, we utilized an image analysis routine previously developed in the Meijer laboratory[19]. This method removes background localizations, identifies the contour of the fibre backbone to study its mechanical properties (Supplementary Fig. 2) and computes the localization density along the polymer. Figure 3c,d display the localization density plots for Cy5- and Cy3-labelled PA, respectively. These profiles contain information about the distribution of the dye-functionalized PA molecules in a nanofibre. As can be clearly observed, the number of localizations shows fluctuations along the fibre. These fluctuations can be attributed to two causes: (i) a heterogeneous distribution of monomers or (ii) the stochastic processes taking place during STORM image acquisition, for example, fluorophore blinking[36] and fluorophore bleaching. Therefore, variations in the density of localizations cannot be unequivocally attributed to changes in local monomer concentration. To address this issue,

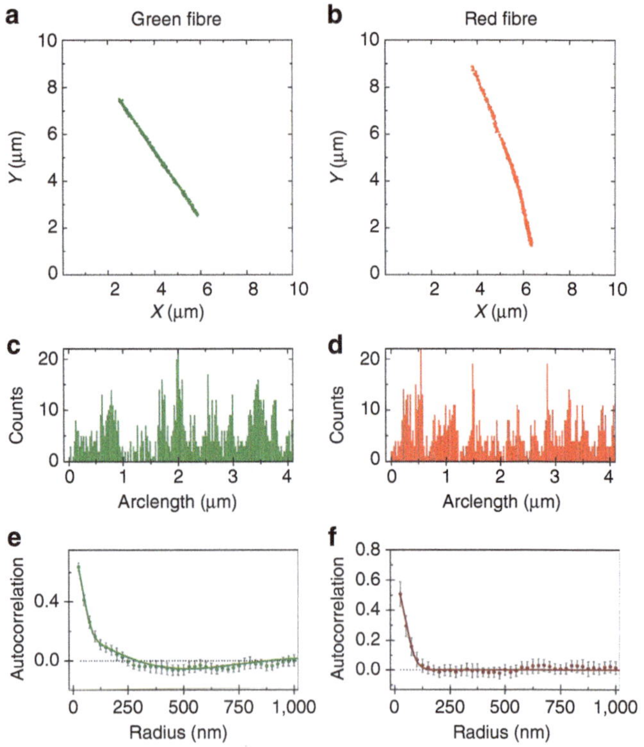

Figure 3 | STORM analysis of single-colour nanofibres. PA nanofibres were labelled with 5 mol% of either Cy3-PA (**a**,**c**,**e**) or Cy5-PA (**b**,**d**,**f**) before the molecular exchange experiment. Localization maps after applying a clustering algorithm, background cleaning and backbone finding (**a**,**b**). Each fibre localization distribution profile was determined along the fibre backbone (arc length) using a bin size of 25 nm (**c**,**d**). Averaged autocorrelation (**e**,**f**) was computed from distribution profiles of a large set of fibre images ($n \geq 19$). Solid lines correspond to model fittings (equation (1) for red channel and equation (2) for green channel) and error bars represent 95% confidence level.

we use a spatial autocorrelation algorithm, a powerful method to investigate the distribution of these localizations. In purely random distributions, localizations at distance r apart are not correlated with each other, by definition giving an autocorrelation $g(r)$ with a value of zero. Positive values of $g(r)$ represent an increased probability that a second localization is found at a certain distance (r) from a first given localization, while negative values indicate a decreased probability. The autocorrelation curves of fibres in single-colour samples are shown in Fig. 3e,f. The autocorrelation of Cy5-labelled nanofibres (Fig. 3f) is strongly positive in the short range (<100 nm) and zero for longer distances. This short-ranged contribution is the typical signature for multiple localizations of the same dye (overcounting) and is present in all STORM images. As previously described, the overcounting contribution in the spatial autocorrelation traces can be modelled using equation (1) (ref. 37),

$$g(r) = A_1 exp(-r^2/4\sigma^2) \quad (1)$$

where the standard deviation (σ) provides an estimate of the resolution achieved. Equation (1) provides a good fit to the autocorrelations of Cy5-labelled single-colour nanofibres (Fig. 3f). Therefore, the localization density fluctuations visible in the STORM images for these PA nanofibres (Fig. 3d) are due to stochastic processes inherent to the technique, while the monomer distribution is homogeneous. On the other hand, the autocorrelation function for the Cy3-channel (Fig. 3e) is not well-described by Equation (1), since the Cy3-functionalized fibres display a consistent anti-correlation ($g(r)<0$) at intermediate distances (~ 500 nm). The autocorrelation behaviour of these fibres was correctly described using a micro-emulsion model, that accounts for the existence of 'microdomains' richer in the fluorescent monomer, according to equation (2) (ref. 37):

$$g(r) = A_1 exp(-r^2/4\sigma^2) + A_2 cos(\pi r/2r_0) exp(-r/\alpha) \quad (2)$$

The first term of this equation is equal to equation (1), since the overcounting phenomenon is still present. In the second term, the parameter r_0 is the average size of the 'microdomains' and α is the coherence length of the domains. As defined by equation (2), 'microdomain' corresponds to an extended fibre region that is enriched in fluorescently labelled molecules compared with the fibre average, thus increasing the likelihood of finding those molecules in that particular region. In this context, 'microdomain' does not mean that molecules are completely clustered and confined to those regions.

An improvement in the goodness of the fit was not observed for the Cy5-labelled nanofibres, therefore it is reasonable to adopt the simpler model with less parameters, which excludes the existence of 'microdomains'. On the other hand, in the case of Cy3-labelled nanofibres a considerable improvement on the χ^2-test (>3.5 times lower) is obtained for the model with the 'microdomain' component, compared with the model that only considers the contribution of the overcounting. A 'microdomain' size of ~ 300 nm and coherence length of ~ 500 nm were obtained from the fit (Fig. 3e), showing a regular fluctuation of localization density in the Cy3-fibres at the submicron length scale. The difference observed between the distributions of Cy3 and Cy5-PAs is surprising, because of the great chemical and structural resemblance of these dyes. These results show that a small change in the PA structure may have noticeable effects on the self-assembly process, and that the STORM-based autocorrelation analysis can be used to investigate structure at the single-aggregate level. In addition, it shows that Cy5-functionalized PA is suitable for use in quantitative exchange experiments, since the Cy5 dye does not affect the molecular distribution of the PA monomers.

Fibre bundling and multicolour imaging. Since the final morphology in supramolecular polymers is highly sensitive to not only the molecular structure, but also the polymerization conditions, it is very important to design a correct sample preparation protocol to obtain the desired structure. A typical problem that may occur is the aggregation of fibres into bundles, a phenomenon that is hard to detect using traditional techniques. Avoiding fibre bundling is crucial when investigating molecular mixing, because bundles of nanofibres with different colours confound the observation of exchanging monomers. In this work we show that STORM-based analysis can be applied to detect PA fibre bundling. When nanofibre preparation was performed at higher PA concentration ($[PA]_{final} > 1\ \mu M$), extremely long and curved fibrillar structures were observed in diffraction-limited fluorescence microscopy (Supplementary Fig. 3). However, the increased resolution attained using STORM imaging revealed finer details of the adsorbed structures, showing extensive fibre bundling. What appeared to be curvature at low resolution was actually the intersection of smaller stiff fibres with different orientations. Two-colour STORM has been performed on PA samples of different concentrations and ionic strengths to unambiguously prove the presence or absence of bundled fibres. Two aqueous solutions of PA nanofibres labelled with either Cy3 or Cy5 were mixed together and immediately adsorbed onto a glass coverslip at room temperature, freezing all kinetics. For this short mixing timescale (<1 min), molecular exchange is negligible according to the FRET measurements shown in Fig. 2b. Therefore, single fibres should be either fully green or fully red. If fibres are adsorbed at high ionic strength and low concentration ($<0.1\ \mu M$), this behaviour was indeed observed (Fig. 4a). On the contrary, as shown in Fig. 4b, fibres adsorbed at concentrations higher than 1 μM can be observed in both channels simultaneously, indicating bundles of PA nanofibres. It is also possible to observe that some of the fibres visible in the green channel are not perfectly aligned with the overlapping fibres visible in the red channel (Fig. 4b). This provides further evidence that at this concentration bundles are observed instead of fully mixed fibres. Therefore, STORM allows us to verify the absence of fibre bundling and therefore select the optimal sample preparation to perform the molecular exchange experiments.

Molecular exchange kinetics and mechanisms. The ability of STORM to image individual nanofibres with high resolution and to study the distribution of different molecular species inside aggregates (*vide supra*), makes it the perfect tool to study exchange in PA samples. We prepared samples for these experiments in similar manner to the FRET measurements. First, two sets of single-colour PA nanofibres with either Cy3-PA (5%) or Cy5-PA (5%) were separately preassembled in aqueous buffer solution, followed by a 16-h aging period. Subsequently the two aged solutions were brought to 37 °C and mixed, allowing molecular exchange. Aliquots of the solution were withdrawn over the course of 48 h and nanofibres were adsorbed onto a glass

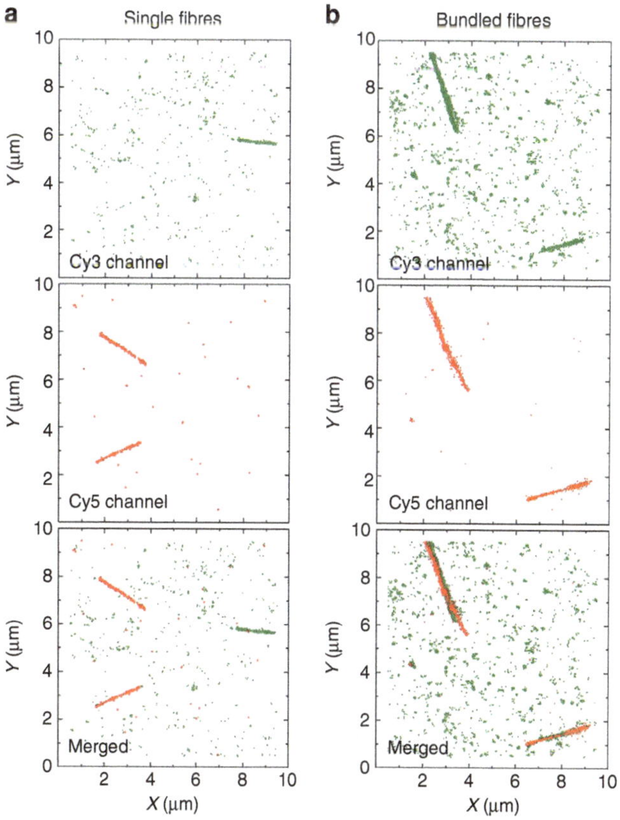

Figure 4 | Single fibre attachment. STORM images (with background) of PA nanofibres attached at a final PA concentration of either (**a**) 0.1 μM or (**b**) 1 μM, showing single fibres or bundled fibres, respectively. Attachment was performed immediately on a glass coverslip after dilution (NaCl 1 M).

coverslip at room temperature to freeze the exchange. Two-colour STORM images of assemblies were then acquired at the different time points. Fig. 5 shows representative PA nanofibres at $t=1$ min and after 48 h of mixing; as clearly shown in the images and in the corresponding localization density plot (Fig. 5b) the fibres initially containing a single label fully mix during the course of the experiment.

We verified that the exchange process is bidirectional by following green fibres incorporating red monomers (Supplementary Fig. 4) and vice versa (Supplementary Fig. 5). The intermediate time points provide insight into the exchange mechanism. Visual inspection of STORM images (Supplementary Fig. 6) showed insertion of labelled monomers over the entire length of the nanofibres throughout the exchange experiment. This finding suggests an exchange mechanism on the basis of expulsion and reinclusion of PA molecules. Polymerization–depolymerization at the nanofibre ends would have resulted in preferential exchange in those regions, while fragmentation–recombination would have resulted in a block-like structure; neither of these mechanisms is consistent with the acquired STORM images.

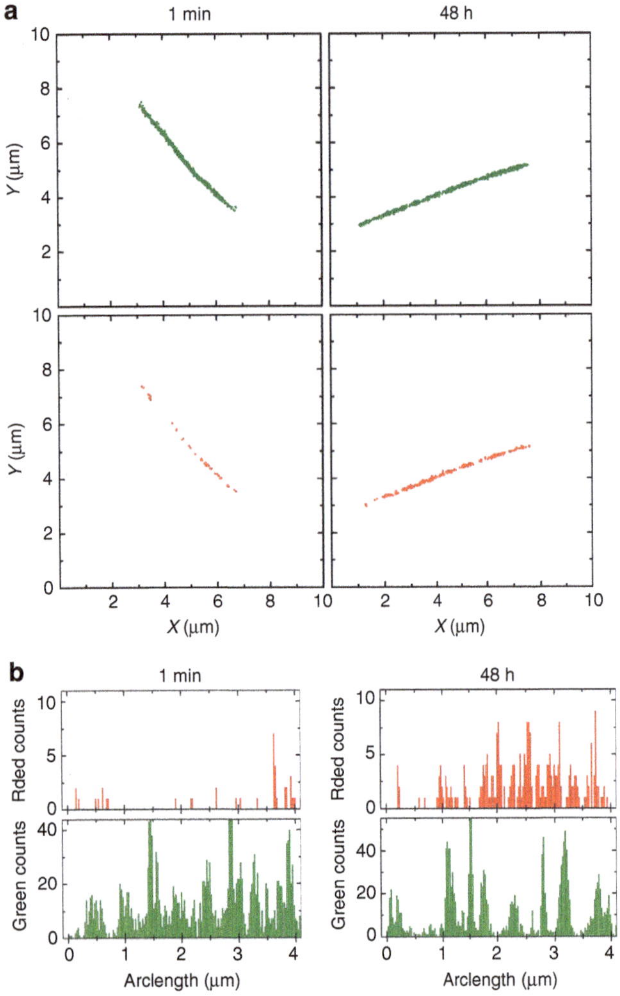

Figure 5 | Quantitative analysis of STORM imaging during a molecular exchange experiment. (**a**) Localization maps of PA nanofibres immobilized on a glass coverslip at different time points, after applying a clustering algorithm, background cleaning and backbone finding. The better reconstruction of the fibres in the green channel reflects the selection of initially Cy3-labelled fibres for this experiment. (**b**) Histograms depict the localization density profiles along the nanofibre backbone (bin size 25 nm).

To confirm this observation and further quantify the exchange, we studied the change in distribution of both dyes with time in initially green-labelled nanofibres, thus tracking the insertion of Cy5-PA. Figure 6 shows the auto- and cross-correlation plots for the green and red channels at the different time points. The green channel displays the previously observed clustered distribution described by equation (2) rendering it less amenable for quantitative analysis. Therefore, we monitored the autocorrelation of the red channel in time to analyse monomer exchange. The red channel autocorrelation, shown in Fig. 6b, can be fit with equation (1) for all time points, suggesting a random exchange of monomers. However, for later time points (that is, after incubation longer than 6 h) the quality of the fit deteriorates, indicating some non-random variations of the Cy5-PA concentration in the nanofibres other than the single decay originated from overcounting. However, the formation of regular 'microdomains' does not seem plausible, because equation (2) did not improve the quality of the fit or yield reasonable fitting parameter values. Since clustering is not observed in single-colour Cy5-PA nanofibres (Fig. 3f), this observation points to a heterogeneous exchange and indicates the presence of structural variations along the fibre or between fibres. This heterogeneity can also be perceived visually by observing single fibre images (Supplementary Fig. 6). The cross-correlation between both channels was computed to assess how the distribution of one labelled PA would affect the distribution of the other, as shown in Fig. 6c. A cross-correlation of zero was found for the entire range, indicating that the distribution of Cy5-PA is not influenced by the presence of Cy3-PA, and vice versa. In other words, the heterogeneous exchange of Cy5-PA is not due to the pre-existing clusters of Cy3-PA, but rather a consequence of structural variation in the PA nanofibre.

The observation of this heterogeneity in exchange, not detected by the FRET experiment, was possible due to the ability of STORM to evaluate the progress of molecular exchange for each individual supramolecular nanofibre. The variability of the exchange progress between different aggregates could be measured, thus allowing us to further probe the heterogeneity in the system during and after molecular exchange. The linear density of localizations, defined as the number of localizations per nanometre arc length, has been computed and could be used as a rough estimate for the ensemble concentration of labelled molecules[19]. The density of Cy5-PA in originally single-colour green fibres was measured as a function of time, showing an increase in the average value as molecular exchange proceeded (Fig. 7a). However, the standard deviation also increased markedly, another indication for the presence of intrinsic structural diversity. In Fig. 7b, it is possible to observe how the distribution of the Cy5-PA linear density gets progressively wider as molecular exchange evolves in time. This can be confirmed visually by inspecting STORM images acquired after 48 h mixing time (Fig. 7c-f), which displayed a range of different behaviours. It was possible to observe a large subset of fibres that had a Cy5-PA density consistent with full mixing (Fig. 7c), as well as fibres comprised of domains of Cy3-PA only (Fig. 7d,e), and both subsets co-existed with a smaller subpopulation of nanofibres that had undergone very little exchange all over their length (Fig. 7f). The existence of regions displaying minimal or no exchange after 48 h implies that full equilibration of these regions will take weeks to months, making them persistent for most practical purposes. This analysis suggests that the population of fibres is considerably more heterogeneous after 48 h than would be expected on the basis of the FRET ensemble measurements.

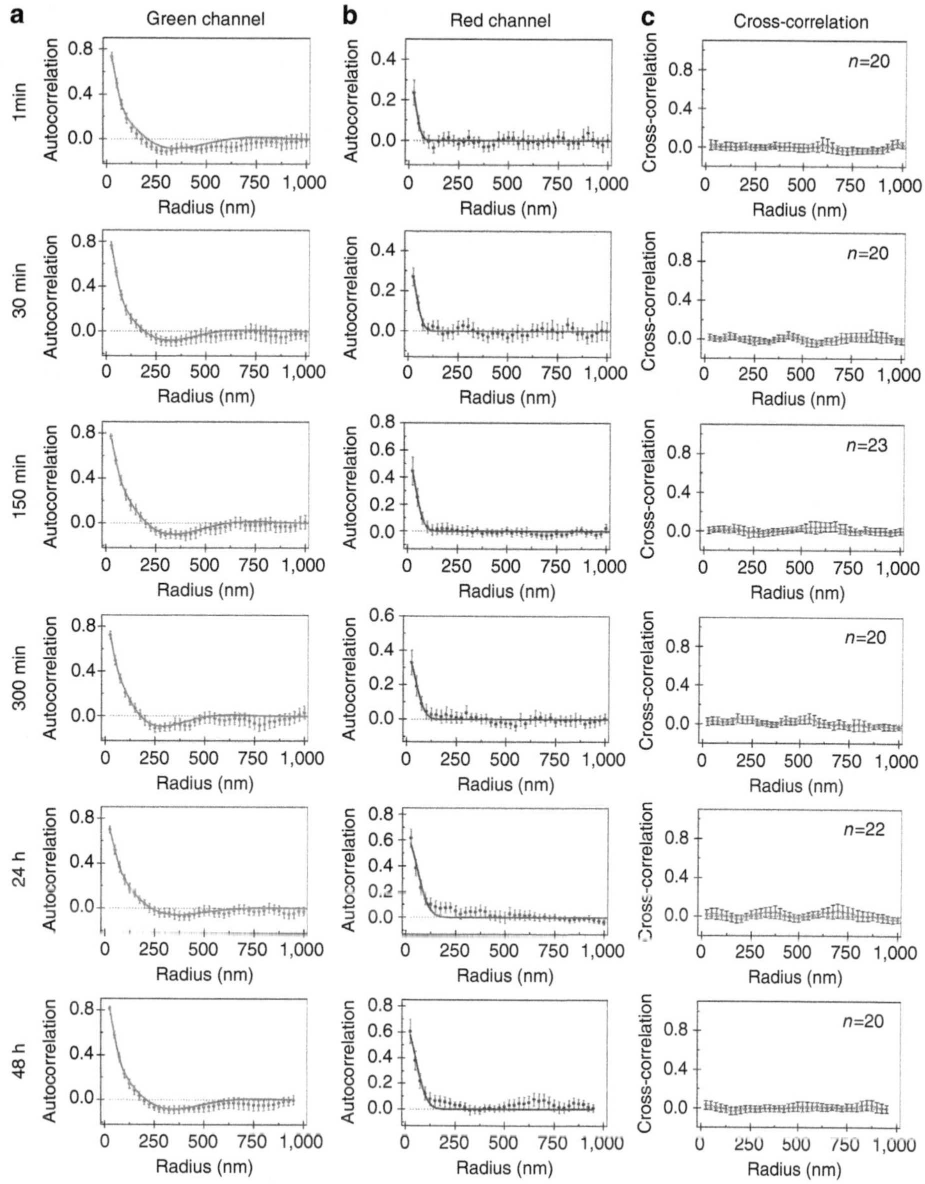

Figure 6 | Correlation of cyanine dyes along PA nanofibres initially only labelled with Cy3 during molecular exchange. Averaged autocorrelation plots for (**a**) Cy3 and (**b**) Cy5-labelled PAs and (**c**) averaged cross-correlation between both dyes. Solid lines correspond to model fittings, namely equation (2) for (**a**) green channel and equation (1) for (**b**) red channel (error bars represent 95% confidence level, $n \geq 20$).

Previous molecular dynamics simulations of a similar PA provide a theoretical framework for this structural diversity[27]. In these simulations, a broad distribution of secondary structures was found in the equilibrated fibre. The heterogeneous molecular exchange pattern observed in our study might stem from this conformational diversity. The key feature that led to the discovery of PA supramolecular nanofibres in the Stupp laboratory was the use of a β-sheet peptide domain to drive one-dimensional self-assembly. Using electron paramagnetic resonance, the β-sheet domain has been recently shown to give rise to locally solid-like behaviour in the interior of nanofibre PA assemblies[32]. On the other hand, the surface moieties in these nanofibres have been identified by the electron paramagnetic resonance experiments as regions where liquid-like dynamics prevail. β-Sheets within the nanofibres are highly cohesive assemblies and therefore the dynamic exchange among supramolecular nanofibres should give rise to a large diversity of supramolecular environments. This is in contrast to supramolecular systems with weaker internal cohesion that could exchange molecules to produce completely homogeneous environments[19]. The formulation of supramolecular systems with high levels of internal order and cohesion such as the PA nanofibres will therefore open new avenues to generate structural diversity for functional purposes. One example will be their ability to adapt structurally to bind important bioactive targets.

Discussion

We have studied the dynamics of PA nanofibres using FRET and super-resolution localization microscopy. While ensemble FRET measurements prove that PAs exchange between different fibres, two-colour STORM imaging reveals a mechanism on the basis of the transfer of monomers and small clusters. Remarkably, the coexistence of fully dynamic and kinetically inactive areas in the aggregate architecture was observed, demonstrating the existence of structural diversity in PA nanofibres. This intriguing dynamic

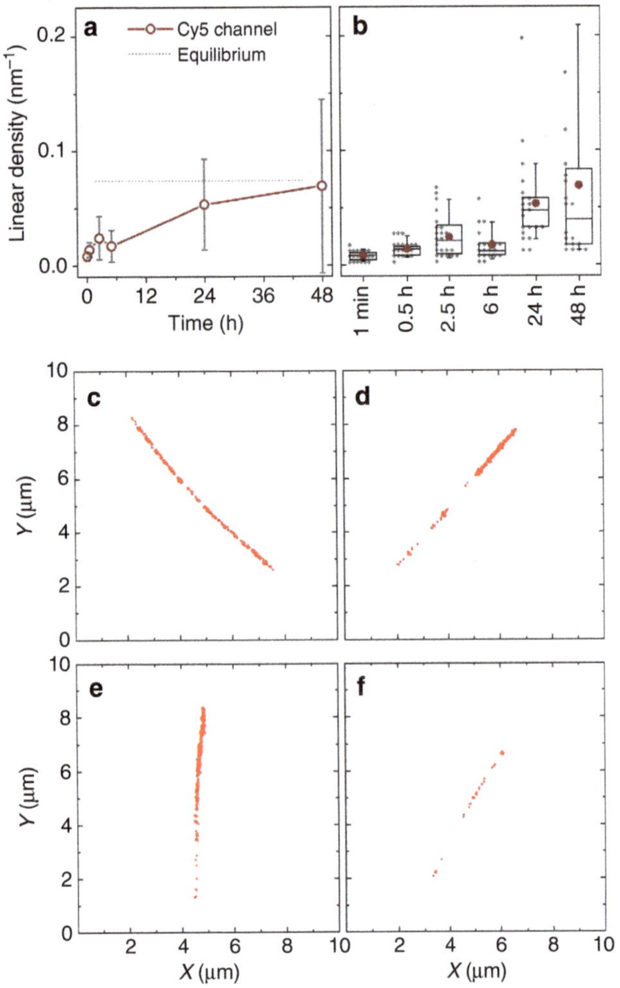

Figure 7 | PA molecular exchange kinetics estimated using the linear density of localizations calculated from STORM data. (**a**) Time course-averaged density of red dye (Cy5) on fibres that were initially only labelled with green dyes (Cy3) ($n \geq 20$, error bars correspond to standard deviation). The equilibrium guideline is estimated from the localization density of original single-colour red fibres. (**b**) Box chart of data represented on **a** depicting average (circle), median (box middle line), quartiles (box edges), 5 and 95% percentiles (error bars) and full data set points (diamond symbols, bin size 5 μm^{-1}); (**c–f**) Examples of localization plots (after background cleaning) of individual fibres at 48 h.

behaviour is foreseen to have important implications in the biological performance of supramolecular systems.

Methods

Peptide synthesis and purification. PAs were synthesized using standard Fmoc solid-phase peptide synthesis. MBHA rink amide resin, Fmoc-protected amino acids and other solid-phase peptide synthesis reagents were purchased from Novabiochem (USA). Cyanine dyes were supplied by Cyandye (USA). ACS-grade solvents were used for synthesis. Water was purified on an EMD Milipore Milli-Q Integral Water Purification System. For each amino acid coupling, 4.1 equiv of Fmoc-protected amino acid was activated for 1 min with 4 equiv of HBTU (O-benzotriazole-*N,N,N′,N′*-tetramethyluroniumhexafluorophosphate) and 6 equiv DIPEA (*N,N*-diisopropylethylamine) in 20 ml DMF. The coupling cocktail was then added to 0.5 mmol of MBHA rink resin and reacted for 1 h. Fmoc removal was performed with 30% piperidine in DMF. Fmoc-Glu(OtBu)-OH (N-α-Fmoc-L-glutamic acid γ-t-butyl ester) (3 ×), and Fmoc-Ala-OH (N-α-Fmoc-L-alanine) (6 ×) were successively coupled to the resin. Finally, the palmitic acid tail (8.1 equiv) was coupled using 8 equiv of HBTU and 12 equiv DIPEA in DMF/DCM (50:50). PAs were cleaved from the resin using a cleavage solution of 95% TFA, 2.5% TIPS (triisopropylsilane) and 2.5% H_2O. Cleavage solution was concentrated by rotary evaporation and the PA precipitated in cold diethyl ether. The crude solid was dissolved in a diluted NH_4OH aqueous solution (~10 mg ml^{-1}) and purified by preparative-scale reversed-phase HPLC on a Varian Prostar 210 HPLC system, using acetonitrile/water gradient containing 0.1% NH_4OH. All eluents and additives were HPLC grade. Separation was achieved using a Phenomenex C18 Gemini NX column (150 × 30 mm) with 5 μm particle size and 110 Å pore size. Product-containing fractions were confirmed by ESI mass spectrometry (Agilent 6510 Q-TOF LC/MS) and combined. ACN was removed by rotary evaporation, lyophilized to yield a white solid product and stored at −30 °C.

Cyanine dye coupling. PAs labelled with cyanine dyes were synthesized first coupling Fmoc-Lys(Mtt)-OH (*N*-α-Fmoc-*N*-ε-4-methyltrityl-L-lysine) to the resin. The rest of the synthetic procedure proceeded as described above. After coupling the palmitic acid tail, Lys(Mtt) was selectively deprotected with 3% TFA, 5% TIPS and 92% DCM for 5 min (3 ×), to expose the side chain amine moiety. Disulfo-cyanine carboxylic acid dyes (Cy3 and Cy5) were dissolved in DMF and stock solutions were stored at -30 °C. (Benzotriazol-1-yloxy)tripyrrolidinophosphonium hexafluorophosphate (PyBOP) was also dissolved in DMF. Cyanine dyes (1 equiv) were activated with PyBOP (1 equiv) and DIPEA (2 equiv) for 1 min and coupled to the lysine-free amine. Cy3- and Cy5-labelled PAs were cleaved and purified as described above. Lyophilization yielded a red powder for Cy3 and blue for Cy5.

Purity. Purity was determined by reversed-phase analytical HPLC (see Supplementary Figs 7–9) using a Phenomenex Gemini C18 column (100 × 4.6 mm), with 5 μm particle size and 110 Å pore size, connected to a HPLC system equipped with an autosampler (Shimadzu SIL-20A XR), degasser (Shimadzu DGU-20A3) and a high-pressure gradient system comprising two LC-20AD XR pumps (Shimadzu). Separation was performed at a flow of 1 ml min^{-1} and using a 20-60% acetonitrile linear gradient in water containing 0.1% NH_4OH. All eluents and additives were HPLC grade. Peptides, cyanine dye-coupled peptides and eventual contaminants were detected using a PDA (Shimadzu SPD-M20A), acquiring a full ultraviolet-visible spectrum between 200 and 750 nm at any time point. Main chromatogram peaks were collected and product mass confirmed by ESI mass spectrometry using a LCQ Deca XP Max (Thermo Finnigan) ion-trap mass spectrometer. For that, samples were manually injected bypassing the column at a flow rate of 0.20 ml min^{-1} and positive ion mass spectra were acquired in standard enhanced mode.

Sample preparation. Stock solutions of non-labelled PA (10 mM), Cy3-labelled PA (1 mM) and Cy5-labelled PA (1 mM) were prepared in Milli-Q water, aliquoted and stored at −30 °C. Similar to other anionic PA systems, these formed hydrogels in the presence of $CaCl_2$. The degree of labelling of the fibres was accurately controlled by simply mixing PA stock solutions at the desired ratio. Solutions of non-labelled PA were mixed with aqueous solutions of either one or both labels at certain mole ratios, flash-frozen in liquid nitrogen and lyophilized. The obtained PA mixture was molecularly dissolved in TFA and readily evaporated under vacuum. The obtained material was redissolved in NH_4OH (aq.), flash-frozen and further lyophilized. These two last steps resembled the cleavage and purification conditions, respectively. These steps were undertaken to provide molecular mixing, as well as to reset PA self-assembly history. Non-labelled PA was treated using the same method before measurements where the fluorescence-labelled molecules were absent. The obtained powders were reconstituted in appropriate buffers before measurements. After the harsh acidic conditions (dissolution in TFA) used to create a homogenous molecular mixture, PAs were checked by analytical HPLC to rule out potential degradation of the original molecules. Analytical HPLC was performed as described above. Supplementary Fig. 10 shows that the PA does not degrade during the harsh procedure of molecular mixing in TFA. Fluorescence measurements showed pronounced FRET for nanofibres labelled with both Cy3 and Cy5. The FRET intensity increased with concentration of acceptor. This indicates that the sample preparation is suitable to assure co-assembly of the three PAs in a single supramolecular object.

Circular dichroism spectroscopy. Circular dichroism spectra were recorded using Jasco Circular Dichroism Spectrometer (model J-715). PA was dissolved in water at 10 mM. Before each measurement, PA stock solution was diluted in appropriate buffers to yield a final PA concentration of 0.1 mM. A quartz cuvette of 1 mm path length was used for the measurements. Each trace represents the average of five scans.

Cryogenic transmission electron microscopy. PA nanofibres were separately labelled with either PA-Cy3 (5%) or PA-Cy5 (5%). Both labelled and non-labelled PA nanofibres were dissolved at 1 mg ml^{-1} in 10 mM HEPES buffer (pH 7.5, NaCl 150 mM). CryoTEM was performed using a JEOL 1230 TEM at an accelerating voltage of 100 kV. Using a Vitrobot Mark IV (FEI) vitrification instrument at 25 °C with 100% humidity, a 6.5-μl drop of the sample was deposited on a 300-mesh copper grid with lacey carbon support (Electron Microscopy Sciences, EMS), blotted twice, plunge-frozen in liquid ethane, then stored in liquid N_2. For imaging, the sample was transferred to a Gatan 626 cryo-holder under liquid N_2, and images were obtained with a Gatan 831 CCD camera.

Critical micelle concentration by Nile Red assay. Nile Red is a hydrophobic solvatochromic dye (water soluble at low concentration) with high affinity to the hydrophobic pocket of micelles. In aqueous environment, the molecule is weakly fluorescent. In hydrophobic environments, fluorescence intensity is around 300 times higher and a pronounced blue shift is observed. Therefore, Nile Red is commonly used to probe the existence of hydrophobic pockets formed by surfactants in aqueous environment[38]. Briefly, a Nile Red 2 mM stock solution was prepared in MeOH. Non-labelled PA was dissolved at 10 mM in 10 mM HEPES buffer (pH 7.5, NaCl 150 mM). PA solution was diluted serially to obtain concentrations in the range 200 nM-10 mM. Nile Red stock solution was diluted in the same buffer to obtain a final concentration of 1 μM. Equal volumes of Nile Red aqueous diluted solution and PA solutions were mixed for each PA concentration. Solutions were aged for 24 h to assure full nanofibre disassembly at concentrations below the critical micelle concentration. Fluorescence-emission spectra (excitation 550 nm) were recorded for an emission range between 570 and 700 nm. The maximum intensity and respective wavelength at maximum intensity were both represented as a function of the logarithm of the PA concentration. At concentrations close to critical micelle concentration, it should be observed a sharp increase in fluorescence intensity and a hypsochromic effect.

Förster resonance energy transfer. Single-colour PA nanofibres labelled with either Cy3 or Cy5 at 1 mol% were assembled in 10 mM HEPES buffer (pH 7.5, NaCl 150 mM). Fluorescence spectra were recorded in a Varian Cary Eclipse Fluorimeter from Agilent Technologies. Emission spectra were acquired using excitation wavelengths of 520 or 615 nm for Cy3 or Cy5, respectively. Excitation spectra were collected using emission wavelengths of 600 or 710 nm for Cy3 or Cy5, respectively. Temperature was kept at 20°C using the in-built peltier system. Supplementary Fig. 11a depicts the spectra of the individually labelled PA nanofibres, showing significant overlap between Cy3 emission (donor) and Cy5 excitation (acceptor). On an independent measurement, PA nanofibres labelled with both Cy3 (1 mol%) and Cy5 (0, 0.1, 0.2 and 0.5 mol%) were assembled in 10 mM HEPES buffer (pH 7.5, NaCl 150 mM) at a final PA concentration of 0.5 mM. Fluorescence-emission spectra of dual-labelled PAs were recorded at the excitation wavelength of Cy3 520 nm, at which direct excitation of Cy5 is negligible. Supplementary Fig. 11b shows considerable energy transfer by FRET, which increases monotonically with acceptor concentration, providing evidence for co-assembly of both Cy3- and Cy5-labelled PAs in the same nanofibres.

Ensemble molecular exchange kinetics. For molecular exchange kinetic experiments, two sets of PA nanofibres were pre-assembled from a mixture of non-labelled PAs and either Cy3 (0.5%) or Cy5 (0.5%). These independently labelled PA solutions were mixed at 1:1 ratio in a 50 μl quartz cuvette and fluorescence was followed over around 24 h (lag time ~30 s). FRET ratio was determined by dividing Cy5 emission at 668 nm by Cy3 emission at 565 nm, at the Cy3 excitation wavelength (520 nm). The contribution of Cy3 emission at 668 nm was first subtracted from the Cy5 emission. At the beginning of the kinetics experiment, Cy3-PA and Cy5-PA are distributed in different nanofibres, being physically separated by distances much higher than the expected FRET radius[35]. Therefore, energy transfer at time zero should be negligible and, consequently, FRET ratio should be zero. A total PA concentration range of 0.5–5 mM was used. Kinetics was recorded at two different temperatures (20 or 37 °C).

Sample preparation for microscopy. Glass microscope coverslips were cleaned by successively immersing in acetone, isopropanol and Milli-Q water. Bath sonication was performed for 10 min with each solvent, followed by drying under N_2 flow. The glass coverslips were then etched with a fresh Piranha solution (3:1 v/v H_2SO_4 (98%):H_2O_2 (30%)) for 30 min. To finish the cleaning procedure, the slides were washed thoroughly with Milli-Q water and rinsed with acetone before drying under N_2 flow. To determine the labelled PA distribution along the nanofibres length, single fibres should be immobilized on a glass surface, and remain still during STORM acquisition. A flow chamber was assembled using a glass slide and a clean coverslip separated by double-sided tape. PA nanofibres were immobilized by adsorption onto the surface of the clean coverslip by flushing diluted PA solutions at high ionic strength (NaCl 1 M). After incubation for 1 min, the unbound nanofibres were washed out of the chamber by flushing with HEPES buffer (2 ×) followed by STORM buffer (2 ×). STORM buffer is composed by 50 mM HEPES buffer (pH = 8.5), 1 M NaCl, an oxygen-scavenging system (0.5 mg ml^{-1} glucose oxidase, 40 μg ml^{-1} catalase, 5 wt% glucose) and 200 mM 2-aminoethanethiol. The high ionic strength conditions were used to screen the electrostatic repulsion between highly charged PA nanofibres and the glass surface. However, repulsion between PA nanofibres is also reduced, causing fibre bundling. To avoid bundling, PA solutions were rapidly diluted at low ionic strength (1 μM). After a final (10 ×) dilution step in NaCl 1M, nanofibres were immediately attached on the glass coverslip. At this concentration isolated nanofibres are mainly observed in the microscopic mounting.

Stochastic optical reconstruction microscopy. STORM images were acquired using a Nikon N-STORM system configured for total internal reflection fluorescence imaging. Excitation inclination was tuned to maximize the signal-to-noise ratio of the glass-absorbed fibres. Cy3- and Cy5-labelled samples were illuminated by 561 and 647 nm laser lines. No activation ultraviolet light was used. Fluorescence was collected by means of a Nikon 100 ×, 1.4NA oil immersion objective and passed through a quad-band pass dichroic filter (97335 Nikon). All time-lapses were recorded onto a 64 × 64 pixel region (pixel size 0.17 μm) of an EMCCD camera (ixon3, Andor). For each channel, 20,000 frames were sequentially acquired. STORM movies were analysed with the STORM module of the NIS Elements Nikon software.

Molecular imaging of fibres during exchange. PA nanofibres were separately labelled with either PA-Cy3 (5%) or PA-Cy5 (5%) at 0.5 mM in 10 mM HEPES buffer (pH 7.5, NaCl 150 mM). Subsequently, 250 μl-aliquots of each solution were mixed in a microcentrifuge tube and gently shaken at 37 °C for different amounts of time. At predefined time points, 2 μl of the solution was withdrawn, diluted and adsorbed on a glass surface as described above. We verified that once adsorbed on the slide, fibres are not able anymore to considerably exchange monomers with the solution in the time frame of the experiment. STORM imaging of Cy5 and Cy3 channels was performed sequentially. Fibres that are initially labelled only with Cy3 and incorporate Cy5 monomers over time were selected observing the fluorescence intensity in the low-resolution images. Original single-colour nanofibres were also observed before mixing as a control.

Image analysis. During STORM imaging, the Nikon software generates a list of localizations by 2D Gaussian fitting of blinking chromophores in the acquired movie of conventional microscopic images. This localizations list was subsequently analysed with our custom-made Matlab scripts, as described in detail elsewhere[19]. Briefly, a first script uses a density-based clustering algorithm to automatically identify the fibres in the image and to remove the background. Examples of output localization maps are shown in Figs 3a,b and 5a. A second script is run on the well-reconstructed fibres obtained in the Cy3-channel to find the fibre backbone. The lower localization density obtained in the Cy5-channel for shorter time periods in the molecular exchange experiment is not enough to fully reconstruct the nanofibres. The polymer backbone coordinates are used to obtain structural information. Next, histograms of localization density profiles along the backbone are generated for both channels, as shown in Figs 3c,d and 5b. The spatial autocorrelation (that is, the average correlation between the localization density at one backbone point in the fibre and the density at another backbone point, as a function of the distance between these points) was also computed using the Matlab-function xcov with unbiased normalization. The spatial cross-correlation of the two channels was calculated in a similar manner. Averaging over multiple fibres is required to obtain accurate correlation decay graphs, such as depicted in Figs 3e,f and 6. Investigation of the average correlation decays yields the distribution of dyes in a fibre, without interference from the stochastic fluctuations inherent to the technique.

References

1. Dominguez, R. & Holmes, K. C. Actin structure and function. *Annu. Rev. Biophys.* **40,** 169–186 (2011).
2. Janke, C. & Kneussel, M. Tubulin post-translational modifications: encoding functions on the neuronal microtubule cytoskeleton. *Trends Neurosci.* **33,** 362–372 (2010).
3. Pollard, T. D. & Borisy, G. G. Cellular motility driven by assembly and disassembly of actin filaments. *Cell* **112,** 453–465 (2003).
4. Pollard, T. D., Blanchoin, L. & Mullins, R. D. Molecular mechanisms controlling actin filament dynamics in nonmuscle cells. *Annu. Rev. Biophys. Biomol. Struct.* **29,** 545–576 (2000).
5. Stournaras, C., Gravanis, A., Margioris, A. N. & Lang, F. The actin cytoskeleton in rapid steroid hormone actions. *Cytoskeleton* **71,** 285–293 (2014).
6. Symons, M. & Rusk, N. Control of vesicular trafficking by Rho GTPases. *Curr. Biol.* **13,** R409–R418 (2003).
7. Albertazzi, L. *et al.* Spatiotemporal control and superselectivity in supramolecular polymers using multivalency. *Proc. Natl Acad. Sci. USA* **110,** 12203–12208 (2013).
8. Marchi-Artzner, V. *et al.* Selective adhesion, lipid exchange and membrane-fusion processes between vesicles of various sizes bearing complementary molecular recognition groups. *Chemphyschem Eur. J. Chem. Phys. Phys. Chem.* **2,** 367–376 (2001).
9. Ferrell, J. E., Lee, K. J. & Huestis, W. H. Lipid transfer between phosphatidylcholine vesicles and human-erythrocytes–exponential decrease in rate with increasing acyl chain-length. *Biochemistry* **24,** 2857–2864 (1985).
10. Lund, R., Willner, L., Richter, D. & Dormidontova, E. E. Equilibrium chain exchange kinetics of diblock copolymer micelles: tuning and logarithmic relaxation. *Macromolecules* **39,** 4566–4575 (2006).
11. Zinn, T., Willner, L., Lund, R., Pipich, V. & Richter, D. Equilibrium exchange kinetics in n-alkyl–PEO polymeric micelles: single exponential relaxation and chain length dependence. *Soft Matter* **8,** 623–626 (2011).

12. Fasting, C. *et al.* Multivalency as a chemical organization and action principle. *Angew. Chem. Int. Ed.* **51,** 10472–10498 (2012).
13. Helms, B. A. *et al.* High-affinity peptide-based collagen targeting using synthetic phage mimics: from phage display to dendrimer display. *J. Am. Chem. Soc.* **131,** 11683 (2009).
14. Silva, G. A. *et al.* Selective differentiation of neural progenitor cells by high-epitope density nanofibers. *Science* **303,** 1352–1355 (2004).
15. Baker, M. B. *et al.* Consequences of chirality on the dynamics of a water-soluble supramolecular polymer. *Nat. Commun.* **6,** 6234 (2015).
16. Bates, M., Huang, B., Dempsey, G. T. & Zhuang, X. Multicolor super-resolution imaging with photo-switchable fluorescent probes. *Science* **317,** 1749–1753 (2007).
17. Shim, S.-H. *et al.* Super-resolution fluorescence imaging of organelles in live cells with photoswitchable membrane probes. *Proc. Natl Acad. Sci. USA* **109,** 13978–13983 (2012).
18. Huang, B., Bates, M. & Zhuang, X. Super-resolution fluorescence microscopy. *Annu. Rev. Biochem.* **78,** 993–1016 (2009).
19. Albertazzi, L. *et al.* Probing exchange pathways in one-dimensional aggregates with super-resolution microscopy. *Science* **344,** 491–495 (2014).
20. Webber, M. J., Berns, E. J. & Stupp, S. I. Supramolecular nanofibers of peptide amphiphiles for medicine. *Isr. J. Chem.* **53,** 530–554 (2013).
21. Boekhoven, J. & Stupp, S. I. 25th anniversary article: supramolecular materials for regenerative medicine. *Adv. Mater.* **26,** 1642–1659 (2014).
22. Stupp, S. I. & Palmer, L. C. Supramolecular chemistry and self-assembly in organic materials design. *Chem. Mater.* **26,** 507–518 (2014).
23. Soukasene, S. *et al.* Antitumor activity of peptide amphiphile nanofiber-encapsulated camptothecin. *ACS Nano* **5,** 9113–9121 (2011).
24. Moyer, T. J. *et al.* Shape-dependent targeting of injured blood vessels by peptide amphiphile supramolecular nanostructures. *Small* **11,** 2750–2755 (2015).
25. Paramonov, S. E., Jun, H.-W. & Hartgerink, J. D. Self-assembly of peptide – amphiphile nanofibers: the roles of hydrogen bonding and amphiphilic packing. *J. Am. Chem. Soc.* **128,** 7291–7298 (2006).
26. Pashuck, E. T., Cui, H. G. & Stupp, S. I. Tuning supramolecular rigidity of peptide fibers through molecular structure. *J. Am. Chem. Soc.* **132,** 6041–6046 (2010).
27. Lee, O.-S., Stupp, S. I. & Schatz, G. C. Atomistic molecular dynamics simulations of peptide amphiphile self-assembly into cylindrical nanofibers. *J. Am. Chem. Soc.* **133,** 3677–3683 (2011).
28. Storrie, H. *et al.* Supramolecular crafting of cell adhesion. *Biomaterials* **28,** 4608–4618 (2007).
29. Stephanopoulos, N., Ortony, J. H. & Stupp, S. I. Self-assembly for the synthesis of functional biomaterials. *Acta Mater.* **61,** 912–930 (2013).
30. Greenfield, M. A., Hoffman, J. R., de la Cruz, M. O. & Stupp, S. I. Tunable mechanics of peptide nanofiber gels. *Langmuir* **26,** 3641–3647 (2010).
31. Newcomb, C. J. *et al.* Cell death versus cell survival instructed by supramolecular cohesion of nanostructuresl. *Nat. Commun.* **5,** 3321 (2014).
32. Korevaar, P. A., Newcomb, C. J., Meijer, E. W. & Stupp, S. I. Pathway selection in peptide amphiphile assembly. *J. Am. Chem. Soc.* **136,** 8540–8543 (2014).
33. Tantakitti, F. *et al.* Energy landscapes and functions of supramolecular systems. *Nat. Mater.* **15,** 469–476 (2016).
34. Levitt, M. Conformational preferences of amino acids in globular proteins. *Biochemistry* **17,** 4277–4285 (1978).
35. Bastiaens, P. I. & Jovin, T. M. Microspectroscopic imaging tracks the intracellular processing of a signal transduction protein: fluorescent-labelled protein kinase C beta I. *Proc. Natl Acad. Sci. USA* **93,** 8407–8412 (1996).
36. Rust, M. J., Bates, M. & Zhuang, X. Sub-diffraction-limit imaging by stochastic optical reconstruction microscopy (STORM). *Nat. Methods* **3,** 793–796 (2006).
37. Veatch, S. L. *et al.* Correlation functions quantify super-resolution images and estimate apparent clustering due to over-counting. *PLoS ONE* **7,** e31457 (2012).
38. Stuart, M. C. A., van de Pas, J. C. & Engberts, J. B. F. N. The use of Nile Red to monitor the aggregation behavior in ternary surfactant-water-organic solvent systems. *J. Phys. Org. Chem.* **18,** 929–934 (2005).

Acknowledgements

Molecular synthesis, purification and characterization were supported by the U.S. Department of Energy, Office of Science, Basic Energy Sciences, under Award # DE-FG02-00ER45810. The work in Eindhoven was supported by was financed by the Dutch Ministry of Education, Culture and Science (Gravity program 024.001.035) and the European Research Council (FP7/2007-2013, ERC Grant Agreement 246829). R.M.P.d.S. and his research activities were funded by the Marie Curie FP7-PEOPLE-2010-IOF program (ref. no. 273295). Compound preparation, purification and characterization were partially performed at the Peptide Synthesis Core at Simpson Querrey Institute. The Biological Imaging Facility (BIF) and Keck Biophysics facilities at Northwestern provided further instrumentation used in this work. Mark Seniw is acknowledged for the preparation of graphics.

Author contributions

R.M.P.d.S., L.A., D.v.d.Z., S.I.S. and E.W.M. conceived the project and contributed to the concept of the manuscript. R.M.P.d.S., L.A., D.v.d.Z. designed the experiments and analysed the results. R.M.P.d.S. performed peptide synthesis, spectroscopic and STORM experiments. S.S.L. performed cryoTEM imaging. R.M.P.d.S., L.A., D.v.d.Z., S.I.S. and E.W.M. wrote the manuscript and contributed to discussions on the project.

Additional information

Competing financial interests: The authors declare no competing financial interests.

22

Rational design of ABC triblock terpolymer solution nanostructures with controlled patch morphology

Tina I. Löbling[1,2], Oleg Borisov[3,4,5], Johannes S. Haataja[2], Olli Ikkala[2], André H. Gröschel[2,†] & Axel H.E. Müller[6]

Block copolymers self-assemble into a variety of nanostructures that are relevant for science and technology. While the assembly of diblock copolymers is largely understood, predicting the solution assembly of triblock terpolymers remains challenging due to complex interplay of block/block and block/solvent interactions. Here we provide guidelines for the self-assembly of linear ABC triblock terpolymers into a large variety of multicompartment nanostructures with C corona and A/B cores. The ratio of block lengths N_C/N_A thereby controls micelle geometry to spheres, cylinders, bilayer sheets and vesicles. The insoluble blocks then microphase separate to core A and surface patch B, where N_B controls the patch morphology to spherical, cylindrical, bicontinuous and lamellar. The independent control over both parameters allows constructing combinatorial libraries of unprecedented solution nanostructures, including spheres-on-cylinders/sheets/vesicles, cylinders-on-sheets/vesicles, and sheets/vesicles with bicontinuous or lamellar membrane morphology (patchy polymersomes). The derived parameters provide a logical toolbox towards complex self-assemblies for soft matter nanotechnologies.

[1] Macromolecular Chemistry II, University of Bayreuth, D-95440 Bayreuth, Germany. [2] Department of Applied Physics, Aalto University School of Science, FIN-02150 Espoo, Finland. [3] Institut Pluridisciplinaire de Recherche sur l'Environnement et les Matériaux UMR 5254 CNRS/UPPA, F-64053 Pau, France. [4] Institute of Macromolecular Compounds, Russian Academy of Sciences, 199004 St Petersburg, Russia. [5] St Petersburg State Polytechnic University, 195251 St Petersburg, Russia. [6] Institute of Organic Chemistry, Johannes Gutenberg-Universität Mainz, D-55099 Mainz, Germany. † Present address: Department of Physical Chemistry, University of Duisburg-Essen, D-45127 Essen, Germany. Correspondence and requests for materials should be addressed to O.I. (email: olli.ikkala@aalto.fi) or to A.H.G. (email: andre.groeschel@uni-due.de) or to A.H.E.M. (email: axel.mueller@uni-mainz.de).

On pursuing increasingly functional materials, one approach aims at bridging the gap between simplistic self-assemblies and the sophisticated level of structural control that is characteristic for biological systems. Therein, multiphase synthetic building blocks that involve a rich interplay of interactions have demonstrated progressively complex self-assembly behaviour[1–5]. In that regard, block copolymers find widespread appreciation in materials science due to their ability to form nanostructures with progressively complex composition, shape and function[6–9]. The solution self-assembly of AB diblock copolymers is comparably straightforward, because solvent selectivity divides the blocks into micellar core and corona. Structural diversity is thereby controlled by block lengths and limited to the geometry that ranges from spheres to cylinders to bilayer sheets and vesicles (polymersomes)[10–12]. For the next degree of complexity, that is, multiphase particles[13,14] or hierarchical assemblies, three or more polymer blocks are required so that at least two blocks phase separate inside the micellar core. Established synthetic techniques provide the necessary building blocks through sequential block extension to linear ABC triblock terpolymers (and beyond). And although some intricate multicompartment nanostructures have been experimentally verified in solution[3,15–18], the multitude of independent interaction parameters between polymer blocks and solvent still complicates rational structuring of terpolymers in solution. As a result, multiphase nanostructures are often perceived as exotic or surprising. So far, only few approaches offer precise control over block positioning in solution self-assembly of ABC triblock terpolymers, creating well-defined core-segmented cylinder micelles and spherical 'patchy' micelles, therein[19–25]. Nanostructured bilayer sheets and vesicles have been observed only in isolated cases[26–29]. Controlling the polymorphism of both micelle geometry and patch morphology still remains a generic challenge.

Here we provide guidelines for the self-assembly of linear ABC triblock terpolymers that allow the separate tuning of micelle geometry and patch morphology. The separate control enables us to construct a library of solution nanostructures, where classical micelle polymorphs such as spheres, cylinders, bilayer sheets and vesicles are further subclassified by their patchy morphology. We identify 13 out of the 16 possible micelle/patch combinations, which are rationalized by a scaling theory that predicts structural transitions and stability regions for each combination.

Results

Self-assembly system. As model system we chose polystyrene-*block*-polybutadiene-*block*-poly(*tert*-butyl methacrylate) (PS-*b*-PB-*b*-PT or SBT) in acetone/isopropanol mixtures (Fig. 1a). Specifics for all polymers are summarized in Supplementary Table 1. Block solubilities and measurements to determine a swelling factor q to correct the PS volume in dependence of the acetone content are summarized in Supplementary Figs 1–4 and Supplementary Tables 2 and 3. In an analogy to lipids and diblock copolymers[30,31], we expect the amphiphilic balance, that is, the volume ratio of corona over core (related to the block PT), to direct self-assembly into spherical micelles, cylinders, and bilayer sheets and vesicles (Fig. 1b). Concurrent microphase separation of the insoluble and immiscible core blocks, PS and PB, then lead to an inner PS core and a PB surface pattern (Fig. 1c)[32,33]. The separate control over micelle geometry and patch morphology allows constructing combinatorial libraries of multicompartment solution nanostructures, for example, spheres-on-cylinders or cylinders-on-sheets, and so on. To target these and other combinations, we synthesized a series of SBT triblock terpolymers with systematically varied block lengths abbreviated as N_S, N_B and N_T (for example, $S_{540}B_{173}T_{137}$).

Self-assembly procedure. In our particular system, that is, PS-*b*-PB-*b*-PT in acetone/isopropanol mixtures, PT is always soluble, PB insoluble and the collapse of PS is controlled by the solvent composition. We use a two-step self-assembly process to direct each block into the desired location within the solution nanostructure[21]. Starting in *N*,*N*-dimethylacetamide (DMAc), the PB middle block selectively collapses to form the core of precursor micelles with a corona composed of PT and PS blocks. Dialysis against acetone/isopropanol mixtures as non-solvent for both PB and PS induces aggregation of the precursor micelles to form spherical multicompartment micelles with a PS core, spherical PB patches stabilized by a PT corona (see Supplementary Fig. 5 for assembly path). Proton nuclear magnetic resonance spectroscopy during dialysis (see Supplementary Fig. 6 for kinetic plot) shows that after 90 min the entire starting solvent had been replaced with the final solvent mixture. During solvent exchange and in the final solvent mixture, all polymer blocks remain in a soft and dynamic state, which reduces the probability of kinetic trapping and facilitates structural rearrangement ($T_{g,PB} < -20\,°C$, acetone plasticises PS). As will be discussed below, depending on the length of the stabilizing corona, the spherical multicompartment micelles undergo a morphological evolution towards higher-order superstructures over time. This slow structural rearrangement occurs on the order of days and is attributed to slow rearrangement of the PS domains. All samples were allowed to equilibrate until no further structural evolution could be detected anymore in TEM (days to weeks depending on the sample). We will first discuss the control over micelle geometry and patch morphology, and characterize the structural features of each identified nanostructure individually, before we go into more detail about the self-assembly kinetics later.

Controlling micelle geometry. We first address the effect of corona length, N_T, on micelle geometry (Fig. 2). Previously, we used scaling arguments to define stability criteria for the special case of spherical micelles with spherical patches[21]. Here we extend this theory to rationalize the polymorphism of patchy micelles to account for morphological transition from spheres-on-spheres to spheres-on-cylinders and further to spheres-on-bilayer sheets/vesicles. Polymorphism could be expected given the delicate balance between the gain in the conformational entropy of the core-forming PS blocks and the penalty in curvature-dependent part of repulsive interactions in the solvated coronal PT domains. These transitions would occur when the radius of the corona becomes smaller than the core radius ($H_{corona} \leq R_{PS}$), that is, when micelles have the crew-cut shape. By acquainting the free energies per chain for different morphologies, we find the binodal (transition) lines as

$$\left(\tilde{N}_T/\left(qN_S^{2/3}\right)\right)^{(\text{sp.}\leftrightarrow\text{cyl.},\text{cyl.}\leftrightarrow\text{lam.})} \simeq \left(\nu_S/\ell^3\right)\left(\gamma_{PS/\text{solvent}}\ell^2\right)^{1/3}\left(\nu_T/\ell^3\right)^{-1/3}, \qquad (1)$$

where numerical coefficients are omitted, $N_{S,B,T}$ are the degrees of polymerization and $\nu_{S,B}$ the volumes of the monomer units of the respective blocks; $\nu_T \sim l^3$ is the excluded volume parameter under good solvent conditions, l the length of the monomer unit (the Kuhn segment), and $k_B T\gamma_{PS/solvent}$ is the surface tension at the PS/solvent interface. The correction factor q accounts for selective swelling of the PS domain (Supplementary Figs 2–4 and Supplementary Note 1). A more detailed and generalized theory is given in Supplementary Note 2 and Supplementary Figs 7 and 8. Unlike in the bulk where the volume fractions of the blocks and incompatibility parameters control the morphology,

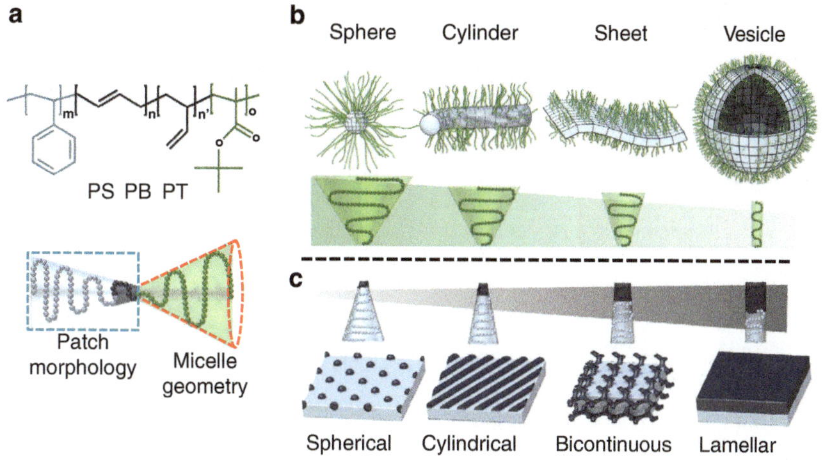

Figure 1 | Parameters controlling micelle geometry and patch morphology. (**a**) Chemical structure of the polystyrene-*block*-polybutadiene-*block*-poly (*tert*-butyl methacrylate) triblock terpolymer (SBT). (**b**) Suggested structural control: the length of the soluble PT corona (green chain) controls the micelle geometry to spheres, cylinders, bilayer sheets and vesicles; (**c**) the block lengths of the insoluble and immiscible blocks, PS (grey) and PB (black), determine the patch morphology to spherical, cylindrical, bicontinuous and lamellar.

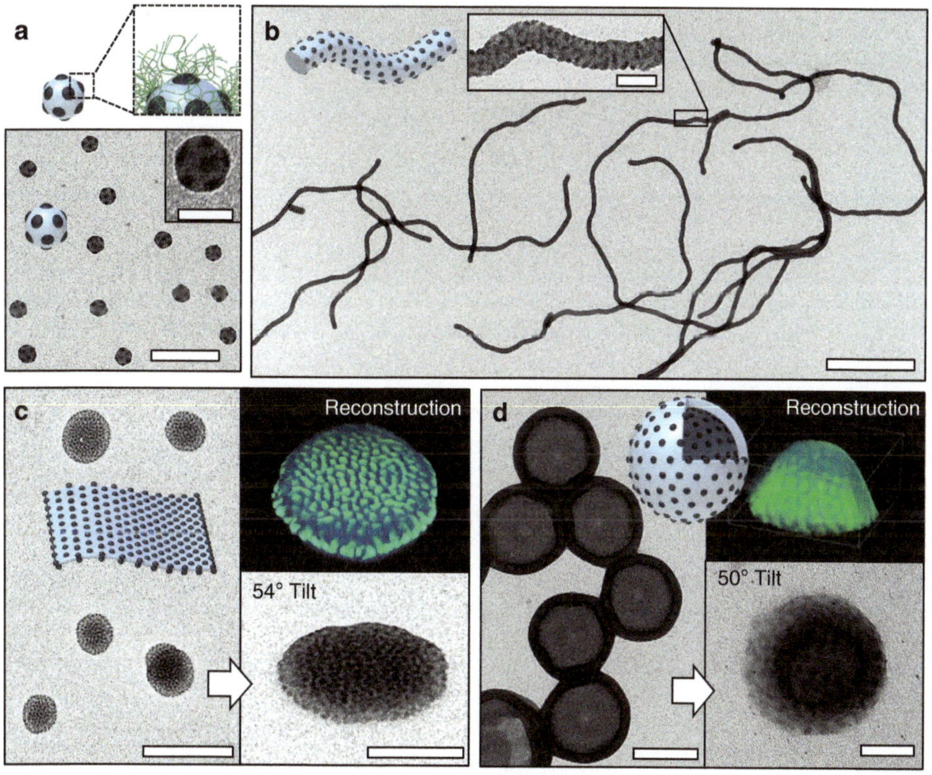

Figure 2 | Micelle polymorphs with spherical patches. Samples were prepared from acetone/isopropanol 60:40 (v/v) unless otherwise noted. (**a**) Spheres-on-spheres of $S_{1105}B_{237}T_{654}$. Schematic of block arrangement shows PS core (grey), PB patches (black) and PT corona (green); this domain sequence applies to all nanostructures and the PT corona always emanates from the dark PB domains. Scale bars, 200 and 50 nm in the inset. (**b**) Spheres-on-cylinders of $S_{540}B_{173}T_{137}$; close-up shows the small PB patches. Scale bars, 1 µm and 100 nm in the inset. (**c**) Spheres-on-sheets of $S_{540}B_{173}T_{137}$ in acetone/isopropanol (85:15 v/v), close-up recorded at 54° tilt angle and tomographic reconstruction showing PB patches in green. Scale bars, 500 and 200 nm in the inset. (**d**) Spheres-on-vesicles of $S_{540}B_{173}T_{88}$; close-up recorded at 50° tilt angle and tomographic reconstruction showing PB patches in green. Scale bars, 200 and 50 nm in the inset.

in solution self-assembly the solvent is the dominating factor. Thus, the length ratio of the blocks, their selectivity towards the solvent and the degrees of swelling become more relevant[31]. Hence, a decrease in the length of the soluble PT blocks (or increase of PS) should lead to successive transitions from spheres-on-spheres to cylinders and further to bilayers, all with PS core. For the micelle geometry, the PB block only enters through the correction term, $\Delta N_T = N_T - \tilde{N}_T$, equal to the number of monomer units in a segment of the PT block protruding from the PB domain up to the inter-patch distance D. As we will see later, this correction term becomes relevant when $N_S v_S \approx N_B v_B$.

We experimentally approach this theory by successively decreasing the parameter $N_T/(qN_S^{2/3})$, here from 1.69 to 0.15, while maintaining constant block lengths of N_S and N_B, which should give spherical PB patches for $N_S v_S \gg N_B v_B$. For $N_T/(qN_S^{2/3}) = 1.69$, we observe spherical micelles in transmission electron microscopy (TEM) with spherical PB patches on top of the PS core (Fig. 2a). The PB phase (dark) was selectively stained with OsO_4 to enhance contrast. This sphere-on-sphere configuration or 'patchy spherical micelle' has been observed before whenever a voluminous corona was present to stabilize the spherical shape of the solvophobic core[19–21,34]. Applying our generalized parameter, $N_C/N_A^{2/3}$ (Supplementary equation (10)), to these literature examples, both literature and our own values suggest a stability region for spheres-on-spheres of $N_C/N_A^{2/3} > 1$. For instance, a $N_C/N_A^{2/3}$ of 1.9 and 6.4 led to the formation of spherical multicompartment micelles in case of polystyrene-*b*-polybutadiene-*b*-poly(2-vinyl pyridine)[34] and polybutadiene-*b*-poly(2-vinyl pyridine)-*b*-poly(*tert*-butyl methacrylate)[20]. Also in a previous publication, we examined a large set of terpolymers without observing structural evolution to patchy cylinder micelles, which we now can explain given their $N_C/N_A^{2/3}$ of 4–10 (ref. 21).

Here, by reducing $N_T/(qN_S^{2/3}) < 1$ (for example, 0.48), we indeed observe the transition to spheres-on-cylinders as suggested by our theory (Fig. 2b). The supramolecular nature of the assemblies and the plasticized core phases (acetone for PS, and $T_{g,PB} < -20\,°C$) facilitate structural rearrangement into this structure. The employed two-step assembly procedure further aids in pre-arranging the polymer chains, reduces the conformational freedom of the participating blocks and promotes reaching equilibrium conditions[21]. Interestingly, the self-assembly proceeds in a hierarchical manner, where first spherical multicompartment micelles form that progressively merge into several micrometre-long patchy cylinder micelles at $N_T/(qN_S^{2/3}) = 0.48$ covered with a dense arrangement of small spherical PB patches.

Reducing $N_T/(qN_S^{2/3})$ further below 0.3, spheres-on-sheets become thermodynamically stable with a PS core of constant thickness and spherical PB patches decorating both sides of the sheets/discs (Fig. 2c). The characteristic arrangement of the patches indicates hexagonal packing, but the close proximity of the nano-sized patches and the overlapping features make visualization challenging in conventional TEM projection. We therefore use transmission electron tomography to resolve the patch morphology (Supplementary Fig. 9 and Supplementary Movie 1). The three-dimensional reconstruction gives an electron density map of high-contrast PB domains (OsO_4 staining, green in inset of Fig. 2c) that indeed shows onset of hexagonal packing of the PB patches and confirms the bilayer structure. The template-free assembly of amorphous polymer into two-dimensional (2D) sheets is rather unusual, because sheets (discs) are not the expected thermodynamic equilibrium morphology and mostly observed for low-entropy motifs above the super-strong segregation limit (for example, liquid-, semi-crystalline and perfluorinated blocks)[35–38]. One other successful approach to form stable 2D polymer disks utilizes the co-assembly of cylinder- and vesicle-forming diblock copolymers (for example, AB + AC). There the two immiscible blocks B/C form the phase-separated core, where the cylinder-forming diblock stabilizes the high-energy edges of the planar phase[16]. In our case, the PB surface pattern seems to stabilize the sheet edges against roll-up. On closer inspection of the reconstruction, we can identify a ring of single-layer PB patches at the sheet edge (Supplementary Fig. 9), whereas the planar part is formed by a double layer of PB patches. The TEM images also support this assumption, because the contrast clearly is lower at the edges reminiscent of a ring surrounding the disc.

At values of $N_T/(qN_S^{2/3}) = 0.15$, we find vesicles (polymersomes) with a compartmentalized shell, where spherical PB patches are located on both sides of the vesicle membrane (Fig. 2d). These patchy vesicles adopt a homogeneous round shape with size distribution typical for block copolymer vesicles. The isotropic spherical PB patches do not influence the vesicle shape. As we will show later on vesicles with other patch morphologies, this is not always the case. Nanostructured vesicles are likewise very intriguing, because a generic concept to control the membrane morphology through self-assembly would impact their application as nanoreactors, drug-delivery vehicles and artificial cell prototypes[39–41].

Controlling the patch morphology. We next explored the possibility to maintain constant micelle geometry, while tuning the PB patch morphology, that is, from spherical to cylindrical, bicontinuous and lamellar, by increasing N_B relative to the corona length, N_T (Fig. 3). Equation (1) applies and spheres-on-cylinders are thermodynamically stable as long as $R_{PB} \leq D \leq H_{corona}$. For shorter soluble PT blocks and/or longer insoluble PB blocks, the size of patches becomes comparable to—or larger than—the extension of the corona (Supplementary Fig. 8). In this regime one can expect shape transformation of the PB domains. This transition is driven by the gain in the conformational entropy of the PB blocks, which is balanced by an increase in the overlap and repulsions between the PT blocks protruding from the surface of the PB domains. The exact numerical factors, which quantify the difference in the conformational entropy of the PB blocks confined in spherical or cylindrical segmental domains are, however, not available. If $N_S v_S >> N_B v_B$, the transition from spherical PB patches to PB cylinders and further to a PB lamella (layered PS/PB) occurs on the surface of quasi-planar PS domains. The length of the PS block has virtually no influence on the position of the transition (tr)

$$\left(N_T/N_B^{2/3}\right)^{(tr)} \simeq \left(v_B/\ell^3\right)\left(\tilde{\gamma}_{BS}\ell^2\right)^{1/3}\left(v_T/\ell^3\right)^{-1/3}, \qquad (2)$$

where $\tilde{\gamma}_{BS}$ accounts for surface tensions on both PB/solvent and PS/PB interfaces. However, at $N_S v_S \simeq N_B v_B$ and for sufficiently short PT block, the PB patches may merge into a cylindrical patch, whereas the PS core still retains a cylindrical shape and hence, a peculiar double-cylindrical patch on a cylindrical core can become thermodynamically stable. Moreover, for nearly symmetrical insoluble blocks one could expect interference of morphological transitions in the PS and PB domains (triggered, for example, by variation in the solvent strength for the soluble PT block).

On cylindrical micelles the patch morphology indeed changes as a function of N_B (Fig. 3). While all cylindrical micelles exhibit comparable diameter ($D_{cyl} \approx 50$–70 nm) as suggested by $N_T/(qN_S^{2/3})$, varying $N_T/N_B^{2/3}$ from 4.41 to 1.15, systematically alters the PB patch morphology (overview images in Supplementary Fig. 10). For $N_T/N_B^{2/3} = 4.41$, we first find small spherical PB patches with diameter $d_{PB} = 12.6$ nm (Fig. 3a) that evolve into larger spherical PB patches with $d_{PB} = 24.2$ nm for $N_T/N_B^{2/3} = 2.33$ (Fig. 3b). Those pack in a dense hexagonal pattern and begin to arrange on a helical trajectory induced by the surface confinement of the cylindrical PS core[42]. Under proper solvent conditions (acetone/isopropanol 80:20 v/v) the large spheres fuse into a continuous PB double helix and both patch morphologies coexist on the same cylinder micelle (Supplementary Fig. 11). As the increasing PB patches (relative to PS) require exceedingly large interfacial area, the spherical domains move closer together, overlap and fuse into the double helix winding around the PS cylinder core (Fig. 3c). Owing to the relatively slow self-assembly kinetics of the micellar building blocks, structural rearrangement (fusion and fission) proceeds over a period of several days until the double helix is well developed. Supplementary Fig. 12 shows

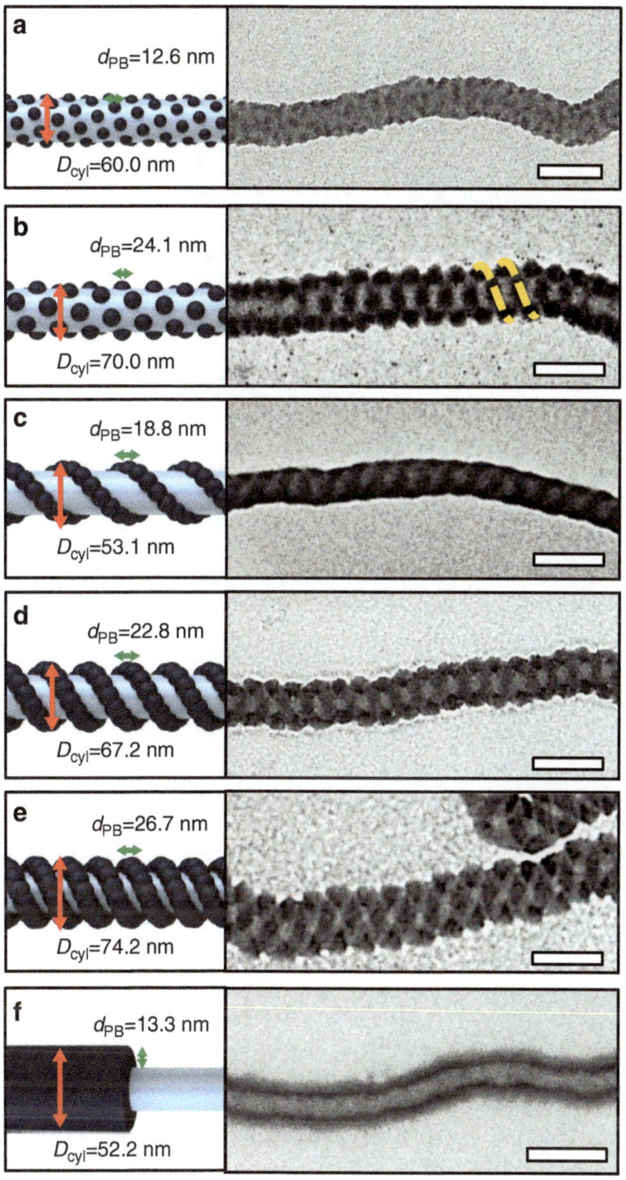

Figure 3 | Cylinder micelles with controlled patch morphology. (**a**) Small PB spheres of $S_{540}B_{173}T_{137}$ in acetone/isopropanol (60:40 v/v) and (**b**) large PB spheres of $S_{510}B_{539}T_{154}$ in acetone/isopropanol (85:15 v/v). (**c-e**) $S_{510}B_{539}T_{154}$ form thin, $S_{307}B_{385}T_{81}$ medium and $S_{307}B_{530}T_{75}$ thick PB double helices all in acetone/isopropanol (60:40 v/v). (**f**) Core-shell-corona for $S_{307}B_{530}T_{75}$ in *n*-hexane/isopropanol (50:50 v/v). Scale bars, 100 nm.

the time-dependent formation of cylindrical micelles decorated with the PB double helical patch. While after 1 and 3 days still precursor or multicompartment micelles are observed, the structure is fully developed after 1 week of ageing. Such long timescales for self-assembly are not entirely surprising as demonstrated before by Liu *et al.*[43] on visually remarkable similar double helix micelles (assembly of 90 days). There, core–shell–corona cylinder micelles intertwine to form the final double helix. These double helices are thus merely visually similar, but with different arrangement of the polymer blocks. Likewise, Wooley and colleagues found polymeric helices in aqueous solution where an organic molecular linker electrostatically drives cylindrical core–shell–corona micelles to curl either into a single-stranded helix or a double helix of two core–shell cylinders (solvent in the centre).[44] At $N_T/N_B^{2/3} = 1.53$, cylindrical core–shell–corona micelles were expected, but instead the PB double helix persists with increased diameter of $d_{PB} = 22.8$ nm (Fig. 3d). Even at $N_T/N_B^{2/3} = 1.15$ the PB double helix rather continues to grow ($d_{PB} = 26.7$ nm) than transforming into the core–shell morphology (Fig. 3e). The delicate interactions of unfavourable interfaces, PS/PB, PS/solvent and PB/solvent most likely suppressed the transition. The slightly repulsive PS/solvent interface (PS swollen) would be replaced on the expense of creating unfavourable PS/PB interface ($\chi_{PS/PB} = 0.06$) and energetically much less-favoured PB/solvent interface ($\chi_{PB/solvent} = 1.6$–2.9). The transition can thus only be realized by decreasing the incompatibility between PB and the solvent, as demonstrated here by exchanging acetone with *n*-hexane (good solvent for PB). In *n*-hexane/isopropanol 50:50 (v/v), PS still remains collapsed in form of cylindrical micelles, but PB now swells with *n*-hexane inducing the transition to cylindrical core–shell–corona micelles (Fig. 3f). These multicompartment nanostructures are merely bound by supramolecular forces, and changes in solvent polarity allow transitions between morphologies, for example, reversible switching from double helical to core–shell cylinders (Supplementary Fig. 13).

Sheets and vesicles with membrane morphology. We also investigated the phase behaviour of patch morphologies on bilayer sheets and vesicle membranes (Fig. 4). As discussed in Fig. 2, by shortening N_T cylindrical micelles become unstable and sheets with a PS core are more favourable. By simultaneously increasing N_B, the patch morphology likewise transforms from spherical to cylindrical and further to bicontinuous and core–shell (lamellar) morphology.

Figure 4a–c illustrates the morphological transition from spheres-on-cylinders to cylinders-on-sheets and vesicles when we reduce the corona length $N_T/(qN_S^{2/3})$ from $0.38 \rightarrow 0.28 \rightarrow 0.18$ and $N_T/N_B^{2/3}$ $4.41 \rightarrow 2.33$. On PS cores with reduced curvature (sheets), the spherical PB patches adapt to the decreasing overall surface area and merge together into cylindrical PB patches (Fig. 4b and Supplementary Movie 2). Tomographic reconstruction clarifies the strictly parallel and equidistant spacing of cylindrical PB patches (displayed in cyan) that locate on top and bottom of the PS sheet (also Supplementary Fig. 14). These cylinders-on-sheets grow larger as compared to the sphere on sheets in Fig. 1, most likely facilitated by the surface pattern that works against the roll-up. These particles have numerous spheres-on-cylinder arms attached to their edges reminiscent of 'jelly-fish' intermediates. A larger selection of these particles is summarized in Supplementary Fig. 15. In a mechanism that is remarkably similar to what is observed for the solution self-assembly of diblock copolymers[35,45], cylinder-on-sheets also grow to a critical size until the edge energy becomes unfavourable and sheets roll up to vesicles. Here the tethered cylinder micelle arms continue to merge into the sheet fuelling membrane growth (Supplementary Fig. 16). Fusion and fission of the soft phases accompanied by equilibration and chain rearrangements facilitate vesicle closure. The cylindrical PB patches on top and bottom of the sheet is thereby transferred to the in- and outside of the vesicle membrane.

So far, only two works utilized the versatility of block terpolymers to implement morphologies into the vesicle membrane through phase separation of solvophobic blocks. Vesicles with a membrane morphology have been shown earlier by Russell *et al.* through rehydration of bulk films[26]. Also the special architecture of miktoarm star terpolymers allows to form compartmentalized vesicles, where the packing frustration of the two solvophobic blocks created laterally structured membranes[27]. However, the decoration of a vesicle membrane with spherical

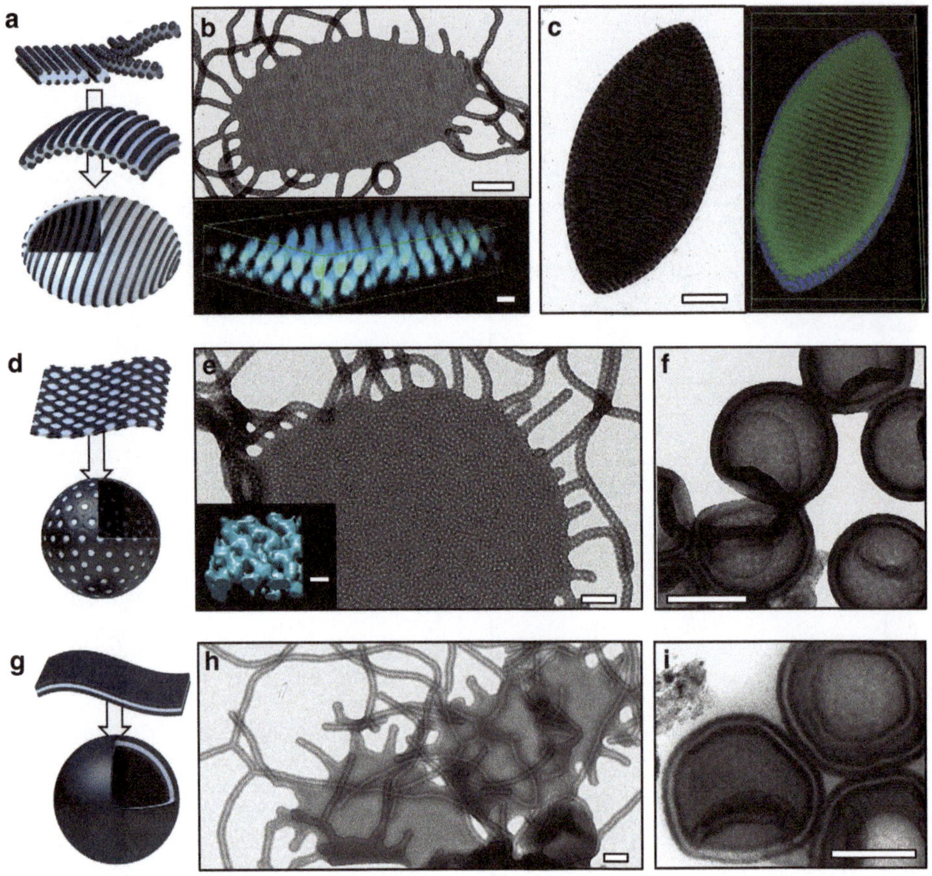

Figure 4 | Polymer sheets and vesicles with defined patch morphology. Samples were prepared from acetone/isopropanol mixtures unless otherwise noted. (**a**) Schematic of spheres-on-cylinders that transform to cylinders-on-bilayer sheets/vesicles. (**b**) TEM image of cylinder-on-sheet of $S_{510}B_{539}T_{154}$ (75:25 v/v), and tomographic reconstruction of PB patches on top and bottom of the PS sheet (PB cyan, PS omitted). (**c**) TEM and tomographic reconstruction of a 'lemon'-shaped vesicle of $S_{510}B_{539}T_{154}$ with PB cylinders on in- and outside of the PS membrane after ageing (85:15 v/v). (**d**) Schematic of sheet and vesicles with bicontinuous membrane. (**e**) TEM and tomographic reconstruction of PB network within the PS sheet of $S_{307}B_{530}T_{75}$ (60:40 v/v) (PB cyan, PS omitted). (**f**) Fully evolved vesicle with bicontinuous membrane of $S_{307}B_{530}T_{63}$ on ageing (60:40 v/v). (**g**) Schematic of the core-shell sheets and (lamellar) vesicle. (**h**) TEM image of a core-shell sheet of $S_{300}B_{756}T_{56}$ (*n*-hexane/isopropanol 50:50 v/v) and (**i**) core-shell (lamellar) vesicle of $S_{300}B_{756}T_{56}$ in (*n*-hexane/isopropanol 35:65 v/v). Scale bars, 200 nm in TEM images and 25 nm in the reconstructions.

and cylindrical patches is reserved for linearly sequenced polymers. The resulting cylinders-on-vesicles are noticeably elongated and adopt a prolate ellipsoidal 'lemon' shape with low curvature in longitudinal direction but pronounced increase in curvature at the tips (Fig. 4c). We interpret the elliptic shape as balance between curvature of the PS membrane and minimization of interfacial energies of the anisotropic PB patches. In atomic force microscopy, the collapsed vesicles display twice the height of the precursor sheet (Supplementary Fig. 17), while not entirely deflated parts of the vesicle are considerable higher, confirming the hollow interior. In vesicular samples, we identify unassembled patchy precursor micelles (spheres and cylinders), all with the same height ($h \approx 50$ nm) as the sheets and the vesicle membrane. This similarity in height further corroborates structural evolution from spherical micelles to cylinder micelles, sheet and finally vesicles. Cryo-TEM imaging of the cylinders-on-vesicles also proves that the 'lemon' shape as well as the striped patch morphology are both present in solution, and are not an effect of drying (Supplementary Fig. 18 and Supplementary Movie 3).

We chose this peculiar solution nanostructure as model to follow the formation of the final lemon-shaped vesicles, as morphological evolution progresses through all other geometries. The slow self-assembly kinetics not only allows us to distinguish single stages of the self-assembly process but also to identify various intermediate structures before reaching a long-term stable nanostructure (assumed from TEM measurements after 2 years of ageing). As already mentioned, the solvent exchange during dialysis is completed after 90 min, while the structural evolution proceeds over the course of weeks in case of striped lemon-shaped vesicles (Supplementary Fig. 19). During the dialysis, first precursor micelles with a PB core and a PS/PT corona are dominant that successively assemble to form patchy spherical micelles (Supplementary Fig. 19a,b). After solvent exchange is completed, the patchy spherical micelles fuse into patchy cylinders, and progressively transform into striped sheets with patchy cylindrical arms after 24 h of ageing (Supplementary Fig. 19c–e). After 3 weeks, a majority of the sheets has rolled up to lemon-shaped vesicles (Supplementary Fig. 19f). The self-assembly from patchy spherical micelles towards striped lemon-shaped vesicles intriguingly demonstrates that fusion and fission processes of soft micellar particles are an integral part of structural evolution. The observed massive structural rearrangements obviously require chain mobility provided by the plasticizing solvent conditions.

We also studied the transition of cylinder micelles with double-helix patches into sheets ($N_T/(qN_S^{2/3})$ from 0.41 to 0.38 and $N_T/N_B^{2/3}$ from 1.53 to 1.15). Here the PB double helices change into a bicontinuous network of PB, which progresses through the PS domain from top to bottom as visualized in cyan in the reconstruction (Fig. 4d,e, PS omitted for clarity

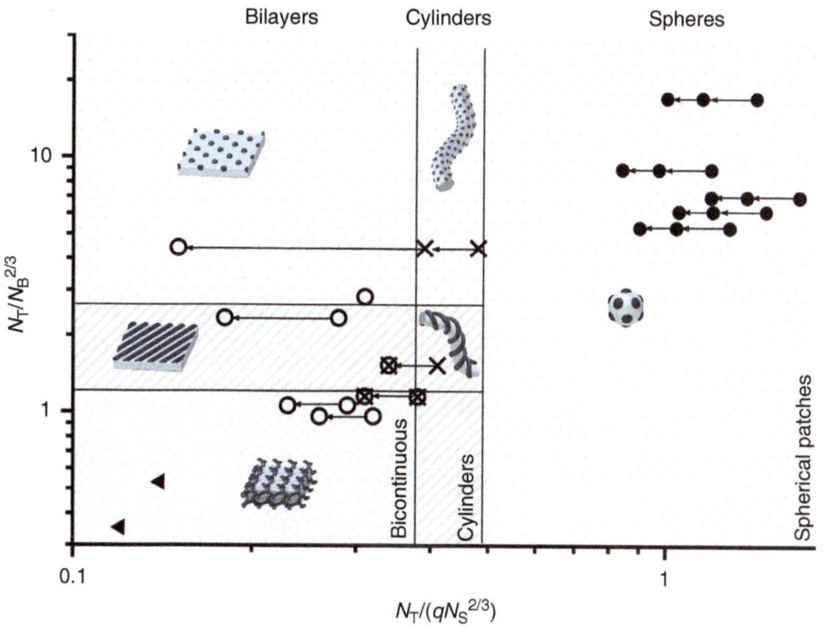

Figure 5 | Experimental phase diagram in dependence of relative block lengths. (●) Spherical micelles with spherical patches. (X) Cylinder micelles with spherical and double-helical patch morphology. (⊗) Mixtures of cylinder micelles and bilayers (structural transition). (○) Bilayer sheets and vesicles with patchy, striped and bicontinuous membrane morphology. (▲) Large terpolymer particles with inverse core morphology. All data points were corrected by swelling factor q (Supplementary Figs 1–4), where arrows (←) indicate same terpolymers, yet in solvents with increasing acetone content.

Supplementary Movie 4). This PS/PB domain arrangement is unexpected given the large unfavourable interface. Then again, interfacial energies are not trivial for the entire system and morphologies under 2D quasi-confinement of the membrane may deviate from the expected case[46]. The bicontinuous sheets are nevertheless long-term stable with double-helical arms protruding from the sheet (Supplementary Fig. 20). The tethered double helices are constrained in motion and we observe self-wrapping to minimize interface with the solvent (intermediate stage between cylinder and sheet). The slowed self-assembly kinetics further allow visualization of the transition from sheets to vesicles with bicontinuous membrane progressing through 'jelly-fish' intermediates with double-helical tentacles (Fig. 4f, Supplementary Fig. 20 and Supplementary Movie 5).

Core-shell-corona sheets and vesicles complete the morphological spectrum (Fig. 4g–i). As discussed for core-shell cylinders, the chosen block sequence and thus sequence of solubilities of PS–PB–PT only allows stable core–shell–corona morphologies under solvent conditions that promote spreading of PB on the PS core ($\chi_{A,solvent} < \chi_{B,solvent} < \chi_{C,solvent}$). For our system, exchange of acetone with n-hexane swells the PB domain and induces the change to a continuous PB shell completely engulfing the PS sheet or vesicle membrane. Core-shell micelles, cylinders and vesicles have been observed before in water[47,48] and organic solvents[49]. Interestingly, in both cases the sequence of polymer blocks also followed $\chi_{A,solvent} < \chi_{B,solvent} < \chi_{C,solvent}$, thus promoting the block arrangement into the core–shell–corona structure.

Experimental phase diagram. To rationalize the observed solution behaviour, we compiled an experimental phase diagram where each data point corresponds to one SBT triblock terpolymer in a specific acetone/isopropanol mixture (Fig. 5). The experimental values are in good agreement with the superimposed phase boundaries as predicted by our theory. Vertical lines represent predicted stability regions for each geometry, that is, transitions from spherical to cylinder micelles and further to bilayer sheets and vesicles. These are governed by the parameter $N_T/(qN_S^{2/3})$, whereas phase boundaries are separated by a factor $(N_T/(qN_S^{2/3}))^{(sph.\leftrightarrow cyl.)}/(N_T/(qN_S^{2/3}))^{(cyl.\leftrightarrow lam.)} = 1.28$ (Supplementary Note 2 and Supplementary Figs 7 and 8). Superposition of SBT terpolymers from different solvent mixtures requires a correction factor that accounts for selective solvent swelling of the polymer domains. Arrows connect identical polymers in different solvents. As stated in the beginning, we chose our system because only PS is affected by the solvent composition. The hydrodynamic radius of PT is unaffected in the employed range of compositions (as verified in DLS measurements on PT multiarm star polymers in acetone/isopropanol mixtures) and the unusually high $\chi_{PB/solvent} = 1.6–2.9$ (Supplementary Table 2) suggests completely collapsed PB blocks in all mixtures, though interfacial tensions for the PB domains are affected by the solvent composition. The solvent dependence thus enters as correction factor, q, only for PS (Supplementary Figs 2–4 and Supplementary Note 1). The parameter $N_T/N_B^{2/3}$ on the other hand controls the resulting patch morphology to spherical, cylindrical and bicontinuous, as represented by the horizontal lines. Core-shell morphologies are omitted from this diagram, as they are only stable under special solvent conditions. Supplementary Table 4 contains all characteristics of the polymers found in the phase diagram and gives an overview over $N_T/(qN_S^{2/3})$ as well as $N_T/N_B^{2/3}$ values and the resulting micellar geometry and patch morphology.

Figure 6 finally summarizes the entire experimental library of multicompartment nanostructures, where classical micellar geometries are divided into subclasses according to the patch morphology (Supplementary Fig. 21 shows a complete schematic library). Characteristics of the polymers in Fig. 6 can be found in Supplementary Table 4 and are highlighted in green. Our combinatorial library allows deducing several trends towards a better understanding of the nanostructure formation of linear ABC triblock terpolymers. While curved spherical cores only support the spherical patch morphology, less-curved cylinder micelles stabilize spherical as well as cylindrical patches, the latter

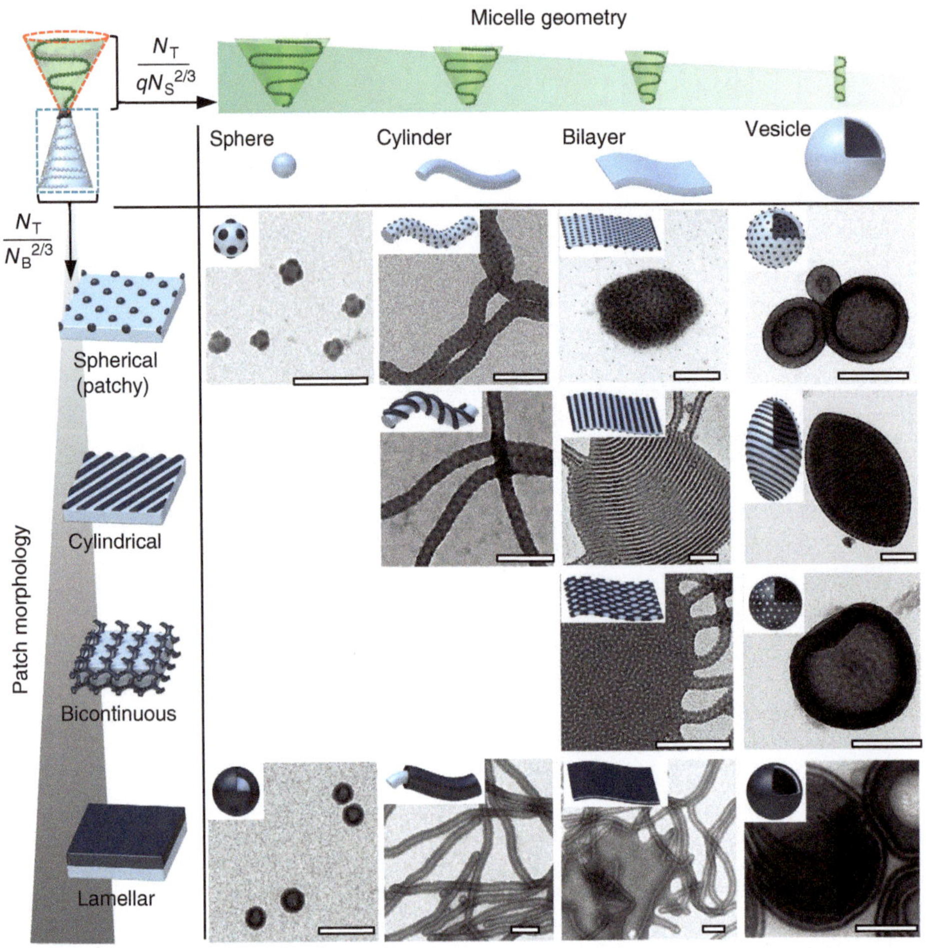

Figure 6 | Experimental combinatorial table sub-classifying micelle geometry by patch morphologies. Sphere-on-spheres, spheres-on-cylinders, spheres-on-bilayer sheets and vesicles; cylinders-on-cylinders (double-helix patch), cylinders-on-bilayer sheets and vesicles; sheets and vesicles with bicontinuous membrane morphology; core-shell micelles, core-shell cylinders, lamellar sheets and vesicles. Solvent composition for each sample is given in Supplementary Table 4. Scale bars, 200 nm.

in form of double helices with a synthetically controllable pitch size. For (quasi-)planar bilayer sheets and vesicles we find close resemblance of the patch morphology to the bulk case, that is, spherical, cylindrical and bicontinuous. Core-shell-corona morphologies are stable on all micellar geometries, but only under specific solvent conditions.

In summary, we have demonstrated how to construct a library of solution nanostructures from a single type of ABC triblock terpolymer, where classical micellar geometries are divided into subclasses according to their patch morphology. It is an interesting and surprising observation that the patch morphologies follow a very similar trend as the bulk morphologies of AB diblock copolymers: from spheres to cylinders to bicontinuous (gyroid) to lamellar microphases, even though the AB blocks are confined to the nanoscale core of the micelle. Given proper block design, these solution nanostructures may find application ranging from templates for nano-optics and -electronics (for example, double helices) to advanced gating and intelligent delivery systems with controlled pharmacokinetic release profiles (for example, vesicles with bicontinuous/porous membrane). In a broader context, mastering the self-assembly of linear ABC triblock terpolymers will provide understanding towards multiblock copolymers (ABCD, ABCDE, ...) with prospect to harvest the potential of their exponentially increasing number of conceivable folding permutations[50].

Methods

Polymer synthesis. A detailed description for the purification of chemicals and monomers involved in synthesis of PS–PB–PT triblock terpolymers is described elsewhere[51]. All SBT triblock terpolymers were synthesized via living anionic polymerization in tetrahydrofuran (THF) at low temperatures in presence of alkoxides. In brief, *sec*-butyllithium (*sec*-BuLi) followed by styrene was added to THF at $-70\,°C$ and allowed to react for 10 min. After 1,3-butadiene was added, the reaction mixture was heated to $-10\,°C$ and stirred for 6.5 h. At $-50\,°C$ a sixfold molar excess (compared to *sec*-BuLi) of diphenylethylene was added and stirred for 1 h. Then, *tert*-butyl methacrylate monomer was added at $-70\,°C$ and the reaction was heated to $-50\,°C$ for 2 h. After complete monomer consumption, 2 ml of degassed methanol was added to the polymer solution to terminate the living chain ends. Characteristics of all synthesized polymers can be found in Supplementary Table 1.

Dialysis procedure. For dialysis, membranes of regenerated cellulose (Spectrum Laboratories, Spectra/Por molecular weight cut-off = 12–14 kDa) were used. After washing with Milliq water the membranes were washed with excess dioxane (Aldrich, analytical grade). DMAc (Aldrich) was p.a. grade and used as received. All other solvents for dialysis experiments were of technical grade and used as received. PS–PB–PT was dispersed in DMAc to give a concentration of $0.1\,g\,l^{-1}$. After annealing at 70 °C over night, dialysis against a certain solvent mixture was performed for 24 h although complete solvent exchange is reached after 90 min. Before analysis, the nanostructures were aged several days to allow full development of nanostructures.

Polymer characterization. Proton nuclear magnetic resonance spectra were recorded on a Bruker Ultrashield 300 machine with a 300-MHz operating frequency using deuterated chloroform as solvent. Size-exclusion chromatography measurements were performed on a set of 30-cm SDV-gel columns of 5-mm particle size having a pore size of 10^5, 10^4, 10^3 and 10^2 Å with refractive index and ultraviolet ($\lambda = 254$ nm) detection. Size-exclusion chromatography was measured at an elution rate of 1 ml min^{-1} with THF as eluent and polystyrene as calibration.

Transmission electron microscopy. TEM was performed either on a Zeiss CEM 902 or a Fei Tecnai 12 electron microscope operated at 80 and 120 kV, respectively. The samples were prepared by placing one drop of the polymer solution onto carbon-coated copper grids. Excess solvent was instantly absorbed by a filter paper. For selective staining of PB, the TEM specimens were exposed to OsO_4 vapour for 3 h.

Cryogenic transmission electron microscopy. Cryo-TEM imaging was carried out using a JEM 3200FSC field emission microscope (Jeol) operated at 300 kV in bright-field mode with an Omega-type zero-loss energy filter. The images were acquired with an Ultrascan 4000 charge-coupled device camera (Gatan) and with Gatan Digital Micrograph software (version 1.83.842), while the specimen temperature was maintained at -187 °C. Vitrified samples were prepared using a FEI Vitribot placing 3 µl of sample solution on 200-mesh holey carbon copper grids under 100% humidity, then blotted with filter paper for 0.5–1.5 s, and immediately plunged into a -170 °C ethane/propane mixture and cryotransfered to the microscope.

Electron tomography. Electron tomography was performed using the Fei Tecnai 12 electron microscope recording a series of projection images at various tilt angles between $\pm \phi_{max}$ with typical values of $\pm 60°$ in 2° increments.

Cryogenic electron tomography. Cryogenic electron tomography was performed with the same transmission electron microscope that was used for cryo-TEM imaging. Electron tomographic tilt series were acquired with the SerialEM software package (version 3.2.2). Samples were tilted between 69° angles with 3° increment steps.

Alignment and reconstruction. The TEM grids were dipped in gold nanoparticle solution before sample deposition ($\delta = 3$–10 nm, stabilized by 11-mercapto-1-undecanol ligand) to ensure proper alignment of captured images with IMOD[52,53]. The fine alignment and cropping was conducted with custom-made Silicon Graphics JPEGANIM software package[54]. The images were binned twice to reduce noise and computation time and maximum entropy method reconstruction scheme was carried out with custom-made programme on Mac or Linux cluster with regularization parameter value of $\lambda = 1.0E^{-3}$ (ref. 54).

Visualization of 3D reconstruction. Volumetric graphics and analyses were performed with the UCSF Chimera package[55] and volume segmentation was carried out using trainable weka segmentation[56].

References

1. Yan, J., Bloom, M., Bae, S. C., Luijten, E. & Granick, S. Linking synchronization to self-assembly using magnetic Janus colloids. *Nature* **491,** 578–581 (2012).
2. Glotzer, S. C. & Solomon, M. J. Anisotropy of building blocks and their assembly into complex structures. *Nat. Mater.* **6,** 557–562 (2007).
3. Gröschel, A. H. *et al.* Guided hierarchical co-assembly of soft patchy nanoparticles. *Nature* **503,** 247–251 (2013).
4. Cademartiri, L. & Bishop, K. J. M. Programmable self-assembly. *Nat. Mater.* **14,** 2–9 (2015).
5. Qiu, H., Hudson, Z. M., Winnik, M. A. & Manners, I. Multidimensional hierarchical self-assembly of amphiphilic cylindrical block comicelles. *Science* **347,** 1329–1332 (2015).
6. Bates, F. S. & Fredrickson, G. H. Block copolymers—designer soft materials. *Phys. Today* **52,** 32–38 (1999).
7. Discher, D. E. & Eisenberg, A. Polymer vesicles. *Science* **297,** 967–973 (2002).
8. Schacher, F. H., Rupar, P. A. & Manners, I. Functional block copolymers: nanostructured materials with emerging applications. *Angew. Chem. Int. Ed.* **51,** 7898–7921 (2012).
9. Hudson, Z. M., Lunn, D. J., Winnik, M. A. & Manners, I. Colour-tunable fluorescent multiblock micelles. *Nat. Commun.* **5,** 3372 (2014).
10. Zhang, L. & Eisenberg, A. Multiple morphologies of 'crew-cut' aggregates of polystyrene-b-poly(acrylic acid) block copolymers. *Science* **268,** 1728–1731 (1995).
11. Jain, S. & Bates, F. S. On the origins of morphological complexity in block copolymer surfactants. *Science* **300,** 460–464 (2003).
12. Discher, B. M. *et al.* Polymersomes: tough vesicles made from diblock copolymers. *Science* **284,** 1143–1146 (1999).
13. Gröschel, A. H. & Müller, A. H. E. Self-assembly concepts for multicompartment nanostructures. *Nanoscale* **7,** 11841–11876 (2015).
14. Moughton, A. O., Hillmyer, M. A. & Lodge, T. P. Multicompartment block polymer micelles. *Macromolecules* **45,** 2–19 (2011).
15. Pochan, D. J. *et al.* Toroidal triblock copolymer assemblies. *Science* **306,** 94–97 (2004).
16. Zhu, J. *et al.* Disk-cylinder and disk-sphere nanoparticles via a block copolymer blend solution construction. *Nat. Commun.* **4,** 2297 (2013).
17. Schacher, F. *et al.* Interpolyelectrolyte complexes of dynamic multicompartment micelles. *ACS Nano* **3,** 2095–2102 (2009).
18. Löbling, T. I. *et al.* Hidden structural features of multicompartment micelles revealed by cryogenic transmission electron tomography. *ACS Nano* **8,** 11330–11340 (2014).
19. Kubowicz, S. *et al.* Multicompartment micelles formed by self-assembly of linear ABC triblock copolymers in aqueous medium. *Angew. Chem. Int. Ed.* **44,** 5262–5265 (2005).
20. Schacher, F., Walther, A., Ruppel, M., Drechsler, M. & Müller, A. H. E. Multicompartment core micelles of triblock terpolymers in organic media. *Macromolecules* **42,** 3540–3548 (2009).
21. Gröschel, A. H. *et al.* Precise hierarchical self-assembly of multicompartment micelles. *Nat. Commun.* **3,** 710 (2012).
22. Cui, H., Chen, Z., Zhong, S., Wooley, K. L. & Pochan, D. J. Block copolymer assembly via kinetic control. *Science* **317,** 647–650 (2007).
23. Fang, B. *et al.* Undulated multicompartment cylinders by the controlled and directed stacking of polymer micelles with a compartmentalized corona. *Angew. Chem. Int. Ed.* **48,** 2877–2880 (2009).
24. Dupont, J. & Liu, G. ABC triblock copolymer hamburger-like micelles, segmented cylinders, and Janus particles. *Soft Matter* **6,** 3654–3661 (2010).
25. Rupar, P. A., Chabanne, L., Winnik, M. A. & Manners, I. Non-centrosymmetric cylindrical micelles by unidirectional growth. *Science* **337,** 559–562 (2012).
26. Zhao, W., Chen, D., Hu, Y., Grason, G. M. & Russell, T. P. ABC triblock copolymer vesicles with mesh-like morphology. *ACS Nano* **5,** 486–492 (2011).
27. Li, Z., Hillmyer, M. A. & Lodge, T. P. Laterally nanostructured vesicles, polygonal bilayer sheets, and segmented wormlike micelles. *Nano Lett.* **6,** 1245–1249 (2006).
28. Brannan, A. K. & Bates, F. S. ABCA tetrablock copolymer vesicles. *Macromolecules* **37,** 8816–8819 (2004).
29. Hu, H. & Liu, G. Miktoarm star copolymer capsules bearing pH-responsive nanochannels. *Macromolecules* **47,** 5096–5103 (2014).
30. Israelachvili, J. N., Mitchell, D. J. & Ninham, W. Theory of self-assembly of hydrocarbon amphiphiles into micelles and bilayers. *J. Chem. Soc. Farady Trans. 2* **72,** 1525–1568 (1976).
31. Zhulina, E. B., Adam, M., Larue, I., Sheiko, S. S. & Rubinstein, M. Diblock copolymer micelles in a dilute solution. *Macromolecules* **38,** 5330–5351 (2005).
32. Khandpurj, A. K. *et al.* Diblock copolymer phase diagram near the order-disorder transition. *Macromolecules* **28,** 8796–8806 (1995).
33. Leibler, L. Theory of microphase separation in block copolymers. *Macromolecules* **1617,** 1602–1617 1980.
34. Ma, Z., Yu, H. & Jiang, W. Bump-surface multicompartment micelles from a linear ABC triblock copolymer: a combination study by experiment and computer simulation. *J. Phys. Chem. B* **113,** 3333–3338 (2009).
35. Antonietti, M. & Förster, S. Vesicles and liposomes: a self-assembly principle beyond lipids. *Adv. Mater.* **15,** 1323–1333 (2003).
36. Rizis, G., van de Ven, T. G. M. & Eisenberg, A. 'Raft' formation by two-dimensional self-assembly of block copolymer rod micelles in aqueous solution. *Angew. Chem. Int. Ed.* **53,** 9000–9003 (2014).
37. Hudson, Z. M. *et al.* Tailored hierarchical micelle architectures using living crystallization-driven self-assembly in two dimensions. *Nat. Chem.* **6,** 893–898 (2014).
38. Kim, J.-K., Lee, E., Lim, Y. & Lee, M. Supramolecular capsules with gated pores from an amphiphilic rod assembly. *Angew. Chem. Int. Ed.* **47,** 4662–4666 (2008).
39. Torchilin, V. P. Recent advances with liposomes as pharmaceutical carriers. *Nat. Rev. Drug Discov.* **4,** 145–160 (2005).
40. Vriezema, D. M. *et al.* Self-assembled nanoreactors. *Chem. Rev.* **105,** 1445–1489 (2005).
41. Christian, D. A. *et al.* Spotted vesicles, striped micelles and Janus assemblies induced by ligand binding. *Nat. Mater.* **8,** 843–849 (2009).
42. Wood, D. A., Santangelo, C. D. & Dinsmore, A. D. Self-assembly on a cylinder: a model system for understanding the constraint of commensurability. *Soft Matter* **9,** 10016–10024 (2013).
43. Dupont, J., Liu, G., Niihara, K., Kimoto, R. & Jinnai, H. Self-assembled ABC triblock copolymer double and triple helices. *Angew. Chem. Int. Ed.* **48,** 6144–6147 (2009).
44. Zhong, S., Cui, H., Chen, Z., Wooley, K. L. & Pochan, D. J. Helix self-assembly through the coiling of cylindrical micelles. *Soft Matter* **4,** 90–93 (2008).

45. Blanazs, A., Madsen, J., Battaglia, G., Ryan, A. J. & Armes, S. P. Mechanistic insights for block copolymer morphologies: how do worms form vesicles? *J. Am. Chem. Soc.* **133,** 16581–16587 (2011).
46. Yabu, H., Higuchi, T. & Jinnai, H. Frustrated phases: polymeric self-assemblies in a 3D confinement. *Soft Matter* **10,** 2919–2931 (2014).
47. Gohy, J., Willet, N., Varshney, S., Zhang, J. & Jerome, R. Core-shell-corona micelles with a responsive shell. *Angew. Chem. Int. Ed.* **40,** 3214–3216 (2001).
48. Lei, L. *et al.* Dependence of the structure of core–shell–corona micelles on the composition of water/toluene mixtures. *Polymer* **47,** 2723–2727 (2006).
49. Stewart, S. & Liu, G. Block copolymer nanotubes. *Angew. Chem. Int. Ed.* **6,** 340–344 (2000).
50. Bates, F. S. *et al.* Multiblock polymers: panacea or Pandora's box? *Science* **336,** 434–440 (2012).
51. Löbling, T. I. *et al.* Bulk morphologies of polystyrene-block-polybutadiene-block-poly(tert-butyl methacrylate) triblock terpolymers. *Polymer* **72,** 479–489 (2015).
52. Raula, J. *et al.* Synthesis of gold nanoparticles grafted with a thermoresponsive polymer by surface-induced reversible-addition-fragmentation chain-transfer polymerization. *Langmuir* **19,** 3499–3504 (2003).
53. Kremer, J. R., Mastronarde, D. N. & McIntosh, J. R. Computer visualization of three-dimensional image data using IMOD. *J. Struct. Biol.* **116,** 71–76 (1996).
54. Engelhardt, P. Electron tomography of chromosome structure. In Encyclopedia of Analytical Chemistry Vol. 6 (eds. Meyer, R.A.) 4948–4984 (John Wiley & Sons, Chichester, England, 2000).
55. Pettersen, E. F. *et al.* UCSF chimera—a visualization system for exploratory research and analysis. *J. Comput. Chem.* **25,** 1605–1612 (2004).
56. Hall, M. *et al.* The WEKA data mining software: an update. *ACM SIGKDD Explor.* **11,** 10–18 (2009).

Acknowledgements

We thank Dr Holger Schmalz for helpful discussions about polymer synthesis and Dr Andreas Hanisch for the synthesis of four SBT terpolymers. We thank Adjunct Professor in Molecular Genetics Peter Engelhardt for help with transmission electron tomography. We acknowledge the Deutsche Forschungsgemeinschaft (DFG) for supporting this work within DFG Mu896/40-1. This work was carried out under the Academy of Finland's Centre of Excellence Programme (2014-2019) and supported by ERC-2011-AdG (291364-MIMEFUN). This work was partially supported by Russian Science Foundation grant #14-33-00003. We further made use of the Aalto University Nanomicroscopy Center (Aalto-NMC) premises.

Author contributions

T.I.L. synthesized the polymers; T.I.L. and A.H.G. performed experiments and collected all data; J.H. helped with the tomograms and commented on the manuscript; O.I. co-designed experiments, discussed results and commented on the manuscript; O.B. developed the theory and commented on the manuscript; T.I.L., A.H.G. and A.H.E.M. designed the experiments, discussed results and wrote the manuscript; O.I., A.H.G. and A.H.E.M. supervised the project.

Additional information

Competing financial interests: The authors declare no competing financial interests.

23

Direct imaging of molecular symmetry by coherent anti-stokes Raman scattering

Carsten Cleff[1], Alicja Gasecka[2,3], Patrick Ferrand[1], Hervé Rigneault[1], Sophie Brasselet[1] & Julien Duboisset[1]

Nonlinear optical methods, such as coherent anti-Stokes Raman scattering and stimulated Raman scattering, are able to perform label-free imaging, with chemical bonds specificity. Here we demonstrate that the use of circularly polarized light allows to retrieve not only the chemical nature but also the symmetry of the probed sample, in a single measurement. Our symmetry-resolved scheme offers simple access to the local organization of vibrational bonds and as a result provides enhanced image contrast for anisotropic samples, as well as an improved chemical selectivity. We quantify the local organization of vibrational bonds on crystalline and biological samples, thus providing information not accessible by spontaneous Raman and stimulated Raman scattering techniques. This work stands for a symmetry-resolved contrast in vibrational microscopy, with potential application in biological diagnostic.

[1] Aix-Marseille Université, CNRS, Centrale Marseille, Institut Fresnel, UMR 7249, Domaine Universitaire de Saint Jérôme, Avenue Escadrille Normandie Niemen, Marseille, F-13397, France. [2] Quebec Mental Health Institute Research Center, Laval University, Québec, Quebec, Canada G1J 2G3. [3] Centre d'Optique, Photonique et Laser (COPL), Laval University, Québec, Quebec, Canada G1V 0A6. Correspondence and requests for materials should be addressed to J.D. (email: julien.duboisset@fresnel.fr).

Obtaining information on the organization of matter on the micrometer-scale with non-destructive methods still remains a challenge in chemical physics and biology. One well-established method for extracting matter organization information is fluorescence microscopy, which uses fluorescent molecules or proteins to tag the sample. However, this technique is limited to the observation of the probe itself, which may differ from the sample organization. Coherent Raman scattering (CRS) microscopy technique has proven to be powerful due to its label-free, three-dimensional, chemical selective and real-time imaging capabilities[1–6]. In coherent anti-Stokes Raman scattering (CARS), two beams of different frequencies interact with the sample to excite a vibrational resonance. A probe beam is used to probe the vibrational excitation by generating a new anti-Stokes frequency shifted beam. Unfortunately, CARS is limited by a non-resonant four wave mixing (FWM) background resulting in reduced image contrast and chemical selectivity. Recently, it has been shown that stimulated Raman scattering (SRS) can provide vibrational spectra without non-resonant background allowing to report vibrational information with high fidelity and high efficiency[7–10] and CARS can benefit from non-resonant background to heterodyne-amplify weak Raman signals[11]. It is well-known that the symmetry properties of matter have a strong influence on its physical properties, for example, in crystalline samples. Similarly, it has been found that in biological environments anisotropic and symmetry properties of tissue are often related to specialized biological functionality. Nevertheless, the organization of molecular bonds in the focal volume, which is of particular interest in numerous situations where the medium is organized, is not contained in the spectral information. However, seminal works from nonlinear optics pioneers have shown that the Cartesian components of the nonlinear susceptibility tensor $\bar{\chi}^{(3)}$ express the vibrational symmetry properties[12,13]. To read the tensor elements, polarization-resolved schemes have been proposed decades ago[14], stimulating more recent developments in microscopy[15–20]. The molecular organization from a sample is usually retrieved by acquiring a stack of images from different polarization angles of the excitation or detection light fields, requiring long acquisition times and time consuming post-processing[21].

In this article, we introduce a label-free microscopy technique that is able to retrieve the individual symmetry orders of molecular organization in a single image acquisition. The symmetry-resolved CARS (SR-CARS) signal not only depends on the presence of molecular bonds, but also on their organization within the focal volume. By switching between combination of left- and right-handed circular polarization states for the involved fields, it is possible to directly image individual symmetry contributions of the sample. Beyond this organization selectivity, we show that our technique can (1) suppress the isotropic background in CARS images and spectra, thus enhancing the contrast by 1–2 orders of magnitude, and (2) retrieve quantitative information without pre-knowledge information on the molecular organization, without post-processing and independently of sample orientation in the transverse plane. SR-CARS provides higher chemical selectivity based on different symmetry characteristics, which are not accessible with regular spontaneous Raman or SRS microscopy.

Results

Light matter symmetry matching. A direct read-out of a specific sample symmetry is possible when the light field tensor probing the sample only consists of the targeted symmetry of the sample[22,23]. In practice, the detected CARS electric field amplitude E_{as} (anti-Stokes) along a specific polarization direction is the projection of a tensor $\bar{F}$, representing all involved light fields, onto the $\bar{\chi}^{(3)}$ susceptibility tensor of the medium

$$E_{as} = \bar{\chi}^{(3)} \cdot \bar{F} = \bar{\chi}^{(3)} \cdot (\hat{e}^*_{as} \otimes \vec{E}_p \otimes \vec{E}^*_s \otimes \vec{E}_{pr}), \qquad (1)$$

where $\otimes$ is the dyadic product and * stands for the complex conjugate, $\vec{E}_p$, $\vec{E}_{pr}$ and $\vec{E}_s$ are the pump, probe and Stokes fields, respectively, and $\hat{e}_{as}$ is the unit vector along the polarization direction of the emitted anti-Stokes field, such that the four vector fields create a rank-four tensor[24].

As our technique targets the investigation of sample's symmetry, it is convenient to choose spherical coordinates (θ, ϕ), for which left- ($E_{\circlearrowleft}$) and right-handed ($E_{\circlearrowright}$) circular polarization and linear z-polarization states ($E_{z\rightarrow}$) can be described by a set of spherical harmonic functions $Y^l_m(\theta, \phi)$ with $l = 1$

$$E_{\circlearrowleft}(\theta, \phi) = aE_{0,\circlearrowleft} \cdot Y^1_1(\theta, \phi) \qquad (2)$$

$$E_{\circlearrowright}(\theta, \phi) = aE_{0,\circlearrowright} \cdot Y^1_{-1}(\theta, \phi) \qquad (3)$$

$$E_{z\rightarrow}(\theta, \phi) = aE_{0,z\rightarrow} \cdot Y^1_0(\theta, \phi) \qquad (4)$$

where $E_{0,\circlearrowleft}$, $E_{0,\circlearrowright}$, $E_{0,\rightarrow}$ are the electric field amplitudes associated to the quantum number $m = 1$, $m = -1$ and $m = 0$, respectively and $a = -\sqrt{4\pi/3}$ is a normalization factor (see Supplementary Note 1).

The advantage of using spherical harmonic functions is that an m-value is a direct reporter of the presence of an m-folded rotational invariant symmetry in the sample plane. In case of the CARS field tensor $\bar{F}$ in equation (1), the multiplication of spherical harmonic functions generates a new set of spherical harmonic functions of fixed m-values, allowing to read m-values higher than the initial individual light fields and to determine the symmetry of the combined light fields

$$\bar{F}(\theta, \phi) \propto Y^1_{m_p} \cdot Y^{1^*}_{m_s} \cdot Y^1_{m_{pr}} \cdot Y^{1^*}_{m_{as}} = \sum_{l=0}^{4} k_l \cdot Y^l_{m_{\bar{F}}}, \qquad (5)$$

with the resulting $m_{\bar{F}}$-value being a summation of the field's m-values

$$m_{\bar{F}} = m_p - m_s + m_{pr} - m_{as}, \qquad (6)$$

where k_l are the Clebsch–Gordan coefficients weighting the different spherical harmonic functions (see Supplementary Note 1) and $Y^{l*}_m = Y^l_{-m}$. Therefore, the light field tensor $\bar{F}(\theta, \phi)$ has a $m_{\bar{F}}$-rotational symmetry in the sample plane defined by the m-values of the individual incident and emitted light fields, see equation (6).

When $\bar{F}$ (light) probes $\bar{\chi}^{(3)}$ (matter) in a CARS process (equation (1)), it can be seen, from the orthogonality of spherical harmonic functions, that $\bar{F}$ only probes parts of the $\bar{\chi}^{(3)}$-tensor with identical rotational invariant symmetries (that is, identical $m_{\bar{F}}$). Thus, engineering of $\bar{F}$ allows the direct read-out of specific sample symmetries, creating a symmetry-based contrast mechanism.

Figure 1a displays the different light fields $\bar{F}(\theta, \phi)$ of symmetry orders $m_{\bar{F}}$ that can be generated using circularly polarized light and degenerated CARS ($m_p = m_{pr}$). As circular polarizations provide only $m = 1$ or $m = -1$ values, sample rotational symmetry contribution of order 0 ($m_{\bar{F}} = 0$), order 2 ($m_{\bar{F}} = 2$) and order 4 ($m_{\bar{F}} = 4$) can be specifically addressed. Using two detectors for the emitted $\circlearrowright$ and $\circlearrowleft$ CARS polarization states allows the independent and simultaneous detection of symmetry contributions of order 0 (isotropic) and 2 (twofold symmetry), respectively, if the pump, Stokes and probe beams have the same circular polarization states (Fig. 1b). If the Stokes circular polarization state is changed from $\circlearrowleft$ to $\circlearrowright$, symmetries of order

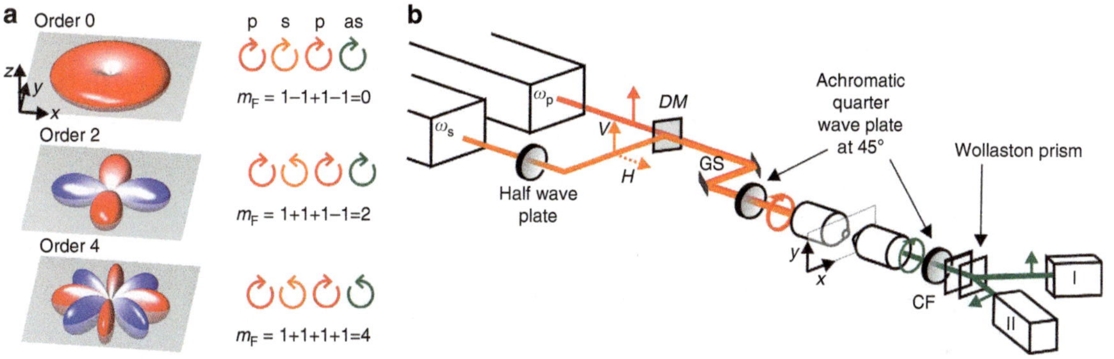

Figure 1 | Symmetry-resolved CARS setup. (**a**) Generated CARS light field $\bar{F}(\theta,\phi)$ with symmetry orders $m_{\bar{F}} = m_p - m_s + m_p - m_{as}$ resulting from the combination of circular polarization states for the pump (red), Stokes (orange) and detected CARS (green) fields; (**b**) schematic of the experimental setup (CF, CARS filter; DM, dichroic mirror; GS, galvo-scanners), switching the Stokes field between *V*- and *H*-polarization states before the first quarter-wave plate allows switching between Stokes field ↺- and ↻-polarization at the sample plane. Using two photomultiplier tubes (I/II) detecting ↺- and ↻-polarized CARS fields allows to target the different symmetry orders $m_{\bar{F}} = 0, 2$ and 4.

2 and 4 (fourfold) are then probed independently and simultaneously. Using circularly polarized light at both excitation and detection paths in CARS then creates a new symmetry-resolved imaging modality that has the unique ability to extract local orientational symmetry information from any sample. The orders $m_{\bar{F}} = 1$ and $m_{\bar{F}} = 3$ are in principle also accessible, but require excitation or detection of *z*-polarization contributions, which is not considered in this work. Note that since the obtained information is invariant on rotation around the *z*-axis, each symmetry order image does not depend on the sample orientation in the sample plane.

Multi lamellar vesicles. To demonstrate the potential of SR-CARS imaging, we applied our scheme to model samples of known symmetry. 1,2-Dipalmitoylphosphatidylcholine (DPPC) multi lamellar lipid vesicles (MLVs) are made of a tight packing of lipid layers, which form a ring of highly ordered matter with twofolded symmetry (see drawing in Fig. 2) and a lipid orientational distribution close to a Gaussian angular shape[20]. Regular CARS imaging of MLVs on C–C stretching vibration at 1,133 cm^{-1} embedded in water, shows a poorly contrasted image due to the non-resonant background, see Fig. 2a. Using SR-CARS imaging, different symmetry contribution can be separately imaged, namely the isotropic ($m_{\bar{F}} = 0$), twofold ($m_{\bar{F}} = 2$) and fourfold ($m_{\bar{F}} = 4$) contributions of the equatorial section of the MLV in the sample plane. Figure 2a shows that the aqueous solution surrounding the MLV is only visible in the order 0 image, due to its purely isotropic nature. The microscopic organization of lipids results in a strong order 2 signal coming from both nonresonant and resonant contributions, allowing background-free imaging of the MLV with a clearly superior contrast respect to regular CARS image. At last, no visible order 4 is present at this frequency, meaning that the lipid C–C stretching vibration at 1,133 cm^{-1} is symmetric, in contrast to anti-symmetric vibrations (see Supplementary Fig. 1). To obtain more quantitative information about the molecular organization, the order 2 image is normalized by the total intensity image (sum over all the order images) pixel by pixel and square rooted, leading to a direct read-out of the order 2 susceptibility tensor contribution. This normalized order 2 is found to be ~0.5 (see Supplementary Fig. 2) and is in good agreement with previous results obtained in similar systems[20].

Zeolite crystal. Since molecular vibrational resonances belong to a variety of symmetry groups, SR-CARS can be used to retrieve the vibration symmetries. As a second model system we selected the cubic crystal octahydrosilasesquioxane $H_8Si_8O_{12}$ crystal (HT8) zeolite, which forms microscopic scale crystals belonging to the O_h crystalline point group, and has a fourfold symmetry for the 932 cm^{-1} O–Si–H vibrational resonance addressed here[25]. Regular CARS imaging of such a crystal embedded in water shows a poorly contrasted image (Fig. 2b) due to the nonresonant background. Employing SR-CARS, the surrounding background disappears in the order 4 image, enhancing the CARS image contrast of the crystal with respect to its isotropic surrounding by a factor of 100. The order 2 image, as expected from a pure fourfold symmetry, shows no visible signal.

SR-CARS can be used to increase spectral contrast and separate close-by or overlapping resonances. Figure 3a shows the vibrational spectrum (obtained by scanning the Stokes beam wavelength) of a HT8 zeolite crystal acquired with spontaneous Raman scattering, SRS and regular CARS. In comparison to spontaneous Raman and SRS spectra, regular CARS is clearly affected by a nonresonant background generated by the crystal itself, which prevents from visualizing all the resonances of HT8 in the scanned spectral range.

Figure 3b displays order 0, order 2 and order 4 spectra acquired with the SR-CARS scheme. The first observation is that order 4 provides a highly contrasted background-free spectrum. It reveals in particular a weak vibrational resonance at 1,120 cm^{-1} that is invisible in regular CARS. Moreover, it clearly separates resonances that are less distinct in Raman or SRS (~2,300 cm^{-1}). The separation of overlapping vibrational resonances with SR-CARS is possible, as the resonances are of different vibrational mode types with different symmetry properties. As a result, they have similar amplitudes for their fourfolded symmetry contributions (thus allowing to separate them) while the total amplitudes of the two resonances are of different orders of magnitude such that they can hardly be distinguished in spontaneous Raman or SRS. While regular CARS of order 0 spectra are clearly distorted by the intrinsic interference with the nonresonant background (which is itself primarily of order 0), the high order symmetry filtering extracts much clearer spectral information (see also the whole spectral scan in Supplementary Movie 1). The order 2 spectrum shows weak signals, which we attribute to residual polarization leakage. The second observation is that all the resonances of the crystal do not have the same symmetry-resolved spectra. The A_{1g} resonance at 2,302 cm^{-1} has mainly order 0 contribution, the order 4 is <1% of the isotropic order and comes from optical leakage. The T_{2g} resonances at 883 cm^{-1}, 897 cm^{-1}, 1,117 cm^{-1}, 2,286 cm^{-1} have only a strong order 4 contribution and the E_g resonance at 932 cm^{-1}

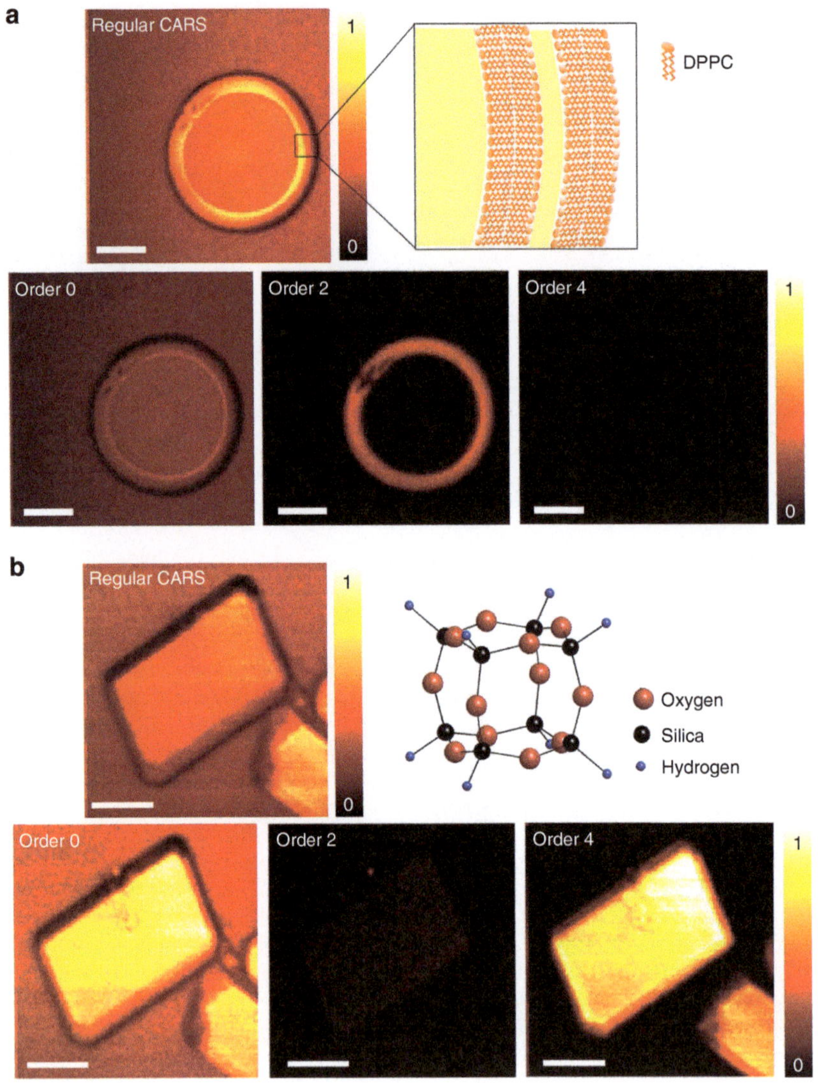

Figure 2 | Symmetry-resolved CARS imaging. (**a**) multi lamellar DPPC vesicle (MLV) imaged with Regular and SR-CARS at 1,133 cm^{-1}. (**b**) HT8 zeolite crystal imaged with Regular and SR-CARS at 932 cm^{-1}. Schematic representation of the membrane structure of a MLV and the cubic unit cell of HT8 zeolite are also shown. Scale bar, 10 μm.

has both order 0 and order 4 contributions. The decomposition of each mode on the symmetry orders is determined by the selection rules and detailed in Supplementary Fig. 3 and Supplementary Note 2. The three different modes of zeolite investigated here have three different signatures on the SR-CARS spectra, allowing an unambiguous experimental identification of each mode.

Myelin sheaths. We now focus on myelin sheaths, that are highly organized multilayer membranes made of lipids and proteins surrounding the axon of neurons. Myelin integrity is essential for the propagation of action potential and CARS has proven to be a powerful label-free technique for myelin imaging, targeting lipid vibrational bonds[26,27]. Polarization-resolved CARS has been reported to describe myelin molecular organization[20], and to relate this organization to demyelination processes[28], however, such experiments take minutes as they require multiple images with different incoming linear polarization states. Here we apply SR-CARS imaging to mice spinal cords and show that it can enhance information and reveal the full molecular lipid organization in a single image acquisition, within biological tissues.

Figure 4a shows a CARS intensity image of an *ex vivo* mouse spinal cord transverse section, using circular excitation on the C–C vibrational bond at 1,099 cm^{-1}. The vibration is chosen for its weak efficiency, as compared with CH_2 stretching modes, to demonstrate the performance of SR-CARS scheme. The order 0 image (Fig. 4b), shows weak contrast due to the presence of lipids bonds surrounding the myelin sheath and an isotropic contribution of the nonresonant background. On the contrary, the order 2 image (Fig. 4c) highlights the ring-shaped myelin around the axons, exhibiting the fact that lipids are highly organized and oriented in such structures. The order 4 image shows very weak signals (not shown), similarly as in MLVs. To provide a quantitative information of molecular order in myelin sheath using SR-CARS, the order 2 image is normalized by the total intensity image (sum of all orders) pixel by pixel, to cancel the molecular density dependency, see Fig. 4d. The molecular order of lipids in myelin sheaths is visibly not homogeneously distributed, with highly organized regions like in MLV can be found (high order 2 values), as well as highly disordered regions with low order 2 (Fig. 2a). This heterogeneity can be attributed to different morphologies and molecular compositions present in

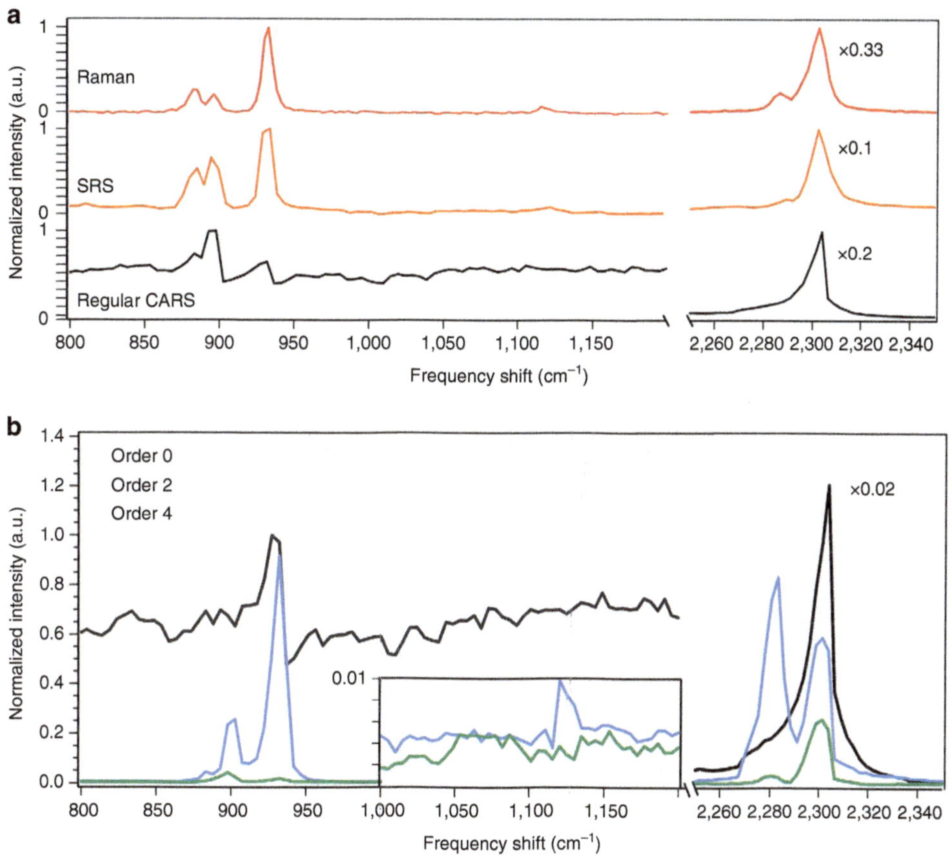

Figure 3 | Symmetry-resolved CARS spectrum. (**a**) Vibrational spectra of a HT8 zeolite crystal obtained with spontaneous Raman, SRS and regular CARS. (**b**) SR-CARS order 0, 2 and 4.

multilayers surrounding axons, which is a topic of current interest to address possible molecular-level neurobiology diagnostics[28]. The symmetry-resolved modality developed here is then able, in a single image allowing for high-speed imaging, to reveal local heterogeneities that are absent from the regular CARS images.

Discussion

This work presents a CARS imaging modality based on symmetry-selectivity, with unprecedented ability to enhance vibrational contrast and to reveal molecular bonds organization symmetry. The use of circular polarization makes this imaging modality independent of the sample orientation in the transverse sample plane, making this contrast enhancement efficient without the need to find an optimal polarization coupling direction. However, the use of polarization makes SR-CARS modality impossible within materials where polarization of light can not be maintained efficiently, such as birefringent materials or scattering media.

We exploited the specific symmetry of molecular bonds assemblies to strongly enhance the CARS image contrast in crystalline and tissue sample. In the past, many other techniques have been presented to improve contrast in CARS, though they mostly aimed to reduce the nonresonant background. A circular excitation approach was already developed in CARS by Upputuri *et al.*[17] to remove the isotropic background and enhance the signal from anisotropic materials. Our SR-CARS scheme not only enhances the non isotropic CARS signal, it also brings additional key information such as the vibration symmetry and the symmetry of the probed molecular assemblies. Such symmetry retrieval is possible because SR-CARS is based on a theoretical background using the irreducible spherical formalism.

Finally, molecular orientational organization imaging can be exploited to achieve a structural contrast in biological samples, which can improve monitoring of biological processes and diseases, as well as diagnostics. This provides quantitative molecular bond symmetry imaging in a single image acquisition without the need for any polarizer rotation nor signal processing. Biochemical studies have shown that molecular organization in myelin sheaths is indeed highly affected in diseases like multiple sclerosis and leukodystrophies; however, no current imaging technique can quantitatively address this issue *in vivo*. Investigating biological functions related to molecular organization in real time would be highly valuable and SR-CARS could be the basis of high-speed diagnostics.

Methods

CARS microscope. CARS imaging was performed on a custom-built setup incorporating a picosecond stimulated Raman optical source[29]. This source is composed of two optical parametric oscillators (OPO1 and OPO2, Emerald, APE) synchronously pumped by a mode-lock frequency doubled Nd:YVO Laser (PicoTrain, HighQLaser) operating at 532 nm. The two mode-locked beams from OPO1 (pump) and OPO2 (Stokes) (pulse duration 5 ps, repetition rate 76 MHz) are overlapped in time and space and sent into a custom made scanning microscope. For spectral scan, the pump wavelength is fixed to 730 nm and the stokes wavelength is scanned from 775 to 790 nm and 870 to 885 nm. SR-CARS signals (order 0, 2 and 4) are detected in the forward direction using PMTs (Hamamatsu, H10682) working in photon-counting regime. Excitation and collection are provided by an numerical aperture = 0.6 objective (Olympus UCPlan FL × 40). Incident powers at the sample plane were 1–4 mW for the pump beam and 1–5 mW for the Stokes beam depending on the samples. Imaging is performed by scanning galvanometric mirrors (typically, pixel dwell time of 50 μs, 100 × 100 pixels and scan range of 30 μm). A dedicated software controls the galvo mirrors, the acquisition card and the OPO wavelength tuning for spectral acquisition[30].

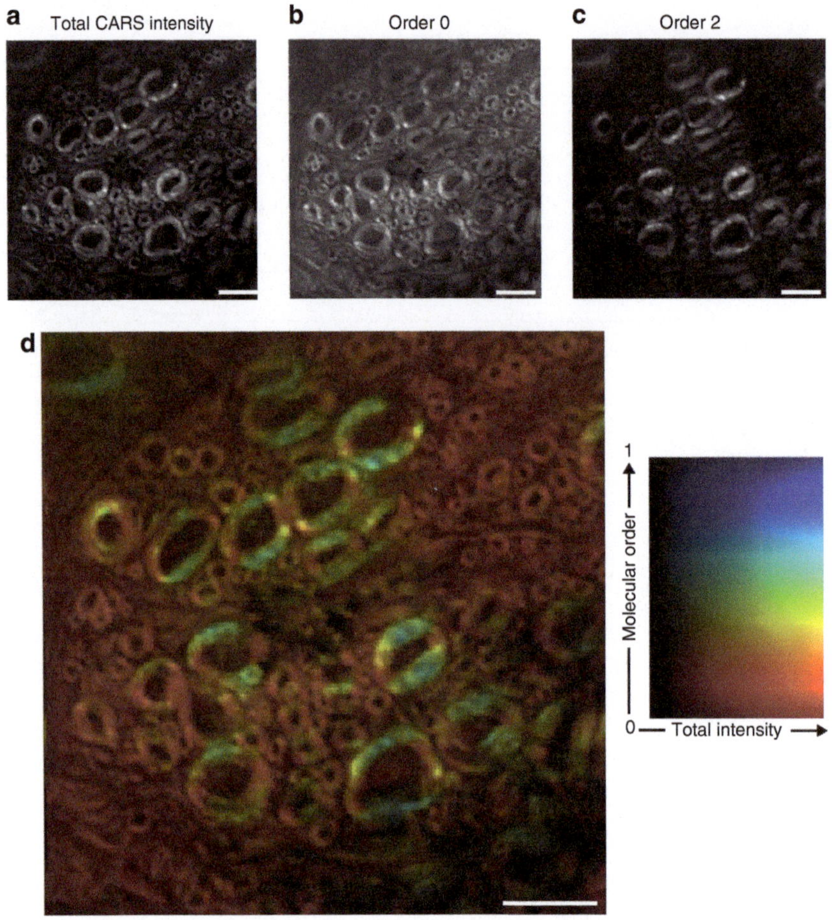

Figure 4 | Symmetry-resolved imaging of *ex vivo* myelin sheaths in mice spinal cord. (**a**) Total CARS intensity (circular excitation and unpolarized detection). (**b**) SR-CARS image of order 0, exhibiting isotropic symmetry contributions in the sample. (**c**) SR-CARS image of order 2, highlighting the twofold symmetry CARS contribution associated to organized lipids. (**d**) Quantification of local molecular order in the myelin sheath using normalized order 2 image with respect to the total intensity image, pixel by pixel. The local molecular order is colour coded, from 0 (for completely disordered) to 1 (for fully oriented), while the total intensity is encoded in the pixel brightness. Scale bar, 5 µm. All data are representative of at least three independant experiments.

To achieve SR-CARS imaging, an achromatic quarter-wave plate was inserted before the focusing microscope objective to excite the sample with circularly polarized light. The linear polarization state (before the quarter-wave plate) of the Stokes beam was switched between V- and H-polarization, resulting in a switching between ↻ and ↺-circular polarization states in the sample plane. The generated CARS signal passed through a second achromatic quarter-wave plate converting circular polarization to linear polarization. Subsequently, a Wollaston prism split the CARS beam into V- and H-polarization, which were detected individually with photomultiplier tubes. Consequently, the two photomultiplier tubes were sensitive to ↻ and ↺-circular polarization states in the sample plane. To ensure that any desired circular polarization state of the excitation is properly delivered to the sample, we used a polarimeter described in ref. 31.

In the SRS operation mode, the pump beam was modulated in amplitude at a frequency of 20 MHz by an acousto-optic modulator (AOM—AAOptoelectronic MT200 A0,2-800). The stimulated Raman gain induced on the Stokes beam was detected in the forward direction by means of a high-speed photodiode and a fast lock-in amplifier (manufactured by APE) in a way similar to ref. 7.

The Raman spectrum was acquired using a HeNe laser at 632.8 nm and a spectrograph (Horiba iHR320) equipped with a Peltier-cooled CCD detector.

MLV samples. MLVs were made from chain 1,2-dipalmitoylphosphatidylcholine (DPPC) lipids and 5% cholesterol. DPPC was hydrated in phosphate-buffered saline (pH 7.4) above the main phase-transition temperature (45 °C) for 1 h, leading to MLVs of 1–30 µm size in a solution enclosed between two spaced coverslips. The MLVs form almost spherical objects made of concentric multilayers of lipids. The images were performed at the equatorial plane of these objects where the distribution of the lipids is expected to be lying along the transverse sample plane.

Zeolite samples. The sample studied in this work is an octahydrosilasesquioxane HT8 crystal, that has cubic symmetry and belongs to the O_h crystallographic point group. The HT8 crystal synthesis can be found in ref. 25. For CARS imaging, micrometric to millimetric size crystals are directly deposited onto a microscope coverslip and surrounded by water.

Myelin samples. All experimental procedures have been performed in accordance with guidelines from the Canadian Council on Animal Care. Mice were anesthetized and perfused intracardially with 4% paraformaldehyde (PFA, Fischer Scientific, Pittsburgh) in 0.1 M phosphate buffer (PB). The whole spinal cord was dissected out from each mouse and placed in 4% PFA overnight. Lumbar spinal segments corresponding to L3–5 level were isolated from the spinal cord, rinsed several times with 0.1 M PB. About 30-µm-thick slices were cut in the sagittal plane using a vibratome (Leica, VT 1000).

References

1. Min, W., Freudiger, C. W., Lu, S. & Xie, X. S. Coherent nonlinear optical imaging: beyond fluorescence microscopy. *Annu. Rev. Phys. Chem.* **62,** 507–530 (2011).
2. Ji, M. *et al.* Rapid, label-free detection of brain tumors with stimulated Raman scattering microscopy. *Science Trans. Med.* **5,** 201ra119 (2013).
3. Chen, X., Grégoire, S., Formanek, F., Galey, J.-B. & Rigneault, H. Quantitative 3D molecular cutaneous absorption in human skin using label free nonlinear microscopy. *J. Control. Release* **200,** 78–86 (2015).
4. Jung, Y., Tam, J., Jalian, H. R., Anderson, R. R. & Evans, C. L. Longitudinal, 3D in vivo imaging of sebaceous glands by coherent anti-stokes Raman scattering

microscopy: normal function and response to cryotherapy. *J. Invest. Dermatol.* **135,** 39–44 (2015).
5. Zumbusch, A., Holtom, G. R. & Xie, X. S. Vibrationnal microscopy using coherent anti-Stokes Raman scattering. *Phys. Rev. Lett.* **82,** 4014–4017 (1999).
6. Potma, E. O. & Xie, X. S. Detection of single lipid bilayers with coherent anti-Stokes Raman scattering (CARS) microscopy. *J. Raman Spectrosc.* **34,** 642–650 (2003).
7. Freudiger, C. W. *et al.* Label-free biomedical imaging with high sensitivity by stimulated Raman scattering microscopy. *Science* **322,** 1857–1861 (2008).
8. Nandakumar, P., Kovalev, A. & Volkmer, A. Vibrational imaging based on stimulated Raman scattering microscopy. *New J. Phys.* **11,** 033026 (2009).
9. Ozeki, Y. *et al.* High-speed molecular spectral imaging of tissue with stimulated Raman scattering. *Nat. Photon.* **6,** 845–851 (2012).
10. Berto, P., Andresen, E. R. & Rigneault, H. Background-free stimulated Raman spectroscopy and microscopy. *Phys. Rev. Lett.* **112,** 053905 (2014).
11. Camp, Jr. C. H. *et al.* High-speed coherent Raman fingerprint imaging of biological tissues. *Nat. Photon.* **86,** 27–634 (2014).
12. Maker, P. D. Spectral broadening of elastic second-harmonic light scattering in liquids. *Phys. Rev. A* **1,** 923–951 (1970).
13. Jerphagnon, J., Chemla, D. & Bonneville, R. The description of the physical properties of condensed matter using irreducible tensors. *Adv. Phys.* **27,** 609–650 (1978).
14. Oudar, J. L., Smith, R. W. & Shen, Y. R. Polarization-sensitive coherent anti-Stokes Raman spectroscopy. *Appl. Phys. Lett.* **34,** 758–760 (1979).
15. Cheng, J.-X., Book, L. D. & Xie, X. S. Polarization coherent anti-stokes Raman scattering microscopy. *Opt. Lett.* **26,** 1341–1343 (2001).
16. Lu, F., Zheng, W. & Huang, Z. Elliptically polarized coherent anti-Stokes Raman scattering microscopy. *Opt. Lett.* **33,** 2842–2844 (2008).
17. Upputuri, P. K. *et al.* Circularly polarized coherent anti-Stokes Raman scattering microscopy. *Opt. Lett.* **38,** 1262–1264 (2013).
18. Munhoz, F., Rigneault, H. & Brasselet, S. High order symmetry structural properties of vibrational resonances using multiple-field polarization coherent anti-stokes Raman spectroscopy microscopy. *Phys. Rev. Lett.* **105,** 123903 (2010).
19. Zimmerley, M., Mahou, P., Debarre, D., Schanne-Klein, M. C. & Beaurepaire, E. Probing ordered lipid assemblies with polarized third-harmonic-generation microscopy. *Phys. Rev. X* **3,** 16 (2013).
20. Bioud, F. Z. *et al.* Structure of molecular packing probed by polarization-resolved nonlinear four-wave mixing and coherent anti-Stokes Raman-scattering microscopy. *Phys. Rev. A.* **89,** 013836 (2014).
21. Duboisset, J. *et al.* Molecular orientational order probed by coherent anti-Stokes Raman scattering (CARS) ans stimulated Raman scattering (SRS) microscopy: a spectral comparative study. *J. Phys. Chem. B* **119,** 3242–3249 (2015).
22. Yuratich, M. A. & Hanna, D. C. Coherent anti-Stokes Raman spectroscopy (CARS). Selection rules, depolarization ratios and rotational structure. *Mol. Phys.* **33,** 671–682 (1977).
23. Brasselet, S. & Zyss, J. Multipolar molecules and multipolar fields: probing and controlling the tensorial nature of nonlinear molecular media. *J. Opt. Soc. Am. B* **15,** 257–288 (1998).
24. Duboisset, J., Rigneault, H. & Brasselet, S. Filtering of matter symmetry properties by circularly polarized nonlinear optics. *Phys. Rev. A.* **90,** 063827 (2014).
25. Marcolli, C., Laine, P., Bhler, R., Calzaferri, G. & Tomkinson, J. Vibrations of H8Si8O12, D8Si8O12, and H10Si10O15 as determined by INS, IR, and Raman experiments. *J. Phys. Chem. B* **101,** 1171–1179 (1997).
26. Wang, H. F., Fu, Y., Zickmund, P., Shi, R. Y. & Cheng, J. X. Coherent anti-Stokes Raman scattering imaging of axonal myelin in live spinal tissues. *Biophys. J.* **89,** 581–591 (2005).
27. Belanger, E. *et al.* Quantitative myelin imaging with coherent anti-Stokes Raman scattering microscopy: alleviating the excitation polarization dependence with circularly polarized laser beams. *Opt. Express* **17,** 18419–18432 (2009).
28. De Vito, G., Tonazzini, I., Cecchini, M. & Piazza, V. RP-CARS: label-free optical readout of the myelin intrinsic healthiness. *Opt. Express* **22,** 13733–13743 (2014).
29. Brustlein, S. *et al.* Optical parametric oscillator-based light source for coherent Raman scattering microscopy: practical overview. *J. Biomed. Opt.* **16,** 021106 (2011).
30. Ferrand, P. GPScan.VI: A general-purpose LabVIEW program for scanning imaging or any application requiring synchronous analog voltage generation and data acquisition. *Comput. Phys. Commun.* **192,** 342–347 (2015).
31. Wang, X., Kress, A., Brasselet, S. & Ferrand, P. High frame-rate fluorescence confocal angle-resolved linear dichroism microscopy. *Rev. Sci. Instrum.* **84,** 053708 (2013).

Acknowledgements

We thank G. Calzaferri (the University of Bern) and P.A. Agaskar (Spherosils LLC, Blacksburg, VA) for providing us with the HT8 crystals. We thank D. Coté (the University of Laval) for providing us with the mice myelin sheath slices. We acknowledge financial support from the Centre National de la Recherche Scientifique (CNRS), Aix-Marseille University A*Midex (noANR-11-IDEX-0001-02), ANR grants France Bio Imaging (ANR-10-INSB-04-01) and France Life Imaging (ANR-11-INSB-0006) infrastructure networks and 'Fondation pour la Recherche Médicale' grant DBS20131128448.

Author contributions

C.C. performed all experiments. A.G. prepared the sample. S.B., H.R. and J.D. designed all experiments and developed the signal-processing methodology and protocols. C.C. and J.D. analysed the results. P.F. developed the acquisition and processing software. C.C., J.D., S.B. and H.R. wrote the manuscript.

Additional information

Competing financial interests: The authors declare no competing financial interests.

24

Semi-metallic Be_5C_2 monolayer global minimum with quasi-planar pentacoordinate carbons and negative Poisson's ratio

Yu Wang[1], Feng Li[1], Yafei Li[1] & Zhongfang Chen[2]

Designing new materials with novel topological properties and reduced dimensionality is always desirable for material innovation. Here we report the design of a two-dimensional material, namely Be_5C_2 monolayer on the basis of density functional theory computations. In Be_5C_2 monolayer, each carbon atom binds with five beryllium atoms in almost the same plane, forming a quasi-planar pentacoordinate carbon moiety. Be_5C_2 monolayer appears to have good stability as revealed by its moderate cohesive energy, positive phonon modes and high melting point. It is the lowest-energy structure with the Be_5C_2 stoichiometry in two-dimensional space and therefore holds some promise to be realized experimentally. Be_5C_2 monolayer is a gapless semiconductor with a Dirac-like point in the band structure and also has an unusual negative Poisson's ratio. If synthesized, Be_5C_2 monolayer may find applications in electronics and mechanics.

[1] College of Chemistry and Materials Science, Jiangsu Key Laboratory of Biofunctional Materials, Nanjing Normal University, Nanjing, Jiangsu 210023, China.
[2] Department of Chemistry, Institute for Functional Nanomaterials, University of Puerto Rico, Rio Piedras Campus, San Juan, Puerto Rico 00931, USA.
Correspondence and requests for materials should be addressed to Y.L. (email: liyafei.abc@gmail.com) or to Z.C. (email: zhongfangchen@gmail.com).

Carbon in known molecules and materials typically has tetrahedral tetracoordination (for example, diamond), planar tricoordination (for example, graphite) or linear dicoordination (for example, ethyne) arrangements. Especially, the tetrahedral preference of tetracoordinate carbon compounds, which was deduced by van't Hoff[1] and Lebel[2] more than a century ago, is one of the major foundations of organic chemistry. In 1968, Monkhorst[3] first discussed an exceptional bonding pattern of carbon, namely planar tetracoordinate carbon (ptC), by proposing a planar methane (D_{4h}), which is actually not a minimum energy structure. Later in 1970, by insightful analysis of ptC bonding in the hypothetical planar methane, Hoffman *et al.*[4] suggested that ptCs can be stabilized electronically by replacing H atoms in planar methane by σ donors (to facilitate electron transfer to electron-deficient σ bonds) or π acceptors (to delocalize the unfavourable lone pair of ptC). Along this line, in 1976 Collins *et al.*[5] theoretically devised the first ptC-containing molecule 1, 1-dilithiocyclopropane. One year later Cotton and Millar[6] synthesized the first ptC-containing molecule, although the unique ptC configuration was not recognized at that time. Ever since, ptC chemistry has been a subject of extensive studies[7–9]. Numerous ptC species have been designed theoretically[10–15] and some global minimum structures, such as CAl_4^+ (ref. 16), CAl_4^{2+} (ref. 17) and CAl_3Si^- (ref. 18) have been observed experimentally. More excitingly, besides ptC, many molecules containing planar pentacoordinate carbon (ppC)[19–22] and hexacoordinate carbon (phC)[23–25] have been designed computationally. The rule-breaking chemical bonding in these planar hypercoordinate carbons can lead to completely new molecules and materials, which are of fundamental importance to chemistry and materials science.

The unique topology and dimensionality may lead to exceptional properties of materials. It is not a surprise to witness the growing interest in designing ptC-containing solids and nanostructures in recent years[26–30]. Especially, stimulated by the extensive studies of graphene[31,32] and inorganic layered materials[33,34], many two-dimensional (2D) materials with rule-breaking chemical bonding have been designed computationally[35–42]. For example, on the basis of ptC molecule CB_4 (ref. 43), Wu *et al.*[35] designed the first ptC-containing 2D material, namely B_2C graphene, which was later confirmed to be only a local minimum[44]. Li *et al.*[36] and Dai *et al.*[37] predicted that Al_2C monolayer in its lowest-energy configuration have all C atoms being ideal ptC. More interestingly, Li *et al.*[45] demonstrated recently that one C atom can bind to six beryllium (Be) atoms in an almost planar manner, yielding a phC-featuring 2D material with a Be_2C stoichiometry. With intriguing structural and electronic properties, these 2D materials are expected to have important applications in some specific fields.

In contrast to ptC and phC, no attempt exists for extending ppC into solids until now. Motivated by the successful design of ptC- and phC-containing materials, we are quite curious whether it is possible to extend ppC into solids. Addressing this issue would deliver not only unique structures but also some fantastic properties, which is of both theoretical and practical importance.

In this work, by means of density functional theory (DFT) computations, we first confirm that the ppC molecule $Be_9C_2^{4-}$ is a local minimum on the potential energy surface. Based on the structural characters of $Be_9C_2^{4-}$, we design a ppC-containing 2D material, namely Be_5C_2 monolayer, in which each C atom binds to five Be atoms to form a quasi-planar pentacoordinate moiety. Our Be_5C_2 monolayer has rather good thermal and kinetic stabilities and is energetically the most favourable isomer in 2D space. Dramatically, the electronic band structure of Be_5C_2 monolayer has a Dirac-like point at the Fermi level, endowing an intriguing semimetallic feature. More interestingly, Be_5C_2 monolayer possesses a negative Poisson's ratio, which is rather rare in nanostructures. These fascinating properties make Be_5C_2 monolayer a promising candidate for future applications in electronics and mechanics.

Results

$Be_9C_2^{4-}$ as the inspiring ppC species for 2D monolayer. Our design of periodic 2D Be_5C_2 monolayer is initially inspired by our finding of the ppC molecule, namely $Be_9C_2^{4-}$, which has singlet ground state and D_{2h} symmetry (Fig. 1a). Computed at B3LYP level of theory with 6–311 + G*(*d*, *p*) basis set, $Be_9C_2^{4-}$ is a local minimum with the lowest vibrational frequency of 158.5 cm^{-1}. In this molecule, each C atom binds to five Be atoms to form a ppC moiety of Be_5C, which has been shown to be a local minima[20]. Structurally, $Be_9C_2^{4-}$ can be viewed as two Be_5C moieties fused by sharing one Be atom. The C–Be bond lengths are in the range of 1.70–1.76 Å, whereas the Be–Be bond lengths are in the range of 2.02–2.39 Å.

According to the natural population analysis, each ppC of $Be_9C_2^{4-}$ possesses a 2.18 |e| negative charge (− 0.40 |e| according to the Hirshfeld charge population analysis) and the natural electron configuration is $2s^{1.42}2p_x^{1.47}2p_y^{1.66}2p_z^{1.60}$, indicating that ppCs in $Be_9C_2^{4-}$ are essentially stabilized through Be σ-donation and delocalization of carbon $2p_z$ electrons. The computed Wiberg bond index for C-Be bonds are 0.52 and 0.61, respectively, resulting in a total Wiberg bond index of 2.81 for each ppC. Moreover, we also scrutinized the molecular orbitals of $Be_9C_2^{4-}$ to get more information on its stabilization mechanism. As shown in Fig. 1b, the highly delocalized π (for example, HOMO-7 and HOMO-8) and σ (for example, HOMO-3,

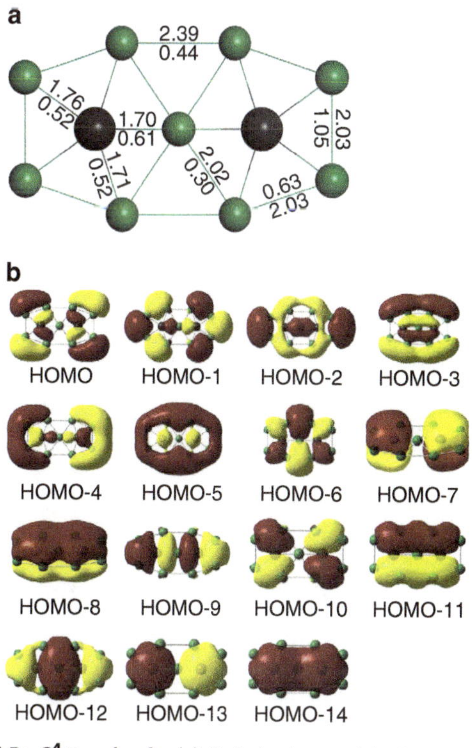

Figure 1 | $Be_9C_2^{4-}$ molecule. (**a**) Optimized geometric structure of $Be_9C_2^{4-}$ molecule. The black and green balls represent C and Be atoms, respectively. The red and black numbers are the Wiberg bond index (WBI) and lengths (in angstroms) of the representative chemical bonds, respectively. (**b**) The canonical molecular orbitals of $Be_9C_2^{4-}$.

HOMO-4 and HOMO-5) orbitals can help maintain the planar configuration.

Geometric properties of Be_5C_2 monolayer. The fusing of two Be_5C moieties to $Be_9C_2^{4-}$ reminds us of the roadmap of fusing benzene rings to polycyclic aromatic hydrocarbons (for example, naphthalene and anthracene) and then to 2D infinite graphene. Inspired by the ppC-containing $Be_9C_2^{4-}$, we designed a new 2D inorganic material, namely Be_5C_2 monolayer by generalizing the structural characters of $Be_9C_2^{4-}$. As shown in Fig. 2a, 1 unit cell of Be_5C_2 monolayer consists of 8 C atoms and 20 Be atoms with the optimized lattice parameters being $a = 8.92$ Å and $b = 9.21$ Å, respectively. Similar to $Be_9C_2^{4-}$, in Be_5C_2 monolayer each C atom binds to five Be atoms to form a ppC moiety of Be_5C and two neighbouring Be_5C moieties share one Be atom (Be_1) to form a Be_9C_2 moiety in the *a* (Be_9C_2-I) or *b* (Be_9C_2-II) direction. Especially, one Be_9C_2-I moiety is fused with four neighbouring Be_9C_2-II moieties by sharing the peripheral Be atoms (Be_2) and *vice versa*, leading to the formation of a 2D network with four Be_9C_2 moieties in one unit cell. The optimized coordinates of Be_5C_2 monolayer are presented in Supplementary Table 1. It is noteworthy that the Be_5C_2 monolayer is remarkably buckled rather than an exactly planar structure (Fig. 2b). The buckling, measured by the vertical distance between the bottommost Be atoms and the uppermost Be atom, is as high as 2.14 Å. Even so, the ppCs in Be_5C_2 monolayer still have a good planarity (Supplementary Fig. 1). According to our computations, the total degrees of five Be–C–Be angles for ppCs in Be_9C_2-I and Be_9C_2-II moieties are 371.44° and 364.84°, respectively, which are a little higher than the ideal 360°. Interestingly, when $Be_9C_2^{4-}$ is neutralized by four protons (H^+), the obtained $Be_9C_2H_4$ molecule is also severely buckled (Supplementary Fig. 2). In Be_5C_2 monolayer, the length of C–Be_1 bonds (1.66 Å, 1.68 Å) is much shorter than that of C–Be_2 bonds (1.73 Å, 1.74 Å). Moreover, our computations revealed that Be_5C_2 monolayer has a non-magnetic ground state, indicating that there are no unpaired electrons in Be_5C_2 monolayer.

We then computed the deformation electronic density of Be_5C_2 monolayer to elucidate its bonding nature. The deformation electronic density is defined as the total electronic density excluding those of isolated atoms. As clearly shown in Fig. 2c, some electrons are extracted from the 2*s* orbitals of Be atoms and well delocalized over C–Be bonds, indicating that C atoms form multicentre covalent bonds with neighbouring Be atoms, which is crucial for stabilizing the ppC moieties. The similar stabilizing mechanism has been found in ptC- and phC-containing 2D materials[35–40,45]. According to the Hirshfeld charge popular analysis, C, Be_1 and Be_2 atoms in Be_5C_2 monolayer possess a −0.32, +0.16 and +0.12 |e| charge, respectively. The buckling of Be_5C_2 monolayer stretches the Be–Be distances and probably helps reduce the otherwise even stronger repulsive interactions between Be atoms.

We also used the recently developed Solid State Adaptive Natural Density Partitioning (SSAdNDP) method[46] to better understand the unique chemical bonding of Be_5C_2 monolayer. According to our results (Supplementary Fig. 3), there is no classical two-centre-two-electron (2c–2e) C–Be bond in Be_5C_2 monolayer. For one unit cell of Be_5C_2 monolayer, the SSAdNDP search revealed twenty-four 3c–2e Be–C–Be σ-bonds (responsible for the bonding within the Be_5C units), four 4c–2e σ-bonds on four Be squares and eight 6c–2e π-bonds over eight Be_5C units, accounting for 72 electrons per unit cell. This bonding pattern is consistent with the symmetry of Be_5C_2 monolayer. Especially, the existence of delocalized σ- and π-bonds could essentially stabilize the ppCs in Be_5C_2 monolayer.

Stability of Be_5C_2 monolayer. Although Be_5C_2 monolayer has rather intriguing structural properties, we are unclear whether it is a stable structure. To assess the stability, we first computed the cohesive energy of Be_5C_2 monolayer, which is defined as: $E_{coh} = (nE_C + mE_{Be} - E_{Be_5C_2})/(n + m)$, in which E_C, E_{Be} and $E_{Be_5C_2}$ are the total energies of a single C atom, a single Be atom and Be_2C monolayer, respectively; *n* and *m* are the number of C and Be atoms in the supercell, respectively. According to our computations, Be_5C_2 monolayer has a cohesive energy of 4.58 eV per atom. As a reference, the cohesive energies of the experimentally realized 2D silicene[47,48] and phosphorene[49,50] are 3.71 and 3.61 eV per atom, respectively. As silicene and phosphorene are composed of covalent bonds, the even higher cohesive energy can ensure that Be_5C_2 monolayer is a strongly connected network.

The stability of Be_5C_2 monolayer can be further confirmed by its phonon dispersion curves. As shown in Fig. 3a, no appreciable

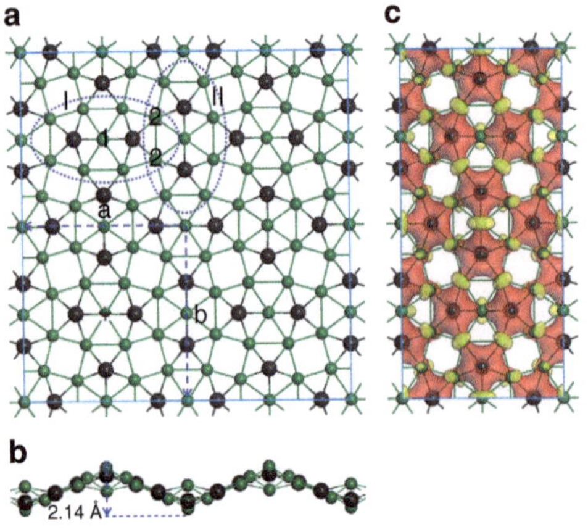

Figure 2 | Geometric and electronic structures of Be_5C_2 monolayer. (**a**) Top and (**b**) side views of geometric structure of Be_5C_2 monolayer. The blue dashed lines denote a unit cell; a and b represent the lattice vectors, 1 and 2 denote the different Be atoms, I and II represent the different Be_9C_2 moieties. (**c**) Deformation charge density of 2D Be_5C_2 monolayer. Red and yellow refer to electron accumulation and depletion regions, respectively.

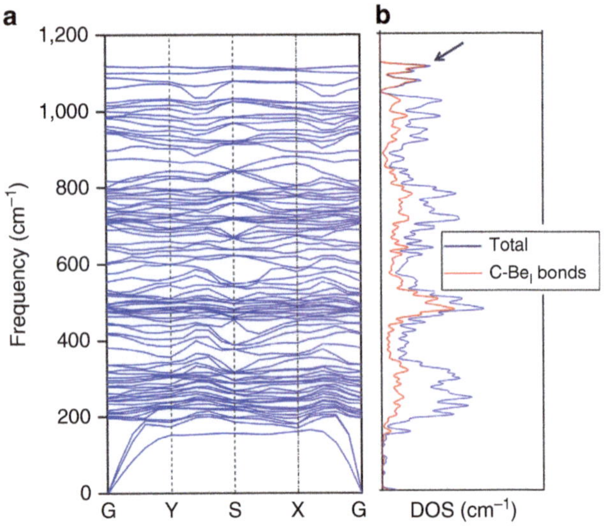

Figure 3 | Kinetic stability of Be_5C_2 monolayer. (**a**) Phonon spectrum and (**b**) phonon DOS of Be_5C_2 monolayer.

imaginary phonon mode is present, suggesting the good kinetic stability of Be_5C_2 monolayer. Remarkably, the highest frequency of Be_5C_2 monolayer reaches up to 1,120 cm^{-1}, which is higher than those of MoS_2 monolayer (473 cm^{-1})[51], silicene (580 cm^{-1})[52] and our recently designed phC-containing Be_2C monolayer (1,020 cm^{-1})[45], indicating robust C–Be bonds in Be_5C_2 monolayer. The analysis of partial phonon density of states (DOS; Fig. 3b) revealed that the highest frequency of Be_5C_2 monolayer is mainly contributed by C–Be_1 bonds.

Moreover, to examine the thermal stability of Be_5C_2 monolayer, we performed first-principles molecular dynamic (FPMD) simulations using a 2 × 2 supercell. Our three simulations at temperature of 1,000, 1,500 and 2,000 K show that Be_5C_2 monolayer can maintain its structural integrity throughout a 10-ps FPMD simulation up to 1,500 K, but is seriously disrupted at 2,000 K, suggesting that Be_5C_2 monolayer has a melting point between 1,500 and 2,000 K (Supplementary Fig. 4). We also performed optimizations for the distorted structures from MD simulations. After full atomic relaxation, those structures from MD simulations at 1,000 and 1,500 K can recover the initial configuration of Be_5C_2 monolayer. The above results demonstrate that Be_5C_2 monolayer has a remarkable thermal stability.

The moderate cohesive energy, all positive phonon modes and good thermal stability can ensure that Be_5C_2 monolayer is at least a local minimum structure on the potential energy surface. However, is the ppC-containing Be_5C_2 monolayer a global minimum? It is noteworthy that the global minimum structure is more likely to be achieved than the local minimum structures experimentally. For example, many isomers of graphene, such as Haeckelite graphene[53], T-graphene[54] and penta-graphene[55] have been designed computationally, but none of them has been realized experimentally. Therefore, we performed a global search for the lowest-energy structure of Be_5C_2 monolayer in the 2D space using first-principles-based particle-swarm optimization (PSO) method as implemented in CALYPSO code. After a comprehensive search, we obtained three stable isomers of 2D Be_5C_2, which are labelled as Be_2C_5-I, Be_2C_5-II and Be_2C_5-III, respectively. As shown in Fig. 4a, Be_2C_5-I is actually the above discussed ppC-featuring Be_5C_2 monolayer. Interestingly, in Be_2C_5-II (Fig. 4b) and Be_2C_5-III (Fig. 4c), each C atom binds to five Be atoms to form a ppC moiety of Be_5C and the Be_9C_2 moieties also can be found in these two isomers. Be_2C_5-I is 50 and 101 meV per atom lower in energy than Be_2C_5-II and Be_2C_5-III, respectively, indicating that Be_2C_5-I is the global minimum structure in the 2D space. Therefore, the ppC-featuring Be_5C_2 monolayer holds great potential to be realized experimentally.

Electronic properties of Be_5C_2 monolayer. With such interesting structural characteristics, does Be_5C_2 monolayer also have intriguing properties? To address this issue, we computed the band structure of the lowest-energy Be_5C_2 monolayer. As shown in Fig. 5a, Be_5C_2 monolayer is gapless or semi-metallic with the conduction band minimum (CBM) and valence band maximum (VBM) meeting at the Fermi level, which is quite similar to that of graphene. However, the meeting point of Be_5C_2 monolayer is located on the path from G (0, 0, 0) point to Y (0, 0.5, 0) point rather than on a high-symmetry point as for graphene. Especially, the conduction and valence bands around the Fermi level exhibit a linear dispersion relation, suggesting that the meeting point of Be_5C_2 monolayer is also Dirac-like. Considering that the PBE functional tend to underestimate the band gap, we recomputed the band structure of Be_5C_2 monolayer using the hybrid HSE06 functional[56] and found that the dispersion of the valence and conduction bands at the Fermi level is similar to that predicted by PBE and no appreciable band gap can be identified (Supplementary Fig. 5). Thus, the gapless property of Be_5C_2 monolayer is solid.

To obtain deeper insight into the electronic properties of Be_5C_2 monolayer, we analysed its DOS. As shown in Fig. 5b, the DOS is zero at the Fermi level, which further supports the presence of the Dirac cone. The partial DOS analysis shows that the VBM and CBM are contributed by both Be-2*p* and C-2*p* states, and the contribution from Be-2*p* states is much more than that from C-2*p* states. Moreover, we plotted the partial charge densities of the VBM and CBM. As shown in Fig. 5c,d, both VBM and CBM are

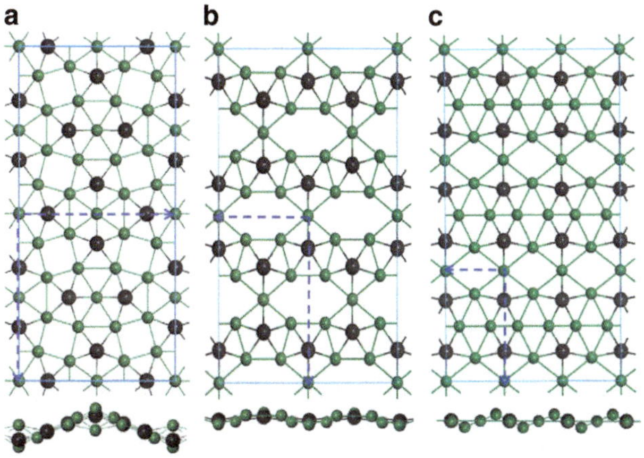

Figure 4 | Low-energy isomers. Geometric structures of Be_5C_2-I (**a**), Be_5C_2-II (**b**) and Be_5C_2-III (**c**). The blue dashed lines denote unit cells.

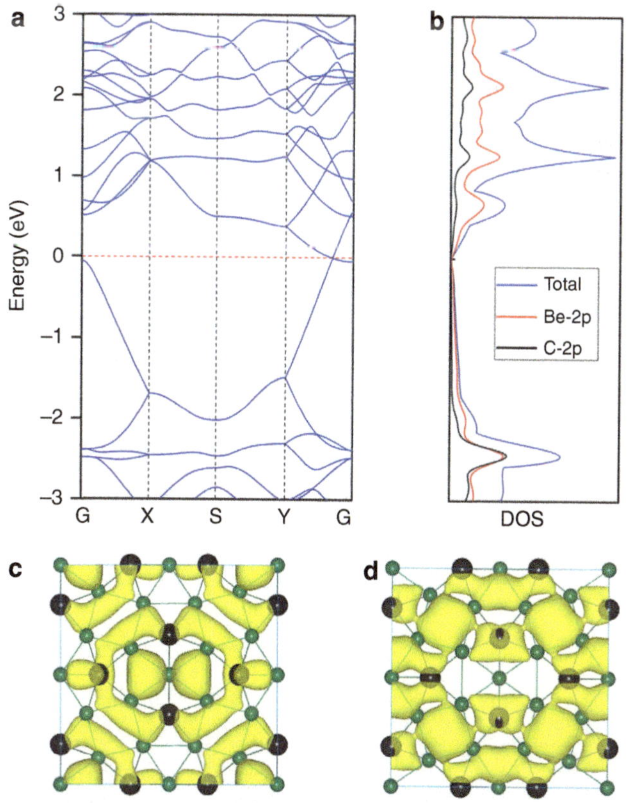

Figure 5 | Electronic properties of Be_5C_2 monolayer. (**a**) Band structure and (**b**) DOS of Be_5C_2 monolayer. The Fermi level is assigned at 0 eV. (**c**,**d**) The isosurfaces of partial charge densities for the (**c**) VBM and (**d**) CBM of Be_5C_2 monolayer. The isovalue is 0.015 e A^{-3}.

mainly originated from the delocalized orbitals of Be atoms and partially from the multicentre bonding between C and Be atoms.

Mechanical properties of Be_5C_2 monolayer. Besides the electronic properties, we also investigated the mechanical properties of Be_5C_2 monolayer by examining its elastic constants. As a validation, the computed elastic constants of graphene are $C_{11}=C_{22}=342.93$ GPa and $C_{12}=C_{21}=62.23$ GPa respectively, which achieve good agreements with experimental measurements[57] and previous computations[58]. For Be_5C_2 monolayer, its elastic constants were computed to be $C_{11}=32.90$ GPa, $C_{22}=130.89$ GPa, $C_{12}=C_{21}=-5.32$ GPa and $C_{66}=48.32$ GPa, which are in agreement with the mechanical stability criteria for a tetragonal 2D sheet ($C_{11}C_{22}-C_{12}^2>0$, $C_{66}>0$)[55]. The in-plane Young's modules along a (Y_a) and b (Y_b) directions, which can be deduced from the elastic constants by $Y_a=(C_{11}C_{22}-C_{12}C_{21})/C_{22}$ and $Y_b=(C_{11}C_{22}-C_{12}C_{21})/C_{11}$, were computed to be 32.68 and 130.03 N m^{-1}, respectively. As Y_a is not equal to Y_b, Be_5C_2 monolayer is mechanically anisotropic. Moreover, computed at the same level of theory, the in-plane Young's modules of Be_5C_2 monolayer are less than those of graphene ($Y_a=Y_b=331.63$ N m^{-1}) but higher than those of phosphorene ($Y_a=25.50$ N m^{-1} and $Y_b=91.61$ N m^{-1}), suggesting that Be_5C_2 monolayer has good mechanical properties.

Remarkably, we noted that Be_5C_2 monolayer has a negative C_{12}, which results in negative Poisson's ratios of -0.041 (C_{12}/C_{22}) and -0.16 (C_{12}/C_{11}) for a and b directions, respectively. It is noteworthy that the Poisson's ratio is defined as the negative ratio of transverse to axial strain. The negative Poisson's ratio indicates that Be_5C_2 monolayer can be compressed or stretched in both two directions at the same time. For a validation, we applied a uniaxial strain of 5% in a and b directions of Be_5C_2 monolayer, respectively. Just as expected, the equilibrium lattice parameters of b and a directions are elongated by ~0.2 and ~0.8%, respectively, confirming that Be_5C_2 monolayer indeed has negative Poisson's ratios.

Discussion

The unusual negative Poisson's ratio may endow Be_5C_2 monolayer with enhanced toughness and shear resistance, as well as enhanced sound and vibration adsorption. Correspondingly, Be_5C_2 monolayer could find some important applications in the fields of mechanics, tissue engineering and national security. It is worth noting that the negative Poisson's ratio is rather peculiar, as in nature nearly all materials have a positive Poisson's ratio, except some so-called auxetic materials[59]. Recently, Jiang et *al.*[60] demonstrated theoretically that single-layer black phosphorus has a negative Poisson's ratio due to the unique puckered configuration. However, the negative Poisson's ratio of phosphorene was observed in the out-of-plane direction, which is different from that of Be_5C_2 monolayer. Remarkably, the Poisson's ratio of Be_5C_2 monolayer in the b direction (-0.16) is much higher than that of phosphorene (-0.027)[60], rendering Be_5C_2 monolayer a more promising candidate for specific application in mechanical devices. No doubt, the negative Poisson's ratio of Be_5C_2 monolayer should be originated from its intriguing structural properties, especially the uniquely oriented chemical bonds. Our results could provide some guidelines for designing materials with a negative Poisson 's ratio.

With so many fascinating properties, it is desirable to synthesize Be_5C_2 monolayer in the laboratory. Considering that there is no layered structure of Be_5C_2 in nature, a promising approach is to grow Be_5C_2 monolayer on the surface of metal or metal oxide via chemical vapour deposition with accurately controlled Be/C ratio, just similar to the growth of silicene[47,48]. It is noteworthy that Be has toxic properties and the chemical vapour deposition synthesis usually requires high temperature; hence, special caution should be given during the experimental realization.

To summarize, inspired by the bonding characters of a ppC-containing molecule, $Be_9C_2^{4-}$, we designed a ppC-containing 2D inorganic material with a Be_5C_2 stoichiometry. Our DFT computations demonstrated that ppCs in Be_5C_2 monolayer are essentially stabilized by the charge transfer from Be ligands. The moderate cohesive energy, absence of imaginary modes in phonon spectrum and high melting point evaluated from FPMD simulations indicated that Be_5C_2 monolayer is experimentally viable. Especially, a global minimum search revealed that Be_5C_2 monolayer is the lowest-energy structure for the Be_5C_2 stoichiometry in 2D space, which endows Be_5C_2 monolayer great possibility to be realized experimentally. Our computations demonstrated that Be_5C_2 monolayer is semi-metallic with a zero band gap in the electronic band structure. More interestingly, Be_5C_2 monolayer has rather intriguing mechanical properties featured with a negative Poisson's ratio. Therefore, Be_5C_2 monolayer is expected to have wide applications in electronics and mechanics. We hope our theoretical studies will promote the experimental realization of this novel material and attract more attentions on investigating nanomaterials with novel chemical bonding.

Methods

DFT computations. For the $Be_9C_2^{4-}$ molecule, geometry optimizations, frequency analyses and electronic structure computations were performed at the B3LYP[61,62] level of theory with the 6-311+G*(d, p) basis set as implemented in Gaussian 03 package[63]. For 2D Be_5C_2 monolayer, DFT computations were performed using the plane-wave technique implemented in Vienna *ab initio* simulation package[64]. The ion–electron interaction was described using the projector-augmented plane wave approach[65,66]. The generalized gradient approximation expressed by PBE functional[67] and a 500-eV cutoff for the plane-wave basis set were adopted in all the computations. The convergence threshold was set as 10^{-4} eV in energy and 10^{-3} eV Å^{-1} in force. We set the x and y directions parallel and the z direction perpendicular to the layer plane, and adopted a supercell length of 15 Å in the z direction. The Brillouin zones was sampled with an $8\times6\times1$ Γ centred k points grid. The phonon spectrum was computed using finite displacement method as implemented in CASTEP code[68]. The elastic constants were also computed using the CASTEP code. The chemical bonding analysis of Be_5C_2 monolayer was done using the SSAdNDP method[46], which can well interpret the chemical bonding in terms of classical lone pairs, two-centre bonds, as well as multi-centre delocalized bonds in bulk solids, surfaces and nanostructures.

Molecular dynamics simulations. The thermal stability of Be_5C_2 monolayer was evaluated by means of first-principles molecular dynamics simulations. The ground-state structure of Be_2C monolayer was annealed at different temperatures. At each temperature, MD simulation in NVT ensemble lasts for 10 ps with a time step of 1.0 fs. The temperature was controlled by using the Nosé-Hoover method[69].

Global minimum structure searches. The PSO method within the evolutionary scheme as implemented in the CALYPSO code[70] was employed to find the low-energy structures of 2D Be_5C_2 monolayer. In our PSO simulation, the number of generation was maintained at 30. Unit cells containing total atoms of 7, 14 and 28 were considered. The structure relaxations during the PSO simulation were performed using Vienna *ab initio* simulation package at PBE level of theory.

References

1. van'tHoff, J. H. On the structure formulas in three-dimensional space. *Arch. Neerl. Sci. Exactes Nat.* 445–454 (1874).
2. LeBel, J. A. Sur les relations qui existent entre les formules atomiques des coprs organique et le pouvoir de leurs dissolutions. *Bull. Soc. Chim. Fr.* **22**, 337–347 (1874).
3. Monkhorst, H. J. Activation energy for interconversion of enantiomers containing an asymmetric carbon atom without breaking bonds. *Chem. Commun.* 1111–1112 (1968).
4. Hoffmann, R., Alder, R. W. & Wilcox, Jr C. F. Planar tetracoordinate carbon. *J. Am. Chem. Soc.* **92**, 4992–4993 (1970).

5. Collins, J. B. *et al.* Stabilization of planar tetracoordinate carbon. *J. Am. Chem. Soc.* **98,** 5419–5427 (1976).
6. Cotton, F. A. & Millar, M. The probable existence of a triple bond between two vanadium atoms. *J. Am. Chem. Soc.* **99,** 7886–7891 (1977).
7. Keese, R. Carbon flatland: planar tetracoordinate carbon and fenestranes. *Chem. Rev.* **106,** 4787–4808 (2006).
8. Merino, G., Mendez-Rojas, M. A., Vela, A. & Heine, T. Recent advances in planar tetracoordinate carbon chemistry. *J. Comput. Chem.* **28,** 362–372 (2007).
9. Yang, L.-M., Ganz, E., Chen, Z., Wang, Z.-X. & Schleyer, P. v. R. Four decades of the chemistry of planar hypercoordinate compounds. *Angew. Chem. Int. Ed.* **54,** 9468–9501 (2015).
10. Krogh-Jespersen, M.-B., Chandrasekhar, J., Wuerthwein, E.-U., Collins, J. B. & Schleyer, P. v. R. Molecular orbital study of tetrahedral, planar, and pyramidal structures of the isoelectronic series BH_4^-, CH_4, NH_4^+, AlH_4^-, SiH_4, and PH_4^+. *J. Am. Chem. Soc.* **102,** 2263–2268 (1980).
11. Boldyrev, A. I. & Simons, J. Tetracoordinated planar carbon in pentaatomic molecules. *J. Am. Chem. Soc.* **120,** 7967–7972 (1998).
12. Merino, G., Méndez-Rojas, M. A. & Vela, A. $(C5M_{2-n})^{n-}$ (M = Li, Na, K, and n = 0, 1, 2). A new family of molecules containing planar tetracoordinate carbons. *J. Am. Chem. Soc.* **125,** 6026–6027 (2003).
13. Merino, G. *et al.* Theoretical analysis of the smallest carbon cluster containing a planar tetracoordinate carbon. *J. Am. Chem. Soc.* **126,** 16160–16169 (2004).
14. Wang, Z. X., Zhang, C. G., Chen, Z. F. & Schleyer, P. v. R. Planar tetracoordinate carbon species involving beryllium substituents. *Inorg. Chem.* **47,** 1332–1336 (2008).
15. Wu, Y. B., Lu, H. G., Li, S. D. & Wang, Z. X. Simplest neutral singlet C_2E_4 (E = Al, Ga, In, and Tl) global minima with double planar tetracoordinate carbons: equivalence of C_2 moieties in C_2E_4 to carbon centers in CAl_4^{2-} and CAl_5^+. *J. Phys. Chem. A* **113,** 3395–3402 (2009).
16. Li, X., Zhang, H.-F., Wang, L.-S., Boldyrev, A. I. & Simons, J. Tetracoordinated planar carbon in the Al_4C^- Anion: a combined photoelectron spectroscopy and *ab initio* study. *J. Am. Chem. Soc.* **121,** 6033–6038 (1999).
17. Li, X., Zhang, H.-F., Wang, L.-S., Geske, G. D. & Boldyrev, A. I. Pentaatomic tetracoordinate planar carbon, $[CAl_4]^{2-}$: a new structural unit and its salt complexes. *Angew. Chem. Int. Ed.* **39,** 3630–3632 (2000).
18. Wang, L.-S., Boldyrev, A. I., Li, X. & Simons, J. Experimental observation of pentaatomic tetracoordinate planar carbon-containing molecules. *J. Am. Chem. Soc.* **122,** 7681–7687 (2000).
19. Wang, Z.-X. & Schleyer, P. v. R. Construction principles of "hyparenes": families of molecules with planar pentacoordinate carbons. *Science* **292,** 2465–2469 (2001).
20. Pei, Y., An, W., Ito, K., Schleyer, P. v. R. & Zeng, X. C. Planar pentacoordinate carbon in CAl_5^+: a global minimum. *J. Am. Chem. Soc.* **130,** 10394–10400 (2008).
21. Jimenez-Halla, J. O. *et al.* CAl_4Be and $CAl_3Be_2^-$: global minima with a planar pentacoordinate carbon atom. *Chem. Commun.* **46,** 8776–8778 (2010).
22. Grande-Aztatzi, R. *et al.* Planar pentacoordinate carbons in CBe_5^{4-} derivatives. *Phys. Chem. Chem. Phys.* **17,** 4620–4624 (2015).
23. Exner, K. & Schleyer, P. v. R. Planar hexacoordinate carbon: a viable possibility. *Science* **290,** 1937–1940 (2000).
24. Ito, K. *et al.* Myriad planar hexacoordinate carbon molecules inviting synthesis. *J. Am. Chem. Soc.* **129,** 1510–1511 (2007).
25. Wang, Z.-X. & Schleyer, P. v. R. Planar hypercoordinate carbons joined: wheel-shaped molecules with C–C axles. *Angew. Chem. Int. Ed.* **41,** 4082–4085 (2002).
26. Pancharatna, P. D., Méndez-Rojas, M. A., Merino, G., Vela, A. & Hoffmann, R. Planar tetracoordinate carbon in extended systems. *J. Am. Chem. Soc.* **126,** 15309–15315 (2004).
27. Yang, L. M., Ding, Y. H. & Sun, C. C. Design of sandwich-like complexes based on the planar tetracoordinate carbon unit CAl_4^{2-}. *J. Am. Chem. Soc.* **129,** 658–665 (2007).
28. Zhang, C. J., Sun, W. X. & Cao, Z. X. Zigzag boron-carbon nanotubes with quasi-planar tetracoordinate carbons. *J. Am. Chem. Soc.* **130,** 5638–5639 (2008).
29. Wu, M.-H., Pei, Y. & Zeng, X. C. Planar tetracoordinate carbon strips in edge decorated graphene nanoribbon. *J. Am. Chem. Soc.* **132,** 5554–5555 (2010).
30. Wu, Y. B., Jiang, J. L., Zhang, E. W. & Wang, Z. X. Computationally designed families of flat, tubular, and cage molecules assembled with "starbenzene" building blocks through hydrogen-bridge bonds. *Chem. Eur. J.* **16,** 1271–1280 (2010).
31. Novoselov, K. S. *et al.* A roadmap for graphene. *Nature* **490,** 192–200 (2012).
32. Tang, Q., Zhou, Z. & Chen, Z. Graphene-related nanomaterials: tuning properties by functionalization. *Nanoscale* **5,** 4541–4583 (2013).
33. Xu, M. S., Liang, T., Shi, M. & Chen, H. Graphene-like two-dimensional materials. *Chem. Rev.* **113,** 3766–3798 (2013).
34. Tang, Q., Zhou, Z. & Chen, Z. Innovation and discovery of graphene-like materials via density functional theory computations. *WIREs Comput. Mol. Sci.* **5,** 360–379 (2015).
35. Wu, X. J., Pei, Y. & Zeng, X. C. B_2C graphene, nanotubes, and nanoribbons. *Nano Lett.* **9,** 1577–1582 (2009).
36. Li, Y., Liao, Y., Schleyer, P. v. R. & Chen, Z. Al_2C monolayer: the planar tetracoordinate carbon global minimum. *Nanoscale* **6,** 10784–10791 (2014).
37. Dai, J., Wu, X., Yang, J. & Zeng, X. Al_xC monolayer sheets: two-dimensional networks with planar tetracoordinate carbon and potential applications as donor materials in solar cell. *J. Phys. Chem. Lett.* **5,** 2058–2065 (2014).
38. Zhang, Z., Liu, X., Yakobson, B. I. & Guo, W. Two-dimensional tetragonal TiC monolayer sheet and nanoribbons. *J. Am. Chem. Soc.* **133,** 19326–19329 (2012).
39. Li, Y., Li, F., Zhou, Z. & Chen, Z. SiC_2 silagraphene and its one-dimensional derivatives: where planar tetracoordinate silicon happens. *J. Am. Chem. Soc.* **133,** 900–908 (2011).
40. Yang, L.-M. *et al.* Two-dimensional Cu_2Si monolayer with planar hexacoordinate copper and silicon bonding. *J. Am. Chem. Soc.* **137,** 2757–2762 (2015).
41. Yang, L.-M. *et al.* Revealing unusual chemical bonding in planar hyper-coordinate Ni_2Ge and quasi-planar Ni_2Si two-dimensional crystals. *Phys. Chem. Chem. Phys.* **17,** 26043–26048 (2015).
42. Yang, L.-M. *et al.* Post-anti-van't Hoff-Le Bel motif in atomically thin germanium ally film. *Phys. Chem. Chem. Phys.* **17,** 17545–17551 (2015).
43. Pei, Y. & Zeng, X. C. Probing the planar tetra-, penta-, and hexacoordinate Carbon in carbon-boron mixed clusters. *J. Am. Chem. Soc.* **130,** 2580–2592 (2008).
44. Luo, X. *et al.* Predicting two-dimensional boron-barbon compounds by the global optimization method. *J. Am. Chem. Soc.* **133,** 16285–16290 (2011).
45. Li, Y., Liao, Y. & Chen, Z. Be_2C monolayer with quasi-planar hexacoordinate carbons: a global minimum structure. *Angew. Chem. Int. Ed.* **53,** 7248–7252 (2014).
46. Galeev, T. R., Dunnington, B. D., Schmidt, J. R. & Boldyrev, A. I. Solid state adaptive natural density partitioning: a tool for deciphering multi-center bonding in periodic systems. *Phys. Chem. Chem. Phys.* **15,** 5022–5029 (2013).
47. Feng, B. J. *et al.* Evidence of silicene in honeycomb structures of silicon on Ag (111). *Nano. Lett.* **12,** 3507–3511 (2012).
48. Fleurence, A. *et al.* Experimental evidence for epitaxial silicene on diboride thin films. *Phys. Rev. Lett.* **108,** 245501 (2012).
49. Li, L. *et al.* Black phosphorus field-effect transistors. *Nat. Nanotechnol.* **9,** 372–377 (2014).
50. Liu, H., Neal, A. T., Zhu, Z., Tomanek, D. & Ye, P. D. Phosphorene: an unexplored 2D semiconductor with a high hole mobility. *ACS Nano* **8,** 4033–4041 (2014).
51. Molina-Sánchez, A. & Wirtz, L. Phonons in single-layer and few-layer MoS_2 and WS_2. *Phys. Rev. B* **84,** 155413 (2011).
52. Cahangirov, S., Topsakal, M., Akturk, E., Sahin, H. & Ciraci, S. Two-and one-dimensional honeycomb structures of silicon and germanium. *Phys. Rev. Lett.* **102,** 236804 (2009).
53. Terrones, H. *et al.* New metallic allotropes of planar and tubular carbon. *Phys. Rev. Lett.* **84,** 1716 (2000).
54. Liu, Y., Wang, G., Huang, Q., Guo, L. & Chen, X. Structural and electronic properties of t graphene: a two-dimensional carbon allotrope with tetrarings. *Phys. Rev. Lett.* **108,** 225505 (2012).
55. Zhang, S. *et al.* Penta-graphene: a new carbon allotrope. *Proc. Natl Acad. Sci. USA* **112,** 2372–2377 (2015).
56. Heyd, J., Scuseria, G. E. & Ernzerhof, M. Erratum: "Hybrid functionals based on a screened Coulomb potential". *J. Chem. Phys.* **124,** 219906 (2006).
57. Lee, C. G., Wei, X. D., Kysar, J. W. & Hone, J. Measurement of the elastic properties and intrinsic strength of monolayer graphene. *Science* **321,** 385–388 (2008).
58. Li, X., Zhang, Q. & Wang, Q. Stability and physical properties of a tri-ring based porous gC_4N_3 sheet. *Phys. Chem. Chem. Phys.* **15,** 7142–7146 (2013).
59. Lakes, R. S. Foam structures with a negative Poisson's ratio. *Science* **235,** 1038–1040 (1987).
60. Jiang, J.-W. & Park, H. S. Negative Poisson's ratio in single-layer black phosphorus. *Nat. Commun.* **5,** 4727 (2014).
61. Becke, A. D. Density-functional thermochemistry. III. The role of exact exchange. *J. Chem. Phys.* **98,** 5648–5652 (1993).
62. Lee, C., Yang, W. & Parr, R. G. Development of the colle-salvetti correlation-energy formula into a functional of the electron density. *Phys. Rev. B: Condens. Mater. Phys.* **37,** 785 (1988).
63. Frisch, M. J. *et al. GAUSSIAN 03, Revision E.02* (Gaussian, Inc., 2004).
64. Kresse, G. & Hafner, J. Ab initio molecular dynamics for liquid metals. *Phys. Rev. B* **47,** 558 (1993).
65. Blöchl, P. E. Projector augmented-wave method. *Phys. Rev. B* **50,** 17953–17979 (1994).
66. Kresse, G. & Joubert, D. From ultrasoft pseudopotentials to the projector augmented-wave method. *Phys. Rev. B* **59,** 1758 (1999).
67. Perdew, J. P., Burke, L. & Ernzerhof, M. Generalized gradient approximation made simple. *Phys. Rev. Lett.* **77,** 3865 (1996).

68. Segall, M. D. *et al.* First-principles simulation: ideas, illustrations and the CASTEP code. *J. Phys. Condens. Matter* **14,** 2717 (2002).
69. Martyna, G. J., Klein, M. L. & Tuckerman, M. E. Nosé-Hoover chains: the canonical ensemble via continuous dynamics. *J. Chem. Phys.* **97,** 2635–2643 (1992).
70. Wang, Y., Lv, J., Zhu, L. & Ma, Y. Crystal structure prediction via particle-swarm optimization. *Phys. Rev. B* **82,** 094116 (2010).

Acknowledgements

Support in China by Natural Science Foundation of China (numbers 21403115 and 21522305) and the NSF of Jiangsu Province of China (number BK20150045), and in the United States by National Science Foundation (Grant EPS-1010094) and Department of Defense (Grant W911NF-12-1-0083) is gratefully acknowledged. The computational resources used in this research were provided by Shanghai Supercomputer Center.

Author contributions

Y.L. and Z.C. conceived the initial idea of this research. Y.W. and F.L. demonstrated the initial idea and collected all the data. Y.W., Y.L. and Z.C. wrote the paper and all authors commented on it.

Additional information

Competing financial interests: The authors declare no competing financial interests.

25

Stabilizing lithium metal using ionic liquids for long-lived batteries

A. Basile[1,2,†], A.I. Bhatt[2] & A.P. O'Mullane[1,3]

Suppressing dendrite formation at lithium metal anodes during cycling is critical for the implementation of future lithium metal-based battery technology. Here we report that it can be achieved via the facile process of immersing the electrodes in ionic liquid electrolytes for a period of time before battery assembly. This creates a durable and lithium ion-permeable solid-electrolyte interphase that allows safe charge-discharge cycling of commercially applicable Li|electrolyte|$LiFePO_4$ batteries for 1,000 cycles with Coulombic efficiencies $>99.5\%$. The tailored solid-electrolyte interphase is prepared using a variety of electrolytes based on the *N*-propyl-*N*-methylpyrrolidinium bis(fluorosulfonyl)imide room temperature ionic liquid containing lithium salts. The formation is both time- and lithium salt-dependant, showing dynamic morphology changes, which when optimized prevent dendrite formation and consumption of electrolyte during cycling. This work illustrates that a simple, effective and industrially applicable lithium metal pretreatment process results in a commercially viable cycle life for a lithium metal battery.

[1] School of Applied Sciences, Applied Chemistry, RMIT University, GPO Box 2476V, Melbourne, Victoria 3001, Australia. [2] Energy Flagship, Commonwealth Scientific and Industrial Research Organisation, Clayton, Melbourne, Victoria 3169, Australia. [3] School of Chemistry, Physics and Mechanical Engineering, Queensland University of Technology, GPO Box 2434, Queensland 4001, Australia. † Present address: Institute for Frontier Materials, Deakin University, Burwood, Victoria 3125, Australia. Correspondence and requests for materials should be addressed to A.I.B. (email: anand.bhatt@csiro.au) or to A.P.O'M. (email: anthony.omullane@qut.edu.au).

The next generations of rechargeable lithium metal anode-based battery technologies such as Li-O_2 and Li-S have specific energies of 3,505 Wh kg^{-1} (Li-O_2) and 2,567 Wh kg^{-1} (Li-S), respectively, which greatly exceed the current graphite anode-based technologies (typically 100–265 Wh kg^{-1}). This order of magnitude increase is the breakthrough needed for applications such as electric vehicles where concerns of 'range anxiety' are negated; in principle, these vehicles could compete with traditional petrol vehicles in terms of drive range per charge. Unlike graphite, lithium metal anodes do not intercalate lithium ions, instead requiring repetitive stripping and plating of the lithium metal surface. This causes volumetric and morphological changes that can lead to cycling instability. The inelastic solid–electrolyte interphase (SEI) formed during cycling cannot compensate for these morphological changes, and as a result is in a continual state of repair. This generally leads to low Coulombic efficiency (CE) and short cycle life because of electrolyte consumption and loss of lithium through cyclic SEI production.

This highlights a significant issue concerning the safe cyclability of lithium metal, that is, difficulty in controlling the thermodynamic instability and reactivity of lithium towards organic electrolytes[1,2]. These factors contribute to the classic literature examples of lithium metal battery failure, which requires avoiding large Li dendrite[3,4] and non-uniform SEI formation that typically results in thermal runaway via the creation of a short circuit[4–7]. Therefore, it is critical that there is not only further development in understanding SEI formation[5,8,9] but also a significant research effort into tailoring the SEI that is transferable into devices that can be practically realized.

Room-temperature ionic liquid (RTIL) electrolytes are potentially a successful alternative to conventional organic solvent-based electrolytes[10] and have shown promising results[11–13]. In particular, RTILs composed of the bis(fluorosulfonyl)imide $[FSI]^-$ anion are attracting significant attention[14,15]. The RTIL comprising the pyrrolodinium cation and *N*-propyl-*N*-methylpyrrolidinium bis(fluorosulfonyl)imide anion, namely ($[C_3mPyr^+][FSI^-]$), possesses low viscosity[16] and has been associated with delayed Al corrosion at the cathode[17,18]. Previously, Budi *et al.* and other reports noted that the SEI formed through chemical pathways using $[C_3mPyr^+][FSI^-]$ provides a smooth morphology that suppresses dendrite formation in symmetrical Li|electrolyte|Li cells, which is a critical prerequisite for secondary cell cycling[19,20].

Ideally, the prospect of tuning the electrolyte to affect a beneficial change in electrode performance[21], rather than modifying the lithium electrode via complicated and tedious surface engineering with nanostructures[22–24], would in principle allow a pathway towards simplified and cheaper battery construction. This approach could readily be translated to existing manufacturing processes and therefore accelerate the development of next-generation energy-storage solutions. We herein report on facile SEI formation via a simple chemical interaction between Li^0 and $[C_3mPyr^+][FSI^-]$ mixtures utilizing LiFSI, $LiPF_6$ and $LiAsF_6$ salts. Remarkably, we observe extensive Li|electrolyte|Li cell cycling at 0.1 and 1.0 mA cm^{-2} for symmetrical cells pretreated using LiFSI and $LiAsF_6$ salts, respectively, dissolved in $[C_3mPyr^+][FSI^-]$, with no evidence of dendrite formation even after 2,500 + hours cycling time. Full Li|electrolyte|$LiFePO_4$ batteries were also evaluated, and these batteries display extensive safe cycling at a rate of 1 C, achieving 1,000 cycles with a CE greater than 99.5% after pretreatment, comparable to or, in certain cases, outperforming current commercial 18,650 Li ion batteries[25]. In addition, this approach is beneficial for cells that are left on the shelf, where a cell that was quiescent for 330 days is shown to cycle effectively with >95% CE.

Results

Time-dependent SEI formation. Our previous work showed that Li reacts with $[C_3mPyr^+][FSI^-]$ to form an SEI layer, and that the application of a potential is not required to observe significant changes on the electrode surface[19]. However, on addition of Li salts to the RTIL there is a dramatic difference in both the reaction rate and composition of the SEI, which has significant implications for the performance of symmetrical and full cells. This initial interaction between the electrolyte and Li anode, occurring on construction of any cell, is a critical factor that can in fact be tailored to increase cyclability, safety and CE, and is the focus of this study that uses $[C_3mPyr^+][FSI^-]$ as the ionic liquid.

The facile lithium pretreatment methodology is shown in Fig. 1a, and the observed morphological changes that occur for the LiFSI/$[C_3mPyr^+][FSI^-]$ electrolyte system after the reaction time of 4 h, 7, 12 and 18 days are illustrated in Fig. 1b–e. Moreover, shown are micrographs of a lithium surface pretreated with $LiPF_6$/$[C_3mPyr^+][FSI^-]$ (Fig. 1f–i) and $LiAsF_6$/$[C_3mPyr^+][FSI^-]$ (Fig. 1j–m) for the same time periods. These micrographs show the diverse topology that can be realized by tuning reaction time and/or electrolyte composition. In all cases there are dramatic changes in the Li surface compared with the relatively smooth pristine metal (Supplementary Fig. 1).

The SEI surface-morphology changes on salt addition are significantly different to that observed for neat $[C_3mPyr^+][FSI^-]$[19]. Interestingly, inclusion of a Li salt results in accelerated changes, which are attributed to the anion. After 4 h of pretreatment with LiFSI/$[C_3mPyr^+][FSI^-]$ (Fig. 1b) the lithium surface contains residual electrolyte with numerous SEI sites. For foils pretreated for 7 days, the surface becomes rough and exfoliated (Fig. 1c and inset). After 12 days, the smoothened surface exhibits ordered SEI deposition sites preferentially arranged into lines of equal spacing of *ca* 0.25–11.0 μm (Fig. 1d). These ordered deposition sites eventually cluster to form ribbons (up to 500 μm in length) after 18 days (Fig. 1e). Honeycomb structures were also observed, dissimilar to those reported for $LiBF_4$/$[C_3mPyr^+][FSI^-]$[14].

The $LiPF_6$/$[C_3mPyr^+][FSI^-]$ electrolyte reacts producing a rough surface with evidence of small deposits on the surface after 4-h interaction (Fig. 1f). Figure 1g shows that additional species (dark areas) dominate the surface and is attributed to an exfoliation process. After 12 days, a thick film of deposition products is formed (Fig. 1h), with rod-like features and surface electrolyte coverage. By 18 days, these resolve into discreet-layered islands of deposition products (Fig. 1i).

For the $LiAsF_6$/$[C_3mPyr^+][FSI^-]$ electrolyte there are three distinct regions formed after 4 h (Fig. 1j), showing deposition sites situated within both the darkest and lightest regions. Medium contrast regions contain adhered residual electrolyte and rough areas are also observed (inset in Fig. 1j). By 7 days of interaction time, there is complete coverage of Li by coral structures (Fig. 1k); unlike the $LiPF_6$/$[C_3mPyr^+][FSI^-]$ pretreatment this structure dominates and alters the entire morphology. Optically, the lithium foil shows a tarnish over the metal (inset in Fig. 1k—typical $LiAsF_6$/$[C_3mPyr^+][FSI^-]$-pretreated foil), whereas all other electrolyte-treated samples retain a reflective surface. After 12 days (Fig. 1l), the surface is dominated by electrolyte adhesion and is presumed to envelop the structures observed after 7 days. After 18 days, the electrolyte and breakdown products become part of the structure. A consolidation of both the deposition products and the electrolyte forms a coral superstructure with complete surface coverage (Fig. 1m). It is clear that the 12-day time point shows a common trend for all electrolytes in that a smoother surface is formed that does not contain as many protrusions and coral-like structures that are

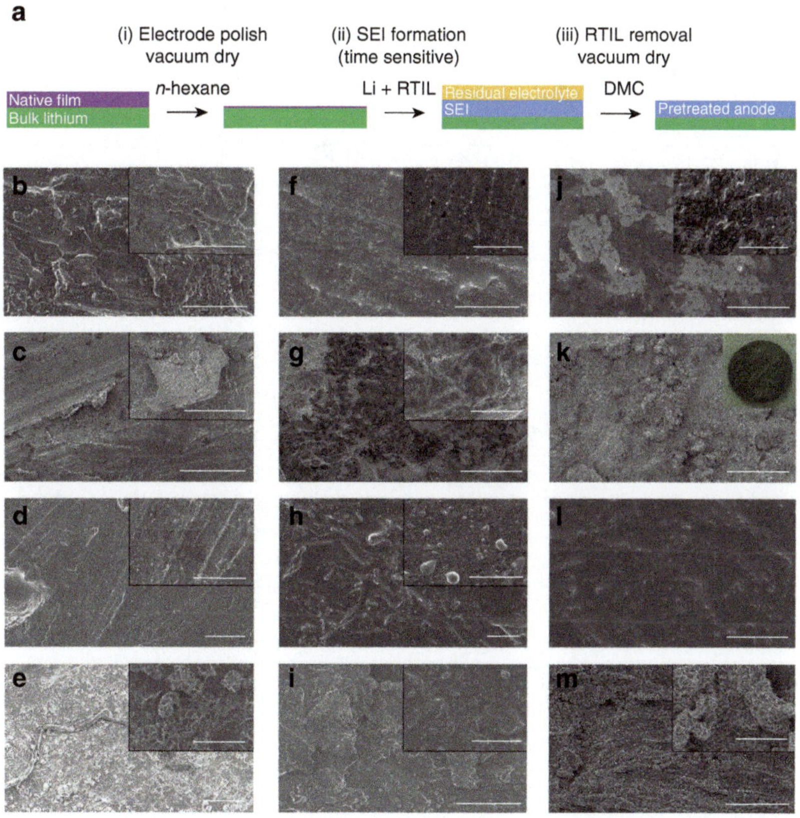

Figure 1 | Scanning electron micrographs of the lithium morphology after pretreatment. (**a**) Facile process for the preparation of SEI materials via chemical pretreatment with ionic liquid electrolyte. Micrographs revealing the dynamic morphology changes at lithium using LiFSI/[C_3mPyr$^+$][FSI$^-$] electrolyte at time periods of (**b**) 4 h, (**c**) 7 d, (**d**) 12 d and (**e**) 18 d. As above, for the $LiPF_6$/[C_3mPyr$^+$][FSI$^-$] electrolyte after (**f**) 4 h, (**g**) 7 d, (**h**) 12 d and (**i**) 18 d, and while using $LiAsF_6$/[C_3mPyr$^+$][FSI$^-$] electrolyte after (**j**) 4 h, (**k**) 7 d, (**l**) 12 d and (**m**) 18 d. Scale bars, 50 μm (inset scale bars, 10 μm). Additional time periods available in Supplementary Fig. 2.

present at other times. Scanning electron microscopy (SEM) images of additional reaction time periods of 4, 10, 14 and 16 days (Supplementary Fig. 2) further illustrate the gradual change of the Li surface. This time-dependent morphological change could have significant implications for battery performance, as this layer will be the initial seed layer that dictates the Li ion ingress/egress during initial cell cycling.

Surface chemistry of SEI via FTIR and XPS. The composition of the diverse SEI structures formed on the lithium surface after pretreatment was identified utilizing a combination of Fourier transform infra-red (FTIR) and X-ray photoelectron spectroscopy (XPS). A full description of the FTIR peak assignments and XPS data is provided in Supplementary Note 1. Table 1 presents the XPS spectral data assignments recorded for lithium foils after pretreatment with each of the electrolytes after 7 days.

In the case of pretreatment with [C_3mPyr$^+$][FSI$^-$], the SEI consists of LiF, LiOH and SO_2F species from breakdown of the RTIL, accompanied by generation of SO_2 gas and surface-bound [C_3mPyr$^+$][FSI$^-$][19]. On addition of LiFSI, there is a rise in the intensity of infrared (IR) bands at *ca.* 1,106 and 1,170 cm^{-1}, suggesting a greater amount of anion breakdown products (Supplementary Table 1). The SEI now changes and consists of LiF, $LiSO_2F$, NSO^- species, LiOH and Li_2S via reduction of the anion (Table 1). Notably, this breakdown process occurs without evolution of SO_2 gas, which illustrates the beneficial properties of a mineral SEI[13]. In addition, the increased concentration of [FSI]$^-$ promotes the breakdown of the cation to propyl pyrrolidine, with reduction taking place via a Hofmann β-elimination (Supplementary Fig. 3), and was observed via FTIR[26–28]. In addition, FTIR measurements reveal that the SEI also contains surface-bound cation and anion species as well as Li_2CO_3 from the native lithium foils.

The conclusion for this mechanism of SEI formation is detailed in the data in Table 1, which illustrates a Li 1*s* peak encompassing the responses for both LiF (56.1 eV) and Li-O (55.3 eV), while the F 1*s* spectrum validates the assignment with a response measured at 685.1 eV. The F 1*s* spectrum also suggests the presence of $LiSO_2F$ with a band at 688.3 eV, associated with the doublet at 170.7 and 169.9 eV (S 2*p*), 55.3 eV (Li 1*s*) and 533.2 eV (O 1*s*). This response is highly likely after further reduction of the stable $^{\bullet}SO_2NSO_2F^-$ radical as validated ref. 13. In that report the oxidation, reduction and subsequent fragmentation of the [FSI]$^-$ at a lithium surface is measured via radiolysis and matrix-isolation electron paramagnetic resonance. That work clarifies the ability of the highly stable $^{\bullet}SO_2NSO_2F^-$ radical ion, which persists after LiF formation, to allow further reduction[13]:

$$Li^0 + N(SO_2F)_2^- \rightarrow LiF + {}^{\bullet}SO_2NSO_2F^- \quad (1)$$

$$Li^0 + {}^{\bullet}SO_2NSO_2F^- \rightarrow LiSO_2F + NSO_2^- \quad (2)$$

This enables SEI formation to take place without the evolution of SO_2 gas, which compromises the beneficial properties of a mineral SEI and is consistent with the report of Philippe *et al.* on using LiFSI at nanosilicon electrodes[29]; however, cleavage of the S–N bond was not described.

Table 1 | Summary of *ex situ* XPS for lithium metal pretreated with [C_3mPyr^+][FSI^-] containing different lithium salts after a 7-day reaction period.

SEI Component	Li 1s	C 1s	N 1s	O 1s	F 1s	S 2p	P 2p/As 3d
	Binding energy (eV)						
$LiFSI/[C_3mPyr^+][FSI^-]$ electrolyte pretreatment							
C-N$^+$		286.3	403.2				
NSO^-			400.3	—		—	
C-C, C-H		285.0					
LiOH	55.2			532.4			
Li_2CO_3	55.2	289.8		533.4			
LiF	56.1				685.1		
Li_2S	—					160.6	
$LiSO_2F$				534.5	688.3	170.5	
						171.9	
$LiPF_6/[C_3mPyr^+][FSI^-]$ electrolyte pretreatment							
C-N$^+$		286.4	403.2				
NSO^-			400.5				
C-C, C-H		285.0					
LiF	—				685.3		
Li_2CO_3	55.0	288.4		533.0			
LiOH	55.0			531.9			
Li_2S						160.5	
PF_x					689.0		136.2
$LiSO_2F$				534.3	688.2	170.2	
						171.5	
$LiAsF_6/[C_3mPyr^+][FSI^-]$ electrolyte pretreatment							
C-N$^+$		286.4	403.2				
NSO^-			400.2				
C-C, C-H		285.0					
Li_2CO_3	55.6	288.3		533.0			
LiF	56.7				685.6		
π-π^*		291.3					
As							—
Li_2S	—					160.2	
$LiSO_2F$					688.7	170.0	
						171.6	

XPS, X-ray photoelectron spectroscopy.

The C 1*s* spectrum contains a small peak at 286.3 eV, confirming the presence of the cation at the lithium surface (C–N$^+$) and was reciprocated in the N 1*s* peak at 403.0 eV. The presence of Li_2CO_3 is also confirmed with peaks at 55.3 eV (Li 1*s*), 289.8 eV (C 1*s*) and 533.2 eV (O 1*s*), which is known to reside in the native lithium foils[30]. The peak observed at 160.6 eV is indicative of Li_2S, and has been detected in other studies of RTIL interaction at lithium metal[31–33].

The reduction of products from equation (2) may take place on lithium as follows[13]

$$NSO_2^- \xrightarrow{Li^0} Li_2O + NSO^- \quad (3)$$

$$NSO_2^- \xrightarrow{Li^0} LiOR + NSO^- \quad (4)$$

Li_2O is not detected via XPS measurements, ruling out equation (3); however, the LiO^- precursor is observed as LiOH-sharing responses with Li_2CO_3 discussed above, giving confidence to equation (4). The thiazate anion may then be assigned to the N 1*s* peak at 400.1 eV (Table 1) as the final [FSI]$^-$ moiety, which is at a similarly low abundance to the cation N 1*s* peak at 403.0 eV.

Similar breakdown products comprise the $LiAsF_6/[C_3mPyr^+][FSI^-]$-pretreated SEI. Namely, propyl pyrrolidine via Hofmann β-elimination of the cation, LiF, $LiSO_2F$, NSO^- and Li_2S from anion breakdown, Li_2CO_3 from the native lithium metal and surface-bound anion and cation. However, the absence of bands in the As 3*d* spectra is indicative of As not taking part in SEI formation. The tarnished surface observed is usually an indication of the well-known arsenic–oxygen polymer 'brown film'[34]. Interestingly, FTIR data (Supplementary Table 2) only show evidence for this at earlier time periods, suggesting that this reaction is not proceeding to completion because of an absence of polymerizing precursors[35].

Finally, for treatment with $LiPF_6/[C_3mPyr^+][FSI^-]$ (Supplementary Table 3), the SEI comprises $LiSO_2F$, NSO^-, surface-bound cations and anions and Li_2S. For this pretreatment, a lower quantity of LiF, LiOH and Li_2CO_3 was observed via energy-dispersive spectroscopy. The most notable difference for this system is the formation of methyl pyrrolidine. The Hofmann β-elimination discussed above may occur via three different pathways. The dominant pathway leads to ring cleavage towards the propyl pyrrolidine species. However, with $LiPF_6$ addition, the less preferred pathway now leads to the formation of methyl pyrrolidine. Finally, breakdown of the PF_6^- anion to form PF_x species is also found. Therefore, the XPS data indicate that the electrolyte containing $LiPF_6$ has a markedly different composition to the other electrolytes, which may have an impact on the performance of any cell that incorporates this electrolyte. Overall, the SEI products formed for each electrolyte system are:

i. $LiFSI/[C_3mPyr^+][FSI^-]$: LiF, LiOH, $LiSO_2F^-$, NSO^-, $[C_3mPyr]^+$, $[FSI]^-$, propyl pyrrolidine, Li_2CO_3 and Li_2S
ii. $LiPF_6/[C_3mPyr^+][FSI^-]$: LiF, PF_x, LiOH, Li_2CO_3, methyl pyrrolidine, NSO^-, $[C_3mPyr]^+$ and $[FSI]^-$
iii. $LiAsF_6/[C_3mPyr^+][FSI^-]$: Li_2CO_3, propyl pyrrolidine, LiF, Li_2SO_2F, NSO^-, $[C_3mPyr]^+$ and $[FSI]^-$

Therefore, the different surface morphologies seen as a result of exposure to different Li salts (Fig. 1) can be attributed to the different species that constitute the SEI. However, it is quite remarkable that the kinetics of the process are similar, given the appearance of a smooth compact SEI for each system at the 10–12-day mark.

Morphological changes after extended cycling of untreated foils. Figure 2 shows the voltage–time (V–t) profiles for Li|electrolyte|Li cells using untreated lithium cycled at 0.1 mA cm^{-2}. The cell containing $LiPF_6/[C_3mPyr^+][FSI^-]$ electrolyte begins with cycling behaviour as reported previously[14,21], with a periodic voltage response due to ambient temperature fluctuation. After 1,300 cycles, a period (*ca.* 250 h) of voltage instability is observed, culminating in cell death. Unpredictably, the cell is re-animated for 100 cycles (*ca.* 3,750–3,850 cycles) after a 'dormancy' of 22 days (550 h) before permanently failing. This unusual behaviour suggests that failure is not due to a dendrite-induced short circuit, as this cannot be undone through continued galvanostatic cycling. The possibility that micro dendrites were formed during cycling, which created short circuits, but broke because of localized heating (that is, high resistances), thereby resurrecting the cell, was discounted because of no evidence of separator piercing. Figure 2b displays the *V–t* plot for a symmetrical cell containing $LiAsF_6/[C_3mPyr^+][FSI^-]$, which cycled for 5,000 cycles reaching a polarization of 13 mV. This is attributed to a resistance decrease through repeated cycling forming a larger, and more porous surface[14]. Signs of voltage instability at various points, *ca.* 3,300, 3,950 and 4,700 cycles, are observed, representing morphology changes during SEI formation[36]. Symmetrical cells containing untreated lithium electrodes and

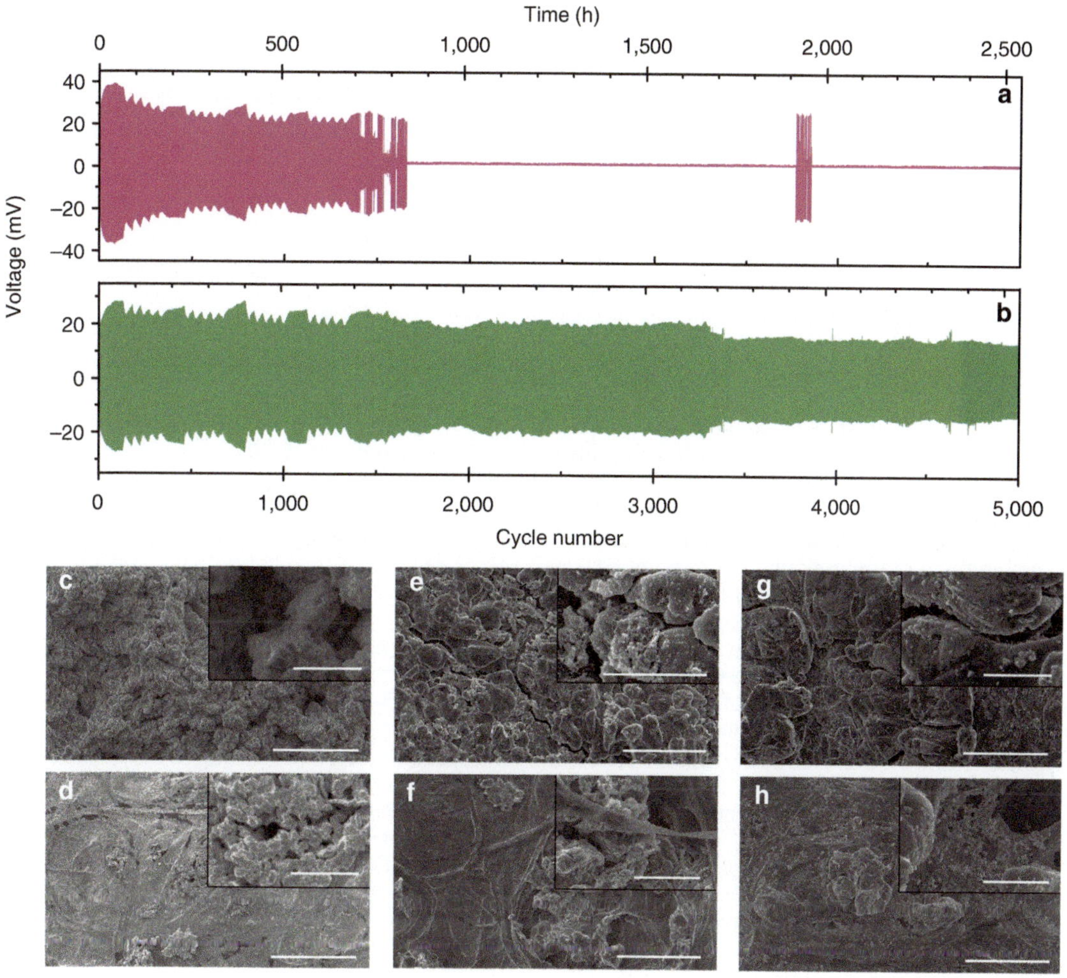

Figure 2 | Characterization of Li-symmetrical cells during 5,000 cycles at 0.1 mA cm^{-2}. Representative voltage-time plot for long-term cycling of a symmetrical cell containing (**a**) $LiPF_6$/[C_3mPyr$^+$][FSI$^-$] or (**b**) $LiAsF_6$/[C_3mPyr$^+$][FSI$^-$] electrolytes. SEM micrographs displaying dendrite suppression for the (**c**) anode and (**d**) separator of the LiFSI/[C_3mPyr$^+$][FSI$^-$] cell; (**e**) anode and (**f**) separator for the $LiAsF_6$/[C_3mPyr$^+$][FSI$^-$] cell; (**g**) anode and (**h**) separator for the $LiPF_6$/[C_3mPyr$^+$][FSI$^-$] cell. Scale bars, 50 µm (inset scale bars, 10 µm).

LiFSI/[C_3mPyr$^+$][FSI$^-$] electrolyte have also been reported to successfully achieve 5,000 cycles[2]. Therefore, these data illustrate that the electrochemical SEI built-up during cycling is undoubtedly electrolyte-dependent. For the case of $LiAsF_6$ and LiFSI as previously reported, the pristine Li surface facilitates stable SEI formation and allows for effective cycling at a low current density of 0.1 mA cm^{-2} for up to 5,000 cycles; however, this is not the case when $LiPF_6$ is used. As discussed later, specific pretreatment of the Li anode before cycling alleviates this phenomenon, and stable cycling can be achieved, which is completely independent of the Li salt used in the electrolyte.

To gain an insight as to why the use of one electrolyte and not another resulted in cell failure, a post-mortem analysis of the anode was undertaken for LiFSI, $LiAsF_6$ and $LiPF_6$ cells. In all cases, SEM images of the lithium electrode (Fig. 2c,e,g) exhibited no signs of dendritic growth of metallic Li from the surface; however, the quantity of surface species is higher than for the chemical pretreatment of the Li surface alone (Fig. 1), resulting in a rougher surface. The electrode surface of the cycled $LiAsF_6$/[C_3mPyr$^+$][FSI$^-$] cell (Fig. 2e) is completely fractured, and fresh Li^0 nucleation now occurs at grain boundaries and steps similar to the LiFSI case (Fig. 2c). This surface-fracturing allows fresh lithium to be exposed, leading to further SEI formation. This is as expected for a mineral SEI, whereby pressure is released from within the SEI because of expansion/contraction via fracturing during lithium-plating/stripping processes. The gradual release of pressure prevents build-up and inherently inhibits dendritic growth[13]. The inset of Fig. 2e shows small nucleation sites that continue to form even near the latter stages of the 2,500-h charge–discharge cycle.

However, given the comparable nature of the SEI and lack of dendrite formation when $LiPF_6$ is used, the brief re-animation indicates that cell failure is due to something other than an effect at the anode. Therefore, an investigation of the separator post cycling was undertaken. Interestingly, the SEM images (Fig. 2d,f,h) show electrolyte-breakdown products that have become detached from the lithium metal surface. For LiFSI (Fig. 2d) and $LiAsF_6$ (Fig. 2f) these detached species are relatively sparse, and the porous nature of the separator is still quite visible, suggesting that the detachment of the SEI occurs at a slow rate. Figure 2d inset in fact shows that these deposits are located within the porous structure. This could suggest that bridging has occurred; however, these cells showed no significant drop in polarization for the experiment duration since the separator area blockage by the insulating SEI species is counteracted by the increase in lithium surface area through morphology changes. However, there is a clear discrepancy when the $LiPF_6$ case is examined (Fig. 2h), as there is a significantly larger quantity of breakdown products observed within the pores that completely dominate the separator. In fact, the pores of the separator are no longer visible (see Supplementary Fig. 4 for SEM image of a pristine separator). Consequently, cell failure occurs because the rate of increase of the lithium surface area is lower than the rate of separator blockage, thus impeding Li^+ flow and causing cell termination.

The FTIR measurements for electrochemically formed SEIs are included in Supplementary Table 4 and show that the chemical composition of the SEI after extensive cycling is relatively consistent with the measurements on the purely chemically formed SEI. However, the presence of cation-breakdown products is noticeably less for the LiFSI-based cell, with negligible responses in the ring-mode region (*ca.* 800–1,000 cm^{-1}) and with fewer bands above 2,800 cm^{-1}. The other chemistries, however, do have a larger abundance of cation products within the SEI layer. In fact, the cell composed of $LiPF_6$ electrolyte exhibits a large response through the ring-mode region, indicating the extensive formation of cation-breakdown products, which may also contribute to the failure of this cell.

Electrochemical impedance spectroscopy (EIS) was undertaken on Li-symmetrical cells before and after 5,000 cycles at a current density of 0.1 mA cm^{-2} to monitor any changes in the Li electrode impedance. EIS has been demonstrated to be highly informative for investigating battery electrodes[37,38]. For the pristine lithium cell using $LiAsF_6$ salt, the interfacial resistance decreases from 251.5 to 51.3 Ω cm^{-2} after 5,000 cycles because of the morphological changes that occur in the SEI. Illustrated in Fig. 3a are typical Nyquist plots obtained for the $LiAsF_6$ system before and after extensive cycling. This phenomenon was also observed for the LiFSI case where the impedance fell from 172.0 to 75.4 Ω cm^{-2} after 5,000 cycles (Supplementary Table 5).

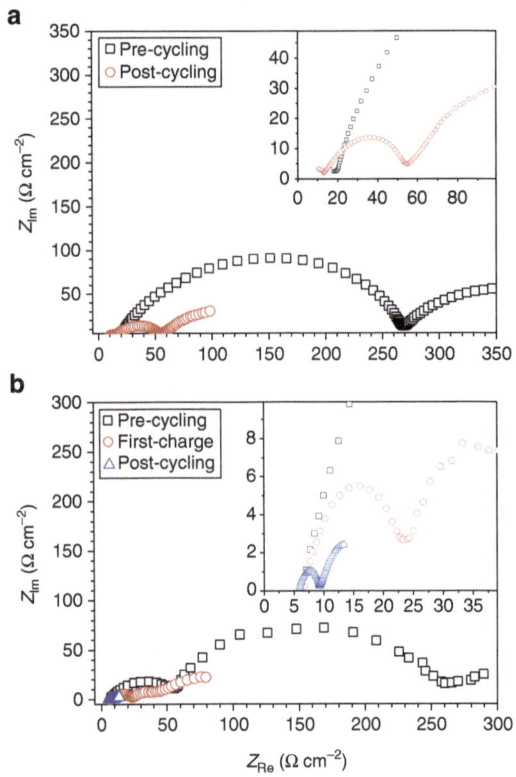

Figure 3 | Electrochemical impedance spectroscopy of pristine and 12-day pretreated Li-symmetrical cells. Nyquist plots for a Li-symmetrical cell containing (**a**) $LiAsF_6/[C_3mPyr^+][FSI^-]$ electrolyte before (black) and after 5,000 cycles (red) at 0.1 mA cm^{-2} and (**b**) 12-day pretreated Li electrode in $LiFSI/[C_3mPyr^+][FSI^-]$ electrolyte before (black) after one charge (red) and after 333 cycles (blue) at 1.0 mA cm^{-2}.

Analysis of pretreated electrodes at higher current density. The symmetrical Li|electrolyte|Li cell-cycling performance of the three pretreated electrodes and the associated electrolyte systems is presented in Fig. 4. The cells were cycled at 1.0 mA cm^{-2} following chemical pretreatment for 12- and 18-day time periods. The most obvious feature is that all the foils pretreated for 18 days begin cycling with large voltage instability, culminating in cell failure within just 6 h. Regardless of electrolyte composition, the SEI that is chemically formed over an 18-day period does not permit effective cycling of symmetrical cells at this current density. However, if the pretreatment is shortened to 12 days, the tailored SEI permits an exceptionally stable *V–t* profile for the entirety of the charge–discharge period. The cycling behaviour of the 12-day pretreated electrodes using $LiFSI/[C_3mPyr^+][FSI^-]$ (Fig. 4a) shows a gradual decrease in overpotential with repeated cycling where these overpotentials are generated via kinetic hindrances within the cell[21]. After a period of 225 cycles (*ca.* 128 h), the cycling behaviour of the symmetrical cell reaches a stable overpotential of 22.5 mV until the end of testing.

Cells of the 12-day pretreated $LiPF_6/[C_3mPyr^+][FSI^-]$ system (failing if untreated, Fig. 2a) undergo a high degree of voltage instability within the first 40 h, reaching a steady and stable overpotential of 21 mV (Fig. 4b). Once this cell reaches a stable voltage, it exhibits the longest stability period for all three systems, displaying how an electrolyte of $LiPF_6$ and $[C_3mPyr^+][FSI^-]$ can be practical for lithium metal batteries. The $LiAsF_6/[C_3mPyr^+][FSI^-]$ system (Fig. 4c) also exhibits a stable overpotential in excess of 300 cycles. The initial *ca.* 50 h of cycling is followed by a *ca.* 30-h period of voltage instability and then by a stable cycling, reaching an overpotential of 28 mV until the end of testing. EIS data obtained for the symmetrical cells with the 12-day pretreated Li anodes for the LiFSI, $LiPF_6$ and $LiAsF_6$ cases also demonstrate a significant drop in the electrode impedance after 300 cycles at this higher current density of 1.0 mA cm^{-2} (Supplementary Table 5 and Supplementary Fig. 6). Typical Nyquist plots are shown for the LiFSI case in Fig. 3b. Promising cycling results for pretreated electrodes suggest that the application of ionic liquids may enable beneficial full-cell cycling. Furthermore, without pretreatment the electrodes do not cycle efficiently. Therefore, it appears that a smooth SEI formed via a chemical interaction of Li with the electrolyte, as indicated in Fig. 1d,h,l facilitates the formation of a stable and well-adhered SEI during electrochemical cycling. If this chemically formed SEI is rough, as shown after 18 days of interaction (Fig. 1e,i,m), it is not a good precursor to the formation of an SEI that is stable during cycling and it results in near-immediate cell failure.

The SEM micrographs of the lithium surface and separator after cycling are shown in Fig. 4d–f. They illustrate that stable SEIs are formed during electrochemical cycling of the pretreated anodes, in this case at a higher current density of 1 mA cm^{-2}. Again, there is no evidence of dendritic growth on the Li surface or within the separator after extended cycling (Fig. 4d,f, respectively). Further evidence is provided in Fig. 4e, which shows a cross-section of the Li electrode that exhibits a smooth morphology devoid of any dendritic growth. SEM micrographs of other areas of the Li surface and separator are shown in Supplementary Fig. 5a,b to illustrate the homogeneity over the entire Li anode after cycling.

Figure 5a shows the *V–t* plot of a symmetrical $Li|LiFSI/[C_3mPyr^+][FSI^-]|Li$ cell prepared with the aforementioned beneficial 12-day pretreatment time frame at a higher current density of 2 mA cm^{-2}. Once again there is a period of larger

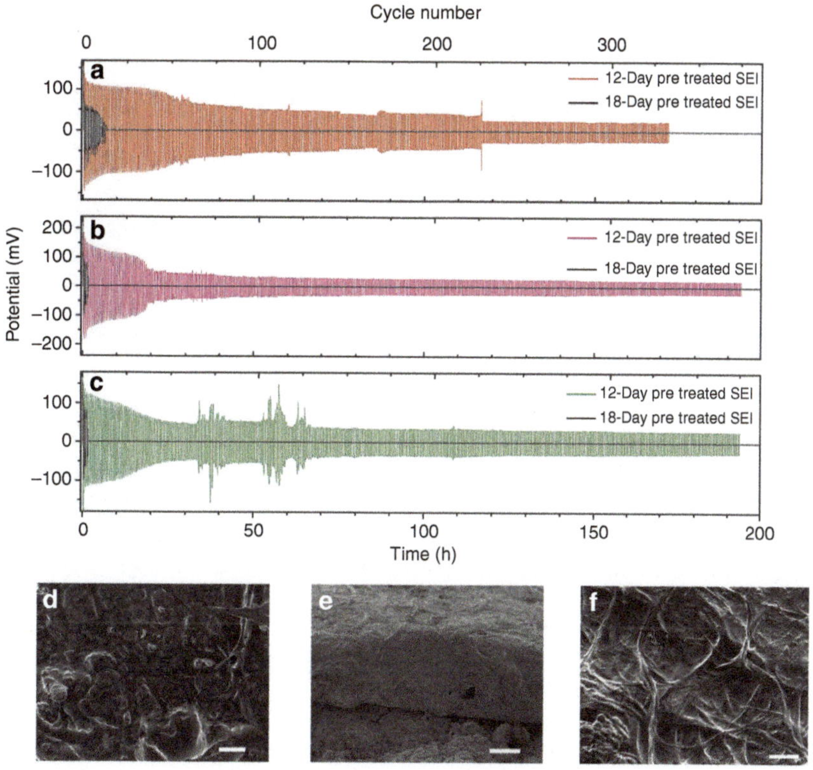

Figure 4 | Comparison of pretreatment time for Li-symmetrical cells cycled at 1.0 mA cm^{-2}. Voltage-time plots of Li-symmetrical cells using (**a**) Li|LiFSI/[C_3mPyr$^+$][FSI^-]|Li, (**b**) $LiPF_6$/[C_3mPyr$^+$][FSI^-]|Li and (**c**) $LiAsF_6$/[C_3mPyr$^+$][FSI^-]|Li, after a period of 12 days (coloured) or 18 days (black). SEM micrographs of the Li|LiFSI/[C_3mPyr$^+$][FSI^-]|Li cell components after 333 cycles highlighting the (**d**) lithium foil, (**e**) cross-section of the foil and (**f**) the separator material. Scale bars, 20 μm (**d**,**f**) and 100 μm (**e**).

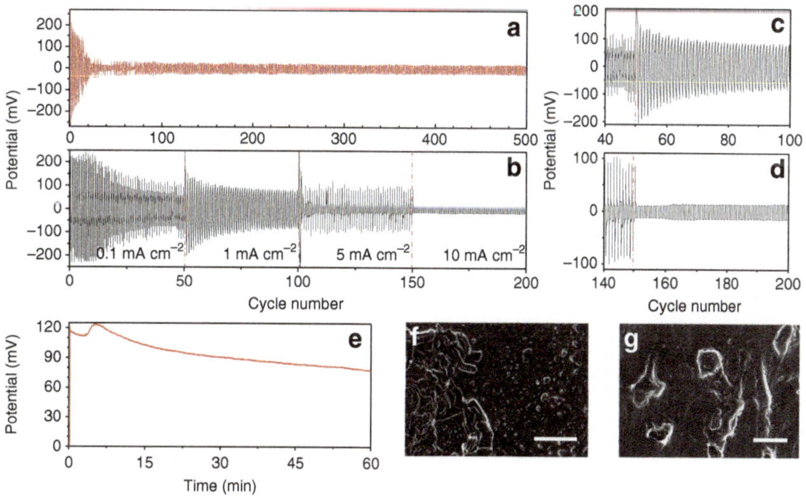

Figure 5 | Characterization of Li|LiFSI/[C_3mPyr$^+$][FSI^-]|Li cells after 12-day pretreatment. Cycled for 500 charge/discharge cycles at (**a**) 2.0 mA cm^{-2}, (**b**) ramping 0.1 to 10 mA cm^{-2} for 50 cycles per current density and (**c**) highlighting the effect of ramping the current density from 0.1 to 1.0 mA cm^{-2} and (**d**) 5.0 to 10 mA cm^{-2}. (**e**) The first charge/discharge of a symmetrical cell and the corresponding SEM images of the Li anode after the initial (**f**) plating (scale bar, 10 μm) and (**g**) stripping of lithium (scale bar, 50 μm).

polarization (>100 mV, 12 cycles) before the cell stabilizes (*ca.* 50 mV after 18 cycles) to 31 mV for the remainder of the cell's cycle lifetime of 150 h. A duplicate cell is cycled to observe the performance of the chemically formed SEI to a stepped current density, whereby the cell is cycled for 50 charge–discharge steps following a protocol of 0.1, 1.0, 5.0 and 10 mA cm^{-2} (Fig. 5b). The same behaviour of decreasing overpotential, after an initial sharp increase in the overpotential, is observed in the initial stages of cycling, and is compounded on the step-up of current density as highlighted within Fig. 5c. However, nearing a cycling time of *ca.* 80 h when stepping the current density to 10 mA cm^{-2} (Fig. 5d), the cell polarization is remarkably low and remains at an overpotential of *ca.* 11 mV. This indicates that there is rapid formation of the SEI on increasing the current density and a high enough supply of Li ions to the electrode that minimizes the possibility of any increases in electrode resistance

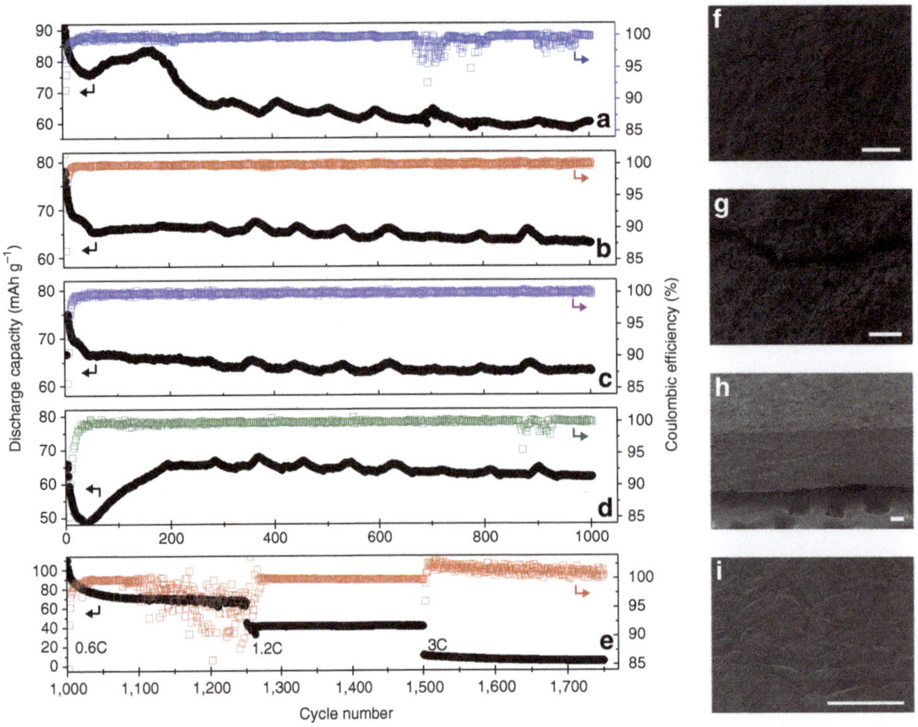

Figure 6 | Electrochemical performance of $LiFePO_4$ versus pristine and tailored lithium anodes after 12-day pretreatment with ionic liquid electrolytes. Discharge capacity and Coulombic efficiencies of Li|electrolyte|$LiFePO_4$ full cells versus cycle number cycle at 1-c rate assembled with **(a)** pristine lithium and LiFSI|[C_3mPyr$^+$][FSI$^-$] electrolyte. Discharge capacity and Coulombic efficiency for cells with LiFSI|[C_3mPyr$^+$][FSI$^-$] electrolyte and pretreated with electrolytes composed of **(b)** LiFSI/[C_3mPyr$^+$][FSI$^-$] **(c)** $LiPF_6$/[C_3mPyr$^+$][FSI$^-$] and **(d)** $LiAsF_6$/[C_3mPyr$^+$][FSI$^-$]. **(e)** Discharge capacity and Coulombic efficiency for the same LiFSI|[C_3mPyr$^+$][FSI$^-$] cell in **(b)** after a shelf-life of 330 days. SEM micrographs of the LiFSI|[C_3mPyr$^+$][FSI$^-$] cell **(f)**, lithium anode **(g)**, higher magnification (scale bar, 10 μm) of **(f)**, **(h)** cross-section of **(f)** and **(i)** separator material. All other scale bars, 100 μm.

as observed previously[11]. To gain more of an understanding of what occurs in the initial stages of electrochemical SEI formation on the pretreated lithium surface, a Li|LiFSI/[C_3mPyr$^+$][FSI$^-$]|Li cell is charged at a current density of 1 mA cm^{-2} for 1 h and the *V–t* profile is shown in Fig. 5e. The SEM micrographs of the lithium electrode after plating and stripping of Li are displayed in Fig. 5f–g. Figure 5f shows the electrode after undergoing lithium plating, which consists of some large lithium deposits in areas where the chemical SEI pretreatment was not entirely effective. However, the areas where ideal chemical pretreatment occurred remain quite smooth. The opposing Li electrode after the stripping process, shown in Fig. 5g, exhibits pitting of the surface where the lithium metal has undergone electrodissolution. Supplementary Fig. 5c–f shows alternate areas of the cell components of this cell, highlighting the consistency of these observations. EIS data indicate that even after one charge the impedance of the Li electrode decreases substantially from 48.8 to 16.6 Ω cm^{-2} (Fig. 3b), which then reduces to 3.4 Ω cm^{-2} after more extensive cycling (Supplementary Table 5).

Performance of full cells utilizing pretreated lithium anodes. Secondary Li|electrolyte|$LiFePO_4$ cells were also cycled (Fig. 6) to determine whether this chemically formed SEI could be beneficial for lithium metal batteries. On the basis of the symmetrical cell cycling, all electrodes were pretreated for 12 days. As a control, cells with pristine, untreated lithium metal were also cycled. The pristine lithium cell (Fig. 6a) displays a CE average of 99.6%; however, the CE plot is constantly undergoing fluctuation and becomes unstable after *ca.* 600 cycles where CE drops to 95.75%. This is not suitable for a commercial battery where technological demands require stability into thousands of cycles. Although the [C_3mPyr$^+$][FSI$^-$] electrolyte has been shown to allow symmetrical cells to cycle effectively[11,12] and for a long period (Fig. 4a) at high current density, this does not translate into stable cycling for full cells containing pristine lithium and LiFSI/[C_3mPyr$^+$][FSI$^-$] electrolyte. The battery capacity after a period of stability decays to 80% of the initial capacity after *ca.* 600 cycles—commonly accepted as cell end of life.

The cells assembled with the pretreated SEI show an initial cell capacity decrease, which may be correlated to the 12-day pretreated symmetrical cell data shown in Fig. 4. The full cells begin cycling with an overpotential that relaxes as the cycling of the lithium lowers the cell resistance to a stable level in the order $LiPF_6$, LiFSI and $LiAsF_6$ after *ca.* 35, 60 and 130 cycles, respectively. This initial capacity fade for the LiFSI/[C_3mPyr$^+$][FSI$^-$] (Fig. 6b) and $LiPF_6$/[C_3mPyr$^+$][FSI$^-$] (Fig. 6c) batteries is due to the anode up until the time periods highlighted above, and all further capacity changes are limited by the cathode (as is the case for the $LiAsF_6$/C_3mPyr$^+$][FSI$^-$] cell (Fig. 6d), reaching a plateau after *ca.* 200 cycles).

The average CEs for each of the pretreated cells are 99.96%, 99.93% and 99.42% for cells containing LiFSI, $LiPF_6$ and $LiAsF_6$, respectively. The minimum measured CE for the pretreated LiFSI/[C_3mPyr$^+$][FSI$^-$] cell is an impressive 99.50%, reinforcing how positive an effect this facile chemical SEI formation method has on battery cycling. After an initial decrease in capacity during the first few cycles, the cell retains 95.4% capacity throughout the entirety of the 1,000 cycles.

The final CE values for the remaining $LiPF_6$/[C_3mPyr$^+$][FSI$^-$]- and $LiAsF_6$/[C_3mPyr$^+$][FSI$^-$]-pretreated cells, which also successfully cycled to 1,000 cycles are 99.02% and 99.89%, respectively. The capacity fade for these cells are *ca.* 5% with retention at 95.2% for the $LiPF_6$ treatment, and 95.0% for the

Table 2 | Comparison and summary of recent literature reports on lithium metal battery cycling utilizing pretreatment methods.

Cell treatment	Cell configuration	Cycle number	Cycling rate	Coulombic efficiency	Capacity fade	Reference
Chemical interaction	Li|$LiFePO_4$	1,000	1C	99.42-99.96%	5.0%	Present study
Carbon nanospheres on copper	[Li/CNS]|Li	150	0.25-1 mA cm^{-2}	99.0%	—	Zheng *et al.*[22]
Graphene or BN layer via CVD and HT	Li|$LiCoO_2$	50	0.5-5 mA cm^{-2}	95.0%	—	Yan *et al.*[32]
PANI-CNT buffer layer	Li | [$LiCoO_2$/CNT]	60	4C	—	14.0%	Zhang *et al.*[23]
Mechanical surface modification	Li|$LiFePO_4$	150	C/2	—	15.0%	Ryou *et al.*[3]
Al_2O_3/polymer CPL	Li|$LiCoO_2$	400	1C	99.8%	12.2%	Lee *et al.*[40]
LiFSI salt in carbonate solvent	Li|$LiCoO_2$	50	0.5/0.2 C	—	7.0%	Han *et al.*[16]
Halogen reinforcement	Li|$Li_4Ti_5O_{12}$	300	1C	—	—	Lu *et al.*[36]
Oxygen removal	$LiTi_2(PO_4)_3$	1,000	C/8	—	10.0%	Luo *et al.*[41]
3D Copper foil	Li|Li-Cu	120	0.5 mA cm^{-2}	98.5%	—	Yang *et al.*[42]
Hybrid POSS solid polymer electrolyte	Li|$LiFePO_4$	50	C/2	>99%	n/a	Pan *et al.*[43]
$LiNO_3$-ternary electrolyte	Li|Li-Cu	75	1 mA cm^{-2}	94%	—	Zhao *et al.*[44]
Cu-graphene 'drum' scaffold	Cu-C|Li	800	2 mA cm^{-2}	93%	—	Zhang *et al.*[45]
Concentrated electrolyte	Li|$LiFePO_4$	200	C/5	>92%	16.7%	Ma *et al.*[46]
Concentrated electrolyte	Li|FeF_2	1,000	140 mA g^{-1}	80 %	—	Gu *et al.*[47]
Concentrated $LiNO_3$ in DMSO	Li|Li	90	0.2 mA cm^{-2}	>80%	—	Togasaki *et al.*[48]
SiO_2 hollow NCSE	Li|$LiFePO_4$	200	C/5	—	<5.0%	Zhou *et al.*[49]
Concentrated electrolyte	Li-Cu	1,000	4 mA cm^{-2}	98.4%	—	Qian *et al.*[50]

3D, three-dimensional; BN, boron nitride; CNT, carbon nanotube; CNS, carbon nanospheres; CPL, composite protective layer; CVD, chemical vapour deposition; DMSO, dimethylsulphoxide; HT, high temperature; NCSE, nanosphere composite solid-electrolyte; PANI, polyaniline; POSS, polyhedral oligomeric silsesquioxane.

$LiAsF_6$ treatment. A recent survey of the best systems reported in the literature (Table 2) demonstrates how this straightforward pretreatment method markedly improves battery-cycling performance. Ionic liquids have been used previously, and a table showing the performance of Li metal-based batteries using ionic liquids is presented as Supplementary Table 6. It is noteworthy that those studies are nowhere near comparable to the data presented here. Indeed, the full cell cycling data presented here are comparable if not better than many commercial Li ion batteries that were recently tested[25].

After a quiescent period of 330 days, the Li|LiFSI/[C_3mPyr^+][FSI^-]|$LiFePO_4$ cell was again cycled, at stepped C-rates, to determine the effect of both the shelf-life and alternate current density (Fig. 6e). Initially, the cell demonstrated a large capacitance that stabilized after *ca.* 70 cycles (including three initial cycles at 0.1 C). Impressively, the full cell retains its charge that is equivalent to that previously measured on initial charge–discharge (Supplementary Table 7). The cell continues to provide close to its original capacity when initially cycled. Although an inoperative time of nearly 1 year required 70 cycles to achieve stability, this is quite comparable to the 40 cycles required for a fresh cell (Fig. 6b). From this point a capacity fade of only 9.6% is measured before stepping to the higher 1.2-C rate (7.3% fade). Figure 6e shows good cycling stability at a 1.2-C rate, with an associated CE greater than 98.5% for the entire 250 cycles, albeit this is reached after a brief period of morphology rearrangement as noted by the instability in the CE. The battery was then dismantled, and the Li anode demonstrated a compact SEI with some evidence of cracking (Fig. 6f,g). From the cross-sectional view (Fig. 6h), the SEI is compact and the thickness was, on average, 124 µm (Supplementary Fig. 7). As with the symmetrical cells, the separator did not exhibit any excessive build-up of material or dendrites, and the porous nature of the separator is still visible.

Discussion

Stable cycling over a commercially acceptable time period of 2,000 h (1,000 cycles) was achieved at a 1-C rate, for lithium metal batteries without the formation of dendrites, thermal runaway or other detrimental characteristics commonly associated with lithium metal anodes. We herein have described a facile pretreatment methodology, which can be easily adopted by researchers and manufacturers with interest in lithium-metal batteries. As cycle life is related to the degree of electrolyte decomposition during cycling, this simple pretreatment of lithium anodes before cell fabrication ensures a reduction in electrolyte breakdown from SEI formation during the 1,000 cycles, thereby increasing cell performance and lifetime. This is highlighted for batteries with pristine, untreated lithium anodes, which do not achieve the same degree of success compared with pretreated cells. By pretreating lithium anodes the SEI must only undergo a rapid equilibration (for example, 35 cycles with $LiPF_6$/[C_3mPyr^+][FSI^-] pretreatment) before reaching unparalleled lithium metal plating/stripping in full-battery scenarios.

The implication of diverse SEI formation on lithium metal anodes using RTIL electrolytes highlights the importance of interaction time for robust SEI formation capable of being used in lithium-metal batteries for extensive and efficient cycling. The salt and RTIL compete during SEI formation, simultaneously contributing to a highly dynamic SEI comprising $[C_3mPyr]^+$, $[FSI]^-$ and salt anion breakdown in the case of $LiPF_6$, or negligible participation, for example, as from $LiAsF_6$. Since adventitious water was minimized, the impact of salt anion impurities (that is, moist $LiPF_6$-producing hydrofluoric acid (HF) was decreased. The major constituents of the mineral SEI formed after pretreatment are LiF, Li_2CO_3, $LiSO_2F$, LiOH as well as cation-breakdown products via a Hofmann elimination mechanism.

Symmetrical cells cycled at 0.1 mA cm^{-2} with pristine lithium highlighted the significant dependence on the salt anion on cyclability and SEI formation during cycling. However, this effect can be completely eliminated in a facile manner via a 12-day pretreatment time, which provides an optimal SEI in all electrolytes permitting the ingress/egress of lithium ions and withstands the associated volume change when cycling at a higher current density of 1.0 mA cm^{-2}. A longer pretreatment time of 18 days did not improve the performance, regardless of salt composition. The formation of a dense and rough SEI is not a beneficial seed layer for further electrochemical SEI formation, indicating the importance of contact time between the Li anode and the electrolyte before cycling of a full battery. These promising symmetrical cell measurements were confirmed in full cell cycling using a commercial cathode material. Each of the electrolyte systems studied herein has accomplished 1,000 charge–discharge cycles in a Li|electrolyte|$LiFePO_4$ configuration, with no lower than 99% efficiency. The one drawback of this approach is that some cycling needs to be performed to achieve a stable response, which is likely to result in some electrolyte consumption, and efforts to reduce this effect are ongoing. However, full cells that were left dormant for nearly 1 year could be effectively cycled at different rates, with a performance comparable to freshly prepared cells. Therefore, this work sheds light on the pre-conditioning that is often utilized for commercial Li ion batteries before roll out. For a battery based on Li|electrolyte|$LiFePO_4$ such pre-conditioning is greatly simplified, whereby Li only needs to be immersed in the relevant electrolyte for 12 days before cell fabrication.

Methods

Pretreatment of lithium anodes. Lithium metal foils (China Energy Lithium, 0.33 mm, 99%) were immersed in $[C_3mPyr^+][FSI^-]$ (Dai-Ichi Kogyo Seigaku, >99%) electrolyte mixtures containing 0.5 mol Kg^{-1} of a lithium salt for various time periods (4 h and 7, 12 and 18 days). These lithium salts include the following: LiFSI (Dai-Ichi Kogyo Seigaku, >95%), $LiPF_6$ (Sigma, 98%) and $LiAsF_6$ (Sigma-Aldrich, 98%). Lithium foils were cleansed of surface contaminants using a nylon brush under *n*-hexane (Merck). The ionic liquid was dried *in vacuo* using standard Schlenk line techniques at 100 °C for 12 h before storage and handling under an Ar atmosphere (less than 5 p.p.m. H_2O and 1 p.p.m. O_2). Karl Fischer titration analysis of the ionic liquids dried using this method indicated less than 10 p.p.m. H_2O content. Lithium salts were dried according to literature reports[39]. On reaching the desired pretreatment time for submerged foils, they were removed and rinsed of any residual electrolyte using dimethyl carbonate (DMC) (Fluka) as reported previously[19].

Coin cell fabrication and cycling. Symmetrical Li|electrolyte|Li coin cells were fabricated using CR2032 metal jackets as reported previously[11]. The symmetrical cells underwent cyclic charge–discharge galvanostatic testing using a Maccor Series 4000 battery tester at ambient temperature (22.2 ± 2 °C) with a low or high current density of 0.1 or 1.0 mA cm^{-2}, respectively. Replicate coin cells were cycled for either hundreds or thousands of cycles as specified in the text. Charge and discharge times were 15 min each, with 30 s rest between cycles (a cycle is defined as a charging step followed by a discharge step). The electrolyte used in symmetrical cells containing pretreated electrodes is kept consistent with that used for the pretreatment. Lithium metal foils used throughout the experimental design were 10 mm in diameter and the amount of electrolyte measured was 60 µl.

Full Li|electrolyte|$LiFePO_4$ cells were fabricated in the same manner as symmetrical cells, with the addition of the cathode material, and underwent galvanostatic testing using an electrochemical pretreatment and subsequent 1-C rate. A composite of $LiFePO_4$ (90 wt%, Phostech), carbon black (2.5 wt%, Timcal Super P) and carboxymethyl cellulose binder (7.5 wt%, Sigma-Aldrich) was coated on aluminium foils using an automated pasting machine (MIT Corporation, Compact tape casting film coater, MSK-AFA-III) and, after drying, cut into disks of 10 mm diameter. Both $LiFePO_4$ and separator materials (Solupor) were dried at 100 °C under vacuum for 12 h before being used in both symmetrical and full cells. All cell-material handling and cell fabrication were performed in an Ar-filled inert atmosphere glovebox operating at less than 5 p.p.m. H_2O and 1 p.p.m. O_2. Duplicate cells were cycled for all systems to ensure reproducibility of data. Before cycling, all cells underwent three formation cycles at a rate of C/10, followed by 1,000 cycles at a rate of 1 C (equivalent lithium current density of 1.25 mA cm^{-2}) at 30 °C temperature. Impedance measurements were carried out using a Solartron 1255B Frequency Response Analyser, a signal amplitude <10 mV, between a frequency range of 100 kHz to 0.1 Hz.

Surface characterization. Surface characterization was performed after rinsing and drying of material with DMC, as stated previously, after pretreatment or cycling was carried out. Dried electrodes were loaded and enclosed in a hermetically sealed attenuated total reflectance stage (Specac Golden Gate) to permit FTIR analysis. After enclosure, the stage was removed from the glove box and spectra were obtained using a spectrum 400 FTIR spectrometer (Perkin-Elmer). All spectra were recorded over 16 scans using 4-cm^{-1} resolution. SEM was carried out using a Philips XL-30 field emission gun scanning electron microscope. A specifically designed brass substrate facilitated imaging of samples transferred to the SEM via a hermetically sealed environmental chamber. Images were recorded with an accelerating voltage of 5 kV under vacuum of 1.7×10^{-5} mbar. A Link ISIS (Oxford Instruments) energy-dispersive X-ray system was used to collect qualitative data of sample surfaces. A Microlab VG310F Spectrometer (Thermo) was used to collect XPS. Samples were introduced to the instrument using a dual layer glove bag (Sigma). Transfer of samples from the glove box to glove bag took place by storing a sample inside a Schott bottle within a heat-sealed aluminium laminate pouch (3 M Film). All XPS measurements were recorded after 7 days of electrolyte pretreatment to ensure that sufficient time had lapsed for SEI reduction products to be detected. It is at this 7-day period that minimal electrolyte could be observed via SEM; thus, any spectral data best describe structural SEI information rather than pore-trapped residues.

References

1. Tarascon, J. M. & Armand, M. Issues and challenges facing rechargeable lithium batteries. *Nature* **414**, 359–367 (2001).
2. Armand, M. & Tarascon, J. M. Building better batteries. *Nature* **451**, 652–657 (2008).
3. Ryou, M.-H., Lee, Y. M., Lee, Y., Winter, M. & Bieker, P. Mechanical surface modification of lithium metal: towards improved Li metal anode performance by directed Li plating. *Adv. Funct. Mater.* **25**, 834–841 (2015).
4. Bhattacharyya, R. *et al.* In situ NMR observation of the formation of metallic lithium microstructures in lithium batteries. *Nat. Mater.* **9**, 504–510 (2010).
5. Bruce, P. G., Freunberger, S. A., Hardwick, L. J. & Tarascon, J.-M. Li-O_2 and Li-S batteries with high energy storage. *Nat. Mater.* **11**, 19–29 (2011).
6. Harry, K. J., Hallinan, D. T., Parkinson, D. Y., MacDowell, A. A. & Balsara, N. P. Detection of subsurface structures underneath dendrites formed on cycled lithium metal electrodes. *Nat. Mater.* **13**, 69–73 (2014).
7. Aurbach, D., Zinigrad, E., Cohen, Y. & Teller, H. A short review of failure mechanisms of lithium metal and lithiated graphite anodes in liquid electrolyte solutions. *Solid State Ionics* **148**, 405–416 (2002).
8. Aryanfar, A. *et al.* Dynamics of lithium dendrite growth and inhibition: pulse charging experiments and monte carlo calculations. *J. Phys. Chem. Lett.* **5**, 1721–1726 (2014).
9. Bieker, G., Bieker, P. M. & Winter, M. Electrochemical *in situ* investigations of the SEI and dendrite formation on the lithium metal anode. *Phys. Chem. Chem. Phys.* **17**, 8670–8679 (2015).
10. Armand, M., Endres, F., MacFarlane, D. R., Ohno, H. & Scrosati, B. Ionic-liquid materials for the electrochemical challenges of the future. *Nat. Mater.* **8**, 621–629 (2009).
11. Bhatt, A. I., Best, A. S., Huang, J. & Hollenkamp, A. F. Application of the N-propyl-N-methyl-pyrrolidinium Bis(fluorosulfonyl)imide RTIL containing lithium Bis(fluorosulfonyl)imide in ionic liquid based lithium batteries. *J. Electrochem. Soc.* **157**, A66 (2010).
12. Best, A. S., Bhatt, A. I. & Hollenkamp, A. F. Ionic liquids with the Bis(fluorosulfonyl)imide anion: electrochemical properties and applications in battery technology. *J. Electrochem. Soc.* **157**, A903–A911 (2010).
13. Shkrob, I. a, Marin, T. W., Zhu, Y. & Abraham, D. P. Why Bis(fluorosulfonyl)imide is a 'magic anion' for electrochemistry. *J. Phys. Chem. C* **118**, 19661–19671 (2014).
14. Basile, A., Bhatt, A. I. & O'Mullane, A. P. A combined scanning electron micrograph and electrochemical study of the effect of chemical interaction on the cyclability of lithium electrodes in an ionic liquid electrolyte. *Aust. J. Chem.* **65**, 1534–1541 (2012).
15. Haskins, J. B. *et al.* Computational and experimental investigation of Li-doped ionic liquid electrolytes: [pyr14][TFSI], [pyr13][FSI], and [EMIM][BF4]. *J. Phys. Chem. B* **118**, 11295–11309 (2014).
16. Han, H. B. *et al.* Lithium bis(fluorosulfonyl)imide (LiFSI) as conducting salt for nonaqueous liquid electrolytes for lithium-ion batteries: physicochemical and electrochemical properties. *J. Power Sources* **196**, 3623–3632 (2011).
17. Kühnel, R.-S., Lübke, M., Winter, M., Passerini, S. & Balducci, A. Suppression of aluminum current collector corrosion in ionic liquid containing electrolytes. *J. Power Sources* **214**, 178–184 (2012).
18. Cho, E. *et al.* Corrosion/passivation of aluminum current collector in bis(fluorosulfonyl)imide-based ionic liquid for lithium-ion batteries. *Electrochem. Commun.* **22**, 1–3 (2012).
19. Budi, A. *et al.* Study of the initial stage of solid electrolyte interphase formation upon chemical reaction of lithium metal and N-methyl-N-propyl-pyrrolidinium-Bis (Fluorosulfonyl) imide. *J. Phys. Chem. C* **116**, 19789–19797 (2012).
20. Grande, L. *et al.* Homogeneous lithium electrodeposition with Pyrrolidinium-based ionic liquid electrolytes. *ACS Appl. Mater. Interfaces* **7**, 5950–5958 (2015).
21. Basile, A., Hollenkamp, A. F., Bhatt, A. I. & O'Mullane, A. P. Extensive charge-discharge cycling of lithium metal electrodes achieved using ionic liquid electrolytes. *Electrochem. Commun.* **27**, 69–72 (2013).
22. Zheng, G. *et al.* Interconnected hollow carbon nanospheres for stable lithium metal anodes. *Nat. Nanotechnol.* **9**, 618–623 (2014).

23. Zhang, D., Yin, Y., Liu, C. & Fan, S. Modified secondary lithium metal batteries with the polyaniline–carbon nanotube composite buffer layer. *Chem. Commun.* **51,** 322–325 (2015).
24. Cheng, X.-B. & Zhang, Q. Dendrite-free lithium metal anodes: Stable solid electrolyte interphase for high-efficiency batteries. *J. Mater. Chem. A* **3,** 7207–7209 (2015).
25. Muenzel, V. *et al.* A comparative testing study of commercial 18650-format lithium-ion battery cells. *J. Electrochem. Soc.* **162,** A1592–A1600 (2015).
26. Markevich, E. *et al. In situ* FTIR study of the decomposition of N-butyl-N-methylpyrrolidinium bis(trifluoromethanesulfonyl)amide ionic liquid during cathodic polarization of lithium and graphite electrodes. *Electrochim. Acta* **55,** 2687–2696 (2010).
27. Lane, G. H. Electrochemical reduction mechanisms and stabilities of some cation types used in ionic liquids and other organic salts. *Electrochim. Acta* **83,** 513–528 (2012).
28. Kroon, M. C., Buijs, W., Peters, C. J. & Witkamp, G.-J. Decomposition of ionic liquids in electrochemical processing. *Green Chem.* **8,** 241 (2006).
29. Philippe, B. *et al.* Improved performances of nanosilicon electrodes using the salt LiFSI: a photoelectron spectroscopy study. *J. Am. Chem. Soc.* **135,** 9829–9842 (2013).
30. Kanamura, K., Tamura, H., Shiraishi, S. S. & Takehara, Z.-I. XPS analysis for the lithium surface immersed in [gamma]-butyrolactone containing various salts. *Electrochim. Acta* **40,** 913–921 (1995).
31. Bhatt, A. I., Kao, P., Best, A. S. & Hollenkamp, A. F. Understanding the morphological changes of lithium surfaces during cycling in electrolyte solutions of lithium salts in an ionic liquid. *J. Electrochem. Soc.* **160,** A1171–A1180 (2013).
32. Yan, K. *et al.* Ultrathin two-dimensional atomic crystals as stable interfacial layer for improvement of lithium metal anode. *Nano Lett.* **14,** 6016–6022 (2014).
33. Howlett, P. C., Brack, N., Hollenkamp, A. F., Forsyth, M. & MacFarlane, D. R. Characterization of the lithium surface in N-methyl-N-alkylpyrrolidinium bis(trifluoromethanesulfonyl) amide room-temperature ionic liquid electrolytes. *J. Electrochem. Soc.* **153,** A595–A606 (2006).
34. Koch, V. R. Reactions of tetrahydrofuran and lithium hexafluoroarsenate with lithium. *J. Electrochem. Soc.* **126,** 181–187 (1979).
35. Aurbach, D., Weissman, I., Schechter, A. & Cohen, H. X-ray photoelectron spectroscopy studies of lithium surfaces prepared in several important electrolyte solutions. A comparison with previous studies by Fourier transform infrared spectroscopy. *Langmuir* **12,** 3991–4007 (1996).
36. Lu, Y., Tu, Z. & Archer, L. A. Stable lithium electrodeposition in liquid and nanoporous solid electrolytes. *Nat. Mater.* **13,** 961–969 (2014).
37. Andre, D. *et al.* Characterization of high-power lithium-ion batteries by electrochemical impedance spectroscopy. I. Experimental investigation. *J. Power Sources* **196,** 5334–5341 (2011).
38. Andre, D. *et al.* Characterization of high-power lithium-ion batteries by electrochemical impedance spectroscopy. II: Modelling. *J. Power Sources* **196,** 5349–5356 (2011).
39. Lane *et al.* Ionic liquid electrolyte for lithium metal batteries. Physical, electrochemical and interfacial studies of *N*-methyl-*N*-butyl morpholinium bis (fluorosulfonyl) imide. *J.Phys. Chem. C* **144,** 21775–21785 (2010).
40. Lee, H., Lee, D. J., Kim, Y.-J., Park, J. & Kim, H. A simple composite protective layer coating that enhances the cycling stability of lithium metal batteries. *J. Power Sources* **284,** 103–108 (2015).
41. Luo, J.-Y., Cui, W.-J., He, P. & Xia, Y.-Y. Raising the cycling stability of aqueous lithium-ion batteries by eliminating oxygen in the electrolyte. *Nat. Chem.* **2,** 760–765 (2010).
42. Yang, C.-P., Yin, Y.-X., Zhang, S.-F., Li, N.-W. & Guo, Y.-G. Accommodating lithium into 3D current collectors with a submicron skeleton towards long-life lithium metal anodes. *Nat. Commun.* **6,** 8058 (2015).
43. Pan, Q., Smith, D. M., Qi, H., Wang, S. & Li, C. Y. Hybrid electrolytes with controlled network structures for lithium metal batteries. *Adv. Mater.* **27,** 5995–6001 (2015).
44. Zhao, C.-Z. *et al.* Li2S5-based ternary-salt electrolyte for robust lithium metal anode. *Energy Storage Mater.* **3,** 77–84 (2016).
45. Zhang, R. *et al.* Conductive nanostructured scaffolds render low local current density to inhibit lithium dendrite growth. *Adv. Mater.* **28,** 2155–2162 (2016).
46. Ma, Q. *et al.* Improved cycling stability of lithium metal anode with novel concentrated electrolytes based on lithium (fluorosulfonyl) (trifluoromethanesulfonyl)imide. *ChemElectroChem.* **6,** 32454–32461 (2016).
47. Gu, W. *et al.* Lithium-iron fluoride battery with *in situ* surface protection. *Adv. Funct. Mater.* **26,** 1507–1516 (2016).
48. Togasaki, N., Momma, T. & Osaka, T. Enhanced cycling performance of a Li metal anode in a dimethylsulfoxide-based electrolyte using highly concentrated lithium salt for a lithium – oxygen battery. *J. Power Sources* **307,** 98–104 (2016).
49. Zhou, D. *et al.* SiO_2 hollow nanosphere-based composite solid electrolyte for lithium metal batteries to suppress lithium dendrite growth and enhance cycle life. *Ad. Energy Mater.* **6,** 1502214 (2016).
50. Qian, J. *et al.* High rate and stable cycling of lithium metal anode. *Nat. Commun.* **6,** 6362 (2015).

Acknowledgements

A.B. acknowledges financial support from CSIRO for the provision of an Office of the Chief Executive (OCE) PhD scholarship. A.P.O'M. gratefully acknowledges support from the Australian Research Council (Future Fellowship FT110100760). A.B. acknowledges Dr John Ward for assistance with SEM measurements. We thank RMIT's Microscopy and Microanalysis Facility for assistance with XPS.

Author contributions

All authors conceived and designed the experiments. A.B. performed the experiments and provided data analysis. All authors discussed the results and co-wrote the manuscript.

Additional information

Competing financial interests: The authors declare no competing financial interest.

26

Integrating a dual-silicon photoelectrochemical cell into a redox flow battery for unassisted photocharging

Shichao Liao[1,2], Xu Zong[1], Brian Seger[3], Thomas Pedersen[4], Tingting Yao[1], Chunmei Ding[1], Jingying Shi[1], Jian Chen[1] & Can Li[1]

Solar rechargeable flow cells (SRFCs) provide an attractive approach for *in situ* capture and storage of intermittent solar energy via photoelectrochemical regeneration of discharged redox species for electricity generation. However, overall SFRC performance is restricted by inefficient photoelectrochemical reactions. Here we report an efficient SRFC based on a dual-silicon photoelectrochemical cell and a quinone/bromine redox flow battery for *in situ* solar energy conversion and storage. Using narrow bandgap silicon for efficient photon collection and fast redox couples for rapid interface charge injection, our device shows an optimal solar-to-chemical conversion efficiency of ~5.9% and an overall photon-chemical-electricity energy conversion efficiency of ~3.2%, which, to our knowledge, outperforms previously reported SRFCs. The proposed SRFC can be self-photocharged to 0.8 V and delivers a discharge capacity of 730 mAh l^{-1}. Our work may guide future designs for highly efficient solar rechargeable devices.

[1] State Key Laboratory of Catalysis, Dalian Institute of Chemical Physics, Chinese Academy of Sciences, Dalian National Laboratory for Clean Energy, iChEM, Dalian 116023, China. [2] University of Chinese Academy of Sciences, Beijing 100049, China. [3] Department of Physics, CINF, Technical University of Denmark, Kongens Lyngby 2800, Denmark. [4] Department of Micro- and Nanotechnology, Technical University of Denmark, Kongens Lyngby 2800, Denmark. Correspondence and requests for materials should be addressed to J.S. (email: jingyingshi@dicp.ac.cn) or to J.C. (email: chenjian@dicp.ac.cn) or to C.L. (email: canli@dicp.ac.cn).

Simultaneous conversion and storage of abundant, but intermittent solar energy has been entering the spotlight as a promising strategy for the controllable utilization of solar energy[1,2]. Specifically, directly storing solar energy in hydrogen (H_2) produced by light-driven water splitting has been regarded particularly attractive, as hydrogen is an efficient and clean fuel for electricity generation[3,4]. Unfortunately, challenges with hydrogen storage and the cost of fuel cells impede wide implementation of solar hydrogen/fuel cell hybrid systems. Moreover, the sluggish half reaction kinetics of water oxidation greatly hamper the improvement of solar energy conversion efficiency in water splitting[5,6].

Alternatively, solar energy can be *in situ* stored in other chemicals by driving non-spontaneous reactions in a photoelectrochemical (PEC) cell[7]. The resulting products can be readily utilized to generate electricity via reversible chemical reactions. On the basis of this principle, efforts to fabricate solar rechargeable cells (SRCs) have been going on for several decades[8]. For instance, solid state electrodes can be integrated in SRCs for storing photogenerated charges[9–12]. However, the storage capacity of SRCs is limited by the physical dimension of solid electrodes. Besides, the sluggish insertion/extraction of the ions into/from the solid electrodes may lead to a poor energy conversion efficiency. This has prompted efforts to integrate SRCs and redox flow batteries (RFBs) with two soluble redox species to improve the storage capacity[13]. Redox couples in RFBs generally present facile electrochemical kinetics, which can be several orders of magnitudes faster than that of water oxidation[14]. Benefiting from the rapid semiconductor/electrolyte interface charge transfer, a higher solar-to-chemical (STC) conversion efficiency based on fast redox species could be expected in comparison with that of solar to hydrogen in water splitting. A conceptual SRC with continuous flow electrolytes was reported by Yang *et al.*, but the overall energy conversion efficiency (<0.1%) is limited by the low conductivity of the inorganic separator between the aqueous and organic electrolytes[15,16]. More recently, Liu *et al.*, proposed to photocharge a vanadium RFB using wide bandgap semiconductors such as WO_3 and TiO_2[17,18]. However, the STC conversion efficiency was low and the discharge performance has not yet been reported. Fabricating an aqueous SRC with high STC conversion efficiency and desirable storage capacity still remains a challenge.

Here we demonstrate a solar rechargeable flow cell (SRFC) integrating a dual-silicon PEC cell into a quinone/bromine RFB for *in situ* solar energy conversion and storage. To attain significant improvements in SRC performance, two important factors in the PEC reactions guide our design. First, a photoelectrode with a narrow bandgap is required to allow for efficient utilization of solar energy. Second, redox couples with fast reaction kinetics and efficient cocatalysts that can catalyse the redox reactions should be employed to expedite semiconductor/electrolyte interface charge transfer kinetics. In this regard, silicon with a favourable bandgap (1.1 eV) is used as the light absorber. Quinones and halogens, which present fast reaction kinetics and excellent electrochemical reversibility[19–21], are used as energy storage media for efficiently capturing photogenerated charges. In our SRFC, the fast PEC reactions of the water-soluble quinone and bromine redox couples on the buried junction and cocatalyst-functionalized dual-silicon absorbers enable the SRFC to achieve an overall photon–chemical–electricity energy conversion efficiency of ~3.2% and deliver a constant discharge voltage of ~0.78 V, which, to our knowledge, are higher than values achieved previously in SRFCs[15–18]. Our work demonstrates aqueous SRFCs with good overall efficiency, high discharge voltage and desirable discharge capacity, and suggests paths for further improvements to allow technological development and use.

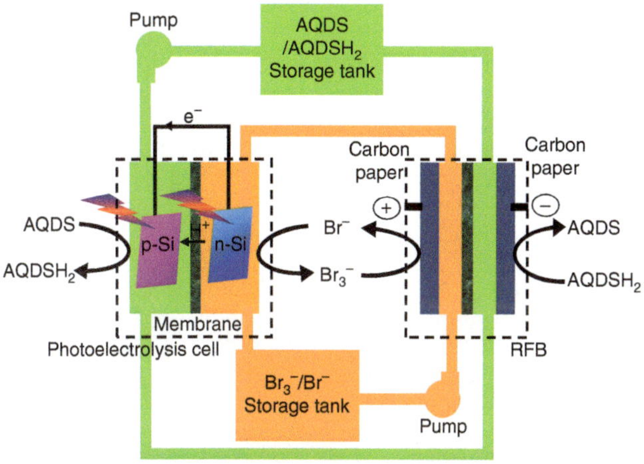

Figure 1 | Schematic configuration of the proposed SRFC. $AQDS/AQDSH_2$ and Br_3^-/Br^- are used as redox couples.

Results

Configuration and working principle of the SRFC. Figure 1 illustrates the configuration of the SRFC used in this work. It consists of a PEC (or photoelectrolysis) cell that deposits solar irradiation into chemical energy and a RFB that converts the as-stored chemical energy into electricity. $AQDS/AQDSH_2$ (9,10-anthraquinone-2,7-disulphonic sodium/1,8-dihydroxy-9,10-anthraquinone-2,7-disulphonic sodium) and Br_3^-/Br^- are used as active redox couples. The PEC cell and RFB are connected through electrolyte circuit loops. During the photocharge process, AQDS is reduced to $AQDSH_2$ on the photocathode and Br^- is oxidized to Br_3^- on the photoanode simultaneously in the PEC cell by short-circuiting the two photoelectrodes under illumination. The resultant $AQDSH_2$ and Br_3^- are then stored in two individual reservoirs that can be readily used by the RFB. A commercial Nafion membrane is used to separate the two compartments in each cell. The involved cell reactions can be expressed as follows:

$$3Br^- \underset{\text{Discharge}}{\overset{\text{Photocharge}}{\rightleftharpoons}} Br_3^- + 2e^- \qquad E^0 = 1.09\text{ V vs. NHE} \tag{1}$$

$$AQDS + 2H^+ + 2e^- \underset{\text{Discharge}}{\overset{\text{Photocharge}}{\rightleftharpoons}} AQDSH_2 \qquad E^0 = 0.20\text{ V vs. NHE} \tag{2}$$

The overall reaction:

$$AQDS + 2H^+ + 3Br^- \underset{\text{Discharge}}{\overset{\text{Photocharge}}{\rightleftharpoons}} AQDSH_2 + Br_3^- \qquad E^0 = 0.89\text{ V} \tag{3}$$

Energy conversion in this SRFC clearly follows a two-step route of solar-chemical-electricity. The overall efficiency is thus decided by the product of the STC efficiency in PEC cell and the chemical-to-electricity efficiency in RFB ($\eta_{overall} = \eta_{STC} \times \eta_{RFB}$). The quinone/bromine flow battery generally demonstrates a high energy efficiency of over 70% (ref. 19). In this way, the STC efficiency during the photocharging process is a critical issue for the as-designed SRFC. Compared with single-photoelectrode system, dual-photoelectrode configuration has the merits of using small bandgap semiconductor materials that allow for efficient light harvesting and two semiconductor/electrolyte liquid junctions for sufficient photovoltage. Crystal silicon, with a small bandgap of 1.1 eV, is a promising material for STC conversion in a p/n dual-electrode PEC cell[22,23]. But so far, its applications are limited due to the instability[24], sluggish surface charge transfer kinetics[25] and insufficient self-driving force[26]. Recent efforts show that the sustainability of crystal silicon in water splitting reaction can be significantly improved to tens of hours by surface passivation and/or cocatalyst[27–33]. Moreover, fabrication of a surface-buried junction (n^+p or p^+n) generally induces a photovoltage of ~0.50 V (refs 22,23,34). Thereby, the dual-silicon (n^+p-Si and p^+n-Si) PEC system is able to offer a photovoltage up to 1.0 V, which is thermodynamically sufficient to drive the above overall reaction for photocharging without the assistance of external bias.

Reduction half reaction during photocharging process. According to the working principle, photocharging of a SRFC involves the PEC reduction of AQDS and PEC oxidation of Br^-. As a proof-of-concept study, PEC reduction of AQDS to produce discharging species of $AQDSH_2$ is firstly studied. The n^+p-Si coated orderly with metal titanium and TiO_2 thin films as a protective layer (hereafter TiO_2/Ti/n^+p-Si) is employed as the photocathode. It is reported that the carbon material serves as an excellent substrate for adsorption and electron transfer for the quinone/hydroquinone redox reaction[35]. Therefore, a carbon film of 5-nm thickness is sputtered onto the TiO_2/Ti/n^+p-Si electrode to achieve a carbon covered surface (hereafter C/TiO_2/Ti/n^+p-Si), as evidenced by the scanning electron microscopy images and X-ray photoemission spectroscopy analysis (Supplementary Fig. 1a,c). Figure 2 shows the current–potential (*I–E*) curves of TiO_2/Ti/n^+p-Si and C/TiO_2/Ti/n^+p-Si electrodes in the electrolyte containing 0.05 M AQDS. Both electrodes generate insignificant currents in the dark. While under illumination, the TiO_2/Ti/n^+p-Si photocathode produces a photocurrent density of $-2.5\,mA\,cm^{-2}$ at 0.1 V versus SCE (the saturated calomel electrode) and has an onset potential of 0.25 V versus SCE (defined as the potential at which a photocurrent exceeds $0.2\,mA\,cm^{-2}$). Upon carbon modification, the photocurrent density significantly increases to $-12.5\,mA\,cm^{-2}$ at the same bias. Meanwhile, the onset potential experiences a positive shift of ~300 mV. These results suggest that carbon is a more favourable substrate for the PEC reduction of AQDS than TiO_2. In sharp contrast, the currents from C/TiO_2/Ti/n^+p-Si photocathodes, both in the dark and under illumination, are negligible in the absence of AQDS (Supplementary Fig. 2a). Thus, the observed photocurrents can be almost completely attributed to AQDS reduction. Since photocurrent spikes grow remarkably as applied potential is decreased under chopped light (Supplementary Fig. 3; Supplementary Note 1), the demonstrated photocurrent is speculated to be limited by the mass transfer of AQDS from bulk solution to the electrode surface.

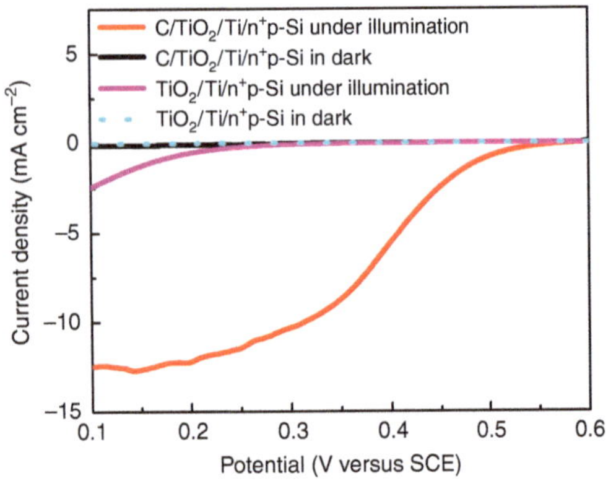

Figure 2 | Photoelectrochemical reduction of AQDS over photocathode. Current-potential curves of the photocathodes in a 0.05 M AQDS + 1.0 M H_2SO_4 solution purged with argon under AM 1.5-G $100\,mW\,cm^{-2}$ illumination.

To investigate the mass transfer effect, controlled experiments were performed in magnetically stirred solution at various speeds and the obtained *I–E* curves are presented in Fig. 3a. A limiting photocurrent appears on each curve and is enhanced with increasing stirring speeds from 0 to 1,000 r.p.m. At 1,000 r.p.m., the saturated photocurrent reaches $-27.0\,mA\,cm^{-2}$, which is about six times higher than that obtained in static electrolyte. The mass transfer limitation is thus confirmed during the AQDS reduction reaction. Actually, the diffusion limitation cannot be fully avoided even at a stirring speed up to 1,000 r.p.m., which is evidenced by the transient current spikes under chopped light as indicated in the inset of Fig. 3a.

To understand the PEC behaviours of the AQDS species, we measured the kinetic parameter for the electrochemical reduction reaction of AQDS to $AQDSH_2$, as well as the diffusion coefficient of AQDS species in the electrolyte via a linear sweep voltammetry method (see Supplementary Fig. 4 and Supplementary Methods for details). The diffusion coefficient (D) of AQDS is determined to be $4.8 \times 10^{-6}\,cm^2\,s^{-1}$, which is in agreement with the previous report[19]. The rate constant (k_0) is evaluated to be $1.05 \times 10^{-2}\,cm\,s^{-1}$. Compared with other redox couples employed in RFBs such as Fe^{3+}/Fe^{2+}, Cr^{3+}/Cr^{2+}, Ce^{4+}/Ce^{3+}, VO_2^+/VO^{2+} and V^{3+}/V^{2+} (Supplementary Table 1), the AQDS/$AQDSH_2$ on carbon electrode exhibits much larger k_0 in combination with a comparable diffusion coefficient. The extremely rapid kinetics might be attributed to little reorganizational energy required in the electrochemical reduction of anthraquinone[20]. However, the transport of AQDS species in the electrolyte is too slow to keep up with the prompt interface charge transfer. These electrochemical properties result in the dominance of mass transfer where it is particularly prominent at more negative applied potentials, which well explains the unusual PEC behaviour of AQDS on C/TiO_2/Ti/n^+p-Si photocathode. The reversible potential for AQDS over a graphite electrode is evaluated to be −0.03 V versus SCE (Supplementary Fig. 2b). On the basis of equation (4) presented below[36], the maximum solar-to-$AQDSH_2$ conversion efficiency can be calculated to be ~6.0% at a bias of 0.30 V versus SCE (Fig. 3b).

Oxidation half reaction during photocharging process. We then moved on to investigate the PEC oxidation of Br^- to Br_3^-, which is the other STC reaction during the storage of solar energy into chemical energy. The p^+n-Si coated with platinum islands as cocatalyst (scanning electron microscopy and X-ray photoemission spectroscopy, respectively, in Supplementary Fig. 1b,d, hereafter Pt/p^+n-Si) is employed as the photoanode. Figure 3c demonstrates the *I–E* curves over Pt/p^+n-Si photoanode in the solution containing 0.2 M HBr and 1.0 M H_2SO_4. Without stirring the electrolyte, the photocurrent is observed to reach a peak value of $22.2\,mA\,cm^{-2}$ at 0.50 V versus SCE and beyond which decreases to a constant value of $19.0\,mA\,cm^{-2}$. In comparison, control experiments show that the currents are negligible without the presence of Br^- species both under light and in the dark (Supplementary Fig. 5a). Therefore, the photocurrent generated is ascribed to the photo-oxidation of Br^-. With magnetic stirring of the solution at 200 r.p.m., the current peak vanishes and the limiting photocurrent promptly elevates to a much higher level of $36.2\,mA\,cm^{-2}$. Unlike the electrochemical behaviour of AQDS during the PEC reduction, increase of the stirring speed to 400–800 r.p.m. does not induce further improvement in the saturated photocurrent. Moreover, no transient current spikes under chopped light irradiation are observed even at more

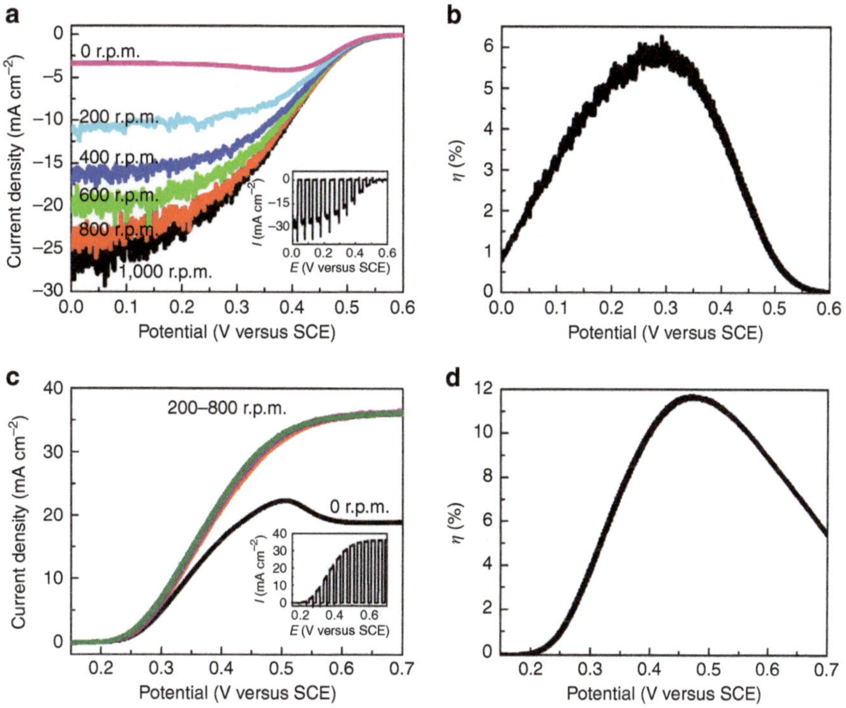

Figure 3 | Photoelectrochemical performance of the photoelectrodes. Constant light current-potential curves of the C/TiO_2/Ti/n^+p-Si photocathode in a 0.05 M AQDS + 1.0 M H_2SO_4 solution (**a**) and Pt/p^+n-Si photoanode in a 0.2 M HBr + 1.0 M H_2SO_4 solution (**c**) under various magnetically stirring speeds. The inset in **a** and **c** shows chopped light current-potential curves of the photocathode at 1,000 r.p.m. magnetic stirring speed and the photoanode at 200 r.p.m. magnetic stirring speed, respectively. Corresponding half-cell STC conversion efficiency for C/TiO_2/Ti/n^+p-Si photocathode at 1,000 r.p.m. stirring speed (**b**) and for Pt/p^+n-Si photoanode at 200 r.p.m. stirring speed (**d**). Light source: AM 1.5-G 100 mW cm^{-2}.

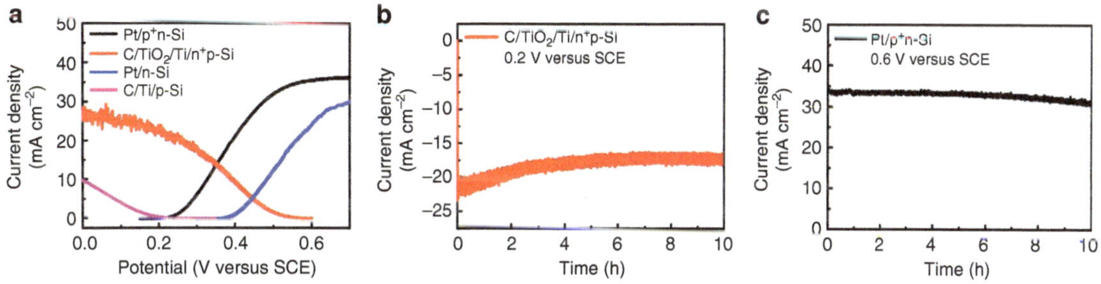

Figure 4 | Photoelectrochemical behaviours of various photoelectrodes. (**a**) Overlaid current-potential curves of the individual photocathodes and photoanodes measured in a three-electrode experiment. C/Ti/p-Si and C/TiO_2/Ti/n^+p-Si in a 0.05 M AQDS + 1.0 M H_2SO_4 solution; while Pt/n-Si and Pt/p^+n-Si/ in a 0.2 M HBr + 1.0 M H_2SO_4 solution. Chronoamperometry of the C/TiO_2/Ti/n^+p-Si photocathode in a 0.05 M AQDS + 1.0 M H_2SO_4 solution (**b**) and the Pt/p^+n-Si photoanode in a 0.2 M HBr + 1.0 M H_2SO_4 solution (**c**). The applied stirring speeds were 1,000 r.p.m. in the AQDS solution and 700 r.p.m. in the HBr solution, respectively. Light source: AM 1.5-G 100 mW cm^{-2}.

positive bias, indicating that most photogenerated holes arriving at the electrode/electrolyte interface participate in the oxidation reaction (inset in Fig. 3c). It is worth noting that the saturation photocurrents in stirred solution are completely dominated by the incident light intensity (Supplementary Fig. 6a,b) and the mass transfer limitation during the PEC oxidation of Br^- thus is fully relieved, which is different from the behaviour of AQDS reduction.

The k_0 for the Br^- oxidation reaction and the D for Br^- ions were also determined (Supplementary Fig. 7) and listed in Supplementary Table 1. Compared with those parameters relative to AQDS, the k_0 (2.82×10^{-2} cm s^{-1}, which is consistent with existing kinetics data[37]) is of the same order while D (6.0×10^{-5} cm^2 s^{-1}) increases by one order of magnitude. The large diffusion coefficient of Br^- allows us to sidestep mass transfer limitations during PEC oxidation reactions. It is also noteworthy that the k_0 for Br^- oxidation is more than twice the value for AQDS reduction and much larger than values for the other redox couples listed in Supplementary Table 1. This ensures rapid charge transfer between the photoelectrode and soluble species, resulting in more effective utilization of the photogenerated carriers. The larger saturated photocurrent is achieved in the case of Br^- oxidation than AQDS reduction (Fig. 3a,c). The reversible potential for Br_3^-/Br^- is estimated to be 0.85 V versus SCE on a Pt electrode (Supplementary Fig. 5b). According to equation (4), the optimal half-cell solar-to-Br_3^- conversion efficiencies is ~11.6% at 0.46 V versus SCE as indicated in Fig. 3d, which is nearly two times as large as that of solar to $AQDSH_2$. In addition, the maximal power conversion efficiency of Pt/p^+n-Si photoelectrode for the electrochemical photovoltaic cell approaches 12.0% (Supplementary Fig. 8), slightly higher than the previously reported value of 10.5% (ref. 38).

The high efficiency further confirms the rapid electrochemical kinetics of Br^- oxidation.

Performance and stability assessment of the PEC cell. After investigating the photoelectrodes in the corresponding electrolytes independently, we can evaluate the performance of the integrated photoanode/cathode PEC cell by overlapping the individual *I–E* data for each photoelectrode as shown in Fig. 4a. The intersection of the two curves indicates the optimal operating current density and the overall STC ($\eta_{O\text{-}STC}$) efficiency in the resulted PEC cell can be calculated according to equation (5) (ref. 5). It can be observed that the intersection is located at 13.5 mA cm^{-2} at 0.35 V versus SCE and the corresponding $\eta_{O\text{-}STC}$ is estimated to be 5.9% for the proposed dual-silicon PEC cell, which is much higher than those of direct solar water splitting, as well as all the reported SRFCs[15–18,39]. Considering that the energy conversion efficiency in this quinone/bromine flow battery is ~70% (Supplementary Fig. 9), the overall energy conversion efficiency of the present SRFC is estimated to be 4.1% ($\eta_{overall} = \eta_{O\text{-}STC} \times \eta_{RFB} = 5.9 \times 70\%$). It should be pointed out that no bias is required to drive the overall photocharge reactions. For comparison, the *I–E* curves for the silicon wafers without a surface doping, C/Ti/p-Si and Pt/n-Si photoelectrodes, are also listed in Fig. 4a. No intersection between these two curves can be observed, indicating that the photocharge reaction in the C/Ti/p-Si‖Pt/n-Si dual-photoelectrode system cannot be realized without additional power input. For the C/Ti/p-Si‖Pt/p$^+$n-Si and C/TiO_2/Ti/n$^+$p-Si‖Pt/n-Si photoelectrolysis cells, the operating photocurrents during solar-charge process are rather low and thus the overall energy conversion efficiencies are far from satisfactory.

Current–time (*I–t*) tests reveal a photocurrent of ~13 mA cm^{-2} for each photoelectrode at the operating potential of 0.35 V versus SCE (Supplementary Fig. 10a,c), which agree with the data obtained from the above *I–E* measurements. This result further confirms the predicted STC efficiency as above. Stability tests illustrate that both photoanode and photocathode can sustain much higher currents for at least 10 h under AM 1.5-G irradiation (Fig. 4b,c). The light absorbance of the catholyte changed after PEC reaction (Supplementary Fig. 11a), suggesting the generation of new species and inevitable absorption losses as a result. Therefore, the appropriate concentration of electrolytes is selected for PEC and SRFC demonstration. To investigate photochemical stability of the electrolytes, experiments were conducted under simulated solar irradiation and visible light illumination (Supplementary Fig. 11b–e). The results show that AQDS is stable under visible light, but it undergoes degradation under ultraviolet irradiation. Br^- is highly stable under the full solar spectrum. When exposed only to visible light to reduce potential photochemical side reactions, the photocathode can run steadily for >100 h and the photoanode for 37 h (Supplementary Fig. 10b,d). The Faradaic efficiencies for photogenerated $AQDSH_2$ and Br_3^- species at the operating potential (0.35 V versus SCE) under visible light illumination are 97% and 99%, respectively (Supplementary Fig. 12; Supplementary Methods), suggesting that the observed photocurrents can be ascribed to the PEC reactions of the electroactive species with negligible side reactions.

Demonstration of a proof-of-concept SRFC. Given the desirable individual performance of both PEC cell and RFB (cycling stability of the RFB is described in Supplementary Fig. 9 and Supplementary Note 2), a prototype SRFC device based on a side-by-side dual-silicon photoelectrode system and the mentioned RFB with dual-carbon discharge electrode is fabricated (for experimental details see Supplementary Methods) as shown in Supplementary Fig. 13. During the photocharge process under AM 1.5-G irradiation, the C/TiO_2/Ti/n$^+$p-Si photocathode and the Pt/p$^+$n-Si photoanode are directly short-circuited without external bias. Figure 5a illustrates the photocharge profiles for the first 2 h. The charging photocurrent for the photocathode starts at ~8.2 mA cm^{-2} and decreases to 4.6 mA cm^{-2}; the corresponding initial and final values for the photoanode are 7.5 and 4.2 mA cm^{-2}. The unequal photocurrent density between the two photoelectrodes is due to a small difference in working area of the electrodes studied herein. The colour of the catholyte in the PEC cell gradually changes from light orange to dark green while that of the anolyte simultaneously varies from colourless to yellow, indicating storage of solar energy. The declined photocurrent during the photocharge is primarily ascribed to the two points: (1) decreased light intensity at the silicon photoelectrode due to the generation of colourful $AQDSH_2$ and Br_3^- species in the electrolyte; and (2) the shifts in Nernst redox potential of the redox couples caused by varied concentration of active species. On the bais of the total exposure areas of photoanode and photocathode, the actual output photocurrent density for the complete photoelectrolysis cell varies from 3.9 to 2.2 mA cm^{-2}, which is lower than the highest photocurrent density 6.8 mA cm^{-2} (half of 13.5 mA cm^{-2}) estimated from the Fig. 4a. Such a deviation may be explained by the inefficient mass transfer of electroactive species in the as-fabricated model system, which is limited by the flow rate of the electrolyte. According to equation (5), the overall STC conversion efficiency during the photocharge process initially achieves 3.4% followed by decrease to the final value of

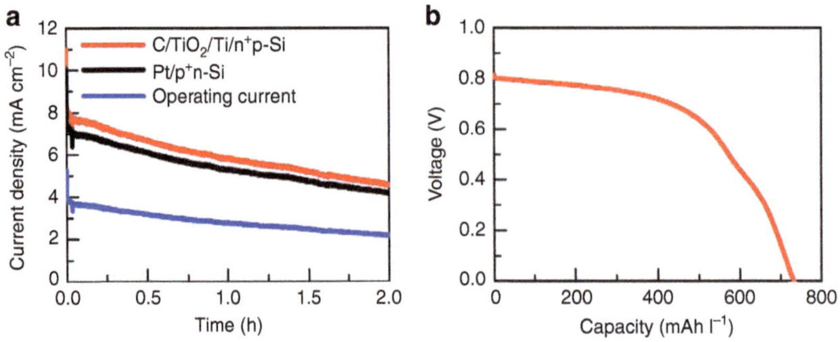

Figure 5 | Electrochemical characterizations of the proposed SRFC. (**a**) Photocharge curves of the SRFC in a 0.2 M HBr + 1.0 M H_2SO_4 solution on the positive side and a 0.05 M AQDS + 1.0 M H_2SO_4 solution on the negative side without the external bias under AM 1.5-G 100 mW cm^{-2} illumination. The surface area of C/TiO_2/Ti/n$^+$p-Si is 0.232 cm^2 and that of Pt/p$^+$n-Si is 0.254 cm^2. The flow rates of the electrolytes are kept at 15 ml min^{-1}. (**b**) The discharge curve of the SRFC at constant current 0.5 mA cm^{-2} under dark conditions.

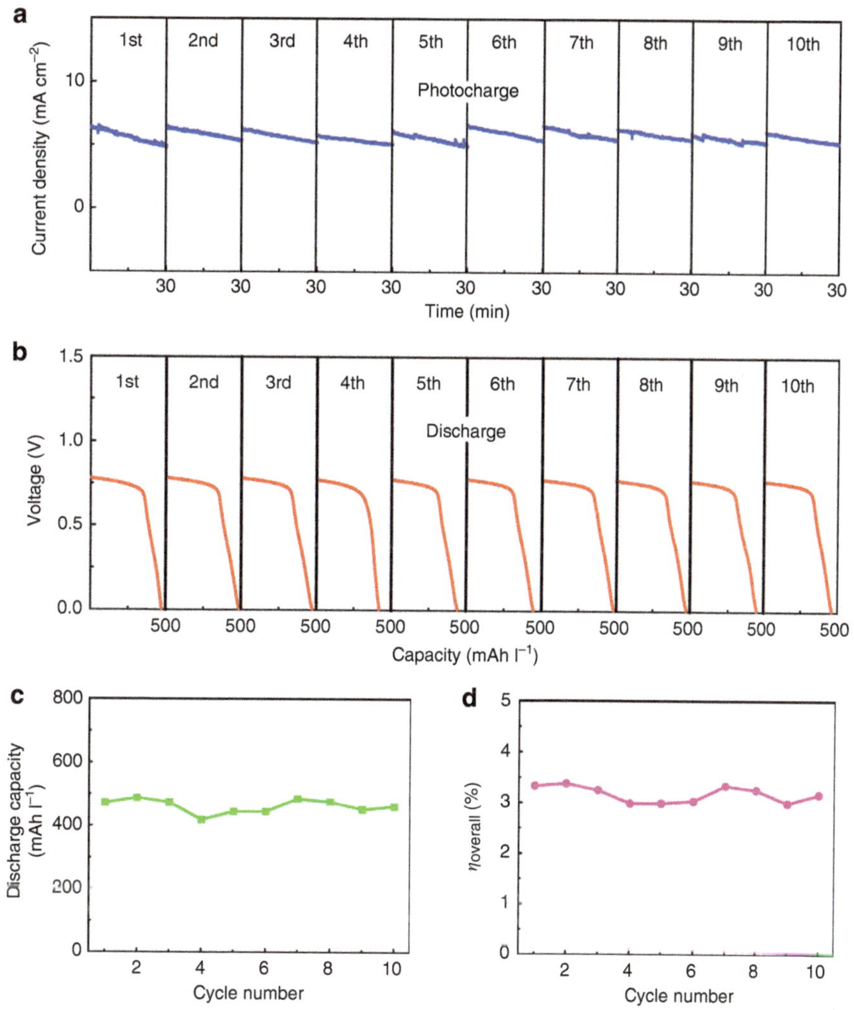

Figure 6 | Cycling performance of the proposed SRFC. (**a**) Photocharge curves of the SRFC in a 0.2 M HBr + 0.005 M Br_2 + 1.0 M H_2SO_4 solution on the positive side and a 0.05 M AQDS + 1.0 M H_2SO_4 solution on the negative side without the external bias under visible light illumination (100 mW cm^{-2}). The surface area of C/TiO_2/Ti/n$^+$p-Si is 0.232 cm^2 and that of Pt/p$^+$n-Si is 0.230 cm^2. The photocurrent density is calculated based on the total illumination areas of photoanode and photocathode. The flow rates of the electrolytes are kept at 100 ml min^{-1}. (**b**) Galvanostatic discharge curves of the SRFC at constant current 2.0 mA cm^{-2} under dark. (**c**) Discharge capacity and (**d**) overall photon–chemical–electricity energy conversion efficiency of the SRFC as a function of the cycle number.

1.9% (Supplementary Fig. 14). The average STC conversion efficiency is calculated to be 2.5% based on equation (6).

Figure 5b shows the galvanostatic discharge curve of the SRFC performed at 0.5 mA cm^{-2}. The initial discharge voltage exceeds 0.80 V and a constant discharge voltage plateau locates at around 0.78 V. It is notable that the discharge capacity is up to 730 mAh l^{-1}. To the best of our knowledge, a combination of such values has not been attained previously by integrated SRFCs. After photocharging, this SRFC can drive an 8 mW electric fan (Supplementary Movie 1). The overall solar–chemical–electricity energy conversion efficiency is calculated to be 1.0% according to equation (7), which is less than half of the above average STC conversion efficiency of 2.5%. This may be ascribed to concentration polarization resulted from the slow flow rate of the electrolytes, evaporation of bromine, potential side reaction of AQDS, partial retention of the electroactive species in the electrolytes after discharge, electrolytes leakage from the pumping system and other parasitic energy loss processes.

While the above SRFC device demonstrates capture and storage of solar energy, its overall solar energy conversion efficiency is far from satisfactory. To address this issue, the device was optimized and photocharged under visible light with a higher flow rate of electrolytes (100 ml min^{-1}). In addition, a very small amount of liquid bromine (0.005 M) was added beforehand to the positive electrolyte to replenish the evaporation loss during operation. The repeated photocharge/discharge curves of the SRFC for the first 10 cycles are exhibited in Fig. 6a,b (voltage curves during photocharge shown in Supplementary Fig. 15), respectively. During photocharge, the voltage approaches 0.8 V and the maximum operating photocurrent density is 6.6 mA cm^{-2}, which is close to the predicted value (6.8 mA cm^{-2}). Almost identical photocharge/discharge curves are presented in the first 10 cycles, implying excellent cycling stability. The SRFC delivers an initial discharge capacity of 471 mAh l^{-1} and maintains a discharge capacity of 462 mAh l^{-1} after 10 photocharge/discharge cycles (Fig. 6c), confirming stable operation. As shown in Fig. 6d, the SRFC possesses highly stable overall photon–chemical–electricity energy conversion efficiency during the 10 photocharge/discharge cycles, with an average efficiency of 3.2%, which to the best of our knowledge is a record-high efficiency attained for the SRFC to date.

Discussion

The SRFC demonstrated in this work exhibits a number of advantages as follows: first, mainly earth-abundant and eco-friendly materials of silicon and carbon are used as photocharge and discharge electrodes, respectively. Second, in comparison to nonaqueous media, the aqueous electrolyte is safer, cheaper and has higher ionic conductivity. Third, the fast electrochemical kinetics of the selected redox couples not only increases the photocharge efficiency, but also improves the discharged power density of the SRFC. Finally, the SRFC can be flexibly designed and optimized by investigating the PEC cell and RFB individually. On the basis of our design, the photocharge and discharge processes can be performed in the same glass-window equipped cell by immersing the photoelectrodes and electrodes into the corresponding half-cell to fabricate a SRFC with a more compact structure. More efficient SRFCs will be enabled by a combination of the following: optimizing the concentration of electroactive species to keep a balance between photocharge capacity and light absorption of the electrolytes; reducing the distance between the photoeletrode and the light incident window; using the photoelectrode as the light incident window that the irradiation can reach the photoabsorber directly without passing through the electrolyte; applying a glass with TiO_2 coating or other ultraviolet-absorption materials as the incident window to filter ultraviolet radiation; and developing more transparent redox couples with larger k_0 and D, and optimizing device fabrication techniques. Investigation into these approaches is underway in our laboratory.

In summary, we successfully designed and fabricated an aqueous SRFC device with the dual function of conversion and storage of solar energy via direct integration of a dual-silicon PEC cell and a RFB using water-soluble AQDS/$AQDSH_2$ and Br_3^-/Br^- redox couples as the energy carriers. The half-cell STC (H-STC) conversion efficiencies achieved for AQDS reduction on C/TiO_2/Ti/n^+p-Si and Br^- oxidation on Pt/p^+n-Si are up to ~6.0% and ~11.6%, respectively. The excellent PEC performance can be attributed to the outstanding light harvesting properties of silicon, active surface cocatalysts, as well as rapid kinetics of redox couples. The maximum STC conversion efficiency during photocharge reaches ~5.9% and the overall photon–chemical–electricity energy conversion efficiency of the as-designed SRFC exceeds 3.0%, record values for an integrated SRFC to the best of our knowledge. In a proof-of-concept SRFC device, the cell voltage can be self-charged to 0.8 V under simulated AM 1.5-G illumination and the discharge capacity is up to 730 mAh l^{-1} after photocharging for 2 h. This aqueous SRFC concept, based on earth-abundant electrodes and fast redox couples, may open new avenues for design and fabrication of highly efficient solar rechargeable batteries towards practical application.

Methods

Electrodes preparation. Commercial p-Si wafers were processed to achieve a thin n^+-doped surface layer followed by successive deposition of 5 nm Ti and 100 nm TiO_2 protective layers according to reported procedures[22]. A thin carbon layer was then deposited onto the surface of wafers by reactive magnetron sputtering of a high-purity graphite target (99.999%). Further experimental details are provided in the Supplementary Methods.

Commercial n-Si wafers were functionalized by surface p^+ doping according to a reported procedure and used as photoanodes[23]. Pt was then sputtered on the surface of wafers as a cocatalyst.

Commercial Toray carbon paper electrodes (200 µm in thickness) were used as the positive and negative electrodes in the SRFC system. The carbon papers were first cleaned under sonication in isopropyl alcohol for 10 min, followed by soaking in a 3:1 (by volume) solution of H_2SO_4 (conc.) and HNO_3 (conc.) for 5 h at 323 K.

Photoelectrochemical measurements. PEC measurements were carried out in a three-electrode system consisting of a photoelectrode as the working electrode, a SCE as the reference electrode and a graphite plate (or platinum plate) as the counter electrode. 1 M H_2SO_4 aqueous solutions were used as supporting or blank electrolytes. An aqueous solution containing 0.05 M AQDS (TCI, 95%) and 1.0 M H_2SO_4 (Sinopharm Chemical, 98%) was used as the catholyte, and an aqueous solution containing 0.2 M HBr (Sinopharm Chemical, 40%) and 1.0 M H_2SO_4 was used as the anolyte. Both electrolytes were deaerated with flowing argon gas during the experiment. Linear sweep voltammetry was performed using a potentiostat (Iviumstat, Ivium Technologies) at a scan rate of 10 mV s^{-1} under illumination. The simulated AM 1.5-G solar light irradiation (100 mW cm^{-2}, Newport Sol 3A, Class AAA Solar simulator) was used as the light source. Chronoamperometry measurements were carried out at various biases under light irradiation.

SRFC device assembly. The SRFC consists of two electrolyte tanks, a p/n photoelectrolysis cell and a common flow battery component. The p/n photoelectrolysis cell was built inside a home-made reactor with quartz windows. The dual absorbers composed of photocanode and photoanode were placed side by side in cathodic and anodic compartments, respectively. Solution of 0.05 M AQDS + 1.0 M H_2SO_4 and 0.2 M HBr + 1.0 M H_2SO_4 was used as initial catholyte and anolyte, respectively. The electrolytes were isolated by a Nafion 115 membrane. During photocharging under AM 1.5-G 100 mW cm^{-2} illumination, the photoelectrodes were directly short-circuited without external bias. The flow rates of the electrolytes were kept at 15 ml min^{-1}. After the photocharge, the produced $AQDSH_2$ and Br_3^- could either be stored in two isolate tanks or flow through the battery to convert the chemical energy to electricity. In the discharge process, the SRFC was performed at a constant current of 0.5 mA cm^{-2}.

SRFC cycling behaviour. Solutions of 0.05 M AQDS + 1.0 M H_2SO_4 and 0.2 M HBr + 0.005 M Br_2 + 1.0 M H_2SO_4 were used as initial catholyte and anolyte, respectively. The individual volume of catholyte and anolyte was 2.5 ml. Before photocharge, galvanostatic charge/discharge of the SRFC was performed under dark condition to remove the dissolved oxygen in the electrolytes, preventing the oxidation of the photogenerated $AQDSH_2$ species. Photocharge of the SRFC in the first 10 cycles was conducted by directly connecting the photoelectrodes without external bias under visible light illumination ($\lambda > 420$ nm, 100 mW cm^{-2}). After the photocharge, the discharge of the SRFC was performed at a constant current of 2.0 mA cm^{-2} under dark condition. The flow rates of the electrolytes were kept at 100 ml min^{-1}.

Calculations. The H-STC energy conversion efficiency ($\eta_{H\text{-}STC}$) of photoelectrode can be estimated according to the following equation (assuming that the Faraday efficiency is 100%) (ref. 36):

$$\eta_{H-STC} = I_{ph} \times |E_{bias} - E^{0\prime}| / P_{in} \times 100\% \tag{4}$$

where I_{ph} represents the photocurrent density of the photoelectrode, E_{bias} stands for the applied bias, $E^{0\prime}$ is the reversible potential of the redox couple and P_{in} denotes the incident solar power (100 mW cm^{-2}).

The expected overall STC ($\eta_{O\text{-}STC}$) efficiency in the PEC cell can be estimated according to[5],

$$\eta_{O-STC} = I_{op} \times |E^{0\prime}_{positive} - E^{0\prime}_{negative}| / P_{in} \times 100\% \tag{5}$$

where I_{op} stands for the operating photocurrent density calculated based on the total exposure areas of the photoelectrodes, $E^{0\prime}$ is the reversible potential of the redox couple and P_{in} denotes incident solar power (100 mW cm^{-2}).

The average STC energy conversion efficiency during photocharge in SRFC system could be roughly estimated according to:

$$\eta_{a-STC} = \left(\int i(t)\mathrm{d}t \times \Delta E^{0\prime}\right) / (P_{in} \times S_{total} \times t) \times 100\% \tag{6}$$

where $i(t)$ represents the photocurrent, $\Delta E^{0\prime}$ (V) denotes the potential difference of the two redox couples at reversible state, P_{in} is the incident solar power (100 mW cm^{-2}), S_{total} stands for the total illumination area of p-Si and n-Si, and t is the illumination time.

The overall solar(photon)–chemical–electricity energy conversion efficiency during the whole photocharge and discharge process in SRFC system could be calculated according to:

$$\eta_{overall} = E_{dis} / (P_{in} \times S_{total} \times t) \times 100\% \tag{7}$$

where E_{dis} represents the discharged electrical energy, P_{in} denotes the incident solar power (100 mW cm^{-2}), S_{total} stands for the total illumination area of p-Si and n-Si, and t is the illumination time.

References

1. Yu, M. *et al.* Aqueous lithium-iodine solar flow battery for the simultaneous conversion and storage of solar energy. *J. Am. Chem. Soc.* **137,** 8332–8335 (2015).
2. Li, N., Wang, Y. R., Tang, D. M. & Zhou, H. S. Integrating a photocatalyst into a hybrid lithium-sulfur battery for direct storage of solar energy. *Angew. Chem. Int. Ed.* **54,** 9271–9274 (2015).

3. Kim, D., Sakimoto, K. K., Hong, D. C. & Yang, P. D. Artificial photosynthesis for sustainable fuel and chemical production. *Angew. Chem. Int. Ed.* **54,** 3259–3266 (2015).
4. Han, H. X. & Li, C. Photocatalysis in solar fuel production. *Natl Sci. Rev.* **2,** 145–147 (2015).
5. Walter, M. G. *et al.* Solar water splitting cells. *Chem. Rev.* **110,** 6446–6473 (2010).
6. Yang, J. H., Wang, D. E., Han, H. X. & Li, C. Roles of cocatalysts in photocatalysis and photoelectrocatalysis. *Acc. Chem. Res.* **46,** 1900–1909 (2013).
7. Fan, F.-R. F., White, H. S., Wheeler, B. L. & Bard, A. J. Semiconductor electrodes. 31. Photoelectrochemistry and photovoltaic systems with n-and p-type tungsten selenide (WSe_2) in aqueous solution. *J. Am. Chem. Soc.* **102,** 5142–5148 (1980).
8. Hodes, G., Manassen, J. & Cahen, D. Photoelectrochemical energy-conversion and storage using polycrystalline chalcogenide electrodes. *Nature* **261,** 403–404 (1976).
9. Hada, H., Takaoka, K., Saikawa, M. & Yonezawa, Y. Energy-conversion and storage in solid-state photogalvanic cells. *Bull. Chem. Soc. Jpn* **54,** 1640–1644 (1981).
10. Licht, S., Hodes, G., Tenne, R. & Manassen, J. A light-variation insensitive high-efficiency solar-cell. *Nature* **326,** 863–864 (1987).
11. Hauch, A., Georg, A., Krašovec, U. O. & Orel, B. Photovoltaically self-charging battery. *J. Electrochem. Soc.* **149,** A1208–A1211 (2002).
12. Liu, P., Yang, H. X., Ai, X. P., Li, G. R. & Gao, X. P. A solar rechargeable battery based on polymeric charge storage electrodes. *Electrochem. Commun.* **16,** 69–72 (2012).
13. Skyllas-Kazacos, M., Chakrabarti, M., Hajimolana, S., Mjalli, F. & Saleem, M. Progress in flow battery research and development. *J. Electrochem. Soc.* **158,** R55–R79 (2011).
14. Calderon, E. H. *et al.* Effectiveness factor of fast (Fe^{3+}/Fe^{2+}), moderate (Cl_2/Cl^-) and slow (O_2/H_2O) redox couples using IrO_2-based electrodes of different loading. *J. Appl. Electrochem.* **39,** 1827–1833 (2009).
15. Liu, P. *et al.* A solar rechargeable flow battery based on photoregeneration of two soluble redox couples. *ChemSusChem* **6,** 802–806 (2013).
16. Yan, N. F., Li, G. R. & Gao, X. P. Solar rechargeable redox flow battery based on Li_2WO_4/LiI couples in dual-phase electrolytes. *J. Mater. Chem. A* **1,** 7012–7015 (2013).
17. Wei, Z., Liu, D., Hsu, C. J. & Liu, F. Q. All-vanadium redox photoelectrochemical cell: an approach to store solar energy. *Electrochem. Commun.* **45,** 79–82 (2014).
18. Liu, D. *et al.* Reversible electron storage in an all-vanadium photoelectrochemical storage cell: synergy between vanadium redox and hybrid photocatalyst. *ACS Catal.* **5,** 2632–2639 (2015).
19. Huskinson, B. *et al.* A metal-free organic-inorganic aqueous flow battery. *Nature* **505,** 195–198 (2014).
20. Yang, B., Hoober-Burkhardt, L., Wang, F., Prakash, G. S. & Narayanan, S. An inexpensive aqueous flow battery for large-scale electrical energy storage based on water-soluble organic redox couples. *J. Electrochem. Soc.* **161,** A1371–A1380 (2014).
21. Zhao, Y. *et al.* A reversible Br_2/Br^- redox couple in the aqueous phase as a high-performance catholyte for alkaliion batteries. *Energy Environ. Sci.* **7,** 1990–1995 (2014).
22. Seger, B. *et al.* Using TiO_2 as a conductive protective layer for photocathodic H_2 evolution. *J. Am. Chem. Soc.* **135,** 1057–1064 (2013).
23. Mei, B. *et al.* Protection of p^+n-Si photoanodes by sputter-deposited Ir/IrO_x thin films. *J. Phys. Chem. Lett.* **5,** 1948–1952 (2014).
24. Chen, Y. W. *et al.* Atomic layer-deposited tunnel oxide stabilizes silicon photoanodes for water oxidation. *Nat. Mater.* **10,** 539–544 (2011).
25. Hou, Y. *et al.* Bioinspired molecular co-catalysts bonded to a silicon photocathode for solar hydrogen evolution. *Nat. Mater.* **10,** 434–438 (2011).
26. Seger, B. *et al.* Hydrogen production using a molybdenum sulfide catalyst on a titanium-protected n^+ p-Silicon photocathode. *Angew. Chem. Int. Ed.* **51,** 9128–9131 (2012).
27. Sun, Y. *et al.* Electrodeposited cobalt-sulfide catalyst for electrochemical and photoelectrochemical hydrogen generation from water. *J. Am. Chem. Soc.* **135,** 17699–17702 (2013).
28. Dai, P. *et al.* Solar hydrogen generation by silicon nanowires modified with platinum nanoparticle catalysts by atomic layer deposition. *Angew. Chem. Int. Ed.* **52,** 11119–11123 (2013).
29. Ding, Q. *et al.* Efficient photoelectrochemical hydrogen generation using heterostructures of Si and chemically exfoliated metallic MoS_2. *J. Am. Chem. Soc.* **136,** 8504–8507 (2014).
30. Dasgupta, N. P., Liu, C., Andrews, S., Prinz, F. B. & Yang, P. Atomic layer deposition of platinum catalysts on nanowire surfaces for photoelectrochemical water reduction. *J. Am. Chem. Soc.* **135,** 12932–12935 (2013).
31. Kenney, M. J. *et al.* High-performance silicon photoanodes passivated with ultrathin nickel films for water oxidation. *Science* **342,** 836–840 (2013).
32. Hu, S. *et al.* Amorphous TiO_2 coatings stabilize Si, GaAs, and GaP photoanodes for efficient water oxidation. *Science* **344,** 1005–1009 (2014).
33. Shaner, M. R., Hu, S., Sun, K. & Lewis, N. S. Stabilization of Si microwire arrays for solar-driven H_2O oxidation to $O_2(g)$ in 1.0 M KOH(aq) using conformal coatings of amorphous TiO_2. *Energy Environ. Sci.* **8,** 203–207 (2015).
34. Benck, J. D. *et al.* Designing active and stable silicon photocathodes for solar hydrogen production using molybdenum sulfide nanomaterials. *Adv. Energy Mater.* **4,** 1400739 (2014).
35. DuVall, S. H. & McCreery, R. L. Control of catechol and hydroquinone electron-transfer kinetics on native and modified glassy carbon electrodes. *Anal. Chem.* **71,** 4594–4602 (1999).
36. Li, Z. S., Luo, W. J., Zhang, M. L., Feng, J. Y. & Zou, Z. G. Photoelectrochemical cells for solar hydrogen production: current state of promising photoelectrodes, methods to improve their properties, and outlook. *Energy Environ. Sci.* **6,** 347–370 (2013).
37. Weber, A. Z. *et al.* Redox flow batteries: a review. *J. Appl. Electrochem.* **41,** 1137–1164 (2011).
38. White, J. R., Fan, F. R. F. & Bard, A. J. Semiconductor electrodes.56. principles of multijunction electrodes and photoelectrosynthesis at texas instruments p/n-Si Solar-Arrays. *J. Electrochem. Soc.* **132,** 544–550 (1985).
39. Jang, J. W. *et al.* Enabling unassisted solar water splitting by iron oxide and silicon. *Nat. Commun.* **6,** 7447 (2015).

Acknowledgements

This project was financially supported by 973 National Basic Research Program of the Ministry of Science and Technology of China (No.2014CB239400) and National Natural Science Foundation of China (No. 21573230).

Author contributions

S.L. and C.L. designed the experiments. S.L. fabricated the RFB and SRFC device, prepared all samples, and performed all electrochemical and PEC measurements. B.S. and T.P. provided the functionalized silicon semiconductor. S.L., J.S., J.C. and C.L. analysed the data and co-wrote the manuscript. X.Z. participated in preparing the manuscript. All authors discussed the results. C.L. conceived the concept and supervised the research project.

Additional information

Competing financial interests: The authors declare no competing financial interests.

27

Mechanistic insights into chemical and photochemical transformations of bismuth vanadate photoanodes

Francesca M. Toma[1,2], Jason K. Cooper[1,2], Viktoria Kunzelmann[1,3], Matthew T. McDowell[4,5,†], Jie Yu[1,2], David M. Larson[1,2], Nicholas J. Borys[6,7], Christine Abelyan[1,2], Jeffrey W. Beeman[1,7], Kin Man Yu[7,†], Jinhui Yang[1,7], Le Chen[1,7], Matthew R. Shaner[4,5], Joshua Spurgeon[4,5,†], Frances A. Houle[1,2], Kristin A. Persson[1,8,9] & Ian D. Sharp[1,2]

Artificial photosynthesis relies on the availability of semiconductors that are chemically stable and can efficiently capture solar energy. Although metal oxide semiconductors have been investigated for their promise to resist oxidative attack, materials in this class can suffer from chemical and photochemical instability. Here we present a methodology for evaluating corrosion mechanisms and apply it to bismuth vanadate, a state-of-the-art photoanode. Analysis of changing morphology and composition under solar water splitting conditions reveals chemical instabilities that are not predicted from thermodynamic considerations of stable solid oxide phases, as represented by the Pourbaix diagram for the system. Computational modelling indicates that photoexcited charge carriers accumulated at the surface destabilize the lattice, and that self-passivation by formation of a chemically stable surface phase is kinetically hindered. Although chemical stability of metal oxides cannot be assumed, insight into corrosion mechanisms aids development of protection strategies and discovery of semiconductors with improved stability.

[1] Joint Center for Artificial Photosynthesis, Lawrence Berkeley National Laboratory, 1 Cyclotron Road, Berkeley, California 94720, USA. [2] Chemical Sciences Division, Lawrence Berkeley National Laboratory, 1 Cyclotron Road, Berkeley, California 94720, USA. [3] Walter Schottky Institut and Physik Department, Technische Universität München, Am Coulombwall 4, 85748 Garching, Germany. [4] Joint Center for Artificial Photosynthesis, California Institute of Technology, 1200 East California Boulevard, Pasadena, California 91125, USA. [5] Division of Chemistry and Chemical Engineering, California Institute of Technology, 1200 East California Boulevard, Pasadena, California 91125, USA. [6] Molecular Foundry, Lawrence Berkeley National Laboratory, 1 Cyclotron Road, Berkeley, California 94720, USA. [7] Materials Sciences Division, Lawrence Berkeley National Laboratory, 1 Cyclotron Road, Berkeley, California 94720, USA. [8] Energy Technologies Area, Lawrence Berkeley National Laboratory, 1 Cyclotron Road, Berkeley, California 94720, USA. [9] Materials Science and Engineering, University of California, Berkeley, 210 Hearst Memorial Mining Building, Berkeley, California 94720, USA. † Present addresses: G.W. Woodruff School of Mechanical Engineering and School of Materials Science and Engineering, Georgia Institute of Technology, Atlanta, Georgia 30332, USA (M.T.M.); Department of Physics and Materials Science, City University of Hong Kong, Kowloon, Hong Kong (K.M.Y.); Conn Center for Renewable Energy Research, University of Louisville, Louisville, Kentucky 40292, USA (J.S.). Correspondence and requests for materials should be addressed to F.M.T. (email: fmtoma@lbl.gov) or to I.D.S. (email: idsharp@lbl.gov).

Photoelectrochemical conversion of solar to chemical energy is an attractive approach for sustainable production of hydrogen and hydrocarbon fuels[1,2]. However, harnessing light to split water involves thermodynamic and kinetic challenges, with major efficiency losses associated with oxygen production[1–3]. Furthermore, material stability is often poor under the harsh conditions in which this reaction is performed. Therefore, much attention has been devoted to the development of *n*-type thin film semiconductors that are chemically stable, can efficiently harvest sunlight and can oxidize water when coupled to appropriate catalysts[3]. Among available candidates for photoanodes, metal oxides are a promising class of potentially durable materials, as they are usually less readily oxidized than water[3,4]. However, many of the most stable metal oxide semiconductors possess wide bandgaps that render them impractical for solar energy conversion. Thus, a central challenge is the discovery and development of new semiconductors that are stable and have bandgaps suitable for efficiently capturing solar energy. In recent times, the quest for such materials has been accelerated by advances in high-throughput computation for prediction of electronic structure and thermodynamic stability of semiconductors[4]. Understanding the present limits of such computational approaches and improving their ability to predict the complex behaviours of real materials requires direct comparison with experiments using known semiconductors.

Although thermodynamic considerations provide a starting point for evaluating (photo)electrochemical stability of materials under the conditions present in solar fuel generators, kinetic factors can also play a dominant role on achievable operational durability. Indeed, some materials exhibit chemical instabilities that cannot be explained by thermodynamic considerations alone. One such material is $BiVO_4$, which has been actively investigated as one of the highest performance oxide photoanodes for water splitting[5]. This material has garnered significant interest, because it absorbs in the visible range, with a bandgap of 2.5 eV (ref. 6), and has conduction and valence band positions—$\sim$4.77 and 7.27 eV below the vacuum level, respectively—that are well-suited for generating large photovoltages and driving the oxygen-evolution reaction (OER)[7]. The calculated oxidation potential of $BiVO_4$ has been predicted to be slightly more positive (1.24 V versus reversible hydrogen electrode, RHE) than the water oxidation potential (O_2/H_2O, 1.23 V versus RHE), suggesting that this material could be resistant to photoinduced corrosion under water splitting conditions[4]. In contrast to this prediction, this material suffers from photoelectrochemical instability in both near-neutral and alkaline pH ranges[8–10]. Although photocorrosion of $BiVO_4$ has been correlated with vanadium leaching from the surface into the electrolyte solution[9,11,12], the mechanisms of photodegradation and the associated chemical transformations have not yet been identified. This poor stability undermines its practical use in large-scale integrated devices[8].

In this work, we establish a methodology for systematically assessing the chemical and photochemical stabilities of semiconductors and apply it to a visible-light-absorbing semiconductor, $BiVO_4$. We investigate chemical and photochemical corrosion pathways using $BiVO_4$ thin films at near-neutral and at high pH, thereby simulating possible operating conditions in solar fuel devices[13]. These measurements demonstrate that $BiVO_4$ is susceptible to chemical corrosion that is accelerated by illumination (photodegradation). Electron microscopy indicates that photodegradation initiates at and propagates from surfaces and grain boundaries, resulting in extensive dissolution of the films. Given the existence of thermodynamically stable bismuth oxides predicted by Pourbaix diagrams under these conditions, formation of a chemically stable surface layer was expected but not observed. Therefore, bulk dissolution of the film cannot be described by thermodynamic arguments. This finding is confirmed by *in situ* electrochemical atomic force microscopy (EC-AFM), which reveals that degradation under operating conditions occurs via dissolution of the film, starting at exposed facets of grains in polycrystalline thin films. To understand this chemical behaviour, we use computational methods to examine both the illuminated and dark stability. This analysis indicates that accumulation of photogenerated holes at the surface leads to structural destabilization and chemical attack via vanadium loss into solution. Although self-passivation of the material with bismuth oxide would be expected, thermodynamically stable phases of this oxide require significant structural transformations from the low-density V-leached state and are not accessible at room temperature. Therefore, photocorrosion results in bulk dissolution of the material. Our measurements reveal that non-equilibrium and kinetic factors dominate chemical and photochemical degradation processes for this commonly studied semiconductor material, and these factors must be considered for prediction of new materials that possess desired combinations of stability and efficiency. Furthermore, the kinetic accessibility of self-passivating surface layers, which broaden the range of available corrosion protection strategies, must be considered for engineering the interfaces of materials possessing lower intrinsic stability. These findings can be used to guide extensions of theory to consider intermediate (photo)chemical states and may provide enhanced predictive power for evaluating and interpreting the stabilities of promising new semiconductors for solar fuel applications. In addition, improved understanding of mechanisms governing photochemical stability will guide development of experimental strategies for stabilizing chemically sensitive semiconductors.

Results

Evaluation of photoelectrochemical stability of bismuth vanadate. Evaluating the stability of materials for photoelectrochemical water splitting requires testing under conditions that are relevant to practical devices. Although near-neutral conditions are often used for testing individual water splitting photoelectrodes and place less severe constraints on material stability relative to acidic and alkaline conditions, current integrated solar water splitting devices require either electrolyte recirculation or more extreme pH environments to eliminate severe pH gradients and ohmic losses[14]. Much work has been reported on the photoelectrochemical properties of $BiVO_4$ at near neutral pH, where stabilization has been realized via integration of OER catalysts, such as cobalt phosphate (CoPi)[15] and iron oxyhydroxide (FeO(OH))[8,16,17]. However, strongly alkaline conditions are often desirable for avoiding hydroxide ion depletion near the electrode and high overpotentials due to ohmic losses in water splitting devices that separate chemical products[14,18]. Fewer reports have addressed the characteristics of $BiVO_4$ under alkaline conditions[9,10]. Notably, a dual layer TiO_2/Ni coating was recently demonstrated to significantly impede degradation at high pH[9]. Under all testing conditions, the long-term stability of this material remains a question and understanding of the factors governing corrosion and photocorrosion may provide a means of predicting lifetime, preventing failure and improving strategies for catalyst integration. Importantly, understanding chemical and photochemical interactions of state-of-the-art materials, such as $BiVO_4$, provides a basis for rational design of new materials by identifying kinetic degradation pathways that can be evaluated computationally.

Because of its different *p*Ka values, phosphate buffer allows for the study of stability at various pH conditions that are relevant for water-splitting applications, while maintaining equal ionic strength. Therefore, we chose to study the stability of $BiVO_4$ using 1 M potassium phosphate (KPi) buffer at pH 6.8 and at pH 12.3 under biased and unbiased dark and illuminated (AM 1.5, 100 mW cm^{-2}) conditions.

For this study, we selected undoped $BiVO_4$ deposited by the spin coating method, because it allows for reproducible fabrication of large area planar thin films. As such, spin-coated $BiVO_4$ is well suited for achieving reliable statistics, drawing comparisons between bulk and surface effects, and analysing the impact of degradation on surface chemistry and morphology. However, to confirm that the findings presented here are generally valid for the material synthesized via other methods, a variety of undoped and doped thin films deposited by different techniques, namely electrodeposition[16], reactive sputtering[17] and spray pyrolysis[19], were also tested under similar conditions and yielded qualitatively similar stability results (Supplementary Figs 1 and 2).

Deposition by spin coating of molecular organic precursors onto fluorine-doped tin oxide (FTO) layers yields pure-phase monoclinic scheelite $BiVO_4$ (Supplementary Fig. 3). Scanning electron microscopy (SEM) shows that the films are characterized by average grain sizes of 153 ± 25 nm (Supplementary Fig. 3). Rutherford backscattering spectrometry (RBS) indicates a film thickness of $\sim 51.6 \pm 3.2$ nm and is in excellent agreement with cross-sectional transmission electron microscopy (TEM) analysis, which reveals a thickness of ~ 54 nm (Supplementary Fig. 3). In addition to thickness, RBS was used to evaluate thin film composition and revealed a stoichiometric ratio of 1:1 Bi:V. Importantly, complementary X-ray photoelectron spectroscopy (XPS) measurements indicate significant differences in surface stoichiometry compared with bulk, with a surface atomic ratio of 1.27:1 of Bi:V and an overall composition of 21.6 at.% Bi, 17.0 at.% V and 61.3 at.% O (compared with the ideal stoichiometric composition of 16.7 at.% Bi, 16.7 at.% V and 66.6 at.% O, respectively; Supplementary Table 1).

As a starting point for the corrosion studies, the performance of $BiVO_4$ thin films was evaluated at different pH values in the presence of 0.1 M sodium sulfite as a hole scavenger. As sodium sulfite oxidation is thermodynamically favoured and is kinetically facile with respect to water oxidation, this approach allows for determination of the charge separation and extraction efficiency in $BiVO_4$, without influence by the intrinsically poor catalytic activity of its surface[8]. Figure 1a show a comparison of photocurrent density versus potential (*J–E*) anodic sweeps of pristine $BiVO_4$ obtained at pH 6.8 and at pH 12.3, respectively. The sample measured at pH 6.8 before its degradation is characterized by an open circuit potential of 0.35 V versus RHE and a current density of 2.2 mA cm^{-2} at 1.23 V versus RHE. The sample measured at pH 12.3 before stability testing shows an identical open circuit potential, but exhibits a significantly reduced fill factor and a smaller photocurrent of 1.8 mA cm^{-2} at 1.23 V versus RHE.

The reduced fill factor and photocurrent density observed under alkaline testing conditions suggest that rapid chemical modification of $BiVO_4$ at high pH (see below) introduces additional recombination centres at the surface and/or an insulating surface layer that increases the resistance to charge transfer across the interface[3,20]. Furthermore, analysis of cyclic voltammograms (CVs) reveals an increase of the *J–E* hysteresis (see Supplementary Fig. 4) at high pH. This hysteresis is consistent with surface state charging and may also be affected by modified interfacial energetics arising from changing occupancies of surface states during illuminated cycling. The pronounced hysteresis under alkaline conditions further suggests that chemical modification at high pH introduces electronically active surface states.

Chronoamperometry measurements show the change of photocurrent from $BiVO_4$ thin films as a function of time at fixed applied potential of 1.23 V versus RHE at both pH 6.8 and 12.3 in the absence of hole scavenger (Fig. 1b). In both cases, these data are characterized by a rapid decrease of the photocurrent during the first ten seconds (inset). This effect is significantly accelerated under alkaline conditions and is consistent with both the reduced fill factor and increased hysteresis relative to near neutral conditions, as described above. The initial fast photocurrent decay is followed by a slower decay over the course of the 60 min (pH 6.8) and 20 min (pH 12.8) measurements.

After the 1 h chronoamperometry test in phosphate buffer at near neutral pH, cyclic voltammetry of degraded $BiVO_4$ was again performed in the presence of sacrificial hole acceptor (Fig. 1c). We observe an anodic shift in the photocurrent onset potential of about 0.1 V, as well as decreases of both the current density and fill factor. Furthermore, analysis of CVs reveals that significant hysteresis is present following sustained operation (Supplementary Fig. 4). This finding suggests that consumption of the films by photocorrosion is also accompanied by the introduction of higher capacitance, probably due to charging of surface defect states, even at moderate pH.

Similarly, the 20 min chronoamperometric stability test at high pH reduces the photoelectrochemical performance of this material, with an anodic shift of the onset potential (~ 0.1 V) and decrease of the current density. Hysteresis remains approximately constant, supporting the finding that chemical transformation of the surface occurs immediately on immersion in the alkaline electrolyte. Interestingly, although the photocurrents were observed to significantly decline after chronoamperometric testing at 1.23 V versus RHE in the absence of sacrificial reagent (by $\sim 75\%$ at pH 6.8 and $\sim 95\%$ at pH 12.3, Supplementary Fig. 5), subsequent *J–E* characterization in the presence of Na_2SO_3 revealed much less pronounced reductions in the photocurrent density (Fig. 1). As neither the local nor bulk pH changes can account for the observed behaviour (Supplementary Note 1), this observation suggests that, in addition to introducing electronic defect states, chemical modification of the surface also reduces the native catalytic activity of the surfaces for water oxidation.

As described below, elemental analysis suggests that photocorrosion may result in enrichment of bismuth oxide on the surface. As oxides of bismuth are known to exhibit poor catalytic activity for water oxidation[21], chronoamperometric testing in pure phosphate buffer yields dramatic performance degradation under sustained operation. However, it is important to note that these measurements convolve catalytic activity with light capture and charge extraction efficiencies. Therefore, comparison of *J–E* curves obtained with a sacrificial hole acceptor, such as those shown in Fig. 1, more accurately represent the effects of photocorrosion and chemical modification on semiconductor performance for solar energy conversion.

Compositional analysis of bismuth vanadate photoanodes. To obtain insights into the mechanism of photodegradation and understand the changes in photoelectrochemical characteristics under operational conditions, we performed a compositional analysis of thin films before and after the chronoamperometric testing. As described above, for the case of pristine $BiVO_4$ thin films, a combination of XPS and RBS was used to analyse differences between the surface and bulk composition, which can be affected by surface chemistry and have an important impact on interfacial charge transport, as well as catalysis.

As summarized in Table 1, RBS indicates that the bulk ratio of Bi:V is stoichiometric, with a ratio of 1:1, and does not change

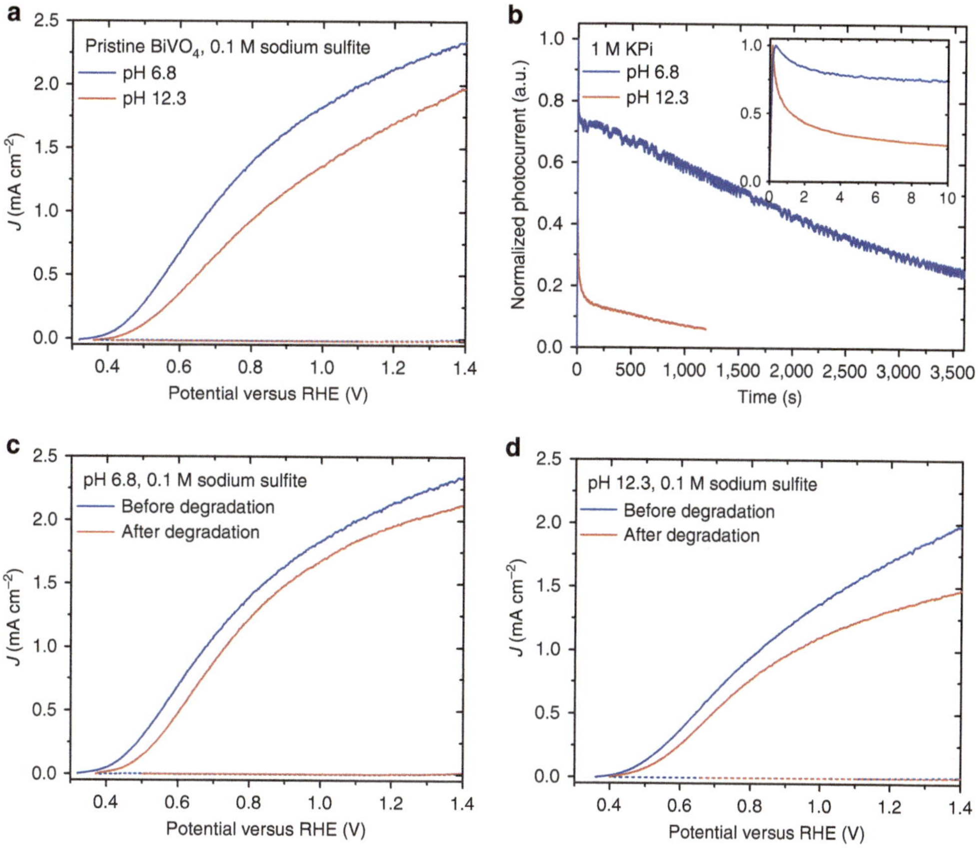

Figure 1 | Photoelectrochemical characterization of $BiVO_4$ electrodes in 1 M KPi buffer. (**a**) Photoelectrochemical *J-E* characterization of pristine $BiVO_4$ thin films in solution containing 0.1 M sulfite at pH 6.8 (blue curve) and pH 12.3 (red curve) under light (solid curve) and dark (dotted curve). (**b**) Chronoamperometry of pristine $BiVO_4$ thin films in pure KPi solution (no sulfite present) at pH 6.8 (blue curve) and pH 12.3 (red curve) at a fixed applied bias of 1.23 V versus RHE. (**c**) *J-E* characteristics of $BiVO_4$ thin films in solution containing 0.1 M sulfite at pH 6.8 under light (solid curve) and dark (dotted curve) before (blue curve) and after (red curve) testing. (**d**) *J-E* characteristics of $BiVO_4$ thin films in solution containing 0.1 M sulfite at pH 12.3 under light (solid curve) and dark (dotted curve) before (blue curve) and after (red curve) testing. All measurements under illumination were performed using simulated AM 1.5 light (Solar Light) adjusted to 100 mW cm^{-2} using a calibrated Si photovoltaic cell (SolarSim calibration, Newport).

Table 1 | Bulk and surface composition analysis.

	Testing condition	Bi (at.%)	V (at.%)	Thickness* (nm)
Ideal†		16.7 (1)	16.7 (1)	
RBS	Pristine	14.3 ± 0.5	14.5 ± 1.3	51.6 ± 3.2
	pH 6.8	14.5 ± 0.7	14.1 ± 0.7	44.1 ± 3.7
	pH 12.3	11.1 ± 1.5	11.1 ± 1.5	37.4 ± 1.8
XPS	Pristine	23.5 ± 1.5	16.3 ± 0.7	
	pH 6.8	25.3 ± 0.7	15.4 ± 0.4	
	pH 12.3	26.7 ± 1.6	14.0 ± 0.5	

RBS, rutherford backscattering spectrometry; XPS, X-ray photoelectron spectroscopy.
*Thickness calculated from the number of atoms detected by RBS, where bulk density of $BiVO_4$ is assumed[48]. Tests were performed for 60 min at pH 6.8 and 20 min at pH 12.3.
†Ideal atomic ratio and ideal stoichiometric ratio in parenthesis. Composition analysis including O at.% is reported in Supplementary Table 1.

during the degradation process. From fitting of RBS data, we calculated the thickness of the pristine $BiVO_4$ layer (51.6 ± 3.2 nm), as well as of the degraded samples (44.1 ± 3.7 nm at pH 6.8 and 37.4 ± 1.8 nm at pH 12.3). The progressive thinning of the layer during sustained operation under illumination indicates photocorrosion of the films over time, with nominal rates of 0.125 nm min^{-1} at pH 6.8 and 0.71 nm min^{-1} at pH 12.3 (7.5 nm h^{-1} at pH 6.8 and 42.6 nm h^{-1} at pH 12.3). Thus, photocorrosion is accelerated under alkaline conditions. However, it must be noted that the RBS model assumes a homogenous corrosion of the film, which is not supported by the SEM and TEM results (*vide infra*). In addition, quantification of bulk oxygen content by RBS in the degraded samples was complicated by the underlying FTO.

In contrast to the constant bulk stoichiometry determined by RBS, important changes of the surface composition are observed by XPS (Fig. 2). Comparison of photoemission spectra obtained before and after photoelectrochemical stability testing reveals no significant changes in the Bi 4*f* or the V 2*p* core level lineshapes. Analysis of the O 1*s* spectral region reveals a main component at 529.4 ± 0.3 eV from oxygen in the thin film. However, a shoulder at 531.7 ± 0.1 eV emerges following illuminated chronoamperometric testing. As discussed below, SEM and TEM analyses prove that the films become discontinuous after stability testing. Therefore, photoelectrons from regions of bare FTO contribute to the overall spectrum. Comparison with a reference FTO standard confirms this and enables assignment of the 531.7 ± 0.1 eV component to oxygen in the FTO substrate. As summarized in Table 1, analysis of XPS core-level intensity changes shows that surfaces become enriched with Bi and slightly depleted of V following chronoamperometric testing at both pH 6.8 and 12.3. Loss of V from the surface is consistent with prior reports of $BiVO_4$ composition modification under aqueous conditions[9,11]. However, the associated enrichment of Bi, on an atomic per cent basis, suggests the possible formation of a terminal surface of bismuth oxide, such as Bi_2O_3 or Bi_4O_7. In

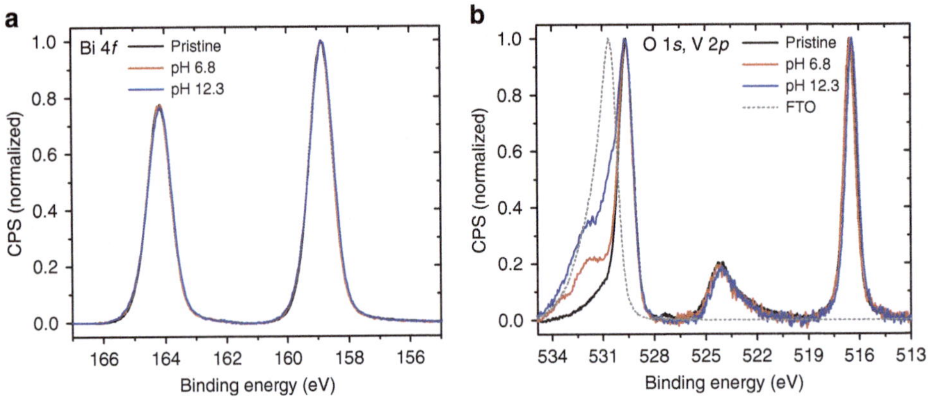

Figure 2 | XPS spectra of pristine and degraded electrodes. (**a**) Bi 4*f*, (**b**) O 1s and V 2*p* core levels of pristine (black) and degraded at pH 6.8 (red) and pH 12.3 (blue) $BiVO_4$ samples compared with FTO (dashed grey).

addition, given that the valence band of Bi_2O_3 is energetically deeper than that of $BiVO_4$, such a layer would be expected to introduce a barrier to photogenerated hole transport across the interface and could explain the anodic shift of the photocurrent onset potential and the reduced fill factor observed from *J–E* curves after stability testing[22]. However, formation of a stable passivating film of bismuth oxide, such as Bi_2O_3, would be expected to impede further chemical corrosion. In contrast, we observe progressive dissolution of the films, with loss of both V and Bi, as measured by RBS. In addition, similar composition results are obtained with RBS and XPS after testing the electrodes under dark conditions at an applied bias of 1.23 V versus RHE (Supplementary Table 1).

Determination of bismuth vanadate etching rates. Although loss of vanadium from the surface of $BiVO_4$ in aqueous environments is consistent with prior reports, our results suggest more extensive (photo)chemical etching, with substantial loss of material from thin films. These findings prompted chemical analysis of the phosphate electrolytes used during stability tests, to detect thin film corrosion products.

Using inductively coupled plasma mass spectrometry (ICP-MS), we analysed the electrolyte solutions from samples treated in different conditions. Table 2 summarizes the estimated degradation rate (nm min^{-1}) on the basis of dissolved Bi and V, calculated assuming homogeneous dissolution. These results are also shown in Fig. 3 and the complete data sets and details of calculations are provided in the Supplementary Information (Supplementary Table 2 and Supplementary Fig. 6). Under all conditions, we observe loss of both Bi and V into solution. We find that $BiVO_4$ degradation is accelerated, in decreasing order, by light, increasing pH and applied bias.

To better understand the role played by solution concentration, as well as by the specific use of phosphate electrolytes, different electrolytes with different concentrations were analysed and compared with the 1 M phosphate buffers (Fig. 3b and Supplementary Information). Notably, $BiVO_4$ thin films soaked in different electrolytes for 72 h tend to solubilize in all of the conditions studied. However, it is clear that the phosphate buffer concentration also plays an important role in determining the degradation rate of the material. At similar pH values, we observe that higher concentration accelerates degradation (Fig. 3b). Although the origin of this concentration dependence is not currently known, trends of stability with illumination, pH and applied bias are mostly preserved (see Supplementary Table 2). Notably, for the case of samples tested in 1 M NaOH (pH 14) at 1.23 V versus RHE and under 1 sun AM1.5 illumination, the

Table 2 | ICP-MS analysis of electrolyte following stability testing.

pH	Light condition	Bias* (V versus RHE)	Time (min)	Degradation rate on V basis (nm min^{-1})	Degradation rate on Bi basis (nm min^{-1})
6.8	Light	1.23	60	0.125	0.122
	Dark	1.23	60	0.03	0.04
	Light	E_{oc}	60	0.06	0.05
	Dark	E_{oc}	60	0.1	Not detectable
12.3	Light	1.23	20	1.24	1.32
	Dark	1.23	20	0.17	0.14
	Light	E_{oc}	20	0.69	0.71
	Dark	E_{oc}	20	0.26	0.21

ICP-MS, inductively coupled plasma mass spectrometry, E_{oc}, open circuit voltage; RHE, reversible hydrogen electrode.

rate of vanadium loss from the films is significantly increased (Supplementary Fig. 6 and Supplementary Table 2). Indeed, under these conditions, we determine the etch rate on the basis of vanadium to be 2.7 nm min^{-1}, corresponding to the entire film thickness after ~20 min. This finding is in agreement with previous reports and is attributed to the extreme chemical instability of vanadium species under water splitting conditions at high pH (refs 8,15).

***Ex-situ* microscopy study of photochemical corrosion.** The effect of photochemical degradation on the $BiVO_4$ film morphology was determined by plan view SEM and cross-sectional TEM (Fig. 4). As seen in Fig. 4a, the average grain size in as-synthesized $BiVO_4$ films is $\sim 153 \pm 25$ nm. Following stability testing at pH 6.8, the grains become rounded and voids form at grain boundaries, suggesting that corrosion initiates in intergrain regions of the polycrystalline films. At high pH, rapid photocorrosion yields large regions of exposed FTO substrate, consistent with XPS observations and the higher corrosion rate determined by ICP-MS and RBS.

TEM images of pristine $BiVO_4$ show a continuous layer of $\sim 54 \pm 13$ nm, in good agreement with the thickness determined by RBS. After stability testing at near-neutral pH for 1 h, the $BiVO_4$ layer remains largely continuous but its thickness is reduced. Inspection of the cross-sectional micrographs reveals evidence of enhanced corrosion rates near the apices and reduced corrosion rates near the valleys of the film. The overall film thickness decreases to 48 ± 13 nm. At high pH, evident signs of degradation appear after stability testing for 20 min. In apex

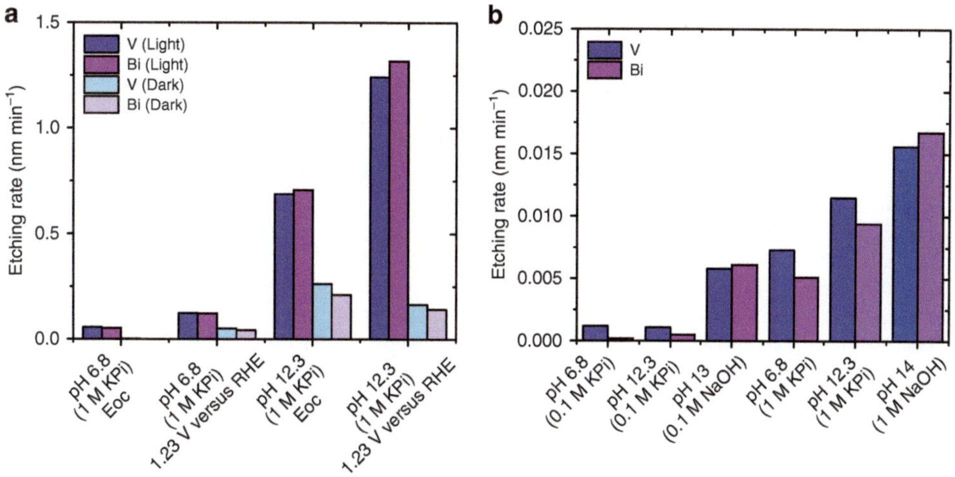

Figure 3 | Etching rates determined by ICP-MS as a function of testing conditions. (**a**) Etching rate on the basis of V (blue) and Bi (purple) in $BiVO_4$ substrates at different potentials and pH conditions under illumination and in the dark (light blue and light purple for V and Bi, respectively). (**b**) Etching rate on the basis of V (blue) and Bi (purple) in $BiVO_4$ substrates soaked in the dark for 72 h in different electrolytes at different concentrations. Homogeneous thinning of films is assumed.

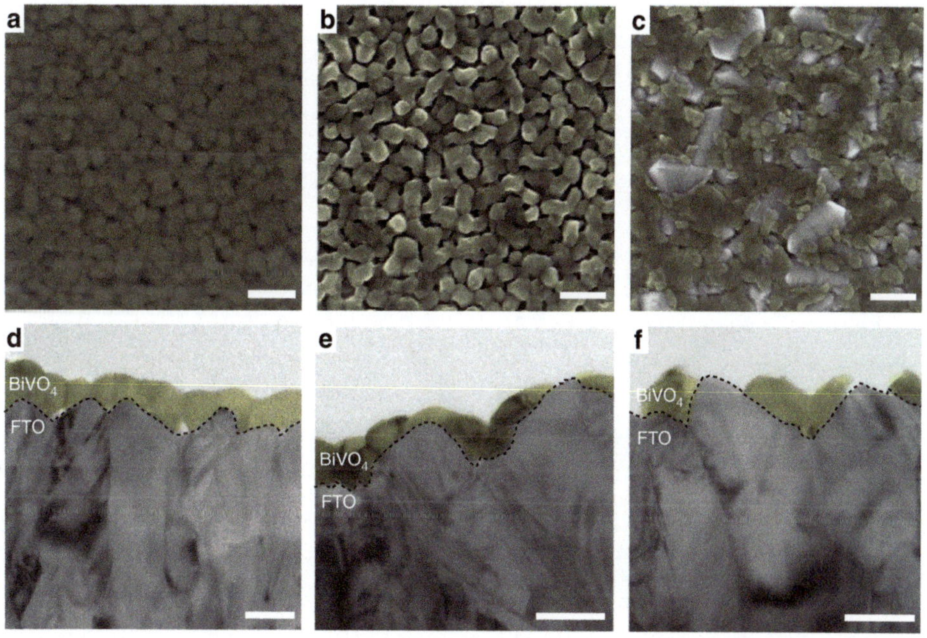

Figure 4 | SEM and TEM characterization. SEM plan view (scale bar, 500 nm) and TEM cross-section (scale bar, 100 nm) images of pristine (**a**,**d**), degraded at pH 6.8 (**b**,**e**) and at pH 12.3 (**c**,**f**) $BiVO_4$ thin films. FTO (transparent blue) and $BiVO_4$ (transparent yellow) are false coloured to better highlight the difference between the two layers.

regions of the substrate, complete loss of $BiVO_4$ is observed, whereas thick islands of $BiVO_4$ remain in valleys.

In-situ **evaluation of chemical stability**. Although poor stability would adversely affect the performance of an operating solar fuel device, identification and enhanced understanding of the morphological, compositional and electronic factors that contribute to corrosion is required to develop strategies for improving durability and efficiency[23–25]. Beyond $BiVO_4$, such mechanistic insights are valuable for evaluating (photo)chemical susceptibilities of other transition metal oxides and approaches to designing next-generation materials that overcome present stability limitations. Therefore, to gain more information on the corrosion mechanism, *in situ* degradation was monitored by EC-AFM. A bias of 1.23 V versus RHE was applied to the $BiVO_4$ working electrode in the electrolyte solution (1 M KPi at pH 12.3) in dark, while monitoring the changes in morphology over time. A total of eight AFM topography scans were collected in the same region, to examine the degradation process over a period of 160 min (Fig. 5a,b and Supplementary Movie 1). Specifically, three different regions within the same scanned area (Fig. 5a) were selected for evaluation of variations in height and surface-to-volume ratio (Fig. 5c,d), as well as the distribution of the height profile. Close inspection of these images reveals that degradation occurs approximately uniformly at crystallographic facets exposed to the electrolyte.

Region 1 exhibits three surfaces (solid/liquid interfaces) in the upper, side and bottom part of the frame that are exposed to the electrolyte. Notably, etching progresses approximately uniformly from all three solid/liquid interfaces. In addition, the height histogram from Region 1 reveals that simultaneous etching occurs from the exposed top surface. In contrast, Region 2 has just one solid/liquid interface, which runs diagonally across the upper left portion of the region, exposed directly to electrolyte. Importantly, solid/solid boundaries that are present within this region do not appear to be specific sites for chemical attack. Analysis of the corresponding height histogram supports this interpretation, in which an approximately bimodal distribution is observed, with low and high values corresponding to the substrate and the $BiVO_4$ grain, respectively. Region 3 exhibits no solid/liquid interfaces and is instead characterized by the presence of solid/solid grain boundaries, along with the exposed top surface of the central grain. Thus, this region is more protected from chemical attack and corrosion occurs almost exclusively from the top surface. As corrosion propagates from the top surface, the distribution of heights shifts monotonically towards smaller values.

Based on these *in-situ* EC-AFM measurements, we can conclude that corrosion of the material initiates predominantly from solid/liquid interfaces. We find no indication for selective etching of localized regions, such as at secondary phases at grain boundaries, or undercutting of the film at its interface to the FTO substrate. However, as these polycrystalline films are not completely compact, the evolution of morphology depends strongly on local geometry. Voids in intergrain regions act as sites for more rapid chemical attack, whereas adjacent grains in intimate physical contact with one another are partially protected. For those regions that are characterized by multiple solid/liquid interfaces, increases of the microscopic surface to volume ratio as etching progresses could lead to accelerated corrosion (Fig. 5d). We note that further advancement of the EC-AFM, currently underway in our laboratories, will allow for implementation of light illumination of the sample under operating conditions. However, the current time-lapse testing procedure enabled *in situ* monitoring of changes over a time scale compatible with the scan rate used for EC-AFM (Supplementary Movie 1).

Computational evaluation of (photo)chemical corrosion. Computational methods are employed to interpret our experimental results and they provide insight into the mechanism of corrosion under dark and illuminated conditions. We first consider the thermodynamic stability of $BiVO_4$ under dark conditions at the open circuit potential, where we observe slow but progressive dissolution of films at all pH values. To this end, we have calculated Bi-V Pourbaix diagrams with ion concentrations of $10^{-5}\,mol\,kg^{-1}$, as implemented in the Materials Project[26] using the hybrid calculated-experimental Pourbaix formalism approach based on the work of Persson *et al.*[27]. To the best of our knowledge, an experimental Pourbaix diagram for $BiVO_4$ has not been established. However, we can gain insight into the processes responsible for photochemical and chemical corrosion of $BiVO_4$ by considering equilibrium thermodynamic stability against decomposition into any combination of solid and/or aqueous species. As shown in Fig. 6a, $BiVO_4$ is stable in a large region with pH values from 1 to 11 and at the electrochemical potentials within the water stability region.

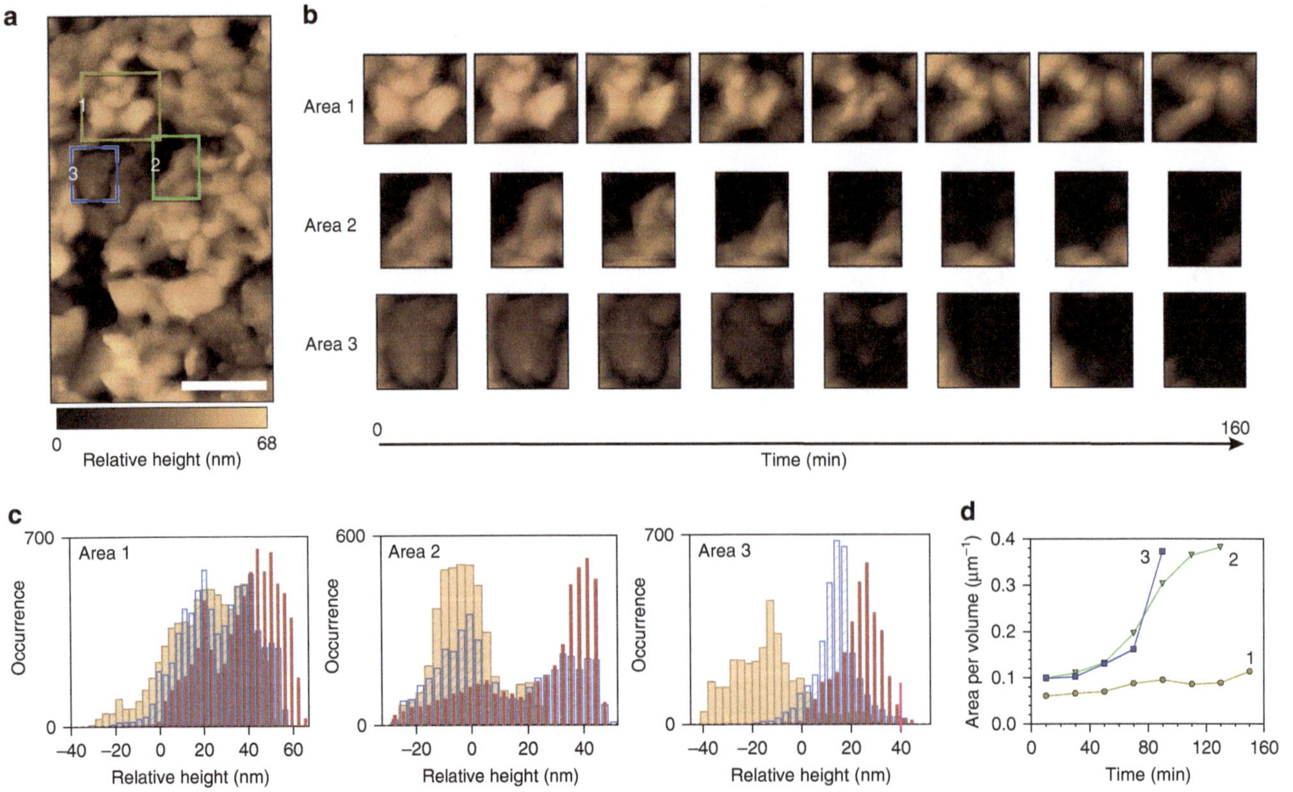

Figure 5 | Monitoring corrosion via *in situ* EC-AFM. (**a**) EC-AFM scan (527 × 898 nm^2; scale bar, 200 nm). The reported height is relative to a point on the underlying FTO substrate that is resolvable in all EC-AFM images. Common points have been used to align eight subsequent 1 × 1 μm^2 scans of the same region, to correct for sample drift and rotation. Three coloured boxes indicate Regions 1 (yellow), 2 (green) and 3 (blue), whose temporal evolution were tracked in detail. (**b**) Aligned EC-AFM scans in the three regions indicated in **a** were used to monitor corrosion-induced changes to $BiVO_4$ morphology at 20 min increments in 1 M KPi (pH 12.3). (**c**) Histograms showing height distributions in each of the three regions at the 10 (red)-, 70 (blue)- and 150 (orange)-min marks demonstrate that the heights of the $BiVO_4$ film decreases over time. Height values <0 nm correspond to the underlying FTO substrate. (**d**) Progression of the surface area to volume ratio for region 1 (yellow), 2 (green) and 3 (blue) over the course of the 160-min test.

Unfortunately, at potentials close to the oxygen evolution potential, $BiVO_4$ is no longer thermodynamically stable. Its constituent V is expected to preferentially dissolve as VO_4^- ions. This is consistent with the single-element V Pourbaix diagram, where solid phases are dissolved and only VO_4^- ions exist in the same region.

From a thermodynamic perspective, it is expected that the $BiVO_4$ would self-passivate via formation of a chemically stable bismuth oxide at its surface. However, although our XPS experiments indicate some Bi enrichment at the surface following extended dark stability testing, we also detect the presence of Bi ions in the supporting electrolyte and conclusively observe that bulk dissolution of the $BiVO_4$ occurs and self-passivation does not impede corrosion. This experimental result indicates that the thermodynamic equilibrium between the solid and liquid is not established. Indeed, a self-passivated surface of $BiVO_4$ would require formation of a more stable Bi–O surface phase (for example, Bi_4O_7 as seen at higher potentials and pH). However, the vanadium to bismuth content in $BiVO_4$ is 1:1, which means that 50% of the cations will dissolve, rendering the resulting Bi–O framework highly defective and unstable. Formation of a stable bismuth oxide or hydroxide would require significant structural rearrangement, which may be kinetically hindered under room-temperature aqueous conditions. Given these considerations, we can visualize the scenario in which structural rearrangement to form a passivating oxide on the solid surface is kinetically unfavourable by artificially removing all solid Bi-oxide phases above the OER potential (in this case, Bi_4O_7 and Bi_2O_3) from the set of phases included in the construction of the Pourbaix diagram. The result is shown in Fig. 6b, where we observe that only ions exist in the aqueous solution in the entire pH range above the OER potential. This result reveals that thermodynamic considerations cannot describe the observed dark stability of $BiVO_4$ thin films. Rather, kinetic limitations associated with

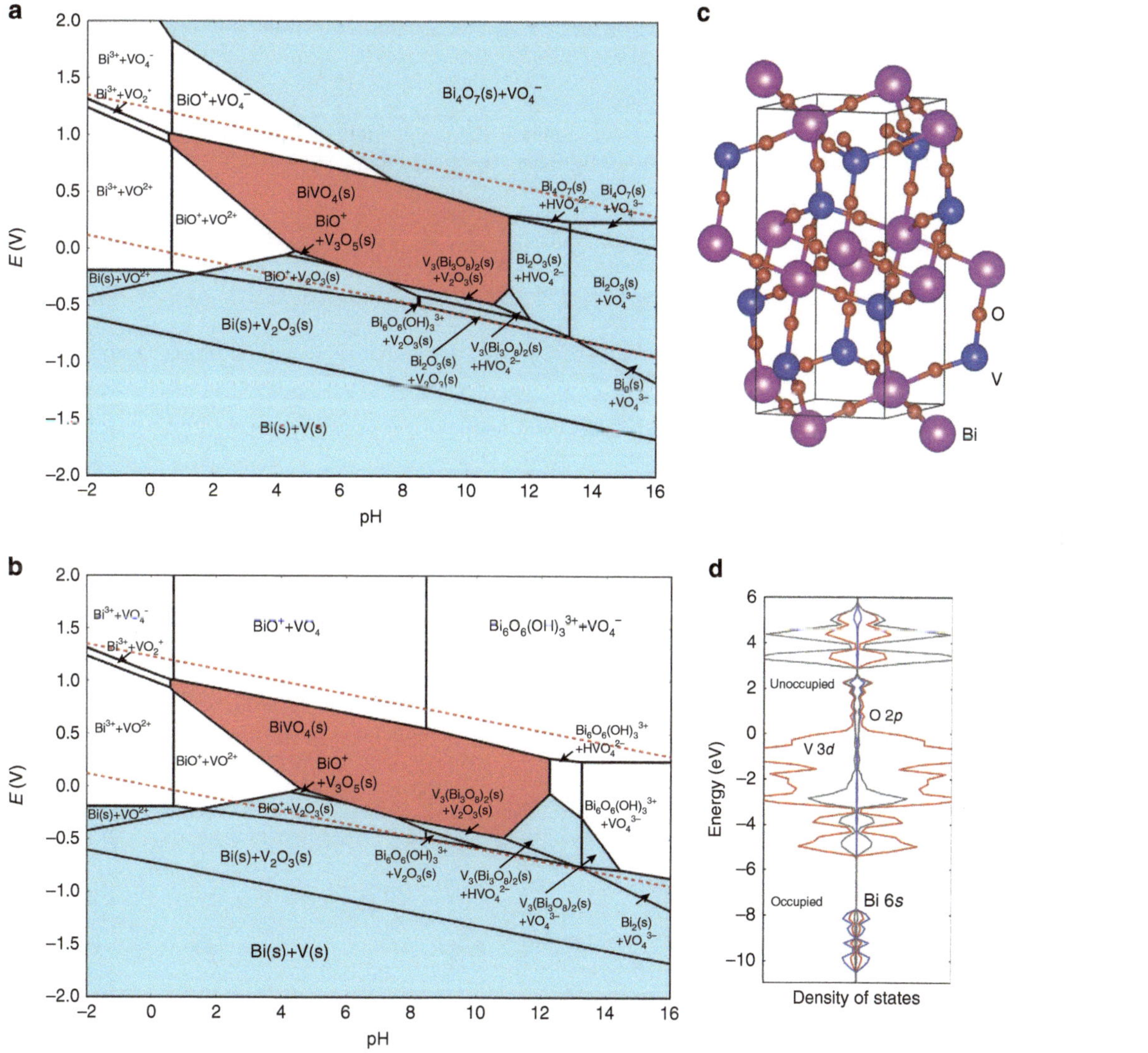

Figure 6 | Evaluation of illuminated and dark stability by computational methods. (**a**) The Materials Project Pourbaix diagram of 50-50% Bi-V system in aqueous solution, assuming a Bi ion concentration at 10^{-5} mol kg^{-1} and a V ion concentration at 10^{-5} mol kg^{-1}, and (**b**) a Pourbaix diagram with Bi_2O_3 and Bi_4O_7 phases removed. The upper red dashed line represents the potential for the oxygen evolution reaction (OER), whereas the lower red dashed line corresponds to the potential for the hydrogen evolution reaction (HER). The cyan regions denote stable solid compounds, while the pink region is $BiVO_4$. In the white regions, only ions are stable in aqueous solution. (**c**) Illustration of predicted $BiVO_5$ structure where blue, magenta and red spheres denote the V, Bi and O atoms, respectively. Thermodynamic evaluation of the stability of this hypothetical compound, which consists of lattice Bi^{5+} and V^{5+}, is used to predict the effect of surface accumulation of photogenerated holes in $BiVO_4$. (**d**) Atom projected density of states of $BiVO_5$. The energy of the Fermi level is set to 0 eV.

room-temperature structural rearrangements of corrosion products must be considered. Such a mechanism for kinetically inaccessible self-passivation is expected to be increasingly important for complex, multi-component oxides. Computational evaluation of possible kinetically hindered surface transformations may be essential for predicting the real world stabilities of novel compounds and for designing strategies for overcoming chemical instabilities.

In addition to the dark chemical instability of $BiVO_4$ thin films, our experimental results indicate that dissolution of the material in water is accelerated by illumination, as well as anodic biasing to 1.23 V versus RHE. These findings suggest that accumulation of photogenerated holes at the surface of $BiVO_4$ promotes photocorrosion. Furthermore, we note that previous studies have found that integration of a catalyst on the surface can suppress photocorrosion and enhance long-term stability of $BiVO_4$ (refs 8,15,16,19,28–31), even for the case of porous and ion-permeable catalysts, such as CoPi and FeO(OH), which do not physically isolate the semiconductor from the electrolyte environment. This suggests that efficient extraction of photogenerated holes from the near-surface region reduces chemical attack of the material.

Given these observations of photoinduced chemical instability, we consider the role of surface-accumulated holes under photoexcitation, as well as the strongly oxidative conditions present with the production of reactive oxygen species (for example, $OH^{\bullet}$ and O^-)[32,33], on the stability of $BiVO_4$. The valence band maximum, where photogenerated holes accumulate, consists of considerable Bi 6*s* orbital character, hybridized with O 2*p* (ref. 7). In prior theoretical work, Kweon and Hwang[34,35] showed that large polaron holes are the stable positive charge carriers within the monoclinic phase of this material, and that, at the surface, holes localize to BiO_6 polyhedra, with 22% of hole charge density on Bi cations and an average of 8% of hole charge density on each of the six O anions in the polyhedron, with balance of charge located outside of the polyhedron. Therefore, we first consider the impact of the modified charge density around Bi ions due to hole localization on stability. To this end, we hypothesize the scenario that brings Bi^{3+} to $Bi^{5+} + 2e^-$ in $BiVO_4$, as Bi^{5+} is a potentially stable oxidation state for Bi, whereas Bi^{4+} is uncommon[36,37]. To obtain a plausible structure representing a Bi^{5+}/V^{5+} compound or film, we use the Materials Project structure predictor[38,39] to predict a new oxide compound containing only Bi^{5+} and V^{5+} valence states. By calculating the stability of an optimized bulk Bi^{5+} and V^{5+} oxide, we can obtain a qualitative picture of the stability of such a configuration in an oxide framework. The structure predictor algorithm searches for the most suitable structure among all known structures found in the ICSD[40] for the target ions and valence states (in this case, Bi^{5+} and V^{5+}) through a data-mining approach[38,39]. The resulting structure is shown in Fig. 6c. From our calculations, the projected density of states show that the Bi 6*s* orbitals are very strongly hybridized with the O 2*p* orbitals, both in the valence band and conduction band (Fig. 6d). Considering that the magnetic moment is zero in the system, it is confirmed that Bi and V in $BiVO_5$ are both in 5+ oxidation states. A new phase diagram for the Bi–V–O system is generated (Supplementary Fig. 7) and it is found that $BiVO_5$ is unstable by 0.215 eV per atom towards oxygen release. This large above-hull energy, defined as the energy of unstable phases, strongly indicates that the thermodynamic driving force for $BiVO_5$ to decompose into $BiVO_4$ and ½ O_2 is significant. One of the reasons for the instability of $BiVO_5$ is that the small ionic radius of Bi^{5+} is not compatible with the BiO_8 structure motif in the crystal. These results suggest that any photo-induced oxidation event on Bi is likely to further destabilize the compound and increase the rate of corrosion under operating conditions, due to an effective Bi–O bond strain in the presence of Bi^{5+}.

In addition to hole localization on Bi ions, we also consider the impact of oxidation of O^{2-} to O^- at the surface of $BiVO_4$. Such a process has been implicated in the photocorroison mechanism of, for example, ZnO, where it has been proposed that degradation proceeds through hole trapping at oxygen surface sites to yield formation of O^- intermediate that results in irreversible loss of oxygen from the lattice as O_2 and consequent release of Zn^{2+} ions in the solution[33]. Although the charge density of surface-accumulated holes has been predicted to be higher at Bi cations than O anions, such a mechanism is not precluded. Furthermore, the energetic position of the valence band maximum of $BiVO_4$, at ~7.27 eV below the vacuum level, renders formation of radical $OH^{\bullet}$ species, as found by Bard and colleagues[33]. Participation of such radicals in the oxygen loss processes described above may serve to further accelerate photocorrosion, in particular at high pH.

Discussion

We demonstrate that polycrystalline $BiVO_4$ thin films are susceptible to chemical and photochemical corrosion. Indeed, chemical attack of the semiconductor is observed under all aqueous testing conditions and is accelerated by illumination, increasing pH and applied anodic bias. Although RBS reveals near-stoichiometric composition in the bulk of as-deposited thin films, XPS indicates a non-stoichiometric composition at the surface that is affected by exposure to electrolyte. Observation of film dissolution in real time by EC-AFM indicates that chemical attack propagates approximately uniformly from exposed solid/liquid interfaces. Therefore, there is no indication of preferential chemical attack at specific regions or at crystal facets of grains, but consumption rate is dependent on local solid/liquid interface morphology. In contrast to thermodynamic predictions, self-passivation of $BiVO_4$ by the formation of a chemically stable Bi oxide surface phase is not observed and bulk dissolution of films occurs on extended exposure to aqueous electrolytes. We attribute the bulk chemical instability to kinetic factors that limit room-temperature structural transformation of the V-deficient degradation product into a stable Bi oxide phase. In addition, photoexcitation results in accumulation of holes at the surface of $BiVO_4$ that destabilize the lattice and increase the dissolution rate.

Mechanistic insights into the chemical and photochemical instability of $BiVO_4$ can be used to guide approaches to stabilization and aid the search for active catalysts and functional interfaces that can improve durability for sustained operation. For example, efficient transfer of holes to catalyst on the surface will dramatically reduce the corrosion rate. However, unless this catalyst layer is dense and impermeable, long-term durability may be limited by the native dark chemical stability of the material and self-passivation cannot be assumed. We note that introduction of surface layers that reduce the corrosion rate and enable time for structural reorganization of the surface before dissolution might promote self-passivation and kinetically stabilize the material. Beyond $BiVO_4$, the methodology presented here can be applied for rigorous investigation of the stabilities of new and emerging functional materials under operating conditions. Identification of kinetic limitations to self-passivation points to a new set of factors that should be considered in the experimental and computational search for stable, visible-light-absorbing semiconductors for photoelectrochemical energy conversion.

Methods

Fabrication of $BiVO_4$ thin film photoelectrodes. Spin-coated bismuth vanadate thin films were prepared using a procedure adapted from the literature[6]. FTO films with nominal resistivity of ~13 Ω/□ (Sigma Aldrich) on 10 × 10 cm^2 glass substrates were thoroughly washed with isopropanol, detergent (Alconex) in deionized water and pure deionized water, dried with a nitrogen gun and treated for 10 min with an ozone cleaner (Jelight Model 42) before deposition of $BiVO_4$ by spin coating. In a typical deposition, 15 ml of a 0.2 M solution of bismuth (III) nitrate pentahydrate (Sigma Aldrich, ≥98%) in acetylacetone (Sigma Aldrich, ≥99%) and 100 ml of a 0.03-M solution of vanadium(IV)-oxy acetylacetonate in acetylacetone were prepared separately and sonicated for 10 min. Then, the two solutions were mixed together and sonicated for an additional 5 min. Approximately 1–1.2 millilitres of the resulting solution millilitres (enough solution to homogeneously cover the whole surface) was filtered with 0.45 μm nylon filters (Thermo Fisher Scientific) and dispensed onto the 10 × 10 cm^2 FTO/glass slide. The substrate was then spun twice in a row at 1,000 r.p.m. for 6 s on a spin coater (Laurell Technologies) with an acceleration rate of 150 r.p.m. s^{-1}. After this spin-coating cycle, the substrate was annealed for 10 min in air at 500 °C in a muffle furnace (Cole-Parmer). This procedure (that is, spin coating followed by short annealing) was repeated nine times. After the last spin-coating cycle, the substrate was annealed for 2 h at 500 °C to achieve a final thickness of ~50 nm.

$BiVO_4$ thin film photoelectrodes were prepared by dicing the $BiVO_4$-coated FTO/glass into ~1 cm × 1.3 cm pieces. To expose FTO to provide electrical connection to external leads, an ~0.3-cm-wide strip of the $BiVO_4$ thin film along the long side of the electrode was etched away using 0.1 M HCl (Sigma Aldrich, ACS reagent, 37%). A Cu wire was connected to the exposed FTO using silver conductive epoxy (Circuit Works, CW2400) and the electrode with the wire was dried for 20 min at 60 °C. Then, the wire was isolated with a glass tube and the Cu wire, silver epoxy and exposed FTO were all sealed together with the glass tube using a non-conductive aqueous resistant epoxy (Loctite 615 Hysol). Exposed surface areas of the assembled photoelectrodes were determined by optical scanning and digital image analysis with Image J. Typical areas were in the range of 0.5–1.0 cm^2.

Photoelectrochemistry and stability measurements. All photoelectrochemical measurements were performed in a three-electrode photoelectrochemical cell with a planar quartz window (5 cm^2) using a BioLogic SP200 potentiostat. Bismuth vanadate photoanodes were configured as the working electrodes, a coiled Pt wire was used as counter electrode and a Ag/AgCl (3 M NaCl, BASI) served as reference electrode. Measurements were performed in electrolyte consisting of 1 M KPi buffer solutions at pH 6.8 or 12.3 with or without 0.1 M sodium sulfite as sacrificial reagent, depending on the particular experiment. In detail, 1 M KPi at pH 6.8 was prepared from KPi monobasic (Sigma Aldrich, ≥99%, 34.0 g, 0.25 mol) and KPi dibasic (Sigma Aldrich, ≥98%, 43.5 g, 0.25 mol) in 500 ml of water. Likewise, 1 M KPi at pH 12.3 was prepared using KPi dibasic (43.5 g, 0.25 mol) and KPi tribasic (Sigma Aldrich, ≥98%, 53.1 g, 0.25 mol) in 500 ml of water. The diffusion constants for HPO_4^{2-} and PO_4^{3-} are reported in Supplementary Note 1 (ref. 41). To evaluate the performance of photoanodes as semiconductor light absorbers, sodium sulfite (Sigma Aldrich, ≥98%, 6.3 g, 0.05 mol) was added to buffer solutions as a sacrificial hole acceptor. For material degradation studies, measurements were performed in the absence of sodium sulfite. Unless otherwise noted, all measurements under illumination were performed using simulated AM 1.5 light (Solar Light) adjusted to 100 mW cm^{-2} using a calibrated Si photovoltaic cell (SolarSim calibration, Newport).

Systematic evaluations of performance and stability at pH 6.8 and 12.3 were carried out using the following procedure. First, baseline photoelectrochemical performance was established by running CVs from the open circuit potential (E_{oc}) to 1.8 V versus RHE at a scan rate of 100 mV s^{-1} in the presence of 0.1 M sodium sulfite as sacrificial reagent. Both dark and illuminated CVs were collected. It is noteworthy that for the sake of clarity, only anodic sweeps obtained from these CVs are shown in Fig. 1. Complete CVs are shown in Supplementary Fig. 4. Second, stability of the same electrodes was tested by performing chrono-amperometry in pure 1 M KPi buffer solutions (pH 6.8 or 12.3) in the dark or under illumination at E_{oc} or at 1.23 V versus RHE. Fresh electrodes were used for each of the illumination and pH conditions. Stability tests at pH 6.8 were run for 60 min, whereas those at pH 12.3 were run for 20 min due to the more rapid degradation. Finally, the post-degradation performance was evaluated by collecting CVs using conditions identical to those used for baseline performance evaluations.

Material characterization. Baseline X-ray diffraction characterization of FTO/glass substrate coated with a spin-cast $BiVO_4$ thin film was obtained with a Rigaku SmartLab X-ray diffractometer. X-ray diffraction patterns were collected with a Cu source at 0.5° with respect to the sample and parallel beam optics. To provide insights into the mechanism and consequences of $BiVO_4$ degradation under photoelectrochemical conditions, comparative measurements before and after stability testing were performed using a variety of complementary spectro-scopic and microscopic methods. To evaluate morphological changes, SEM was performed on a Zeiss Gemini Ultra-55 analytical field emission SEM and on an FEI Quanta 250 FEG. In addition, cross-sectional TEM images were obtained on an FEI Tecnai F30, with a 300-kV acceleration voltage.

The near-surface chemical composition and valence band structure were determined by XPS using a monochromatized Al Kα source (hν = 1486.6 eV), operated at 225 W, on a Kratos Axis Ultra DLD system at a takeoff angle of 0° relative to the surface normal and a pass energy for the narrow scan core level and valence band spectra of 20 eV. Spectral fitting was conducted using Casa XPS analysis software. Spectral positions were corrected using adventitious carbon by shifting the C 1*s* core level position to 284.8 eV and curves were fit with quasi-Voigt lines following Shirley background subtraction.

Complimentary bulk composition measurements were obtained by RBS, which also allowed for determination of nominal film thickness. He^+ ions were accelerated to 3,040 keV on a 5SDH pelletron tandem accelerator manufactured by National Electrostatics Corporation with a sample tilt of 30–60° and backscattered ions were detected at an angle of 165° using a Si surface barrier detector. Data fitting was performed with SIMNRA software. RBS analysis may allow for calculation of an effective film thickness, assuming a planar and homogeneous $BiVO_4$ overlayer deposited on a flat surface. Although our films exhibit good coverage with a reduced roughness on the surface, the underlying FTO has a similar roughness and we have to account that the model is affected by these parameters.

Analysis of corrosion products in electrolyte solution was conducted by ICP-MS with an Agilent 7,900 using a standard configuration consisting of Micromist nebulizer and quartz sample introduction system of Scott-type with spray chamber and one piece torch with a 2.5-mm inner diameter injector; Ni cones were used throughout. Internal standards were added on-line via the sample delivery peristaltic pump. The instrument was optimized using default autotune conditions directly from the MassHunter Workstation for U-HMI. Samples were measured in helium collision mode for all analytes.

EC-AFM was performed on a Bruker Dimension Icon AFM system, using an *in situ* electrochemical cell as sample holder, equipped with a $BiVO_4$ working electrode, a coiled Pt wire as counter electrode and a Ag wire as pseudo-reference. All the measurements were performed in pH 12.3 and a BioLogic SP200 potentiostat was used to apply a 1.23 V versus RHE bias. Topography AFM images were acquired in Scanasyst mode with Si tips (Bruker Scanasyst—Air). It is worth noting that no degradation of the Si tip was observed over the measurement period.

Evaluation of illuminated and dark stability by theoretical methods. The first-principles calculations are based on density functional theory using the Vienna software package (VASP)[42]. We choose the PAW pseudopotentials[43] with the generalized gradient approximation as implemented by Perdew, Burke and Ernzerhoff (PBE[44]). The PBE + U method is used to address the on-site Coulomb interactions in the localized *d* orbitals by adding an additional Hubbard-type U-term, with the value of 3.25 eV for V as reported for binary metal oxides[45]. A 6 × 6 × 6 Monkhorst–Pack *k*-point mesh for the integrations over the Brillouin zone is used. The energy cutoff is 450 eV and the spin polarization is included in all calculations.

The analysis of the Bi-V ternary Pourbaix diagram is based on the work of Persson *et al.*[46], where solid and dissolved species are combined in a single-phase diagram to determine the stable species (solids and/or aqueous ions) as a function of pH and potential[36]. Following ref. 46, the theoretical data obtained with the PBE + U method are used for the total energies of solid phases and the experimental data are used for the dissolved ions.

To construct the phase diagram, we use the phase diagrams app[47] from the Materials Project database. By calculating the energies of all known compounds in a given chemical system (for example, Bi–V–O), we can determine the phase diagram for that system at 0 K and 0 atm.

References

1. Maeda, K. & Domen, K. Photocatalytic water splitting: recent progress and future challenges. *J. Phys. Chem. Lett.* **1**, 2655–2661 (2010).
2. McKone, J. R., Lewis, N. S. & Gray, H. B. Will solar-driven water-splitting devices see the light of day? *Chem. Mater.* **26**, 407–414 (2013).
3. Sivula, K. Metal oxide photoelectrodes for solar fuel production, surface traps, and catalysis. *J. Phys. Chem. Lett.* **4**, 1624–1633 (2013).
4. Chen, S. & Wang, L.-W. Thermodynamic oxidation and reduction potentials of photocatalytic semiconductors in aqueous solution. *Chem. Mater.* **24**, 3659–3666 (2012).
5. Park, Y., McDonald, K. J. & Choi, K.-S. Progress in bismuth vanadate photo-anodes for use in solar water oxidation. *Chem. Soc. Rev.* **42**, 2321–2337 (2013).
6. Cooper, J. K. *et al.* Indirect bandgap and optical properties of monoclinic bismuth vanadate. *J. Phys. Chem. C* **119**, 2969–2974 (2015).
7. Cooper, J. K. *et al.* Electronic structure of monoclinic $BiVO_4$. *Chem. Mater.* **26**, 5365–5373 (2014).
8. Kim, T. W. & Choi, K.-S. Nanoporous $BiVO_4$ photoanodes with dual-layer oxygen evolution catalysts for solar water splitting. *Science* **343**, 990–994 (2014).
9. McDowell, M. T. *et al.* Improved stability of polycrystalline bismuth vanadate photoanodes by use of dual-layer thin TiO_2/Ni coatings. *J. Phys. Chem. C* **118**, 19618–19624 (2014).
10. Lichterman, M. F. *et al.* Enhanced stability and activity for water oxidation in alkaline media with bismuth vanadate photoelectrodes modified with a cobalt oxide catalytic layer produced by atomic layer deposition. *J. Phys. Chem. Lett.* **4**, 4188–4191 (2013).

11. Ding, C. *et al.* Visible light driven overall water splitting using cocatalyst/$BiVO_4$ photoanode with minimized bias. *Phys. Chem. Chem. Phys.* **15,** 4589–4595 (2013).
12. Abdi, F. F., Firet, N., Dabirian, A. & van de Krol, R. Spray-deposited Co-Pi Catalyzed $BiVO_4$: a low-cost route towards highly efficient photoanodes. *Mater. Res. Soc. Symp. Proc.* **1146,** 7–12 (2012).
13. Zhai, P. *et al.* Net primary energy balance of a solar-driven photoelectrochemical water-splitting device. *Energy Environ. Sci.* **6,** 2380–2389 (2013).
14. Modestino, M. A. *et al.* Robust production of purified H_2 in a stable, self-regulating, and continuously operating solar fuel generator. *Energy Environ. Sci.* **7,** 297–301 (2014).
15. Abdi, F. F. *et al.* Efficient solar water splitting by enhanced charge separation in a bismuth vanadate-silicon tandem photoelectrode. *Nat. Commun.* **4,** 2195 (2013).
16. Seabold, J. A. & Choi, K.-S. Efficient and stable photo-oxidation of water by a bismuth vanadate photoanode coupled with an iron oxyhydroxide oxygen evolution catalyst. *J. Am. Chem. Soc.* **134,** 2186–2192 (2012).
17. Chen, L. *et al.* Mo-doped $BiVO_4$ photoanodes synthesized by reactive sputtering. *ChemSusChem.* **8,** 1066–1071 (2015).
18. Haussener, S. *et al.* Modeling, simulation, and design criteria for photoelectrochemical water-splitting systems. *Energy Environ. Sci.* **5,** 9922–9935 (2012).
19. Abdi, F. F., Firet, N. & van de Krol, R. Efficient $BiVO_4$ thin film photoanodes modified with cobalt phosphate catalyst and W-doping. *ChemCatChem.* **5,** 490–496 (2013).
20. Smith, W. A., Sharp, I. D., Strandwitz, N. C. & Bisquert, J. Interfacial band-edge energetics for solar fuels production. *Energy Environ. Sci.* **8,** 2851–2862 (2015).
21. Hardee, K. L. & Bard, A. J. Semiconductor electrodes: X. Photoelectrochemical behavior of several polycrystalline metal oxide electrodes in aqueous solutions. *J. Electrochem. Soc.* **124,** 215–224 (1977).
22. Myung, N. *et al.* Tailoring interfaces for electrochemical synthesis of semiconductor films: $BiVO_4$, Bi_2O_3, or composites. *J. Phys. Chem. C* **115,** 7793–7800 (2011).
23. Warren, S. C. *et al.* Identifying champion nanostructures for solar water-splitting. *Nat. Mater.* **12,** 842–849 (2013).
24. Howell, S. L. *et al.* Spatial mapping of efficiency of GaN/InGaN nanowire array solar cells using scanning photocurrent microscopy. *Nano Lett.* **13,** 5123–5128 (2013).
25. Economou, N. J., Mubeen, S., Buratto, S. K. & McFarland, E. W. Investigation of arrays of photosynthetically active heterostructures using conductive probe atomic force microscopy. *Nano Lett.* **14,** 3328–3334 (2014).
26. Jain, A. *et al.* Commentary: the materials project: a materials genome approach to accelerating materials innovation. *APL Mater.* **1,** 011002 (2013).
27. Persson, K. A., Waldwick, B., Lazic, P. & Ceder, G. Prediction of solid-aqueous equilibria: scheme to combine first-principles calculations of solids with experimental aqueous states. *Phys. Rev. B* **85,** 235438 (2012).
28. Jeon, T. H., Choi, W. & Park, H. Cobalt-phosphate complexes catalyze the photoelectrochemical water oxidation of $BiVO_4$ electrodes. *Phys. Chem. Chem. Phys.* **13,** 21392–21401 (2011).
29. Park, Y., Kang, D. & Choi, K.-S. Marked enhancement in electron-hole separation achieved in the low bias region using electrochemically prepared Mo-doped $BiVO_4$ photoanodes. *Phys. Chem. Chem. Phys.* **16,** 1238–1246 (2014).
30. Zhong, D. K., Choi, S. & Gamelin, D. R. Near-complete suppression of surface recombination in solar photoelectrolysis by 'Co-Pi' catalyst-modified W:$BiVO_4$. *J. Am. Chem. Soc.* **133,** 18370–18377 (2011).
31. Han, L. *et al.* Optimization of amorphous silicon double junction solar cells for an efficient photoelectrochemical water splitting device based on a bismuth vanadate photoanode. *Phys. Chem. Chem. Phys.* **16,** 4220–4229 (2014).
32. Gerischer, H. Electrochemical behavior of semiconductors under illumination. *J. Electrochem. Soc.* **113,** 1174–1182 (1966).
33. Park, H. S., Leonard, K. C. & Bard, A. J. Surface interrogation scanning electrochemical microscopy (SI-SECM) of photoelectrochemistry at a W/Mo-$BiVO_4$ semiconductor electrode: quantification of hydroxyl radicals during water oxidation. *J. Phys. Chem. C* **117,** 12093–12102 (2013).
34. Kweon, K. E. & Hwang, G. S. Structural phase-dependent hole localization and transport in bismuth vanadate. *Phys. Rev. B* **87,** 205202 (2013).
35. Kweon, K. E. & Hwang, G. S. Surface structure and hole localization in bismuth vanadate: a first principles study. *Appl. Phys. Lett.* **103,** 131603 (2013).
36. Pourbaix, M. Atlas of electrochemical equilibria. in *Aqueous Solutions* (National Association of Corrosion Engineers, 1974).
37. Mazin, I. I., Liechtenstein, A. I. & Ivanov, O. V. Breathing instability and disproportionation of Bi^{4+} ions in $BaBiO_3$. *Ferroelectrics* **164,** 169–175 (1995).
38. Hautier, G., Fischer, C., Ehrlacher, V., Jain, A. & Ceder, G. Data mined ionic substitutions for the discovery of new compounds. *Inorg. Chem.* **50,** 656–663 (2011).
39. Jain, A. *et al.* A high-throughput infrastructure for density functional theory calculations. *Comp. Mater. Sci.* **50,** 2295–2310 (2011).
40. Bergerhoff, G. & Brown, I. D. in *Crystallographic Databases.* (eds Allen, F. H. *et al.*) (Hrsg) (International Union of Crystallography, 1987).
41. Buffle, J., Zhang, Z. & Starchev, K. Metal flux and dynamic speciation at (bio)interfaces. Part I: critical evaluation and compilation of physicochemical parameters for complexes with simple ligands and fulvic/humic substances. *Environ. Sci. Technol.* **41,** 7609–7620 (2007).
42. Kresse, G. & Furthmuller, J. Efficiency of *ab initio* total energy calculations for metals and semiconductors using a plane-wave basis set. *Comp. Mater. Sci.* **6,** 15–50 (1996).
43. Blochl, P. E. Projector augmented-wave method. *Phys. Rev. B* **50,** 17953–17979 (1994).
44. Perdew, J. P., Burke, K. & Ernzerhof, M. Generalized gradient approximation made simple. *Phys. Rev. Lett.* **77,** 3865–3868 (1996).
45. Wang, L., Maxisch, T. & Ceder, G. Oxidation energies of transition metal oxides within the GGA + U framework. *Phys. Rev. B* **73,** 195107 (2006).
46. Persson, K. A., Waldwick, B., Lazic, P. & Ceder, G. Prediction of solid-aqueous equilibria: scheme to combine first-principles calculations of solids with experimental aqueous states. *Phys. Rev. B* **85,** 235438 (2012).
47. Ong, S. P., Wang, L., Kang, B. & Ceder, G. Li-Fe-P-O_2 phase diagram from first principles calculations. *Chem. Mater.* **20,** 1798–1807 (2008).
48. Sleight, A. W., Chen, H. Y., Ferretti, A. & Cox, D. E. Crystal growth and structure of $BiVO_4$. *Mater. Res. Bull.* **14,** 1571–1581 (1979).

Acknowledgements

This study is based on work performed at the Joint Center for Artificial Photosynthesis, a DOE Energy Innovation Hub, supported through the Office of Science of the U.S. Department of Energy under Award Number DE-SC0004993. Imaging work at the Molecular Foundry was supported by the Office of Science, Office of Basic Energy Sciences, of the U.S. Department of Energy under contract number DE-AC02-05CH11231. The EC-AFM part of this work was supported in part by the Laboratory Directed Research and Development Program of Lawrence Berkeley National Laboratory under U.S. Department of Energy contract number DE-AC02-05CH11231. V.K. and F.M.T. acknowledge support from the BaCaTeC programme, project number 2015-1. Craig Jones from Agilent Technologies, Inc. is greatly acknowledged for his help with ICP-MS measurements.

Author contributions

F.M.T., F.A.H. and I.D.S. conceived of and designed this study. F.M.T., J.K.C., V.K., M.T.M., D.L., C.A., J.W.B. and I.D.S. performed the experiments. F.M.T., J.K.C., N.J.B., K.M.Y., J. Yang, F.A.H. and I.D.S. analysed and interpreted the data. J. Yu and K.A.P. performed the theoretical calculations and associated analysis. L.C., M.R.S. and J.S. provided some of the materials used for preliminary testing. All authors contributed to the final version of the manuscript.

Additional information

Competing financial interests: The authors declare no competing financial interests.

Permissions

All chapters in this book were first published in NC, by Nature Publishing Group; hereby published with permission under the Creative Commons Attribution License or equivalent. Every chapter published in this book has been scrutinized by our experts. Their significance has been extensively debated. The topics covered herein carry significant findings which will fuel the growth of the discipline. They may even be implemented as practical applications or may be referred to as a beginning point for another development.

The contributors of this book come from diverse backgrounds, making this book a truly international effort. This book will bring forth new frontiers with its revolutionizing research information and detailed analysis of the nascent developments around the world.

We would like to thank all the contributing authors for lending their expertise to make the book truly unique. They have played a crucial role in the development of this book. Without their invaluable contributions this book wouldn't have been possible. They have made vital efforts to compile up to date information on the varied aspects of this subject to make this book a valuable addition to the collection of many professionals and students.

This book was conceptualized with the vision of imparting up-to-date information and advanced data in this field. To ensure the same, a matchless editorial board was set up. Every individual on the board went through rigorous rounds of assessment to prove their worth. After which they invested a large part of their time researching and compiling the most relevant data for our readers.

The editorial board has been involved in producing this book since its inception. They have spent rigorous hours researching and exploring the diverse topics which have resulted in the successful publishing of this book. They have passed on their knowledge of decades through this book. To expedite this challenging task, the publisher supported the team at every step. A small team of assistant editors was also appointed to further simplify the editing procedure and attain best results for the readers.

Apart from the editorial board, the designing team has also invested a significant amount of their time in understanding the subject and creating the most relevant covers. They scrutinized every image to scout for the most suitable representation of the subject and create an appropriate cover for the book.

The publishing team has been an ardent support to the editorial, designing and production team. Their endless efforts to recruit the best for this project, has resulted in the accomplishment of this book. They are a veteran in the field of academics and their pool of knowledge is as vast as their experience in printing. Their expertise and guidance has proved useful at every step. Their uncompromising quality standards have made this book an exceptional effort. Their encouragement from time to time has been an inspiration for everyone.

The publisher and the editorial board hope that this book will prove to be a valuable piece of knowledge for researchers, students, practitioners and scholars across the globe.

List of Contributors

Sarah L. Perry, Dimitrios Priftis, Derek Wong, Ryan A. Klein, Charles F. PierceIII, Khatcher O. Margossian and Jian Qin
Institute for Molecular Engineering, University of Chicago, Chicago, Illinois 60637, USA

Lorraine Leon, Matthew J. Kade, Juan J. de Pablo, Matthew Tirrell and Jonathan K. Whitmer
Institute for Molecular Engineering, University of Chicago, Chicago, Illinois 60637, USA
Argonne National Laboratory, Argonne, Illinois 60439, USA

Kyle Q. Hoffmann
Institute for Molecular Engineering, University of Chicago, Chicago, Illinois 60637, USA
Department of Chemical Engineering, University of Wisconsin, Madison, Wisconsin 53706, USA

Katie A. Black
Department of Bioengineering, University of California at Berkeley, Berkeley, California 94720, USA

Akihiko Hirata and Akeshi Fujita
WPI Advanced Institute for Materials Research, Tohoku University, Sendai 980-8577, Japan

Shinji Kohara
Quantum Beam Unit, National Institute for Materials Science (NIMS), 1-1-1 Kouto, Sayo, Hyogo 679-5148, Japan
Information Integrated Materials Research Unit, Research Center for Information Integrated Materials, NIMS, 1-2-1 Sengen, Tsukuba, Ibaraki 305-0047, Japan
Division of Rsearch & Utilization , Japan Synchrotron Radiation Research Institute, Hyogo 679-5198, Japan
Schools of Materials Science, Japan Advanced Institute of Science and Technology, Nomi, Ishikawa 923-1291, Japan
JST, PRESTO, 4-1-8 Honcho, Kawaguchi, Saitama 332-0012, Japan

Toshihiro Asada, Masazumi Arao, Chihiro Yogi and Hideto Imai
Device-functional Analysis Department, NISSAN ARC Ltd., 1 Natsushima, Yokosuka 237-0061, Japan

Yongwen Tan
WPI Advanced Institute for Materials Research, Tohoku University, Sendai 980-8577, Japan
State Key Laboratory of Metal Matrix Composites and School of Materials Science and Engineering, Shanghai Jiao Tong University, Shanghai 200030, China

Mingwei Chen
WPI Advanced Institute for Materials Research, Tohoku University, Sendai 980-8577, Japan
State Key Laboratory of Metal Matrix Composites and School of Materials Science and Engineering, Shanghai Jiao Tong University, Shanghai 200030, China
JST, CREST, 4-1-8 Honcho, Kawaguchi, Saitama 332-0012, Japan

Sahika Inal, Eleni Stavrinidou, Xenofon Strakosas and George G. Malliaras
Department of Bioelectronics, Ecole Nationale Supérieure des Mines, CMP-EMSE, MOC, Gardanne 13541, France

Jonathan Rivnay
Department of Bioelectronics, Ecole Nationale Supérieure des Mines, CMP-EMSE, MOC, Gardanne 13541, France
PARC, a Xerox Company, 3333 Coyote Hill Road, Palo Alto, California 94304, USA

Dean M. Delongchamp
Material Science and Engineering Division, National Institute of Standards and Technology (NIST), Gaithersburg, Maryland 20899, USA

Brian A. Collins
Material Science and Engineering Division, National Institute of Standards and Technology (NIST), Gaithersburg, Maryland 20899, USA
Department of Physics and Astronomy, Washington State University, Pullman, Washington 99164, USA

Michele Sessolo
Instituto de Ciencia Molecular, Universidad de Valencia, Paterna 46980, Spain

Christopher Tassone
Stanford Synchrotron Radiation Lightsource (SSRL), SLAC National Accelerator Laboratory, Menlo Park, California 94025, USA

Zhaochuan Fan, Li-Chiang Lin, Wim Buijs and Thijs J.H. Vlugt
Process and Energy Department, Delft University of Technology, Leeghwaterstraat 39, 2628CB Delft, The Netherlands

Marijn A. van Huis
Soft Condensed Matter, Debye Institute for Nanomaterials Science, Utrecht University, Princetonplein 5, 3584CC Utrecht, The Netherlands

Jason Potticary, Lui R. Terry and Simon R. Hall
Complex Functional Materials Group, School of Chemistry, University of Bristol, Bristol BS8 1TS, UK

Christopher Bell
School of Physics, HH Wills Physics Laboratory, Tyndall Avenue, Bristol BS8 1TL, UK

Alexandros N. Papanikolopoulos, Peter C.M. Christianen and Hans Engelkamp
High Field Magnet Laboratory (HFML-EMFL), Radboud University, Toernooiveld 7, 6525 ED Nijmegen, The Netherlands

Andrew M. Collins
Bristol Centre for Functional Nanomaterials, HH Wills Physics Laboratory, Tyndall Avenue, Bristol BS8 1TL, UK

Simon Crampin and Enrico Da Como
Department of Physics, University of Bath, Claverton Down, Bath BA2 7AY, UK

Claudio Fontanesi
Department of Physics, University of Bath, Claverton Down, Bath BA2 7AY, UK
Dipartimento di Ingegneria Enzo Ferrari, Universita' di Modena e Reggio Emilia, Via Vivarelli 10, 41125 Modena, Italy

Gabriele Kociok-Köhn
Department of Chemistry, University of Bath, Claverton Down, Bath BA2 7AY, UK

Meital Eckshtain-Levi, Eyal Capua, Shinto P. Mathew and Ron Naaman
Department of Chemical Physics, Weizmann Institute of Science, Rehovoth 76100, Israel

Sivan Refaely-Abramson, Soumyajit Sarkar and Leeor Kronik
Department of Materials and Interfaces, Weizmann Institute of Science, Rehovoth 76100, Israel

Yaakov Levy and Yulian Gavrilov
Department of Structural Biology, Weizmann Institute of Science, Rehovoth 76100, Israel

Yossi Paltiel
Department of Applied Physics and Center for Nano Science and Nanotechnology, The Hebrew University, Jerusalem 91904, Israel

C.S. Lehmann
Argonne National Laboratory, Argonne, Illinois 60439, USA
Present address: Fachbereich Chemie, Philipps-Universität Marburg, Marburg, Germany

A. Picón, G. Doumy, P.J. Ho, E.P. Kanter, B. Krässig, A.M. March, D. Moonshiram, L. Young, S.T. Pratt and S.H. Southworth
Argonne National Laboratory, Argonne, Illinois 60439, USA

C. Bostedt
Argonne National Laboratory, Argonne, Illinois 60439, USA
Department of Physics and Astronomy, Northwestern University, Evanston, Illinois 60208, USA

A. Rudenko
J.R. Macdonald Laboratory, Department of Physics, Kansas State University, Manhattan, Kansas 66506, USA

A. Marinelli, T. Osipov, K.R. Ferguson and T. Gorkhover
SLAC National Accelerator Laboratory, Menlo Park, California 94025, USA

D. Rolles
J.R. Macdonald Laboratory, Department of Physics, Kansas State University, Manhattan, Kansas 66506, USA
Deutsches Elektronen-Synchrotron (DESY), Hamburg 22607, Germany

N. Berrah
Department of Physics, University of Connecticut, Storrs, Connecticut 06269, USA

C. Bomme and B. Erk
Deutsches Elektronen-Synchrotron (DESY), Hamburg 22607, Germany

M. Bucher
Argonne National Laboratory, Argonne, Illinois 60439, USA
SLAC National Accelerator Laboratory, Menlo Park, California 94025, USA

J. Krzywinski and A.A. Lutman
SLAC National Accelerator Laboratory, Menlo Park, California 94025, USA

D. Ray
SLAC National Accelerator Laboratory, Menlo Park, California 94025, USA
Lawrence Berkeley National Laboratory, Berkeley, California 94720, USA

Pei-Xi Wang and Mark J. MacLachlan
Department of Chemistry, University of British Columbia, 2036 Main Mall, Vancouver, British Columbia, Canada V6T 1Z1

Wadood Y. Hamad
FPInnovations, 2665 East Mall, Vancouver, British Columbia, Canada V6T 1Z4

Shuya Wei, Shaomao Xu, Akanksha Agrawral, Snehashis Choudhury and Lynden A. Archer
School of Chemical and Biomolecular Engineering, Cornell University, Ithaca, New York 14853, USA

Yingying Lu
College of Chemical and Biological Engineering, Zhejiang University, Hangzhou 310027, China

Zhengyuan Tu and Lin Ma
Department of Material Science and Engineering, Cornell University, Ithaca, New York 14853, USA

Yueqi Li, Limin Xiang and Julio L. Palma
Center for Bioelectronics and Biosensors, Biodesign Institute, Arizona State University, Tempe, Arizona 85287-5801, USA
School of Molecular Sciences, Arizona State University, Tempe, Arizona 85287-5801, USA

Yoshihiro Asai
Research Center for Computational Design of Advanced Functional Materials (CD-FMat), National Institute of Advanced Industrial Science and Technology (AIST), Central 2, Umezono 1-1-1, Tsukuba, Ibaraki 305-8568, Japan

Nongjian Tao
Center for Bioelectronics and Biosensors, Biodesign Institute, Arizona State University, Tempe, Arizona 85287-5801, USA
School of Chemistry and Chemical Engineering, Nanjing University, Nanjing, Jiangsu 210093 China

Qi Liu, Zhe-Fei Li, Yadong Liu and Jian Xie
Department of Mechanical Engineering, Purdue School of Engineering and Technology, Indiana University-Purdue University, Indianapolis, Indiana 46202, USA

Hangyu Zhang and Yun Zhou
School of Materials Engineering, Purdue University, West Lafayette, Indiana 47907, USA

Yang Ren and Cheng-Jun Sun
X-ray Science Division, Advanced Photon Source, Argonne National Laboratory, 9700 South Cass Avenue, Lemont, Illinois 60439, USA

Wenquan Lu
Chemical Science and Engineering Division, Argonne National Laboratory, 9700 South Cass Avenue, Lemont, Illinois 60439, USA

Lia Stanciu
School of Materials Engineering, Purdue University, West Lafayette, Indiana 47907, USA
Weldon School of Biomedical Engineering, Purdue University,West Lafayette, Indiana 47907, USA

Eric A. Stach
Center for Functional Nanomaterials, Brookhaven National Laboratory, Upton, New York 11973, USA

Long Chen, Xiaoli Dong, Yonggang Wang and Yongyao Xia
Department of Chemistry, Shanghai Key Laboratory of Molecular Catalysis and Innovative Materials, Institute of New Energy, iChEM (Collaborative Innovation Center of Chemistry for Energy Materials), Fudan University, Shanghai 200433, China

Guilhem P. Baeza
Foundation for Research and Technology - Hellas (FORTH), Institute of Electronic Structure and Laser, Heraklion, Crete 70013, Greece

Claudia Dessi, Salvatore Costanzo and Dimitris Vlassopoulos
Foundation for Research and Technology - Hellas (FORTH), Institute of Electronic Structure and Laser, Heraklion, Crete 70013, Greece
Department of Materials Science and Technology, University of Crete, Heraklion, Crete 71003, Greece

Dan Zhao and Sanat K. Kumar
Department of Chemical Engineering, Columbia University, New York, New York 10027, USA

Ralph H. Colby and Shushan Gong
Department of Materials Science and Engineering, Pennsylvania State University, University Park, Pennsylvania 16802, USA

Angel Alegria
Departamento de Física de Materiales y Centro de Física de Materiales CSIC-UPV/EHU, Universidad del País Vasco UPV/EHU, San Sebastian E-20018, Spain

Michael Rubinstein
Department of Chemistry, University of North Carolina, Chapel Hill, North Carolina 27599-3290, USA

Jeong Kuk Shon
Samsung Advanced Institute of Technology, Samsung Electronics Co., Ltd., Suwon 443-803, Republic of Korea
Department of Chemistry, Sungkyunkwan University, Suwon 440-746, Republic of Korea

Hyo Sug Lee, Gyeong Su Park, Jae-Man Choi, Hyuk Chang and Seokgwang Doo
Samsung Advanced Institute of Technology, Samsung Electronics Co., Ltd., Suwon 443-803, Republic of Korea

Gwi Ok Park and Ji Man Kim
Department of Chemistry, Sungkyunkwan University, Suwon 440-746, Republic of Korea
Department of Energy Science, Sungkyunkwan University, Suwon 440-746, Republic of Korea

Jeongbae Yoon and Won-Sub Yoon
Department of Energy Science, Sungkyunkwan University, Suwon 440-746, Republic of Korea

Hansu Kim and Eunjun Park
Department of Energy Engineering, Hanyang University, Seoul 133-791, Republic of Korea

Soo Sung Kong
Department of Chemistry, Sungkyunkwan University, Suwon 440-746, Republic of Korea

Mingshi Jin
Key Laboratory of Natural Resource of the Changbai Mountain and Functional Molecular (Yanbian University), Ministry of Education, Department of Chemistry, Park Road 977, Yanji City, Jilin Province 133002, China

Chanho Pak
Fuel Cell Group, Corporate R&D Center, Samsung SDI Co., Ltd., Yongin 446-577, Republic of Korea

Galen D. Stucky
Department of Chemistry and Biochemistry, University of California, Santa Barbara, California 93106, USA

Shigeki Kawai
International Center for Materials Nanoarchitectonics, National Institute for Materials Science, 1-1, Namiki, Tsukuba, 305-0044 Ibaraki, Japan
Department of Physics, University of Basel, Klingelbergstrasse 82, CH-4056 Basel, Switzerland
Precursory Research for Embryonic Science and Technology (PRESTO), Japan Science and Technology Agency, 4-1-8, Honcho, Kawaguchi, 332-0012 Saitama, Japan

Adam S. Foster
COMP, Department of Applied Physics, Aalto University, PO Box 11100, FI-00076 Aalto, Finland
Division of Electrical Engineering and Computer Science, Kanazawa University, Kanazawa 920-1192, Japan

Torbjörn Björkman
COMP, Department of Applied Physics, Aalto University, PO Box 11100, FI-00076 Aalto, Finland
Physics/Department of Natural Sciences, Åbo Akademi University, 20500 Turku, Finland

Sylwia Nowakowska and Ernst Meyer
Department of Physics, University of Basel, Klingelbergstrasse 82, CH-4056 Basel, Switzerland.

Jonas Björk
Department of Physics, Chemistry and Biology (IFM), Linköping University, Linköping 58183, Sweden

Filippo Federici Canova
COMP, Department of Applied Physics, Aalto University, PO Box 11100, FI-00076 Aalto, Finland
Aalto Science Institute, Aalto University, PO Box 15500, FI-00076 Aalto, Finland

Lutz H. Gade
Anorganisch-Chemisches Institut, Universität Heidelberg, Im Neuenheimer Feld 270, 69120 Heidelberg, Germany

Thomas A. Jung
Department of Physics, University of Basel, Klingelbergstrasse 82, CH-4056 Basel, Switzerland
Laboratory for Micro- and Nanotechnology, Paul Scherrer Institute, CH-5232 Villigen, Switzerland

Xinyong Tao
College of Materials Science and Engineering, Zhejiang University of Technology, Hangzhou 310014, China
Department of Materials Science and Engineering, Stanford University, Stanford, California 94305, USA

JianguoWang and Qiuxia Cai
College of Chemical Engineering, Zhejiang University of Technology, Hangzhou 310014, China
Chong Liu, HaotianWang, Hongbin Yao, Guangyuan Zheng, ZhiWei Seh, Weiyang Li, Guangmin Zhou and Chenxi Zu
Department of Materials Science and Engineering, Stanford University, Stanford, California 94305, USA

Yi Cui
Department of Materials Science and Engineering, Stanford University, Stanford, California 94305, USA
Stanford Institute for Materials and Energy Science, SLAC National Accelerator Laboratory, Menlo Park, California 94025, USA

Ke Fan
Department of Chemistry, Organic Chemistry, KTH Royal Institute of Technology, 10044 Stockholm, Sweden
State Key Laboratory of Advanced Technology for Materials Synthesis and Processing,Wuhan University of Technology,Wuhan 430070, China

Hong Chen, Quentin Daniel and Fusheng Li
Department of Chemistry, Organic Chemistry, KTH Royal Institute of Technology, 10044 Stockholm, Sweden

Yongfei Ji and Yi Luo
Division of Theoretical Chemistry and Biology, School of Biotechnology, KTH Royal Institute of Technology, SE-106 91 Stockholm, Sweden

Hui Huang and Per Martin Claesson
Department of Chemistry, Surface and Corrosion Science, KTH Royal Institute of Technology, SE-10044 Stockholm, Sweden

Bertrand Philippe and Håkan Rensmo
Department of Physics and Astronomy, Uppsala University, Box 516, 751 20 Uppsala, Sweden

Licheng Sun
Department of Chemistry, Organic Chemistry, KTH Royal Institute of Technology, 10044 Stockholm, Sweden
State Key Laboratory of Fine Chemicals, DUT-KTH Joint Education and Research Center on Molecular Devices, Dalian University of Technology (DUT), Dalian 116024, China

Xu Peng, Huili Liu, Qin Yin, Junchi Wu, Pengzuo Chen, Guangzhao Zhang, Guangming Liu, Changzheng Wu and Yi Xie
Hefei National Laboratory for Physical Sciences at the Microscale, iChEM (Collaborative Innovation Centre of Chemistry for Energy Materials), Hefei Science Center (CAS), and CAS Key Laboratory of Mechanical Behavior and Design of Materials, University of Science and Technology of China, Hefei, Anhui 230026, China

Bin Feng, Akihito Kumamoto and Naoya Shibata
Institute of Engineering Innovation, The University of Tokyo, Tokyo 113-8656, Japan

Tatsuya Yokoi
Department of Adaptive Machine System, Osaka University, Osaka 565-0871, Japan

Masato Yoshiya
Department of Adaptive Machine System, Osaka University, Osaka 565-0871, Japan
Nanostructures Research Laboratory, Japan Fine Ceramics Center, Nagoya 456-8587, Japan

Yuichi Ikuhara
Institute of Engineering Innovation, The University of Tokyo, Tokyo 113-8656, Japan
Nanostructures Research Laboratory, Japan Fine Ceramics Center, Nagoya 456-8587, Japan
WPI advanced Institute for Materials Research, Tohoku University, Sendai 980-8577, Japan

Xueguang Ren
Max-Planck-Institut für Kernphysik, Saupfercheckweg 1, 69117 Heidelberg, Germany
Physikalisch-Technische Bundesanstalt, Bundesallee 100, 38116 Braunschweig, Germany

Elias Jabbour Al Maalouf and Stephan Denifl
Institut für Ionenphysik und Angewandte Physik, Universität Innsbruck, Technikerstrasse 25, 6020 Innsbruck, Austria

Alexander Dorn
Max-Planck-Institut für Kernphysik, Saupfercheckweg 1, 69117 Heidelberg, Germany

Ricardo M.P. da Silva
Simpson Querrey Institute for BioNanotechnology (SQI), Northwestern University, Chicago, Illinois 60611, USA
Laboratory of Macromolecular and Organic Chemistry and Institute for Complex Molecular Systems, Eindhoven University of Technology, Eindhoven MB 5600, The Netherlands
Craniofacial Development & Stem Cell Biology, King's College London, London, SE1 9RT, UK

E.W. Meijer and Daan van der Zwaag
Laboratory of Macromolecular and Organic Chemistry and Institute for Complex Molecular Systems, Eindhoven University of Technology, Eindhoven MB 5600, The Netherlands

Lorenzo Albertazzi
Laboratory of Macromolecular and Organic Chemistry and Institute for Complex Molecular Systems, Eindhoven University of Technology, Eindhoven MB 5600, The Netherlands
Nanoscopy for Nanomedicine Group, Institute for Bioengineering of Catalonia (IBEC), Barcelona 08028, Spain

Sungsoo S. Lee
Department of Materials Science and Engineering, Northwestern University, Evanston, Illinois 60208, USA

Samuel I. Stupp
Simpson Querrey Institute for BioNanotechnology (SQI), Northwestern University, Chicago, Illinois 60611, USA
Department of Materials Science and Engineering, Northwestern University, Evanston, Illinois 60208, USA
Department of Chemistry, Northwestern University, Evanston, Illinois 60208, USA
Department of Medicine, Northwestern University, Chicago, Illinois 60611, USA
Department of Biomedical Engineering, Northwestern University, Evanston, Illinois 60208, USA

Tina I. Löbling
Macromolecular Chemistry II, University of Bayreuth, D-95440 Bayreuth, Germany
Department of Applied Physics, Aalto University School of Science, FIN-02150 Espoo, Finland

Oleg Borisov
Institut Pluridisciplinaire de Recherche sur l'Environnement et les Matériaux UMR 5254 CNRS/UPPA, F-64053 Pau, France
Institute of Macromolecular Compounds, Russian Academy of Sciences, 199004 St Petersburg, Russia
St Petersburg State Polytechnic University, 195251 St Petersburg, Russia

Johannes S. Haataja and Olli Ikkala
Department of Applied Physics, Aalto University School of Science, FIN-02150 Espoo, Finland

André H. Gröschel
Department of Applied Physics, Aalto University School of Science, FIN-02150 Espoo, Finland
Department of Physical Chemistry, University of Duisburg-Essen, D-45127 Essen, Germany

Axel H.E. Müller
Institute of Organic Chemistry, Johannes Gutenberg-Universität Mainz, D-55099 Mainz, Germany

Carsten Cleff, Patrick Ferrand, Hervé Rigneault, Sophie Brasselet and Julien Duboisset
Aix-Marseille Université, CNRS, Centrale Marseille, Institut Fresnel, UMR 7249, Domaine Universitaire de Saint Jérôme, Avenue Escadrille Normandie Niemen, Marseille, F-13397, France

Alicja Gasecka
Quebec Mental Health Institute Research Center, Laval University, Québec, Quebec, Canada G1J 2G3
Centre d'Optique, Photonique et Laser (COPL), Laval University, Québec, Quebec, Canada G1V 0A6

Yu Wang, Feng Li and Yafei Li
College of Chemistry and Materials Science, Jiangsu Key Laboratory of Biofunctional Materials, Nanjing Normal University, Nanjing, Jiangsu 210023, China

Zhongfang Chen
Department of Chemistry, Institute for Functional Nanomaterials, University of Puerto Rico, Rio Piedras Campus, San Juan, Puerto Rico 00931, USA

A. Basile
School of Applied Sciences, Applied Chemistry, RMIT University, GPO Box 2476V, Melbourne, Victoria 3001, Australia
Energy Flagship, Commonwealth Scientific and Industrial Research Organisation, Clayton, Melbourne Victoria 3169, Australia
Institute for Frontier Materials, Deakin University, Burwood, Victoria 3125, Australia

A.I. Bhatt
Energy Flagship, Commonwealth Scientific and Industrial Research Organisation, Clayton, Melbourne Victoria 3169, Australia

A.P. O'Mullane
School of Applied Sciences, Applied Chemistry, RMIT University, GPO Box 2476V, Melbourne, Victoria 3001, Australia
School of Chemistry, Physics and Mechanical Engineering, Queensland University of Technology, GPO Box 2434, Queensland 4001, Australia

Shichao Liao
State Key Laboratory of Catalysis, Dalian Institute of Chemical Physics, Chinese Academy of Sciences,
Dalian National Laboratory for Clean Energy, iChEM, Dalian 116023, China
University of Chinese Academy of Sciences, Beijing 100049, China

Xu Zong, Tingting Yao, Chunmei Ding, Jingying Shi, Jian Chen and Can Li
State Key Laboratory of Catalysis, Dalian Institute of Chemical Physics, Chinese Academy of Sciences,
Dalian National Laboratory for Clean Energy, iChEM, Dalian 116023, China

Brian Seger
Department of Physics, CINF, Technical University of Denmark, Kongens Lyngby 2800, Denmark

Thomas Pedersen
Department of Micro- and Nanotechnology, Technical University of Denmark, Kongens Lyngby 2800, Denmark

David M. Larson, Francesca M. Toma, Jason K. Cooper, Jie Yu, Christine Abelyan, Frances A. Houle and Ian D. Sharp
Joint Center for Artificial Photosynthesis, Lawrence Berkeley National Laboratory, 1 Cyclotron Road, Berkeley, California 94720, USA
Chemical Sciences Division, Lawrence Berkeley National Laboratory, 1 Cyclotron Road, Berkeley, California 94720, USA

Viktoria Kunzelmann
Joint Center for Artificial Photosynthesis, Lawrence Berkeley National Laboratory, 1 Cyclotron Road, Berkeley, California 94720, USA
Walter Schottky Institut and Physik Department, Technische Universität München, Am Coulombwall 4, 85748 Garching, Germany

Nicholas J. Borys
Molecular Foundry, Lawrence Berkeley National Laboratory, 1 Cyclotron Road, Berkeley, California 94720, USA
Materials Sciences Division, Lawrence Berkeley National Laboratory, 1 Cyclotron Road, Berkeley, California 94720, USA

Matthew T. McDowell and Joshua Spurgeon
Joint Center for Artificial Photosynthesis, California Institute of Technology, 1200 East California Boulevard, Pasadena, California 91125, USA
Division of Chemistry and Chemical Engineering, California Institute of Technology, 1200 East California Boulevard, Pasadena, California 91125, USA
G.W. Woodruff School of Mechanical Engineering and School of Materials Science and Engineering, Georgia Institute of Technology, Atlanta, Georgia 30332, USA (M.T.M.); Department of Physics and Materials Science, City University of Hong Kong, Kowloon, Hong Kong (K.M.Y.); Conn Center for Renewable Energy Research, University of Louisville, Louisville, Kentucky 40292, USA (J.S.).

JeffreyW. Beeman, Jinhui Yang and Le Chen
Joint Center for Artificial Photosynthesis, Lawrence Berkeley National Laboratory, 1 Cyclotron Road, Berkeley, California 94720, USA
Materials Sciences Division, Lawrence Berkeley National Laboratory, 1 Cyclotron Road, Berkeley, California 94720, USA

Kin Man Yu
Materials Sciences Division, Lawrence Berkeley National Laboratory, 1 Cyclotron Road, Berkeley, California 94720, USA
G.W. Woodruff School of Mechanical Engineering and School of Materials Science and Engineering, Georgia Institute of Technology, Atlanta, Georgia 30332, USA (M.T.M.); Department of Physics and Renewable Energy Research, University of Louisville, Louisville, Kentucky 40292, USA (J.S.).
Materials Science, City University of Hong Kong, Kowloon, Hong Kong (K.M.Y.); Conn Center for

Matthew R. Shaner
Joint Center for Artificial Photosynthesis, California Institute of Technology, 1200 East California Boulevard, Pasadena, California 91125, USA
Division of Chemistry and Chemical Engineering, California Institute of Technology, 1200 East California Boulevard, Pasadena, California 91125, USA

Kristin A. Persson
Joint Center for Artificial Photosynthesis, Lawrence Berkeley National Laboratory, 1 Cyclotron Road, Berkeley, California 94720, USA
Energy Technologies Area, Lawrence Berkeley National Laboratory, 1 Cyclotron Road, Berkeley, California 94720, USA
Materials Science and Engineering, University of California, Berkeley, 210 Hearst Memorial Mining Building, Berkeley, California 94720, USA

Index